Deep Decarbonization of Modern Energy Networks for Climate–Energy–Sustainability Nexus

IEEE Press
445 Hoes Lane
Piscataway, NJ 08854

Deep Decarbonization of Modern Energy Networks for Climate–Energy–Sustainability Nexus

Edited by

Mohammadreza Daneshvar
Laboratory of Multi-Carrier Energy Networks Modernization, Faculty of Electrical and Computer Engineering,
University of Tabriz
Tabriz, Iran

Amjad Anvari-Moghaddam
Department of Energy (AAU Energy), Aalborg University
Denmark

Mahdieh S. Sadabadi
Department of Electrical and Electronic Engineering,
The University of Manchester
Manchester, UK

IEEE Press Series on Power and Energy Systems
Ganesh Kumar Venayagamoorthy, Series Editor

IEEE*Press*

WILEY

Published by John Wiley & Sons, Inc., Hoboken, New Jersey.

Library of Congress Cataloging-in-Publication Data:

Names: Daneshvar, Mohammadreza Editor | Anvari-Moghaddam, Amjad Editor | Sadabadi, Mahdieh S. Editor
Title: Deep Decarbonization of Modern Energy Networks for Climate–Energy–Sustainability Nexus/edited by Mohammadreza Daneshvar, Amjad Anvari-Moghaddam, and Mahdieh S. Sadabadi.
Description: [Hoboken, New Jersey]: [John Wiley & Sons, Inc.], [2027] | Series: IEEE Press Series on Power and Energy Systems | Includes bibliographical references and index.
Identifiers: LCCN 2026025473 | ISBN 9781394394760 cloth | ISBN 9781394394777 ePub | ISBN 9781394394791 adobe pdf
Subjects: LCSH: Power resources—Environmental aspects | Carbon dioxide mitigation | Renewable energy sources
Classification: LCC TD195.E49 D44 2027
LC record available at https://lccn.loc.gov/2026025473

Cover Design: Wiley
Cover Image: ima/stock.adobe.com

Contents

List of Contributors

Habib A. Aalami
Electrical & Computer Department
Eyvanekey University
Tehran, Iran

Mehdi Abapour
Energy System Research Institute
(ESRI), Faculty of Electrical and
Computer Engineering
University of Tabriz
Tabriz, Iran

Hamdi Abdi
Electrical Engineering Department
Razi University
Kermanshah, Iran

Prince Aduama
Smart Operation Research (OR) Lab,
Advanced Power and Energy Center,
Department of Electrical Engineering
Khalifa University
Abu Dhabi, UAE

Ahmed Al Durra
Advanced Power & Energy Center,
EECS Department
Khalifa University
Abu Dhabi, UAE

Ameena S. Al-Sumaiti
Smart Operation Research (OR) Lab,
Advanced Power and Energy Center,
Department of Electrical Engineering
Khalifa University
Abu Dhabi, UAE

Mohammad Mehdi Amiri
Faculty of Electrical Engineering
Shahid Beheshti University
Tehran, Iran

Alireza Zare Ansaroudi
Energy System Research Institute
(ESRI), Faculty of Electrical and
Computer Engineering
University of Tabriz
Tabriz, Iran

Hamidreza Arasteh
Center for Research on Microgrids
(CROM), Huanjiang Laboratory
Shaoxing, Zhejiang, China

Power Systems Operation and
Planning Research Department
Niroo Research Institute
Tehran, Iran

Akram Avami
Energy Systems Engineering Group,
Department of Energy Engineering
Sharif University of Technology
Tehran, Iran

Negash Begashaw
Department of Computer Science,
Physics and Engineering
Benedict College
Columbia, SC, USA

Marwan Bikdash
Department of Computational Data
Science and Engineering
North Carolina Agricultural and
Technical State University
Greensboro, NC, USA

Gurcan Comert
Department of Computational Data
Science and Engineering
North Carolina Agricultural and
Technical State University
Greensboro, NC, USA

Mohammadreza Daneshvar
Laboratory of Multi-Carrier Energy
Networks Modernization, Faculty of
Electrical and Computer Engineering
University of Tabriz
Tabriz, Iran

Ali Darvishi
Clinical-Laboratory Center of Power
System & Protection, Faculty of
Intelligent Systems Engineering and
Data Science
Persian Gulf University
Bushehr, Iran

Rahman Dashti
Clinical-Laboratory Center of Power
System & Protection, Faculty of
Intelligent Systems Engineering and
Data Science
Persian Gulf University
Bushehr, Iran

Hadi Ghaebi
University of Mohaghegh Ardabili
Ardabil, Iran

Reza Gharibi
Clinical-Laboratory Center of Power
System & Protection, Faculty of
Intelligent Systems Engineering and
Data Science
Persian Gulf University
Bushehr, Iran

Josep M. Guerrero
Center for Research on Microgrids
(CROM), Huanjiang Laboratory
Shaoxing, Zhejiang, China

Seyed Mohsen Hashemi
Power Systems Operation and
Planning Research Department
Niroo Research Institute
Tehran, Iran

Hossein Hosseinalibeiki
DISA-MIS Department
University of Salerno
Fisciano, Campania, Italy

Rajesh Kumar
RAMAN Lab, Department of Electrical Engineering
Malaviya National Institute of Technology
Jaipur, India

Department of Human Anatomy and Physiology, Faculty of Health Sciences
University of Johannesburg
Johannesburg, South Africa

Mohammad Reza Mansouri
Clinical-Laboratory Center of Power System & Protection, Faculty of Intelligent Systems Engineering and Data Science
Persian Gulf University
Bushehr, Iran

Behnam Mohammadi-Ivatloo
Department of Electrical Engineering, School of Energy Systems
LUT University of Technology
Lappeenranta, South Karelia, Finland

Sina Nabighadim
Energy System Research Institute (ESRI), Faculty of Electrical and Computer Engineering
University of Tabriz
Tabriz, Iran

Ehsan Naderi
Department of Electrical Engineering, College of Engineering and Computer Science
Arkansas State University
Jonesboro, AR, USA

Asma Nasiri
Energy System Research Institute (ESRI), Faculty of Electrical and Computer Engineering
University of Tabriz
Tabriz, Iran

Nima Nasiri
Resilient Smart Grids Research Lab, Electrical Engineering Department
Azarbaijan Shahid Madani University
Tabriz, Iran

Mehrdad Setayesh Nazar
Faculty of Electrical Engineering
Shahid Beheshti University
Tehran, Iran

Justice Adjei Owusu
Department of Computational Data Science and Engineering, North Carolina Agricultural and Technical State University
Greensboro, NC, USA

Anand Rajendran
Center for Research on Microgrids (CROM), Huanjiang Laboratory
Shaoxing, Zhejiang, China

Ali Akbar Rezazadeh
Energy Systems Engineering Group, Department of Energy Engineering
Sharif University of Technology
Tehran, Iran

Mahdieh S. Sadabadi
Department of Electrical and Electronic Engineering
The University of Manchester
Manchester, UK

Vikash Kumar Saini
Smart Operation Research (OR) Lab,
Advanced Power and Energy Center,
Department of Electrical Engineering
Khalifa University
Abu Dhabi, UAE

Reza Sepehrzad
Electrical and Computer Engineering
Department
University of Canterbury
Christchurch, New Zealand

Heresh Seyedi
Energy System Research Institute
(ESRI), Faculty of Electrical and
Computer Engineering
University of Tabriz
Tabriz, Iran

Alireza Sheikhi-Fini
Power Systems Operation and
Planning Research Department
Niroo Research Institute
Tehran, Iran

Pierluigi Siano
Department of Management &
Innovation Systems
University of Salerno
Fisciano, Salerno, Italy

Elahe Soleymani
University of Mohaghegh Ardabili
Ardabil, Iran

Ayobami Abiodun Taiwo
Department of Computational Data
Science and Engineering
North Carolina Agricultural and
Technical State University
Greensboro, NC, USA

Hamed Tajalli
Energy System Research Institute
(ESRI), Faculty of Electrical and
Computer Engineering
University of Tabriz
Tabriz, Iran

Theresa Thompson
Department of Computer Science,
Physics and Engineering
Benedict College
Columbia, SC, USA

Sajjad Tohidi
Energy System Research Institute
(ESRI), Faculty of Electrical and
Computer Engineering
University of Tabriz
Tabriz, Iran

Behrooz Vahidi
Electrical Engineering Department
Amirkabir University of Technology
Tehran, Iran

Zahra Yousefi
Department of Chemical and Polymer
Engineering
Yazd University
Yazd, Iran

Kazem Zare
Energy Systems Research Center
Khazar University
Baku, Azerbaijan

Energy System Research Institute
(ESRI), Faculty of Electrical and
Computer Engineering
University of Tabriz
Tabriz, Iran

About the Editors

Mohammadreza Daneshvar, PhD, SMIEEE, is an Assistant Professor, founder and head of the Laboratory of Multi-Carrier Energy Networks Modernization at the Faculty of Electrical and Computer Engineering, University of Tabriz, Tabriz, Iran. Prior to that, he was a postdoctoral research fellow in the field of modern multi-energy networks at the Smart Energy Systems Lab of the University of Tabriz for two years. He obtained his MSc and PhD degrees in electrical power engineering from the University of Tabriz, all with honors. He has (co)authored more than 50 technical journal and conference articles, 10 books, 30 book chapters, and 10 national and international research projects in the field. Dr. Daneshvar is a member of the Full Editorial Board of the *Sustainable Cities and Society* and the *Energy and Built Environment* journals. He also served as the guest editor for the *Sustainable Cities and Society* and *Sustainable Energy Technologies and Assessments* journals. Moreover, he serves as an active reviewer for more than 130 top journals of the IEEE, Elsevier, Springer Nature, Wiley, Taylor & Francis, and IOS Press, and was ranked among the top 1% of reviewers in engineering and cross-field based on Publons global reviewer database. His research interests include smart grids, transactive energy, energy management, renewable energy sources, integrated multi-energy systems, grid modernization, energy storage systems, sustainable cities and society, microgrids, energy hubs, machine learning and deep learning, digital twin, and optimization techniques and AI.

Amjad Anvari-Moghaddam is a Full Professor and leader of the iGRIDS Research Group at the Department of Energy (AAU Energy), Aalborg University, Denmark, where he is the coordinator of the Integrated Energy Systems Laboratory (IES-Lab). His research interests include planning, control, and operation management of microgrids, renewable/hybrid power systems, and integrated energy systems with appropriate market mechanisms. He serves as an alternate member of Council for Energy Efficient Transition at the Danish Ministry of Climate, Energy and Utilities, a member of Reference Group 5 "Climate, Energy and Mobility" for Horizon Europe (HE) at the Danish Agency for Education and Research (UFS), a board member of Energy Mission at Aalborg University, and the chair of IEEE Denmark Section.

He has (co)authored more than 300 technical articles, 11 books, and 20 book chapters in the field. Prof. Anvari-Moghaddam is the Editor-in-Chief of *Academia Green Energy* journal and serves as the associate editor of several leading journals. He is a member of IEC SC/8B Working Groups (WG 3 and WG 6) as well as technical committee member of several IEEE PES/IES/PELS and CIGRE WGs.

Mahdieh S. Sadabadi received the PhD degree from École Polytechnique Fédérale de Lausanne (EPFL), Switzerland, in 2016. She is currently an Assistant Professor in the Department of Electrical and Electronic Engineering at The University of Manchester, UK. Previously, she held academic positions at Queen Mary University of London and the University of Sheffield. She was a postdoctoral research associate in the Department of Engineering at the University of Cambridge, affiliated with Trinity College, and a postdoctoral fellow in the Division of Automatic Control at Linköping University, Sweden. She has held visiting positions at École Polytechnique de Montréal (Canada), LAAS-CNRS (France), HHMI Janelia Research Campus (United States), and KTH Royal Institute of Technology (Sweden). She is a Senior Member of IEEE, a Humboldt Fellow, and a Fellow of the Higher Education Academy (FHEA). Her research interests include microgrids, cyber-physical energy systems, and the stability and control of power-electronics-rich power systems.

Preface

Recent evolutions in the technology domain have directly affected people's lifestyles, driving them from past traditional forms toward digitalized and most energy-dependent fashions. Similar rapid transformations have significantly raised the need for various carriers of energy, exacerbating the energy crisis. This alarm has highlighted the necessity for increasing energy generation potential to continuously meet the new pattern of energy consumption in its different forms. Herein, the negative impacts of fossil-fuel-powered generation units on the clean environment and climate change, as well as the limited sources of such fuels, have driven the energy infrastructure to avail carbon-free solutions in the energy production process. As renewable energy sources (RESs) are known as a multi-beneficial option for compensating for energy shortages in line with the digitalization trend of the world, their inherent nature in delivering uncontrollable energy has made remarkable challenges in procuring secure and sustainable energy to end users. Although uncertainty in RESs' output has enhanced the risk of fully relying on them in energy generation, future modern energy networks (MENs) are scheduled to highly utilize the potential of RESs for clean energy production under deep decarbonization frameworks. In this respect, deep decarbonization not only emphasizes a high/full usage of RESs in the energy production sector, but also refers to a set of efforts in the vast area of energy interactions from energy conversion to storage, trading, and management. However, considering all the effects of such strategies on climate change, energy security, and sustainability in energy delivery is one of the most prominent steps in successfully reaching the Sustainable Development Goals adopted by the United Nations. Nevertheless, how deep decarbonization strategies can be effectively implemented in energy structures while taking effective factors of climate change and grid sustainability into account is a key question that needs to be addressed in deep detail. As the demand for multi-carrier energy is rapidly growing and global concerns are rising due to the unacceptable release of carbon emissions, a great need is felt for deep decarbonization of multi-energy networks by focusing on the climate–energy–sustainability nexus (CESN) frameworks. This book covers multifarious efforts for deep decarbonization of MENs and states creative ways for developing new solutions for developing green grid structures in line with CESN.

As a pioneering book that presents the fundamental theories, technologies, and solutions for real-world problems in the deep decarbonization of MENs by relying on the nexus of climate, energy, and sustainability, this book covers a conceptual introduction to deep decarbonization strategies, highlights potential technologies and managing solutions, and discusses requirements for realizing CESN in the deep decarbonization of hybrid energy grids. In the initial chapters, conceptual requirements with more emphasis on the definition, structure, features, and challenges of developing deep decarbonization frameworks are studied. The subsequent chapters of the book cover the potential technologies, paradigms, and systems, including powering technologies for resilience of decarbonized communities with energy diversification, AI techniques for carbon removal, intelligent cyber-physical energy systems (CPESs) for CESN, cybersecure digital twin architectures and quantum-enhanced analytics for smart CPESs, innovative carbon capture cycles for zero-emission resilient power generation, policymaking models for carbon removal, as well as real-world case studies for CESN in deep decarbonization of MENs.

As any research achievement may not be free of gaps, the editors kindly welcome any suggestions and comments from respectful readers to improve the quality of this work. Interested readers can share their valuable comments with the editors via mohammadrezadaneshvar95@gmail.com.

1

Transition Requirements in the Need for Deep Decarbonization of Modern Energy Networks

Mohammadreza Daneshvar[1]* and Mahdieh S. Sadabadi*[2]

[1] *Laboratory of Multi-Carrier Energy Networks Modernization, Faculty of Electrical and Computer Engineering, University of Tabriz, Tabriz, Iran*
[2] *Department of Electrical and Electronic Engineering, The University of Manchester, Manchester, UK*

*Corresponding author: m.r.daneshvar@tabrizu.ac.ir

List of Acronyms	
CESN	Climate–energy–sustainability nexus
EVs	Electric vehicles
MENs	Modern energy networks
RESs	Renewable energy sources

1.1 Introduction

In recent decades, industrialization has accelerated the growth of energy consumption by automating most services as well as improving the quality of life with energy-dependent devices [1]. As such changes denote the positive impacts of recent evolutions, another side of this transformation has brought tremendous challenges for the secure operation of power systems [2]. The stable and secure exploitation of the electrical power grid requires a dynamic establishment of balance in the amount of power production and consumption. The initial step for the above-mentioned balance is to ensure that the power generation matches its consumption. However, relying on fossil fuels with environmental emissions is no longer acceptable for modern societies, given advanced technologies in clean energy generation and governments' net-zero targets. As the main backbone of energy generation in modern energy networks (MENs), renewable energy sources (RESs) are considered as the overarching option for implementing carbon-free plans in the energy world [3]. However, their impacts are not limited to delivering clean energy, and intermittencies in their outputs threaten the sustainable development of the system [4]. This is while achieving a sustainable structure for MENs is further prominent to make the system able to provide continuous energy dispatch. On the other hand, supplying energy without interruptions is a necessity for serving high-quality services for end users. RESs challenge this task by fluctuating power, which creates uncertainties in the energy generation sector. Modern management schemes and capable energy strategies are indispensable to increasing the reliability index of energy supply. Hence, in addition to the sustainability factor, different energy services should also be taken into account in manipulating modern functionalities for future energy infrastructures. Nonetheless, the world witnesses substantial changes in the climate that trace back to human activities. Herein, how energy grids need to be developed directly affects the intensification of climate change. Recent experiences have discovered the negative impacts of constructing traditional power plants with fossil fuels on climate change factors. However, new policies emphasize relying on environmentally friendly ways for procuring the energy requirements of societies. As severe changes in climate are possible due to the rapid advancements in various technologies, accounting for climate change factors in modernizing energy grids is critical for reaching a zero-emission globe. Therefore, the nexus of climate, energy, and sustainability is vital in grid modernization. For this aim, this chapter covers the different dimensions of the need for the deep decarbonization of MENs for the climate–energy–sustainability nexus (CESN). It provides a general overview of this need, concentrating on clarifying the transition requirements.

The current book focuses on bridging gaps for the deep decarbonization of MENs for the climate-energy-sustainability nexus. It consists of fourteen chapters. Chapter 1 assesses the transition requirements for deep decarbonization of MENs. It also studies how deep decarbonization can facilitate reaching a carbon-free world with zero-emission energy structures. Chapter 2 analyzes the tight nexus of climate change, energy, and sustainable development by using the key role of demand response programs. This chapter also proposes a comprehensive framework and intends to support progressive development by developing a hierarchical process of activities and strategies. Chapter 3 delves into the impact of deep decarbonization on digitalization and industrialization of modern energy societies. It also examines the effects of digitalization on enabling real-time monitoring, applying control mechanisms for emission reduction, easing incorporation of advanced systems, and increasing efficiency. Chapter 4 inspects the creative practices, changing mindsets, and interaction of developing technologies for the climate-energy-sustainability nexus. It also explores the demand-side flexibility and promotes a unified strategy for sustainable and resilient net-zero energy grids.

Chapter 5 reveals energy justice and security by assessing nexus frameworks of climate-energy-sustainability to realize low-carbon societies. It also evaluates the impacts of key strategies like smart homes, electric vehicle deployment, and renewable energy adoption on energy justice and security. Chapter 6 concentrates on investigating energy storage pathways for the nexus of climate-energy-sustainability to facilitate deep decarbonization. It also examines the decline in operational emissions under the support of storage usage in the energy grid. Chapter 7 scrutinizes practical strategies for carbon dioxide removal in line with the realization of deep decarbonization. This chapter also presents several practical and implemented examples to discuss the effectiveness of the mentioned strategies. Chapter 8 covers investigations of technological innovations in the resilient operation of decarbonized communities. It also assesses energy diversification metrics to give applicable insights for policymakers. Chapter 9 accommodates AI techniques for carbon removal of MENs. It also explores AI's role in deep decarbonization and its indirect and direct applications, as well as covers carbon sequestration, utilization, and capture. Chapter 10 analyzes intelligent cyber-physical systems in deep decarbonization of MENs. It also highlights the creation of smart decision-support units by incorporating large language models and digital twins. Chapter 11 argues how quantum-augmented and cybersecure digital twin architectures can participate in deep decarbonization by using smart cyber-physical energy systems. It also contributes to comprehensive threat analysis and mitigation, as well as secure optimization of smart grids. Chapter 12 evaluates sustainable deep decarbonization of MENs by innovating carbon capture cycles for the power system with resilient and zero-emission features. It also evaluates exergy and energy for discovering thermodynamic losses and clarifies potential improvements in energy exchangers. Chapter 13 presents policymaking models to remove carbon under the CESN structure. It also reviews the capability of various methods such as game theory, agent-based modeling, mathematical programming, data-driven approaches, and life cycle assessment for CESN in the MEN. Finally, Chapter 14 assesses real-world case studies for CESN in the deep decarbonization of the MEN. It also examines how CESN can drive sustainability and climate in low-carbon transitions.

1.2 The Necessity for Deep Decarbonization

Nowadays, decarbonization activities are more sensible in practice due to the growing trend of deploying RESs and the relevant supportive technologies [5]. However, the breadth of energy-consumed activities has highlighted the essentiality of not restricting decarbonization objectives to specific levels and sectors [6]. Such an approach drives decarbonization endeavors to much deeper dimensions of effective parts. The deep decarbonization in the energy area forces substantial shifts of the system in both energy demand-side profiles and energy production portfolios [7]. The new policies also strengthen this shift by incentivizing RESs usage, aiming to mitigate the carbon intensity of energy generation [8]. Achieving the significant alleviation of climate change is also another key motivation for transitioning towards the deep decarbonization of MENs [9]. To reach deep decarbonization of the energy world, different solutions are widely assessed and implemented in practice, yet, which indicate multiple aspects facing this trend in the research community. The main one implies the high deployment of RESs and benefits various types of them for clean energy harvesting. However, intermittent power generation, particularly with an uncertainty feature, has made RESs as a challenging option for sensitive goals. RESs impose more operational costs on the energy system for securely integrating the grid, as energy balancing should be done in their presence [10]. Herein, RESs, such as onshore and offshore wind energy as well as solar, have a role as predominant sources for supplying carbon-free energy and play a central role in managing renewables fluctuations under different decarbonization constraints [11]. However, endeavors for decarbonization are not limited to RESs alone, and multidimensional pathways are being developed around the globe to ease the reach of deep decarbonization goals. Some of the prominent pathways are demonstrated in Figure 1.1 [12].

As shown in Figure 1.1, different urgent pathways are considered for Canada to reach deep decarbonization goals. Grid flexibility enhancement is one of the most crucial ones that concentrates on tight flexible ways, like energy storage systems, for tackling the intermittencies in renewables' outputs. However, energy storage technologies are not just intended for the grid-level applications, and their vehicle-scale versions are targeted to be developed in line with decarbonization principles. As one of the effective sectors in the grid, the transportation area is expected to undergo changes in using carbon-free fuels that mostly rely on hydrogen and advanced biofuels. In the decarbonization path, exerting low-carbon techniques should not be restricted to special parts, and advanced extraction techniques can be applied for oil sands, such as solvent extraction. These approaches can also be developed for various carbon functionalities to enable the modern grid for carbon capture, production, extraction, storage, conversion, and utilization. Given that the industry is at the point of huge energy consumption, such carbon functionalities can be adopted with other practical green technologies to reduce the portion of the industry in CO_2 release. Herein, CO_2 process sinks can be developed in a way to be optimally integrated with diverse carbon functionalities for affordably optimizing processes. A set of the mentioned tools empowered the energy network to shape zero-emission processes to harness the carbon release of various areas.

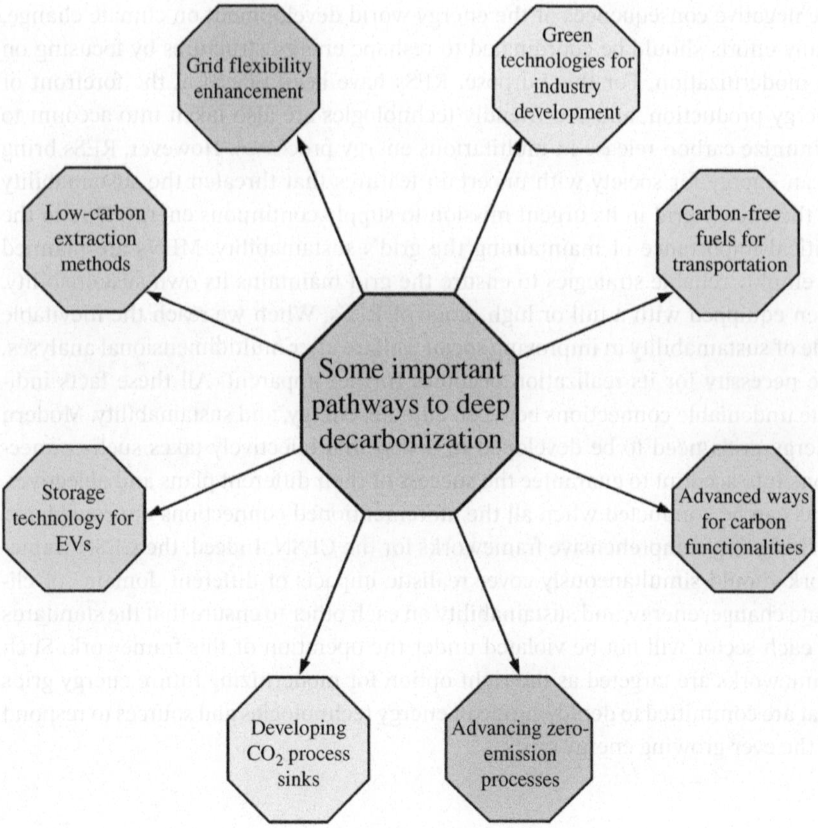

Figure 1.1 Some of the prominent pathways for deep decarbonization.

1.3 CESN Framework and Connections

Nowadays, climate change has been accelerated, particularly under human-origin activities. The increasing need for energy has been highlighted in the present style of human life that mostly depends on energy-consuming equipment. Such evolutions in the energy area have driven this sector to experience huge changes in its different parts to be able to dynamically meet the energy consumption. This transformation has occurred from energy generation to energy conversion, storage, and management. The impact of such changes in the energy landscape has not influenced the energy area alone, and one of its considerable effects has been on the climate. Indeed, growing traditional energy systems, especially fossil-fuel-based units, have substantial impacts on climate change. To mitigate the negative consequences of the energy world development on climate change, many efforts should be coordinated to reshape energy structures by focusing on its modernization. For this purpose, RESs have been placed at the forefront of energy production, and eco-friendly technologies are also taken into account to minimize carbon release in multifarious energy processes. However, RESs bring clean energy for society with uncertain features that threaten the sustainability of the energy grid in its urgent mission to supply continuous energy. Due to the critical importance of maintaining the grid's sustainability, MENs are planned to employ reliable strategies to ensure the grid maintains its own sustainability, even equipped with a full or high range of RESs. When we reach the inevitable role of sustainability in improving social welfare after multidimensional analyses, the necessity for its realization becomes further apparent. All these facts indicate undeniable connections between climate, energy, and sustainability. Modern energy grids need to be developed in a way that effectively takes such connections into account to guarantee the success of their different plans and objectives. This can be conducted when all the aforementioned connections are considered in designing comprehensive frameworks for the CESN. Indeed, the CESN framework should simultaneously cover realistic impacts of different domains of climate change, energy, and sustainability on each other to ensure that the standards of each sector will not be violated under the operation of this framework. Such frameworks are targeted as the right option for modernizing future energy grids that are committed to deploying clean energy technologies and sources to respond to the ever-growing energy crisis.

1.4 Conclusions

Today, the world has witnessed ambitious changes in different fields that have directly affected styles of human lives and have made it dependent on various forms of energy more than ever before. Due to this, accessing clean, affordable, and sustainable energy has become a desirable and strategic request among human societies. Given the essentiality of this need, this chapter delves into assessing transition requirements for deep decarbonization of MENs. The chapter begins by discussing the urgent need for deep decarbonization of energy systems. Due to the direct impacts of decarbonization activities on climate change, energy accessibility, and the grid's sustainability, the CESN frameworks and connections are more revealed to provide the correct vision of frameworks for modernizing future MENs. By discussing such foundational concepts, the chapter prepares readers for the detailed analyses in subsequent chapters.

1.5 Future Directions and Challenges

In the modern energy landscape, deep decarbonization of MENs for CESN paves forming the set of endeavours to correctly drive the integrated system towards a net-zero emission structure. However, some critical challenges can negatively affect the complete success of the relevant plans. Technology limitation is one of the most remarkable ones that makes the smooth implementation of some beneficial and scheduled plans difficult. On the other hand, technology complexity alleviates the feasibility of developed paradigms and makes them hard to form in practice. Decarbonization-oriented plans are often hierarchical and interdependent in nature, which work coordinately with each other to achieve specific goals. Such mechanisms make the process costly, and their affordability will be questionable. This issue can lead to a significant decline in participation of different players in grid interactions. Human activities are always another effective dimension in environmental consequences and climate change. The lack of alignment of such activities typically hinders the effectiveness of deep decarbonization schemes. Another key challenge is that energy demand is dramatically surging, leading to an undeniable increase in energy generation and further challenges for the deep decarbonization of societies.

The aforementioned challenges are only some of the central difficulties ahead of employing deep decarbonization plans. From a futuristic perspective, each of them designs new roads for researchers to make them solvable by finding applicable solutions. Thus, future directions can be constructed on the described central challenges. Advancing the science of technologies can not only assist the system to address technology limitations but also can mostly ease the complexity involved in the process. Adopting emerging sciences such as multi-energy paradigms and digitalization platforms can support the integrated structure to reach the mentioned goals. On the other hand, further deploying mature technologies and nature-supported solutions can foster cost-effective mechanisms for implementing decarbonization frameworks, allowing for further reduction in energy costs. Improving human awareness of the different dimensions of decarbonization efforts and providing appropriate training for aligned actions can reduce the adverse effects of human activities to some extent. Such measures can also be expanded to the field of energy management that drives consumers to the correct energy consumption patterns, resulting in a decline in energy usage and achieving expected demand behaviors. All the mentioned items can be considered in future efforts of researchers to effectively open the ways of deep decarbonization for MENs.

References

1 G. Xiu, "Analyzing the impact of supply chain and digitalization in the Chinese economy: What is the role of energy consumption, government expenditure, and industrialization in environmental pollution?," *Energy Economics*, vol. 145, p. 108412, 2025.

2 M. Daneshvar, B. Mohammadi-ivatloo, S. Asadi, and S. Galvani, "Short term optimal hydro-thermal scheduling of the transmission system equipped with pumped storage in the competitive environment," *Majlesi Journal of Electrical Engineering*, vol. 14, no. 1, pp. 77–84, 2020.

3 M. Daneshvar, B. Mohammadi-ivatloo, and A. Anvari-Moghaddam, "Risk-oriented operational model for fully renewable cooperative prosumers in a modern water-energy nexus structure," *IEEE Transactions on Sustainable Energy*, vol. 14, no. 4, pp. 2006–2019, 2023.

4 L. Zhang, R. Zhang, X. Zhang, B. Xie, and X. Chen, "China's development path for intermittent renewable energy power generation: The influence of diffusion factors," *Chinese Journal of Population, Resources and Environment*, vol. 23, no. 2, pp. 193–204, 2025.

5 Z. Wang, Y. Sun, H. Kong, and C. Xia-Bauer, "An in-depth review of key technologies and pathways to carbon neutrality: Classification and assessment of decarbonization technologies," *Carbon Neutrality*, vol. 4, no. 1, p. 15, 2025.

6 M. Daneshvar, B. Mohammadi-ivatloo, K. Zare, and A. Anvari-Moghaddam, "Risk-aware stochastic scheduling of hybrid integrated energy systems with 100% renewables," *IEEE Transactions on Engineering Management*, vol. 71, pp. 9314–9324, 2023.

7 Z. Zhou, T. Levin, and A. Botterud, "Evolution of grid services for deep decarbonization: The role of hydropower," *Current Sustainable/Renewable Energy Reports*, vol. 11, no. 4, pp. 145–155, 2024.

8 M. Arbabzadeh, R. Sioshansi, J. X. Johnson, and G. A. Keoleian, "The role of energy storage in deep decarbonization of electricity production," *Nature Communications*, vol. 10, no. 1, p. 3413, 2019.

9 J. H. Williams et al., "The technology path to deep greenhouse gas emissions cuts by 2050: The pivotal role of electricity," *Science*, vol. 335, no. 6064, pp. 53–59, 2012.

10 A. Mileva, J. Johnston, J. H. Nelson, and D. M. Kammen, "Power system balancing for deep decarbonization of the electricity sector," *Applied Energy*, vol. 162, pp. 1001–1009, 2016.

11 R. Khorramfar, M. Santoni-Colvin, S. Amin, L. K. Norford, A. Botterud, and D. Mallapragada, "Cost-effective planning of decarbonized power-gas infrastructure to meet the challenges of heating electrification," *Cell Reports Sustainability*, vol. 2, no. 2, p. 100307, 2025.

12 C. Bataille, D. Sawyer, and N. Melton, *Pathways to Deep Decarbonization in Canada*. Sustainable Development Solutions Network, 2015.

2

The Important Role of Demand Response in the Tight Nexus of Climate Change, Energy, and Sustainable Development

Hamidreza Arasteh[1,2], Seyed Mohsen Hashemi[2], Habib A. Aalami[3], Alireza Sheikhi-Fini[2], Anand Rajendran[1], and Josep M. Guerrero[1]*

[1] *Center for Research on Microgrids (CROM), Huanjiang Laboratory, Shaoxing, Zhejiang, China*
[2] *Power Systems Operation and Planning Research Department, Niroo Research Institute, Tehran, Iran*
[3] *Electrical & Computer Department, Eyvanekey University, Tehran, Iran*

*Corresponding author: hr.arasteh@ieee.org

List of Acronyms	
AI	Artificial intelligence
AMI	Advanced metering infrastructure
CBL	Customer baseline load
CEF	Carbon emission flow
DER	Distributed energy resource
DG	Distributed generation
DR	Demand response
DSL	Digital sharing line
DSM	Demand-side management
DSR	Demand-side response
DTLR	Dynamic thermal line rating
EE	Energy efficiency
EMS	Energy management system
ESG	Environmental, social, and governance
EV	Electric vehicle

List of Acronyms	
GHG	Greenhouse gas
HAN	Home area network
ICT	Information and communication technology
IoT	Internet of Things
LAN	Local area network
ML	Machine learning
PLC	Power line communication
RESs	Renewable energy sources
SDG	Sustainable Development Goal
V2G	Vehicle-to-grid
WAN	Wide area network
WPS	Wind power spillage

2.1 Introduction

In conventional power systems, generation resources are the main active agents in system operation procedures. The system operator forecasts the electrical load during the short-term and long-term horizons and determines desired decisions, respectively, to expand and operate the generation resources. In other words, by increasing *X*% in the annual peak load, the generation resources should be extended by about *X*% to deal with the load growth. Different programs of demand-side management (DSM) were developed by advancing the smart grids and activating the demand-side resources in the system operation procedure, different programs of DSM were developed. Two main subsets of DSM are energy efficiency (EE) and demand response (DR) [1].

The power system, one of the most crucial energy-supplying infrastructures, is experiencing a new structural change. High penetration of renewable energy technologies, distributed generation (DG), and energy storage units is significantly changing the power generation portfolio [2, 3]. Energy management system (EMS) could be employed for the optimum scheduling of the system, including different generation units and demand-side resources [4–7].

The increasing dependence of customers on electrical power and the development of digital and intelligent technologies are changing the power consumption portfolio [8]. The high capability of the active loads to participate in the system operation procedure makes it necessary to predict the future condition of DR and explain its related concepts. Several entities could play roles in the future smart grids to provide different services, such as DR [9–12]. DR incorporates different domains, e.g., technical, economic, social, and cultural aspects. This chapter aims to examine diverse implementation factors of DR. In order to complete this objective, it evaluates the specifications of DR's stakeholders, as well as their challenges and needs. The procedure to employ DR programs is typically specified by recognizing the most critical problems faced by the players, e.g., consumers, investors, and distribution companies. This chapter not only identifies the main research directions in DR activation but also develops real-world solutions to overcome the relevant challenges. From a technical viewpoint, advanced metering infrastructure (AMI) and communication systems are pivotal technologies to implement DR programs. Also, the employment of analyzing tools empowers the DR planners or aggregators to extract valuable information about the consumption pattern of the loads and select the most appropriate DR programs for them. From the economic viewpoint, the cost of implementing DR programs, including the investment and administration costs, should be appropriately analyzed compared to their values. According to different uncertainties, there may be economic risks that the DR planners should consider. There are different stakeholders in implementing DR programs. Different legislative requirements should be considered to regulate their roles and relations. Some rules should be passed for developing the necessary infrastructures, providing financial resources, and determining the incentive or punitive policies. In addition to the legislation, the successful implementation of DR programs needs to train human resources. Moreover, it may be necessary to change the organizational structure of utility companies to define new managerial tasks for better implementation of DR programs.

As above-mentioned, there are several challenges to implement DR programs. Some of the challenges are from the consumer's side, such as their knowledge about DR programs and the available technologies required to implement DR. Another challenge is the consumers' reluctance to follow the DR programs because of the complexity of the DR programs. Also, DR incentives may not be exciting enough to incentivize consumers to follow them. There are challenges from the DR service providers' side, such as the investment recovery of DR programs or their conflict of interest in reducing power consumption. Other challenges, such as economic, social, and technical, are studied in this chapter. Also, the chapter provides appropriate solutions to these challenges and requirements. As is shown in Figure 2.1, the successful activation of DR includes satisfying certain needs and overcoming different challenges using effective solutions for achieving profits for the customers and the network. Moreover, this chapter proposed a framework to

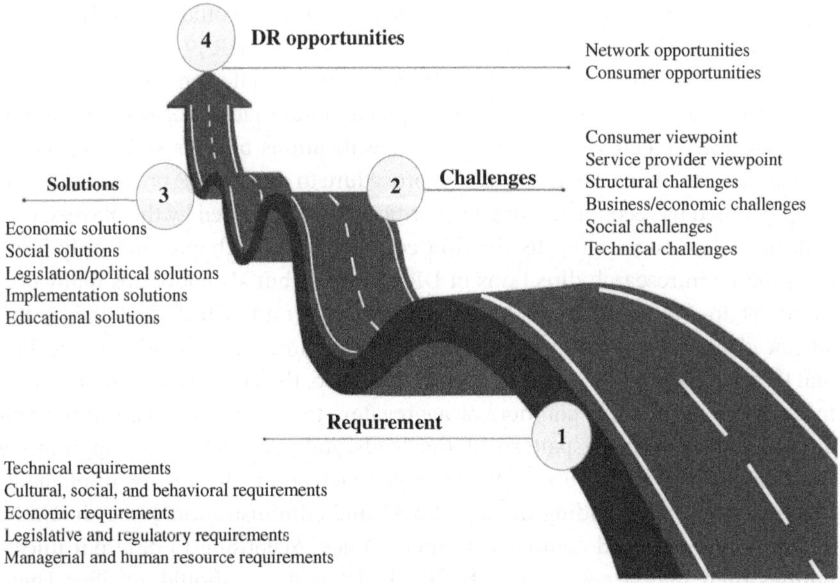

Figure 2.1 Unlocking the potential of DR: meeting requirements, tackling challenges, and implementing effective solutions.

implement DR by introducing three generations of DR. The goal of the proposed concept is to provide an organized procedure for tailoring specific sets of actions essential for unlocking the capacity of DR. Proposing these DR generations not only accounts for planning horizons concerning the employment of DR, but also takes into account the current situation of the system and the readiness to include various DR models.

In addition to the classification of DR into three generations, an inclusive procedure is proposed for activating and maximizing the profits of DR. It highlights the necessity of identifying the system's status and different time horizons, allowing decision makers and operators to evaluate the system's readiness to encompass various DR models. Then, the selection of the appropriate DR generation requires a set of efforts to efficiently activate DR and proceed toward a perfect employment. Also, this chapter investigates the vital criteria to evaluate DR. Additionally, it meticulously analyses the present condition of a test system, with the aim of assessing its potential to implement DR. Investigation of the DR specifications serve as a basic concept to determine a system's capacity to host different generations of DR. Indeed, using the presented framework, it would be possible to understand the system's readiness to embrace diverse tiers or phases of DR. This procedure not only reveals the system's capacities, but also clarifies its strengths and potential areas needing improvement. So, it offers an inclusive comprehension of the system's potential to successfully support diverse DR strategies and progress. In fact, the investigations of the test system's status offer deep understanding into the real-world applications of DR.

Lastly, the chapter develops the concepts in order to discover the strategic role of DR in the global climate–energy–sustainability nexus. It underscores how cutting-edge paradigms, such as carbon-aware DR and digital orchestration, can redefine DR as a key element for deep decarbonizing, integrating renewables, and transitioning toward resilient and just systems. This forward-thinking study highlights the important role of DR as a foundation of sustainable grids and its important contribution to achieving wider climate and sustainability objectives.

Briefly, the main contributions of the chapter are:

- Representation of various DR opportunities for customers and stakeholders.
- Identification and classification of DR needs, including technical, economic, social, and cultural aspects.
- Analysis of DR challenges from the DR players' points of view.
- Proposing efficient solutions for coping with the DR challenges.
- Proposing three generations of DR, as well as an organized applicable process for unlocking the actual DR capacity.

Therefore, the main aims of this chapter are: (1) to comprehensively investigate the technical, economic, social, cultural, and legislative needs for efficient DR employment; (2) to recognize main challenges faced by several players and propose possible solutions; (3) to propose three generations of DR and offer a hierarchical procedure for the progressive development; and (4) to underscore the strategic role of DR to support decarbonization, high renewable penetration, and sustainability within future power systems.

2.2 DR Opportunities

DR programs not only provide financial profits but also have the capability to provide technical, environmental, and market-related benefits. This section classifies these opportunities.

2.2.1 Opportunities for Customers

Economic opportunities could be achieved as the result of consumers' participation in DR, duo to the changes in the time or value of energy usage for reducing the billing costs [13]. Incentive-based DR also offers incentives for customers to participate in DR. From the technical point of view, the participation of consumers in DR can improve the reliability level of the system [14–16], and decrease the risk of energy curtailments. In addition, customers who have participated in DR may enjoy from knowing that their responses help to avoid energy curtailments for others, even if they have not received any direct monetary rewards. This sense of efficacy can be considered as motivation for customers to respond to DR calls.

2.2.2 Opportunities for System Operators

Price-based DR is a tool to introduce real-time pricing, where customers can face the prices reflecting actual power system conditions [17]. Incorporating DR into the real-time operation of the system can alleviate the risk of sudden price spikes in electricity markets [18, 19]. DR empowers customers to adjust the energy usage patterns in response to energy prices. This responsiveness is a tool to react against the misuse of market power. When producers try to manipulate the energy prices, DR can modify the energy consumption in response to high energy prices, and therefore limit the ability of producers to increase market prices [20].

Additionally, DR plays a vital role in deferring power generation and network expansion planning by effectively decreasing peak power values [21–24]. By alleviating the constraints and requirements for the expansion plans, DR can postpone the need for network expansion.

The reserve capacity is a backup for the online generation capacity and can be used in the contingency conditions. DR also can be considered as the reserve capacity to improve the reliability of the network [25]. Moreover, DR can reduce the peak load level and prevent the synchronization of inefficient power plants characterized by high fuel consumption. Therefore, such the reductions in peak load levels also decrease the emission rates and contribute to environmental advantages. DR can further improve the environmental benefits when joined with other distributed energy resources (DERs), e.g., electric vehicles (EVs) [26–28].

2.2.3 Opportunities for the Grids' Sustainability

Through the coordination of energy usage patterns with renewable generation, DR can play an important role in promoting the sustainability of modern energy systems. By mitigating the peak load problems and facilitating the integration of renewable energy sources (RESs), e.g., wind and solar, DR can minimize the reliance on fossil-based power plants and thereby reduce greenhouse gas (GHG) emissions. It also has the capability to optimize resource usage and mitigate stress on generation and network assets, and so extend their operational lifespan and defer the need for costly capacity expansions. Through these mechanisms, DR programs contribute to more efficient, environmentally responsible, and economic grid operations.

Moreover, DR improves grid resilience and adaptability by activating flexible load management during contingencies, equipment failures, or climate-induced disruptions. The aggregation of DERs and active and responsive customers fosters a decentralized grid structure supporting reliability, fairness, and local self-sufficiency. Combined with emerging digital technologies, e.g., artificial intelligence (AI), Internet of Things (IoT), and blockchain, DR programs empower real-time coordination, transparency, and carbon-linked decision-making, key components to advance the long-term sustainability and stability of future smart grids.

2.3 DR Requirements

For the successful implementation of DR, different requirements should be considered. This section provides complementary explanations regarding the essential needs.

2.3.1 Technological Requirements

Information and communication technologies (ICTs) are crucial requirements for a successful implementation of DR [29]. Infrastructures like AMI, communication platforms, and data analyzing tools are the leading technologies needed to implement DR. AMI is an important part of DR that enables the measurement and management of data through different timescales. The key application of AMI is metering, gathering, and analyzing the energy-related data [30]. AMI includes several software and hardware systems, e.g., communication systems, controllers, thermostatic equipment, smart meters, and data centers. In most DR programs, consumers should consider real-time electricity prices to make actions, such as load reduction. Also, in some DR programs, consumers can participate in the power markets by providing ancillary services [31]. According to these requirements and considering the system security standards, an efficient metering and communication system should be used.

The energy consumption procedure is a continuous action that should be continuously metered. In a smart grid, the energy flow should be metered and reported to the data analysis servers [32]. Smart meters record the flow of energy consumption in near real-time and send it to data centers regularly. They use two-way communication.

Monitoring and storing energy consumption data is essential in the smart grid. However, it should be completed by a robust data analysis module. This module can extract valuable datasets from the gathered data [33]. The imported data from the smart meters should be checked, validated, and processed before any professional analysis.

Communication systems have essential roles to facilitate the successful implementation of DR by enabling seamless connectivity between several program elements [34]. The scale of communication networks, including local area network (LAN), wide area network (WAN), and home area network (HAN), determines the scope and coverage of connectivity within DR systems. Moreover, a range of communication technologies like power line communication (PLC), digital sharing line (DSL), optical fiber, Ethernet, Bluetooth, low-power radio frequency, cellular networks, 5G technology, and WiMAX are available that offer important benefits and applications within DR contexts. The integration of IoT allows interconnected physical devices to exchange data, which enables remote monitoring and control of household appliances through smartphones or tablets [35–37]. However, it is crucial to ensure cybersecurity within DR programs. Since DR comprises the collection and analysis of customers' data, a secure communication framework is required. Standards like IEC61968 by the IEC have been developed to secure data transmitted through power distribution systems, which emphasize the cruciality of cybersecurity in DR employment.

High penetration of DR programs needs to analyze large amounts of data. In fact, after checking the data and constructing a dataset, a data analyzing module should use it to generate appropriate information for DR implementation. Data science techniques can typically be implemented to preexisting datasets with various objectives, such as knowledge extraction or pattern discovery. Data mining, in particular, is a widely used subset of data science that provides a potent means to extract valuable insights from vast amounts of data. Data mining is a technology to discover patterns from large datasets. It applies different statistics and machine learning (ML) techniques to translate the data into practical information and rules [38]. Using data mining technology, DR providers discover the consumption pattern of the consumers and their dependence on different economic and environmental factors. It helps the DR provider to design appropriate DR programs for different consumers.

It should be noted that data mining requires an extensive dataset to discover the relationships properly. Combining data mining methods and computational technologies can create a powerful, intelligent tool for power consumption management [39]. In a broader view, AI plays an essential role. AI receives the data from the environment and acts through the actuators on the environment [40], and includes different methods of statistical learning [41, 42], soft computing [43], logic-based and knowledge-based systems [44, 45], etc. AI can be utilized in different problems, such as estimating the customer baseline load (CBL) based on the historical data [46–49], forecasting the market price for managing the loads [50], load forecasting to apply the DR programs, and designing the customer pricing methods [51, 52]. Figure 2.2 shows the applications of AI in DR.

2.3.2 Cultural, Social, and Behavioral Requirements

It is critical to discover theoretical underpinnings, practical applications, and statistical insights in order to examine the key cultural, social, and behavioral domains within DR. The subsequent explanations delineate these aspects, shedding light on the intricate interplay between cultural, social, and behavioral elements in the successful implementation of DR programs. The following offers an inclusive insight into the theoretical frameworks, real-world implications, and empirical findings.

Theoretical concepts: DSM theoretical concepts include some significant aspects influencing its successful employment. Customer awareness of DSM is one of them. The implementation of DSM programs can have important economic and technical profits. Electric companies can educate customers about proper usage behaviors and encourage their adoption.

Cultural and lifestyle factors also meaningfully influence consumption patterns. Consumption behaviors are closely linked with lifestyles, which are vital in a successful implementation of DR. Existing studies have widely studied the connection between lifestyles and consumption behaviors [53–55].

Environmental knowledge is another vital factor. The 21st century has observed a significant increase in environmental pressures resulting from

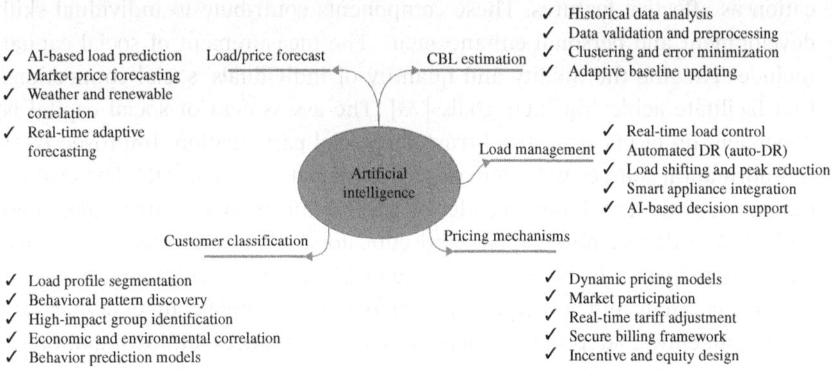

Figure 2.2 Applications of AI in DR.

human actions. Attitudes and beliefs are serious determinants of human behavior, and different studies have been developed to understand their role in predicting environmental behaviors. Knowledge is a crucial element to facilitate behavioral changes [56]. Improving environmental knowledge is necessary to successfully implement DR that aims to reduce human impact on the environment.

People's psychological and behavioral impressionability also plays a key role in shaping electricity consumption. Electricity price is an influential factor in consumers' energy consumption. Many mathematical models have been presented to calculate the loads' sensitivity to price based on the concept of price–demand elasticity [57]. The elasticity is the sensitivity of the consumer's load to the electricity prices [58]. It can be modeled by the linear and nonlinear mathematical models [57, 59]. During a daily horizon, the electricity's value may vary by time, so the elasticity can be calculated at different times of the day [60, 61]. Elasticity can forecast the load in real-time applications [62]. Consumers' behaviors are generally interconnected with financial rewards [63]. Also, in the context of psychology, nonfinancial programs can affect the consumption pattern of consumers [64]. Although both the reward and penalty can affect the short-term behaviors of the consumers [65], the reward-based programs make a habit in the consumers and result in the long-term effects [53, 66]. Based on studies in the context of human behaviors, different consumers react differently to the penalties and rewards [67]. Nevertheless, some general principles can be considered, including consumers' loss aversion [68]. This principle suggests that consumers prioritize avoiding losses over maximizing profits [69–71].

Practicality aspects: in connecting theoretical aspects to the practical implementation of DR, this part addresses several solutions. For assessing cultural capital, some researches such as [72] suggested to evaluate habits and education as affective features. These components contribute to individual skill development and personal enhancement. The measurement of social capital includes gauging the quality and quantity of individuals' social connections that facilitate achieving their goals [73]. The assessment of social capital is dependent on facets, e.g., social trust and social participation. Improved trust in information sources will increase the participation rate in DR. The evaluation of social trust includes considering several information sources, like radio and TV broadcasts, newspapers, and educators, as well as expert opinions. Social participation involves conscious engagement in specific programs with clear goals. Analyzing participation in energy management courses, adherence to the guidelines, adoption of energy-efficient appliances, and installation of multi-time meters offer insights into social participation. Moreover, economic capital contains financial assets of people, family income, and the number of electrical appliances owned, all of which contribute to measuring economic capital [74].

Statistical studies: identification and usage of common practical DR programs, e.g., the employment of efficient appliances, switching off surplus lights, or reducing the utilization of high-consumption devices during peak hours, provide visions for prevailing practicality of these programs [75]. Assessing the credibility of information sources regarding DR through statistical analysis aids decision makers in employing the most effective educational and promotional plans [76]. Furthermore, the segmentation of customers based on their responsiveness to DR calls allows prioritizing DR programs and policies [18], allowing decision makers to tailor initiatives better.

2.3.3 Economic Requirements

Overall economic aspects of DR programs are illustrated in Figure 2.3, and can be categorized into: DR value or pricing, investment recovery, and risk management.

DR value or pricing: a successful implementation of DR is dependent on different aspects, particularly the relationship between electricity prices and their real values [77]. Programs relying on energy prices should guarantee that prices properly reflect the energy values for incentivizing customers' responses. Incentive-based programs could be facilitated using competitive frameworks such as the DR exchange market, allowing customers to offer their DR services, ensuring proper financial settlement approaches are in place [78, 79]. Moreover, customers' income levels meaningfully affect DR implementation results. Generally, lower-income people show more sensitivity to electricity prices and therefore, higher participation in DR programs [77]. In addition, decision makers should carefully take into account the implementation costs of DR initiatives

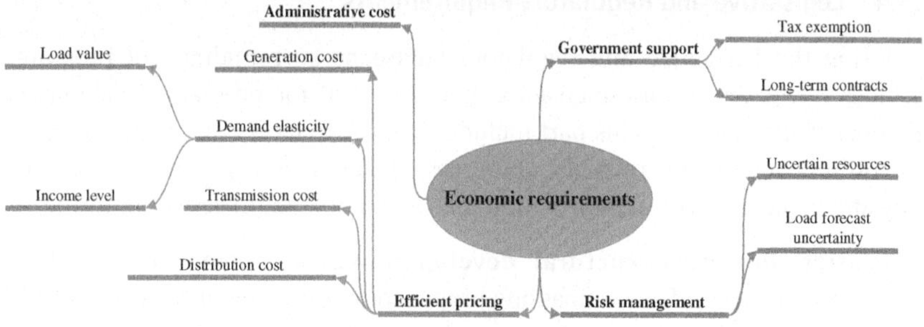

Figure 2.3 Overall economic aspects of DR programs.

[80], looking for governmental financial support for enhancing investment attractiveness and mitigating potential income reductions for power distribution companies. It should be noted that the implementation of DR may decrease the income of the power distribution companies, since responsive customers can modify their consumption patterns and reduce their electricity bills. Also, the assessment of DR influences on operational and investment costs within power systems is vital, since it serves as a metric for prioritizing DR initiatives [81, 82]. Cost-effective DR initiatives that yield significant decreases in financial costs must take precedence [83].

Investment recovery: implementing DR programs often demands substantial investment in measurement, communication, and control equipment that emphasizes the requirements for robust economic frameworks to recuperate these expenses [13]. To ensure economic viability and incentivize participation, mid- or long-term contracts can be established, offering guarantees for investment recovery and enhancing the attractiveness of DR. Moreover, providing tax exemptions can further improve the willingness to these programs [84]. In long-term electricity markets, such as capacity markets, responsive customers or aggregators can have participation in DR by presenting their price-capacity offers. The development of these frameworks to include DR appropriately provides financial resources [85–87].

Risk management: although DR offers mitigating financial and technical risks within energy systems, all providers and customers participating in DR programs are still susceptible to different risks. The management of these risks needs an inclusive structure that covers several pertinent subjects, such as risk management instruments [88], dealing with uncertainties of DR [89, 90], and the management of DR characterized by high levels of intermittent generation capacity [91, 92].

2.3.4 Legislative and Regulatory Requirements

Exploring the legislative and regulatory landscape surrounding DR programs includes navigating multidimensional needs crucial for program development and operational success. This part includes legislative facets that focus on infrastructural development, financial resources, stakeholder engagement, incentivizing and penalizing rules, information management, and process monitoring.

Legislation for infrastructural developments: developing DR initiatives requires a smart grid, encompassing proper communication infrastructure and smart elements [93]. Hence, the legislation and regulation requirements of the smart grid should be created [94, 95].

Legislation for financial resources and developing the business environment in the power distribution system: it is critical to the successful implementation and expansion of DR programs. One essential aspect is attracting private investors. To encourage their participation, appropriate legislative support, such as tax exemption [84] and payment guarantees, is essential.

Another important factor is fostering competition among different DR service providers. DR programs can be provided in a competitive environment where different DR providers try to increase service quality and decrease prices.

In other words, different business environments should be formed to provide better DR services [96, 97]. As DR services are integrated into the energy management systems, the provided business environments have to consider possible technical limitations of these systems [98]. In order to support and expand this business, appropriate rules should be created by the legislative body.

Furthermore, cooperation with startup companies, universities, and research institutes is crucial, especially given the relative novelty of DR in many regions. These institutions can play a pivotal role in identifying efficient, innovative, and context-appropriate solutions for DR implementation [99, 100]. For this purpose, appropriate rules should support financial contracts [101].

Facilitating rules to employ the existing potentials of the system: it is essential for advancing DR programs. One such potential lies in the self-production capabilities of large industries. In many cases, there are several load-side resources in the system that the system operator can employ to implement DR programs. For example, generation resources in large industries can be used to serve the loads during peak hours. In order to encourage the loads to offer their generation resources, supportive rules should be created in different aspects, such as financial rules.

Additionally, promoting stakeholder engagement in policy-making and planning processes is crucial. To successfully implement DR programs, stakeholders should be involved in the related policy-making actions. In other words, a legal structure should be provided for all stakeholders to participate in the policy-making procedures [102].

Legislation for incentivizing and punitive policies: it plays a pivotal role in ensuring their effectiveness and sustainability. In addition to the needed rules for the implementation of DR programs, appropriate rules should have existed to control and manage the implementation of DR programs. The accurate models for penalty and reward [103, 104] can be calculated based on the consumers' properties. Mainly, it will be more critical when no appropriate metering equipment and smart grid infrastructure exist.

Another important consideration is the inclusion of satisfaction criteria in the design and execution of DR programs. The implementation of DR programs should be performed considering the satisfaction aspects of the consumers and the planners. In other words, several satisfaction criteria should be selected on different sides of the DR programs, which highly encourage people to utilize DR programs [105–109].

Equally important is the motivation of employees within power distribution networks, who are among the key actors in the operationalization of DR initiatives. Well-designed motivational approaches must be suggested for enhancing their performance and commitment. Such approaches are mainly valuable in the long term since they contribute to the sustained employment and continuous development of DR initiatives [110].

Legislation for information management: this is an important element to ensure a secure and efficient operation of DR. DR providers gather participant data, like consumption patterns, equipment performance, and other metrics. This data is vital for the successful implementation of DR, enabling providers to forecast demand, optimize load balancing, and detect potential problems before they become critical. However, gathering this data also creates a risk for security and transparency breaches. As an instance, if unauthorized parties have access to this data, they might utilize it for a cyberattack, commit fraud, or other malicious actions. DR providers must employ stringent security tools to reduce these risks for protecting the collected data. This includes protocols, e.g., data encryption, access controls, and regular vulnerability testing.

Moreover, DR providers should be transparent about their data collection and use practices. This means providing clear information to participants about what data is being gathered, how it will be utilized, and who will have access to it. In this way, providers can create trust and confidence among DR participants and stakeholders and guarantee the success of the DR initiatives.

Furthermore, implementing a robust process monitoring system is essential to guarantee that DR initiatives are efficiently harmonized with other energy resources, like energy storage, and RESs [111, 112]. The high level of RESs, energy storages, and DR can improve the reliability of the system and alleviate the uncertainties. It is critical to provide proper legislative instructions for distribution system operators to efficiently control and monitor the operational procedures in the presence of these sources [111].

2.3.5 Human Resource and Managerial Requirements

It is crucial to explore different integral aspects to understand the key human resource and managerial requirements within the realm of DR. This part navigates through critical components covering staff training, organizational restructuring, defining new managerial tasks, and program prioritization.

Training the staff: this is a basic prerequisite for implementing DR successfully within energy systems. Educational development can be studied as the core facet of organizational development, as it comprises diverse pillars of the development of a company. Power distribution companies should have enough attention to training their staff about DR programs [113]. It should be noted that, from an economic viewpoint, educational costs can be considered an investment, as they increase staff performance to employ DR programs appropriately.

Beyond technical knowledge, education should also aim to reshape the mindset and attitudes of employees. One of the main goals of education is to create a proper way of thinking which encourages people to follow the goals of the related companies appropriately.

Changing the organizational structure: it is essential to support the effective implementation of DR programs. One key step is establishing dedicated teams responsible for monitoring and evaluating the performance of DR initiatives. This office has different tasks, such as identification of the load potentials for peak-shaving, providing programs for seasonal load management, and budget estimation for implementation of DR programs. Several types of DR programs have different performances in different conditions [114] that should be monitored and analyzed.

Defining new managerial tasks and programs: it is essential for integrating DR into the broader operational and strategic framework of power distribution systems. Although DR programs are implemented in the short-term horizons, they can have considerable effects on the mid-term and long-term planning of the power system [115, 116]. The interaction between different economic aspects of the power system and DR programs can significantly influence long-term expansion decisions [117].

To optimize the deployment of DR programs, it is necessary to develop advanced mechanisms that can identify consumers with high DR potential. Although some general DR programs can be employed in all power grids and for different energy consumers, some power distribution systems may have particular potential to implement special DR programs [118–120]. Consequently, proper frameworks have to be shaped to recognize and analyze these potentials.

Furthermore, efficient employment of DR needs the design of motivational approaches to motivate all players, like customers and DR providers. These approaches should be carefully created to improve the responsiveness of the customers, and ensure that DR reach its full capacity in contributing to an optimal operation of the system [121].

Prioritizing the DR programs: different DR initiatives have different efficiencies within different power distribution systems. It is vital to prioritize DR initiatives regarding their efficiencies in the system under study [122].

2.4 DR Challenges

Evaluating the inherent barriers and challenges for implementing DR needs multidimensional analyses, taking into account the aspects including customers, DR providers, structural aspects, business and economic dimensions, social implications, and technical complications. This part accurately investigates the various challenges, classifying them into distinct subsections. Through the examination of these challenges, the goal of this part is to discover the challenges hindering the perfect employment of DR.

2.4.1 Challenges from Consumers' Viewpoint

Consumers' knowledge and information feed: the promotion of proper energy usage patterns is essential, as customers usually lack enough information regarding their energy usage and standardized patterns. Improving customers' awareness about their usage encourages their active participation in adapting to standardized consumption patterns. Moreover, clearer comprehension of the relationship between usage patterns and energy costs is vital. Customers' inadequate knowledge regarding power markets presents dual challenges: hindering a quick employment of DR and impeding the formation of real-time markets, which are vital platforms to foster DR programs [13].

Available technologies: the ability for customers to monitor real-time electricity prices is critical for efficient participation in DR initiatives. Customers need access to real-time prices and the essential tools and infrastructure in order to react to varying prices and modify their consumption patterns accordingly. Implementing DR includes careful consideration of several technology-related facets [123, 124].

Moreover, the cost implications of technology and financing should be assessed. Employing such technologies on a large-scale for many numbers of customers can significantly enhance costs. Motivating private investors to provide these investments could be a feasible policy, but it needs adequate support and incentives to motivate them. Offering incentives can be vital to attract and maintain investors' involvement in such initiatives.

Consumers' interest in following DR programs: the interests of customers for participating in DR initiatives are affected by some behavioral, economic, and technical reasons. Response fatigue is one of the main challenges that arises in dynamic tariff-based DR programs where customers need to frequently modify their usage in response to price changes. Although these models offer better energy consumption behaviors, they can lead to fatigue over time, particularly in long-term scenarios, since customers may feel tired because of continuous monitoring and decision-making. In such cases, many number of customers may prefer to revert to the simplicity of fixed electricity tariffs [13, 125, 126].

The next limiting aspect is the perceived value of potential savings. In several countries, electrical energy costs a small portion of household expenditures. Consequently, investing in the implementation of price-based DR programs may have no proper efficiency.

Another concern is the utilization of smart meters, which are crucial to monitor customers' reactions in DR initiatives. Although these devices are essential for precise billing and system management, their implementation includes significant costs and also increases privacy concerns. As smart meters collect detailed energy usage data, distribution companies should access personal consumption details, which some consumers perceive as intrusive [127, 128].

Finally, the satisfaction level of customers with changing consumption patterns is a vital factor for the successful implementation of DR. Most end users prefer simplicity in managing their electricity consumption. Although the utilization of complicated pricing models may provide some benefits for customers, they are still not pleased to participate in DR calls which significantly change their usage behaviors. The utilization of initiatives with prepayment in which the smart meters monitor and report the consumption statistics can be taken into account as an effective technique to overcome these concerns.

2.4.2 Challenges from Service Providers' Viewpoint

Investment recovery: investments in the employment of DR face with some uncertainties. Thus, it is needed to persuade private investors to invest in these programs. The government must provide supportive policies, like tax exemption, for the investors [13].

2.4.2.1 Conflict of Interest

The conflict between managerial incentives and DR initiatives causes a substantial challenge. The decision makers in companies generally derive their incentives from the company's revenues. However, DR often aims to decrease energy usage during peak hours, potentially decreasing the total income of the company.

This conflict of interest might reduce managerial motivation for an active implementation and support of these programs [129].

Furthermore, the employment of DR may inadvertently lead to unintended effects on load and voltage profiles. The main aim of DR is to modify the electrical load profile. However, the implementation of different load management programs could yield to adverse influences, like the creation of new peak levels, thereby yielding to price spikes in the electricity market. These economic implications could further result in technical problems related to voltage profiles, causing extra concerns to the successful operation of DR [130, 131].

2.4.3 Structural Challenges

Numerous structural challenges hinder the successful employment of DR. Compatibility of DR initiatives with different customer segments is one of the key problems. DR initiatives have various performances considering different properties of customers. Properties like geographical features, as well as education and income levels, are some of the important factors that would raise such concerns [13, 132].

Another key challenge is the improper structure of electricity markets, which often do not recognize DR as a formal market commodity. In many current systems, electricity markets are primarily centered on energy and capacity as tradable goods, while DR remains unintegrated as a standard product. This lack of recognition limits its value and restricts consumer participation. According to [133–135], the participation of the consumers in the DR programs has a direct relation with the market structure. The value of DR programs is not the same in different power networks and may have a higher value in systems with high penetration of intermittent generation resources. Hence, in many cases, DR cannot recover its implementation costs. Strong dependence of the incentive-based DR programs on smart meters: implementing the price-based DR programs requires smart meters to monitor the performance of the consumers. Replacing conventional meters with smart ones may require long-term programs and high costs. Therefore, the smart meter requirement can be considered an important challenge.

Lastly, there is the issue of consumer capability to adjust consumption behavior. In the price-based DR programs, the electricity price is dynamic, and consumers should change their consumption patterns. However, they may not be able to change their consumption pattern completely [136].

2.4.4 Business/Economic Challenges

The implementation of DR programs is accompanied by several business and economic challenges that must be addressed to ensure their success. One major obstacle is the behavioral uncertainty surrounding DR participants. Lack of exact information about the DR providers is an effective challenge. This challenge may have different reasons, such as the consumers' load uncertainty and their potential to follow DR programs [137].

Another issue is the incomplete market structure, which can lead to inefficiencies such as free-riding. In cases where property rights have not been correctly defined, unfair incomes or costs may be created, and big differences will be seen between the individual benefits and the social ones. Phenomena such as free-riding can be considered as the result of these problems [138].

Furthermore, the costs associated with developing new market structures for DR can be considerable. Although trading DR in the market-based structures facilitates the trading of DR, it may result in some extra expenses, such as administrative costs. These costs will be greater in the pool-based DR trading structures. The structure of the DR market has been analyzed in some research, such as [139–141] in which linear or nonlinear models have been developed for DR programs.

Market imperfections, such as limited competition or concentrated market power, also present challenges. Each market structure is exposed to undesired conditions or phenomena such as market power. The regulators should monitor any market structure for DR programs to guarantee fairness and competitiveness.

Finally, the economic attractiveness of DR has an important role in driving investment and contribution. In comparison with conventional expansion planning alternatives, DR may not always appear cost-effective, mainly because of the expenditures involved in the installation of ICTs, smart meters, and control systems. If the efficiency gains and financial returns of DR do not outweigh these upfront investments, their adoption may be meaningfully hindered [21].

2.4.5 Social Challenges

The successful implementation of DR is often hindered by several social challenges. Inefficiency of officials and managers in the execution of DR is one of the key problems. If environmental values and economic considerations are overlooked within organizational structures, managers may struggle to appropriately employ and promote DR initiatives.

A significant barrier to consumer participation is distrust toward DR providers or implementation companies. When the consumers do not trust the DR providers or the related implementation companies, they will not participate in DR programs.

There is also an issue of incompatibility between DR programs and the values or beliefs of consumers. DR providers have to pay attention to their consumers' values and beliefs as they attempt to maximize their welfare [142–144]. Reducing their welfare or engaging them to respond to many numbers of signals will reduce their interest in participating in DR programs.

Another challenge lies in behavioral resistance. In many cases, although consumers are aware of the economic benefits of participating in DR programs, they resist changing their habits since they think that DR programs violate their privacy [145–147].

Furthermore, personal limitations like income level, age, or educational background may reduce customers' participation. Individuals from low-income households may lack access to the essential technologies. In addition, elderly or less-educated customers might find the programs complicated [148]. Coping with these problems needs tailored communication and education efforts to effectively consider the unique needs and concerns of different consumers. Moreover, the provision of incentives and support for responsive customers can help to overcome these challenges and encourage broader acceptance of these programs.

2.4.6 Technical Challenges

DR initiatives cope with several technical problems that should be handled to ensure an effective employment of these programs. The utilization of proper metering infrastructure, actuators, and reliable communication systems is one of the most fundamental requirements. In the electricity markets, the market-clearing processes are performed in short-term periods (5–15 minutes). Therefore, there should be appropriate metering and communication infrastructure to accurately meter data and send it to the metering centers [149].

Another critical technical challenge is the availability of robust data analysis tools. Gathering the consumption data of the loads may lead to large amounts of data that would not be useful without an appropriate data analysis tool. Using data mining tools to study the data is very useful for identifying the behavior patterns of consumers [150, 151].

Furthermore, there is a pressing need for standardization of DR programs to ensure compatibility with other emerging smart grid technologies. By the extensive extension of DR resources and renewables, the centralized power networks will be replaced by decentralized microgrids that contain different infrastructures, such as IoT and cloud computing tools. In order to coordinate with different technologies, DR programs should be standardized to be compatible with the other smart grid technologies [152–154].

2.5 Solutions to Address DR Challenges

Due to the complex challenges associated with DR implementation, it is essential to identify effective solutions. This section closely examines a various of strategic solutions that designed to address these challenges. From economic barriers to social, technical, legislative, implementation, and educational hurdles, distinct subsections outline targeted solutions geared toward addressing each facet. Various economic solutions are viable consisting of improving DRX markets, developing innovative contract frameworks, and implementing mechanisms to reduce market power, which yield promoting investment support. Addressing social challenges requires the implementation of educational initiatives focused on DR concepts, the development of smart energy management systems, the establishment of trust-building mechanisms, and the promotion of environmental awareness. Technical solutions which encompass the advancements in data analysis, utilizing effective communication technology, reinforcement in the data security aspect, standards development, and skill enhancement. The challenges are involved with legislative and political approaches concerning the revision and auditing of tax policy, the dynamic energy system regulations, and coordinating consumer electricity pricing with wholesale energy market prices. Further, solutions to tackle implementation challenges and educational necessities focused on real-time pricing and public education initiatives, respectively. This section focuses on various effective solutions to create a pathway toward overcoming barriers and taking necessary actions to successfully implement the DR initiatives.

2.5.1 Solutions to Deal with Economic Challenges

To effectively deal with the economic challenges related to DR implementation, the development of a DRX market, as a DR exchange platform, provides a viable solution. This market mechanism improves demand-side flexibility through the integration of DR into a well-defined economic framework, allowing consumers and service providers to participate more efficiently in energy markets [155]. Enabling customized DR contracts based on the priority of consumer choices is an essential strategy for active participation. Different consumers have different goals for participating in DR programs. In other words, using a unique framework for all consumers is impossible. DR providers should use various forms of contracts for the consumers' attraction.

Furthermore, dealing with market power is essential for maintaining efficiency and neutrality, especially in localized or regional markets with less competition. Strong regulatory oversight is necessary to monitor market dynamics and ensure equitable access and pricing. Developing novel trading structures, such as peer-to-peer framework, can be another solution to deal with the market power [156, 157].

Besides, mechanisms for investment support, including low-interest loans and subsidies, might increase the attraction of private sector investment in DR infrastructure. At the same time, subsidies can be used to offset hidden or unforeseen costs that may arise during DR deployment, thereby enhancing the financial viability and scalability of these programs [158, 159].

2.5.2 Solutions to Deal with Social Challenges

Addressing the social challenges associated with DR programs requires a multifaceted approach centered on education, trust-building, and cultural change. Public education plays a fundamental role in promoting DR by raising awareness and influencing consumer behavior [160]. It should be noted that such solutions require enough time to make appropriate cultures. Also, specialized training courses may be required to educate some energy-related concepts such as energy management, energy audit, and future of energy. Also, implementing some psychological techniques, such as displaying the energy consumption meters in the view of customers, can provide effective energy feedback [161]. This approach can increase awareness of energy use and encourage customers to make better choices regarding their energy consumption. Customers obtain real-time feedback on their energy usage and allowing them to adjust their behavior and minimize their energy consumption.

Ensuring consumer comfort is also vital to the successful implementation of DR programs. Smart energy management systems, such as automation technologies like smart thermostats, optimize energy usage without compromising comfort levels [142]. These technologies will pave the way for consumers to participate in DR programs without any inconvenience.

Establishing mutual trust between DR providers and consumers is a significant driver in encouraging participation. Using trustworthy outside companies to handle DR programs can help consumers feel more secure about their personal data, encouraging them to participate more actively.

Lastly, promoting environmental values within public attitudes helps to motivate consumer participation in DR initiatives. Promoting a culture of public awareness and emphasizing the significance of environmental issues in daily life can effectively encourage people to participate in DR programs [162].

2.5.3 Solutions to Deal with Technical Challenges

In order to address the technical challenges of DR programs, it is important to use feasible advanced technologies to establish robust support frameworks. A key solution is using highly sophisticated data analysis and data mining methods. The vast amounts of data generated through DR initiatives need to be thoroughly analyzed, and advances in data mining and cloud computing have the potential to overcome many technical barriers associated with managing and interpreting this data effectively [163, 164].

The advancement of communication technologies, especially the IoT, provides promising solutions pertinent to large-scale data transmission. A key problem is the lack of standardized communication protocols capable of managing huge data flows.

Creating universal standards for data formats, transfer methods, and other functions are important for achieving effective communication between devices and systems. IoT technology can reduce communication gaps, which makes way for seamless connectivity and interoperability in terms of implementing DR [154, 165, 166].

Ensuring data network security and protecting consumer privacy are also critical. Privacy violation is a barrier to promoting DR programs. Developing popular methods to protect consumers' data reduces their concerns about their privacy (methods such as using VPNs, firewalls, authentication, data tagging, and physical coding could be utilized). Also, using the blockchain framework is a technical facility that improves privacy and security [167–169].

Developing integrated standards for the various technologies involved in smart grids is necessary for efficient data exchange and system operation. Smart grids include interdisciplinary aspects apart from electricity, which is associated with multiple classifications and standards that ensure interoperability and effective coordination.

Lastly, reinforcing technical support teams is essential to manage the complexity of smart grid technologies, including IoT. Operating and maintaining these systems demands specialized skills. It is advised to make use of some services, which can be outsourced. At the same time, it is important to have structured protocols for recruiting, training, and retaining skilled personnel to provide continuous and effective technical support.

2.5.4 Solutions to Deal with Legislation/Political Challenges

It is important to address legislative and political challenges that promote the broad use of DR programs across consumers. Reforming tax policies is capable of creating economic incentives that encourage participation in DR initiatives [170]. For instance, tax credits or rebates can be provided to individuals who reduce their energy use during peak demand times. The new tax policies can encourage the development of new technologies and infrastructure for DR programs, which include smart grid systems and energy storage solutions. By creating a favorable economic environment for DR, tax policies can help to accelerate the transition to a more flexible and sustainable power system.

There is a need to review and update the rules for energy systems to make sure that producers and consumers have legitimate roles in the market. Market regulations should consider all costs encountered by producers and consumers. This will allow for equal and transparent participation in energy trading and demand management.

Additionally, configuring end-user electricity prices based on wholesale market prices can enhance demand-side responsiveness. When consumers are exposed to the price fluctuations of the energy market, they can modify their consumption patterns based on their price sensitivity [171, 172].

2.5.5 Solution to Deal with Implementation Challenges

It is important to address the challenges of real-time pricing. This approach updates energy prices in real time, which can promote incentive benefits for consumers in order to encourage active participation in the power distribution system management procedure [173].

2.5.6 Solutions for Educational Requirements

Meeting educational requirements necessary for successful DR program implementation requires ample public outreach and customized training strategies. The strategic use of public media is an effective approach. Television, newspapers, news agencies, journals, billboards, and other public platforms can serve as powerful tools to raise awareness and educate the public about the goals and benefits of DR programs [174]. The mentioned public media resources are tools for public education of DR programs. Using the potential of municipalities and governorates can be considered an effective solution to educate DR programs publicly.

Another important educational strategy is the establishment of specific educational courses with practical training. These courses should address through significant socioeconomic variables like education level, income potential, and lifestyle to be communicated effectively with consumers about the significance and advantages of DR participation. Promoting DR programs in financially secure regions may need a different strategy compared to low-income areas.

Government organizations are able to support and expand DR programs in a wider way to attain cost benefit at both ends. The specialized training is needed for employees and departments in these organizations which makes it adopt more sustainable energy practices internally and promotes broader public participation. These training programs can help further develop the organization's commitment to EE and environmental responsibility. In addition, introducing basic energy-saving concepts in schools is crucial to embedding energy-conscious behavior from an early age. As the future pillars of society, children and teenagers play a crucial role in DR within their communities. It is important to provide proper training in DR programs from a young age. Incorporating DR education into the curriculum of schools can achieve this. Providing them with the essential knowledge and skills enables the younger generations to take part in creating a safer and more resilient future for everyone.

2.6 Evolution of DR Generations and Hierarchical Solutions for Real-world Implementation

This part proposes a method constructed based on the current state of systems and their different time horizons. Therefore, three successive generations of DR are outlined here.

First generation of DR (short-term package): the first generation of DR represents a solution adaptable to the existing system conditions, requiring negligible or no adjustments, so creating the foundational necessities to trade DR. Considering the minimum alterations involved, this first generation of DR is envisioned as short-term, spanning a time horizon around one year. To gain more accurate visions and anticipate potential issues, this initial generation is recommended for employment as a pilot in nominated regions.

Second generation of DR (mid-term package): the second generation of DR offers a transitional stage, progressing from the first generation's minimum potential toward the envisioned situation of an optimal DR market. This phase needs wider adjustments and facilitates a higher capacity of DR. While this stage presents intermediate requisites to improve the DR level, realizing these potential improvements requires meeting more substantial requirements. Envisioned within a mid-term horizon, approximately around two to five years, the aim of this package is to bridge the gap between the initial and perfect DR implementation.

Third generation of DR (long-term package): the third generation offers the perfect and most beneficial condition of DR. In this stage, the potential of DR activation is expected to achieve its maximum possible value, including a vast number of stakeholders. Constructing this advanced market needs satisfying both present and future necessities, envisioning a wide array of players and advanced infrastructures within the perfect DR market. Given the need for an inclusive understanding of this advanced market's dynamics and the formation of necessary infrastructure, many of which require long-term employment, this package embodies a long-term outlook, projected over 5–10 years.

Consumers' accessibility is one of the key facets, referring to the ease of participation in DR. Another important aspect is the presence of a technological platform, encompassing the presence of AMI and the availability of dynamic pricing mechanisms, enabling real-time monitoring and flexible pricing.

The presence of smart networks and devices also has a critical role in supporting automated responses and seamless communications between customers and operators. Moreover, the specified budget for DR initiatives affects the scope and quality of services, while the participation rates in DR reflect the program's outreach and appeal. Several DR programs, offer different options to customers and can increase engagement and satisfaction among the participants.

The effectiveness of measurement and verification systems guarantees that DR actions are precisely evaluated and rewarded. It should be noted that timely calculation of incentives and penalties is also crucial that reinforces participation and compliance. The maturity and structure of the DR market, along with main performance indicators like maximum peak load reduction and load factor improvements, offer further insight into the successful implementation of DR.

Diverse criteria could be defined to assess DR programs/markets [175]. Table 2.1 presents two important criteria for the United States, Canada, and China. Based on the initial assessment of these criteria, several American states stand out as potential models. To successfully employ DR in a real world, assessing and prioritizing these criteria are vital. Decision makers should focus on the most critical criteria aligned with their policies to improve DR execution. As an example, Table 2.2 details a typical test system's current status in terms of these criteria.

Investigating the features shown in Table 2.2 offers critical insights into the readiness of the test system to embrace different generations of DR. The current state of the system specifies numerous limitations and insufficiencies, hindering its appropriateness for hosting advanced DR packages. Low accessibility for responsive consumers and lack of access to the retail markets are few of the main limitations. These limitations restrict their participation and engagement in market-driven DR programs. Additionally, the system supports an average penetration level of AMI, with approximately 40–50% load monitoring, showing a moderate technological readiness, but still leaving room for expansion

Table 2.1 Evaluation criteria for different regions.

Country	Provinces/states	Customers' participation	Measurement and verification
America	New York	High	Advanced
	New England	High	Advanced
	California	High	Advanced
	Texas	High	Advanced
	Hawaii	High	Advanced
	Arizona	Average	> Average
	Florida	High	Advanced
	Illinois	< Average	Average
	Kentucky	< Average	< Average
	Massachusetts	< Average	< Average
	Minnesota	Average	Average
Canada	All provinces, except where DR programs are not carried out	> Average	Average
China	Only in Shanghai, Beijing, Tangshan, Foshan, and Suzhou provinces	< Average	Advanced

Table 2.2 The current states of DR indicators in the test system.

Description	Current status	Criteria
Customers can access DR programs, but not the market (no retail-side electricity market)	< Average	Customers' accessibility
Approximately 450,000 smart meters, mainly for large-scale customers; 40–50% load monitoring	Average	Penetration level of AMI
Implementation of three-tariff time-of-use pricing for smart meter owners	Average	Presence of dynamic rates
Poor penetration rates for smart networks and devices	Low	Smart network and device integration
Inadequate budget compared to the cost of curtailments (DR value); no incentives for implementation	Low	Allocated budget
Notably low participation (since the potential of DR in different sectors has not been assessed, it is impossible to explain this criterion accurately. However, because of the low-participation of residential and small-scale commercial customers, it is clear that the participation rate is low)	< Average	Level of customers' participation
Approximately five implemented DR programs	Average	Variety of programs, offering customer choice
No specific reported complaints	> Average	Effectiveness of measurement and verification
Usually paid at year-end	> Average	Timely processing of incentives and penalties
Absence of a DR market	Low	DR market
Reported peak reduction around 7% (is acceptable compared to developed countries), potential for higher reduction	> Average	Maximum peak load reduction or growth
Very low annual load factor (about 65%)	Low	Load factor

and inclusion of smaller-scale customers. The presence of average dynamic rates and a few applied DR mechanisms shows an incomplete and not fully integrated pricing mechanism and limited program diversity for consumers to select from, hindering an inclusive engagement. Moreover, the limited budget, low participation rates, poor smart network integration, absence of a DR market, and remarkably low annual load factor together highlight the system's unreadiness for advanced DR packages. Consequently, taking into account the features of this test system, it is obvious that the system is currently unprepared to host the second or third generations of DR. These challenges show that the system is better suited for the implementation of the first generation of DR. Efforts should be focused on rectifying these shortages and foster higher consumer participation by integrating smart devices, assigning enough budgets, and forming an operational DR market. It is critical to overcome these deficiencies to build a robust foundation for the next generations of DR. In addition, a tailored package of actions needs to be planned to handle the first stage of DR development. However, alongside the first-generation DR package, extra actions should be done to proceed toward achieving the optimum state of DR. The goal of these additional actions is to pave the way for the system's evolution toward a perfect DR infrastructure.

To achieve this goal, strategic themes and strategic goals should be accurately defined. Then, effective strategies should be developed to achieve the strategic goals. These strategies should first be defined with an emphasis on realizing the strategic goals and then be organized into clusters based on different areas, including legal, infrastructure and technology, expert manpower, financial and economic, educational and research, as well as social and cultural aspects. Furthermore, strategies should be classified considering different generations of DR. The employment actions could be separately investigated for these strategies and aspects. Moreover, introducing the important projects in the DR field establishes the final stage. Figure 2.4 provides a representation of the hierarchical procedure within the set of strategic plans, clarifying main pathways for managers, policymakers, and stakeholders to enhance the performance of the system toward the desired state of DR markets.

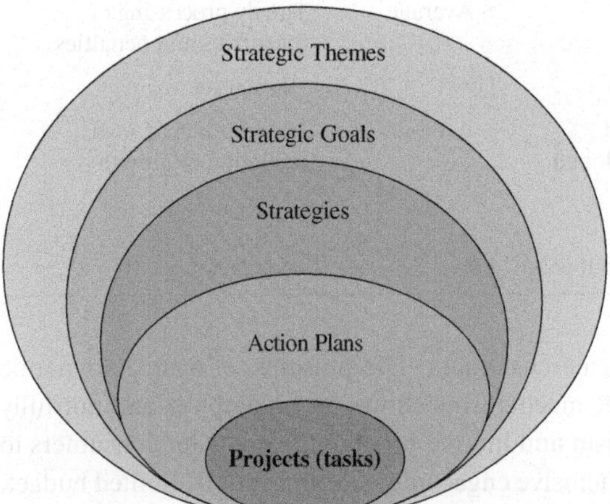

Figure 2.4 The hierarchical structure of strategic plans [176–178].

2.7 Summary of Studies and Analytical Findings

Table 2.3 accurately classifies the referenced researches investigated across different parts of this chapter. This inclusive categorization helps to clarify different aspects explored within the realm of DR. In fact, Table 2.3 is an organized review of the various domains included in the field of DR execution. It comprises main research subjects like the definition of DR, its application in short-term operation and long-term planning, viewpoints from entities and customers, influences on system reliability and power markets, risk management instruments, evaluating potentials, executive necessities, data science applications, and customer psychological aspects. Each category is also detailed using sub-topics and references, providing a systematic breakdown of the wide literature survey conducted in this research. Consequently, this detailed classification not only enables an organized comprehension of the existing literature, but also performs as a comprehensive resource for readers looking for deeper understanding within this specific area of knowledge.

Table 2.3 Classification of the DR researches.

Main subject	Sub-topics	References
Introduction of DR	Coordination of DR and other DSM programs	[1, 75]
Application in short-term operation	EMS	[4–7]
	Coordination with DERs	[8, 112]
	Providing flexibility	[9–12]
Application in long-term planning	Prioritization of DR based on investment and operational cost	[81–83, 115–117]
	Deferring generation and transmission expansion	[2, 3, 21–24]
	Participation in capacity markets	[85–87]
DR in utilities perspective	Implementation cost	[80]
	Revenue reduction as a result of DR programs	[129]
	Effects on operational variables	[130, 131]
DR in consumer's perspective	Economic motivations (earned profit)	[13]
	Economic motivations (tax exemptions)	[84, 170]
	Electricity price vs. electricity value	[77]

(Continued)

Table 2.3 (Continued)

Main subject	Sub-topics	References
Effects on system reliability	Application as reserve services	[14–16, 25, 31]
	Deal with intermittency of renewables	[91, 92]
Effects on power market	Sensitivity to real-time prices and mitigating price spikes	[17–19, 171–173]
	Reducing market power	[20, 156, 157, 179]
	Effects on the market model	[114, 139–141]
	Participation in the power market and DRX market	[78, 79, 133–135, 155]
Risk	Instruments to manage risks	[88]
	Uncertainty of DR programs	[89, 90]
Assessing potentials and prioritization of DR	Strategic plans to implement DR	[176–178]
	DR potential considering system conditions and DR efficiency	[111, 118–120, 122, 132]
	Comparison criteria of DR programs	[175]
Executive requirements	Business environment	[96–98]
	Legislation and regulation	[93–95, 101, 102]
	Technological aspects (general ICT technologies)	[29, 34, 149]
	Technological aspects (AMI)	[30, 32, 33]
	Technological aspects (IoT)	[35–37, 123, 124, 165, 166]
	Technological aspects (blockchain)	[167–169]
	Technological aspects (compatibility of different technologies)	[152–154]
	Education and collaboration with effective agents	[99, 100, 113]
Data science application	Methods and algorithms (data mining and its subsets)	[38, 39, 41–45, 150, 151, 163, 164]
	Methods and algorithms (AI)	[40]
	Estimation of CBL	[46–49]
	Price forecast for load management	[50]
	Designing a consumer pricing mechanism	[51, 52]

Main subject	Sub-topics	References
Customer psychological factors	Effect of lifestyle	[53–55]
	Effect of consumers knowledge	[56, 148, 160, 174]
	Price sensitivity	[57–62]
	Reward and penalty	[63, 65–67, 103, 104]
	People's attention to the environment	[26–28, 162]
	Effects of nonfinancial programs	[64]
	Consumers' loss aversion	[68–71]
	Cultural capital	[72, 73]
	Family income	[74]
	Credibility of information sources	[76]
	Consumer's satisfaction and comfort	[105–109, 136, 137, 142–144]
	Consumer fatigue	[13, 125, 126]
	Consumer privacy	[127, 145–147]
	Individual benefits or the social ones	[138]
	Providing feedback on energy consumption for consumers	[161]

2.8 The Strategic Role of DR in the Climate–Energy–Sustainability Nexus

DR programs have progressed beyond a means for adjusting the consumption patterns to a key enabler of the climate–energy–sustainability nexus, incorporating carbon reduction, energy security, and justice-oriented development with modern power system planning. Through the provision of flexible demand-side resources, DR allows energy grids to better host high penetration of variable RESs, e.g., wind turbines and PVs, thus reducing the intermittency challenges. This flexibility assists in decreasing the need for load curtailment, increasing the stability of the system during peak load or unpredictable outages, and improving the resilience of the system against technical or climate-related troubles. Moreover, DR is able to contribute to avoiding climate emissions and reducing the dependencies on fossil-based power plants through adjusting the energy consumption, which makes it a key element for low-carbon transition plans. Also, DR offers wider benefits through incorporating social and economic goals, encompassing energy affordability and just access to energy services. So, DR has an essential role not only in enhancing system efficiency but also in fostering deep decarbonization and incorporating the United Nations Sustainable Development Goals (SDGs), like "climate action" (SDG 13), and "affordable and clean energy" (SDG 7), locating it as a vital concept toward equitable and sustainable energy systems.

2.8.1 Innovative Paradigms for Deep Decarbonization

As energy systems move toward net-zero emission concepts, different DR models can integrate flexibility with decarbonization actions. By redefining DR programs as dynamic carbon-responsive mechanisms, such models are able to combine real-time energy consumption with carbon intensity and renewable availability.

2.8.1.1 Carbon-aware DR and Renewable Integration

Advanced DR paradigms could be joined with carbon-intensity alarms resulting from renewable generation predictions. "Green DR" models manage flexible loads to coincide with the low-emission times. These carbon-aware DR programs transform customers into carbon-manager players. AI-based predictive controllers and digital twins enhance this coordination, enabling consumers to minimize their carbon footprint dynamically. Huang et al. [180] investigated a problem with the aim of achieving a low-carbon and cost-efficient solution through a demand-side management framework. Through the integration of the carbon emission flow (CEF) theory with the Shapley value method, they ensure fair distribution of carbon emission responsibilities between several aggregators. A price-based DR strategy and an incentive-based carbon trading framework are also combined to motivate active contribution to decrease emissions. Fawzy et al. [181] proposed a market-clearing model to minimize wind power spillage (WPS) using DR and DTLR (dynamic thermal line rating), considering the uncertainties of wind generation and demand levels. The proposed model improves the flexibility level and renewable usage through the coordination of DR with real-time thermal line capacities.

2.8.1.2 Advanced Coordination of EVs and DR

The growing penetration of EVs is representing a dual challenge and opportunity for energy systems. The grid stress could be increased due to the uncoordinated charging behavior, especially in peak periods. But, using the advanced and efficient coordination of EV behaviors and DR activation, such huge distributed energy storages could be transformed into flexible energy resources. A corrective strategy is introduced in [182], in order to coordinate DR programs and EVs to improve the utilization of renewables in the system. Du et al. [183] used practical data and simulation models based on trip-chain analysis to capture both temporal and spatial charging dynamics and presented a low-carbon DR framework for EVs. The suggested method introduced dynamic carbon emission factors as guidance signals for influencing user charging and battery-swapping behaviors, to encourage EV owners to modify their consumption patterns with the times of high renewable generations. This paper offers a method to coordinate fast and slow charging behaviors, by modeling various charging modes and user profiles. Experimental results from Changzhou confirmed that EVs could achieve 73% lower driving emissions compared to conventional vehicles. Also, the results show that the employment of low-carbon DR approaches could further decrease total charging-related emissions by up to 15% by 2035. This study highlights the potential of integrating carbon-aware DR into EV management systems to advance China's broader decarbonization and clean energy goals.

Smart coordination of EVs and DR can manage the timing and power of charging sessions, and shift the demand to other hours with higher renewable generation or lower load levels. Additionally, vehicle-to-grid (V2G) technologies make it possible to discharge stored energy back to the grid that provides significant ancillary services such as peak shaving and frequency regulation. This enhances EVs from passive consumers into active players, who can have a role to enhance grid stability and support higher renewable integration, providing a new revenue stream for EV owners. Therefore, the coordinated DR and EV deployment is a key concept to create a sustainable, green, and resilient energy system.

2.8.1.3 Data-driven Framework and Digitalization

The growth of AI-driven platforms and blockchain-secured peer-to-peer DR markets presents decentralized flexibility frameworks. DR providers and aggregators can manage community-scale DR portfolios. Blockchain technologies guarantee clear interactions, carbon accounting, and traceable incentives [184]. Such digital platforms enhance the concept of DR as a main part of cyber-physical platforms in sustainable systems.

Additionally, data-oriented DR uses the growing availability of data from smart devices and sensors, as well as AMI for activating predictive and automated control of demand-side flexible resources. Through the employment of ML, data mining, and digital twins, system operators can recognize load patterns, assess flexibility potentials, and improve scheduling algorithms [185]. These data-driven methods transform DR from static, rule-based initiatives into dynamic programs with abilities such as self-learning and continuous improvements. Furthermore, the combination of various data, e.g., weather, calendar, behavior of participants, and market signals, improves the accuracy and responsiveness, and leads to more efficient and customer-oriented DR programs.

2.8.1.4 Just and Resilience Energy Systems

DR could play a role to promote the energy systems toward resilience and equitable systems. Under the extreme climate events, DR can mitigate the stress on energy systems and help recovery processes. Therefore, DR is promoted to be considered as an operational and social backbone of the fair energy transitions [186, 187]. The critical intersection of DSR and energy fairness in the transition toward decarbonized, decentralized, and digitalized energy systems is explained in [188]. Demand-side response (DSR) can adjust residential electricity usage in response to signals from the system, and may yield uneven and variegated results for customers. Therefore, a question can be raised regarding the winners and losers of these adjustments. This research investigated the opportunities for increasing justice, as well as the risks of regressive effects. The results highlight the importance of programs' designs and policies to achieve fairness through DSR. Also, an efficient implementation can reduce GHG emission without decreasing the welfare levels of customers.

2.8.2 Critical Effects and Challenges Toward Sustainability

Although DR programs offer huge capacity in sustainable development, achieving their advantages faces multifaceted challenges across environmental, economic, and social aspects.

2.8.2.1 Environmental Aspect

From the environmental point of view, DR alleviates fossil-based generation to provide peak demands and can also increase the penetration level of RESs. However, rebound effects, which are post-DR recoveries, may increase total energy usage and therefore offset some parts of the benefits. Hou et al. [189] presented a low-carbon DR approach for buildings that clearly considers the load rebound problem. Integrating buildings into carbon markets and DR programs is vital for sustainable energy management, since they contribute notably to both peak loads and carbon emissions (because of the large and synchronized loads). Building demands are first classified in this paper based on their schedulable capabilities and a relevant DR model is constructed. Then, through the identification of diverse rebound effects exhibited by various kinds of loads, a load rebound optimization problem is developed for representing and controlling the rebound effects.

2.8.2.2 Economic Aspect

Economically, DR supports cost-efficient decarbonization by postponing the expansion plans of the system [82]. However, inconsistent pricing of flexibility, market fragmentation, and insufficient carbon valuation affect its investment requirements. In order to cope with these challenges, coordinated markets for carbon and flexibility are needed, where DR programs are used not only for consumption pattern modifications but also for verified CO_2 abatement. Frameworks such as carbon-aware DR credits could internalize environmental value into DR pricing models.

2.8.2.3 Social Aspect

From the social point of view, DR should guarantee inclusivity. Digital divides, data-privacy concerns, and cultural heterogeneity affect the involvement of customers in DR. Justice-aware DR can include levelized incentives and prioritize low-income individuals and communities. Musilek and Hussain [190] introduced a fairness-aware power-sharing paradigm which includes DR for promoting fair access to RESs in mixed communities (renewable energy and conventional homes). The aim of this research is to decrease disparities to access the green energies across various residential customers, considering that RESs are unevenly distributed and not equally affordable. It first applied a DR approach for conventional homes to encourage them to shift electricity usage to periods of RES surplus, and therefore maximize local renewable utilization. Then, a proportionally just resource distribution approach is suggested to distribute excess renewable equitably between customers.

2.8.2.4 System and Policy-relevant Challenges

At the system level, sustainable DR can help to balance different goals, such as decarbonization, flexibility, and reliability. Renewable-dominated energy systems need real-time coordination among different energy sources that requires interoperable data standards and cross-sector regulation. As explained in [191], a pilot project in New Delhi, India, showed the feasibility of large-scale behavioral DR without monetary incentives. The "Urja Arpan" program, led by Tata Power Delhi Distribution Limited, engaged more than 90,000 residential and commercial-industrial customers, achieving a flexibility potential of 560 MW by 2024. This community-based, incentive-free approach highlights the effectiveness of social motivation and environmental awareness as drivers for demand flexibility. Policymakers should adopt integrated assessment frameworks in order to link DR performance to sustainability indicators, e.g., emission intensity and energy justice. Such algorithms can quantify all the profits and guide incentive design. Integrating DR into green taxonomies and environmental, social, and governance (ESG) investment criteria can accelerate capital inflows to flexible infrastructure.

2.8.3 Policy Implications and Future Outlook

To fully harness DR within the climate–energy–sustainability nexus, future efforts should aim to:

- Leverage digital technologies: AI, IoT, and blockchain technologies could be used to enhance the autonomy, transparency, and security of large-scale DR ecosystems.
- Consider DR in decarbonization pathways: national and regional climate policies should clearly identify DR as an option, supported by quantified carbon indexes.
- Foster inclusive contribution: policies should guarantee fair access to DR opportunities through digital literacy programs, micro-incentives, and social innovation mechanisms.
- Create carbon-responsive market mechanisms: future electricity markets must consider multi-attribute trading platforms where flexibility, emissions, and equity are co-optimized.
- Establish sustainability-linked evaluation frameworks: integrating DR metrics into SDG goals and corporate ESG disclosures will enhance accountability and investment alignment.

Consequently, strategic redefinitions of DR programs as climate-aware, carbon-responsive, and socially inclusive resources position them as the key component for the transitions toward sustainable energy systems. As energy systems move toward deep decarbonization, DR programs must evolve from short-term operational tools to long-term systemic enablers of global sustainability objectives.

2.9 Conclusions

This chapter addressed the evolution of DR from a short-term operational concept to a strategic pillar of the climate–energy–sustainability nexus. DR programs demonstrate that DR not only is able to improve grid flexibility but also can drive decarbonization, resilience, and energy justice by aligning energy usage behaviors with renewable generation. In this chapter, DR is comprehensively analyzed through various technical, economic, social, cultural, and legislative aspects. One of the main achievements of this chapter was the development of an organized framework and proposing three generations of DR, and also presenting a hierarchical procedure encompassing strategic themes, goals, and implementation actions considering technical and nontechnical aspects. It also recognized and investigated key challenges and opportunities faced by different players in adopting DR, which provided insights to overcome practical barriers. Additionally, this research proposed multidimensional evaluation criteria to assess DR readiness in practical systems and highlighted advanced paradigms, encompassing carbon-responsive DR, sector coupling, and digitalization, redefining the future direction of DR. These achievements construct an inclusive road map to advance DR as a key player of the sustainable energy systems.

References

1 H. Arasteh, V. Vahidinasab, M. S. Sepasian, and A. Ghaderi, "Integrated generation, transmission and energy efficiency planning," in *2019 International Power System Conference (PSC)*, IEEE, 2019, pp. 571–578.

2 H. Arasteh, M. Kia, V. Vahidinasab, M. Shafie-khah, and J. P. Catalão, "Multiobjective generation and transmission expansion planning of renewable dominated power systems using stochastic normalized normal constraint," *International Journal of Electrical Power & Energy Systems*, vol. 121, p. 106098, 2020.

3 V. Vahidinasab, M. Tabarzadi, H. Arasteh, M. I. Alizadeh, M. M. Beigi, and H. R. Sheikhzadeh, "Overview of electric energy distribution networks expansion planning," *IEEE Access*, vol. 8, pp. 34750–34769, 2020.

4 M. Kia, M. Shafiekhani, H. Arasteh, S. M. Hashemi, M. Shafie-khah, and J. P. S. Catalão, "Short-term operation of microgrids with thermal and electrical loads under different uncertainties using information gap decision theory," *Energy*, vol. 208, p. 118418, 2020.

5 S. M. Hashemi and V. Vahidinasab, "Energy management systems for microgrids," in *Microgrids: Advances in Operation, Control, and Protection*, Springer, 2021, pp. 61–95.

6 R. Darshi, S. Shamaghdari, A. Jalali, and H. Arasteh, "Decentralized energy management system for smart microgrids using reinforcement learning," *IET Generation, Transmission & Distribution*, vol. 17, no. 9, pp. 2142–2155, 2023.

7 B. Chegari, M. Tabaa, E. Simeu, and M. El Ganaoui, "Optimal energy management of a hybrid system composed of PV, wind turbine, pumped hydropower storage and battery storage to achieve a complete energy self-sufficiency in residential buildings," *IEEE Access*, vol. 12, pp. 126624–126639, 2024. https://doi.org/10.1109/ACCESS.2024.3454149

8 S. M. Hashemi, H. Arasteh, M. Shafiekhani, M. Kia, and J. M. Guerrero, "Multi-objective operation of microgrids based on electrical and thermal flexibility metrics using the NNC and IGDT methods," *International Journal of Electrical Power & Energy Systems*, vol. 144, p. 108617, 2023.

9 H. Arasteh et al., "A system-of-systems planning platform for enabling flexibility provision at distribution level," in *Flexibility in Electric Power Distribution Networks*, CRC Press, 2021, pp. 41–65.

10 A. Akbari, V. Vahidinasab, H. Arasteh, and E. Kazemi-Robati, "Rural and residential microgrids: Concepts, status quo, model, and application," in *Residential Microgrids and Rural Electrifications*, Elsevier, 2022, pp. 131–161.

11 H. Song, G.-S. Seo, and D. Won, "Pricing strategy of electric vehicle aggregators based on locational marginal price to minimize photovoltaic (PV) curtailment," *IEEE Access*, vol. 13, pp. 11232–11247, 2025. https://doi.org/10.1109/ACCESS.2025.3528626

12 H. R. Massrur, M. Fotuhi-Firuzabad, P. Dehghanian, and F. Blaabjerg, "Fog-based hierarchical coordination of residential aggregators and household demand response with power distribution grids—part II: Data transmission architecture and case studies," *IEEE Transactions on Power Systems*, vol. 40, no. 1, pp. 99–112, 2025. https://doi.org/10.1109/TPWRS.2024.3392018

13 J.-H. Kim and A. Shcherbakova, "Common failures of demand response," *Energy*, vol. 36, no. 2, pp. 873–880, 2011.

14 I. Arafat, E. C. Guerra, and J. Meng, "Loads mismatch and network voltage behavior for future planning of demand response with customer satisfaction," *IEEE Access*, vol. 13, pp. 14471–14484, 2025. https://doi.org/10.1109/ACCESS.2025.3531300

15 Z. M. Zenhom, S. H. A. Aleem, E. E. D. A. El Zahab, Z. M. Ali, A. Almalaq, and T. A. Boghdady, "Synergistic hierarchical bi-level optimization of demand response programs and smart inverter volt/var control for maximizing hosting capacity of power systems with renewable-based distributed generators," *IEEE Access*, vol. 13, pp. 15390–15409, 2025. https://doi.org/10.1109/ACCESS.2025.3530557

16 S. M. Hashemi, V. Vahidinasab, M. S. Ghazizadeh, and J. Aghaei, "Valuing consumer participation in security enhancement of microgrids," *IET Generation, Transmission & Distribution*, vol. 13, no. 5, pp. 595–602, 2019.

17 X. Yan, Y. Ozturk, Z. Hu, and Y. Song, "A review on price-driven residential demand response," *Renewable and Sustainable Energy Reviews*, vol. 96, pp. 411–419, 2018.

18 D. H. Vu, K. M. Muttaqi, A. P. Agalgaonkar, and A. Bouzerdoum, "Customer reward-based demand response program to improve demand elasticity and minimise financial risk during price spikes," *IET Generation, Transmission & Distribution*, vol. 12, no. 15, pp. 3764–3771, 2018.

19 V. Subramanian and T. K. Das, "A two-layer model for dynamic pricing of electricity and optimal charging of electric vehicles under price spikes," *Energy*, vol. 167, pp. 1266–1277, 2019.

20 T.-H. Yoo, W. Ko, C.-H. Rhee, and J.-K. Park, "The incentive announcement effect of demand response on market power mitigation in the electricity market," *Renewable and Sustainable Energy Reviews*, vol. 76, pp. 545–554, 2017.

21 S. Ghaderi, H. Shayeghi, and Y. Hashemi, "Impact of demand response technique on hybrid transmission expansion planning and reactive power planning," *Journal of Operation and Automation in Power Engineering*, vol. 9, no. 1, pp. 1–10, 2021.

22 H. Arasteh, M. Sepasian, and V. Vahidinasab, "Toward a smart distribution system expansion planning by considering demand response resources," *Journal of Operation and Automation in Power Engineering*, vol. 3, no. 2, pp. 116–130, 2015.

23 H. Arasteh, M. S. Sepasian, and V. Vahidinasab, "Incorporation of distribution system reconfiguration and expansion planning problems by considering the role of demand response resources," *Journal of Electrical Systems*, vol. 3, no. 1, pp. 23–36, 2015.

24 H. Arasteh, M. S. Sepasian, and V. Vahidinasab, "An aggregated model for coordinated planning and reconfiguration of electric distribution networks," *Energy*, vol. 94, pp. 786–798, 2016.

25 M. D. Kamruzzaman and M. Benidris, "A reliability-constrained demand response-based method to increase the hosting capacity of power systems to electric vehicles," *International Journal of Electrical Power & Energy Systems*, vol. 121, p. 106046, 2020.

26 H. Liang, H. Zhang, X. Zhang, J. Huang, and D. Zhang, "Role of demand response in the decarbonisation of China's power system," *Environmental Impact Assessment Review*, vol. 104, p. 107313, 2024.

27 J. Duan, Y. Xia, R. Cheng, Q. Gao, and F. Liu, "Low carbon and economic optimal operation of electricity-gas integrated energy system considering demand response," *Sustainable Energy, Grids and Networks*, vol. 38, p. 101290, 2024.

28 T. Yi and W. Ren, "Low carbon economy scheduling of integrated energy system considering the mutual response of supply and demand," *Sustainable Energy, Grids and Networks*, vol. 38, p. 101279, 2024.

29 European Commission (2020). A Strategy for Smart, Sustainable and Inclusive Growth. https://eur-lex.europa.eu/LexUriServ/LexUriServ.do?uri=COM:2010:2020:FIN:en:PDF

30 D. F. Ramírez and S. Céspedes, "Routing in neighborhood area networks: A survey in the context of AMI communications," *Journal of Network and Computer Applications*, vol. 55, pp. 68–80, 2015.

31 J. MacDonald, P. Cappers, D. Callaway, and S. Kiliccote, "Demand response providing ancillary services: A comparison of opportunities and challenges in US wholesale markets," 2023, Accessed: Jan. 21, 2024. [Online]. Available: https://escholarship.org/uc/item/1ct1g5j9

32 M. H. Rashid, "AMI smart meter big data analytics for time series of electricity consumption," in *2018 17th IEEE International Conference on Trust, Security and Privacy in Computing and Communications/12th IEEE International Conference on Big Data Science and Engineering (TrustCom/BigDataSE)*, IEEE, 2018, pp. 1771–1776.

33 M. Martinez-Pabon, T. Eveleigh, and B. Tanju, "Smart meter data analytics for optimal customer selection in demand response programs," *Energy Procedia*, vol. 107, pp. 49–59, 2017.

34 European Commission (2011). Communication from the Commission to the European Parliament, the Council, the European Economic and Social Committee and the Committee of the Regions Youth Opportunities Initiative. https://eur-lex.europa.eu/legal-content/EN/TXT/?uri=celex:52011DC0933

35 F. Moazzen, O. Shahhoseini, H. Arasteh, S. M. Mirsadeghi, and F. Jabari, "Emerging business models for IoT-based smart distribution systems," in *Synergy Development in Renewables Assisted Multi-Carrier Systems*, Springer, 2022, pp. 461–495.

36 M. S. Dildar, A. S. Khan, I. A. Abbasi, R. Shaheen, K. Al Ruqaishi, and S. Ahmed, "End-to-end security mechanism using blockchain for Industrial Internet of Things," *IEEE Access*, vol. 13, pp. 20584–20598, 2025. https://doi.org/10.1109/ACCESS.2025.3535821

37 H. Arasteh et al., "IoT-based smart cities: A survey," in *2016 IEEE 16th International Conference on Environment and Electrical Engineering (EEEIC)*, IEEE, 2016, pp. 1–6.

38 S. Chakrabarti et al., "Data mining curriculum: A proposal (version 1.0)," in *Intensive Working Group of ACM SIGKDD Curriculum Committee*, vol. 140, 2006, pp. 1–10.

39 S. Zhang, D. Zhang, Y. Zhang, J. Cao, D. Gao, and J. Pang, "The research on smart power consumption technology based on big data," in *2016 International Conference on Smart Grid and Clean Energy Technologies (ICSGCE)*, IEEE, 2016, pp. 12–18.

40 S. Russell and P. Norvig, *Artificial Intelligence: A Modern Approach*. Prentice Hall, 2002.

41 G. James, D. Witten, T. Hastie, and R. Tibshirani, *An Introduction to Statistical Learning*, vol. 112. Springer, 2013.

42 P. Norvig, "On Chomsky and the two cultures of statistical learning," in *Berechenbarkeit der Welt?*, Springer, 2017, pp. 61–83.

43 D. Kim, *Soft Computing in Artificial Intelligence*, vol. 270. Springer, 2014.

44 B. Buchanan, G. Sutherland, and E. A. Feigenbaum, "Heuristic DENDRAL: A program for generating explanatory hypotheses," *Organic Chemistry*, vol. 30, pp. 209–254, 1969.

45 C. Green, "Theorem proving by resolution as a basis for question-answering systems," *Machine Intelligence*, vol. 4, pp. 183–205, 1969.

46 M. Sun, Y. Wang, F. Teng, Y. Ye, G. Strbac, and C. Kang, "Clustering-based residential baseline estimation: A probabilistic perspective," *IEEE Transactions on Smart Grid*, vol. 10, no. 6, pp. 6014–6028, 2019.

47 J. Oyedokun, S. Bu, Z. Han, and X. Liu, "Customer baseline load estimation for incentive-based demand response using long short-term memory recurrent neural network," in *2019 IEEE PES Innovative Smart Grid Technologies Europe (ISGT-Europe)*, IEEE, 2019, pp. 1–5.

48 S. Lei, D. Hong, J. L. Mathieu, and I. A. Hiskens, "Baseline estimation of commercial building HVAC fan power using tensor completion," *Electric Power Systems Research*, vol. 189, p. 106624, 2020.

49 H. Arasteh, N. Moslemi, and S. M. Hashemi, "Demand response measurement and verification approaches: Analyses and guidelines," in *Industrial Demand Response: Methods, Best Practices, Case Studies, and Applications*, IET, 2022, pp. 133–161.

50 R. Lu, S. H. Hong, and M. Yu, "Demand response for home energy management using reinforcement learning and artificial neural network," *IEEE Transactions on Smart Grid*, vol. 10, no. 6, pp. 6629–6639, 2019.

51 R. Lu, S. H. Hong, and X. Zhang, "A dynamic pricing demand response algorithm for smart grid: Reinforcement learning approach," *Applied Energy*, vol. 220, pp. 220–230, 2018.

52 K. Aurangzeb, S. Aslam, S. M. Mohsin, and M. Alhussein, "A fair pricing mechanism in smart grids for low energy consumption users," *IEEE Access*, vol. 9, pp. 22035–22044, 2021.

53 P. T. Baboli, M. Eghbal, M. P. Moghaddam, and H. Aalami, "Customer behavior based demand response model," in *2012 IEEE Power and Energy Society General Meeting*, IEEE, 2012, pp. 1–7.

54 A. Naeem, A. Shabbir, N. U. Hassan, C. Yuen, A. Ahmad, and W. Tushar, "Understanding customer behavior in multi-tier demand response management program," *IEEE Access*, vol. 3, pp. 2613–2625, 2015.

55 J. Lin, J. Dong, D. Liu, Y. Zhang, and T. Ma, "From peak shedding to low-carbon transitions: Customer psychological factors in demand response," *Energy*, vol. 238, p. 121667, 2022.

56 F. G. Kaiser, S. Wölfing, and U. Fuhrer, "Environmental attitude and ecological behaviour," *Journal of Environmental Psychology*, vol. 19, no. 1, pp. 1–19, 1999.

57 H. A. Aalami, M. P. Moghaddam, and G. R. Yousefi, "Demand response modeling considering interruptible/curtailable loads and capacity market programs," *Applied Energy*, vol. 87, no. 1, pp. 243–250, 2010.

58 D. S. Kirschen and G. Strbac, *Fundamentals of Power System Economics*. John Wiley & Sons, 2018.

59 H. A. Aalami, M. P. Moghaddam, and G. R. Yousefi, "Modeling and prioritizing demand response programs in power markets," *Electric Power Systems Research*, vol. 80, no. 4, pp. 426–435, 2010.

60 M. A. Domínguez-Garabitos, V. S. Ocaña-Guevara, F. Santos-García, A. Arango-Manrique, and M. Aybar-Mejía, "Demand-shifting strategies to optimize the performance of the wholesale electricity market: A Dominican Republic study case," *IEEE Access*, vol. 10, pp. 44362–44383, 2022.

61 G. Kansal and R. Tiwari, "Elasticity modelling of price-based demand response programs considering customer's different behavioural patterns," *Sustainable Energy, Grids and Networks*, vol. 36, p. 101244, 2023.

62 A. J. Conejo, J. M. Morales, and L. Baringo, "Real-time demand response model," *IEEE Transactions on Smart Grid*, vol. 1, no. 3, pp. 236–242, 2010.

63 T. Santarius, "Energy efficiency, human behavior, and economic growth: Challenges to cutting energy demand to sustainable levels," *AIP Conference Proceedings*, vol. 1652, pp. 70–81, 2015.

64 H. Allcott and S. Mullainathan, "Behavior and energy policy," *Science*, vol. 327, no. 5970, pp. 1204–1205, 2010.

65 J. Luft, "Bonus and penalty incentives contract choice by employees," *Journal of Accounting and Economics*, vol. 18, no. 2, pp. 181–206, 1994.

66 J. D. Dodson, "Relative values of reward and punishment in habit formation," *Psychobiology*, vol. 1, no. 3, p. 231, 1917.

67 Z. Carmon and D. Ariely, "Focusing on the forgone: How value can appear so different to buyers and sellers," *Journal of Consumer Research*, vol. 27, no. 3, pp. 360–370, 2000.

68 S. Mohajeryami, P. Schwarz, and P. T. Baboli, "Including the behavioral aspects of customers in demand response model: Real time pricing versus peak time rebate," in *2015 North American Power Symposium (NAPS)*, IEEE, 2015, pp. 1–6.

69 A. Tversky and D. Kahneman, "Advances in prospect theory: Cumulative representation of uncertainty," *Journal of Risk and Uncertainty*, vol. 5, no. 4, pp. 297–323, 1992.

70 A. Tversky and D. Kahneman, "Loss aversion in riskless choice: A reference-dependent model," *The Quarterly Journal of Economics*, vol. 106, no. 4, pp. 1039–1061, 1991.

71 D. Kahneman and A. Tversky, "Choices, values, and frames," in *Handbook of the Fundamentals of Financial Decision Making: Part I*, World Scientific, 2013, pp. 269–278.

72 J. H. Turner and P. R. Turner, *The Structure of Sociological Theory*. Dorsey Press, 1978.

73 A. Portes, "Social capital: Its origins and applications in modern sociology," *Annual Review of Sociology*, vol. 24, no. 1, pp. 1–24, 1998.

74 M. Shekari, H. Arasteh, A. Sheikhi Fini, and V. Vahidinasab, "Demand response requirements from the cultural, social, and behavioral perspectives," *Applied Sciences*, vol. 11, no. 23, p. 11456, 2021.

75 C. Goldman (2010). Coordination of Energy Efficiency and Demand Response. Lawrence Berkeley National Laboratory.

76 C. Zabin and K. Chapple (2011). California Workforce Education and Training Needs Assessment for Energy Efficiency, Distributed Generation, and Demand Response. Institute for Research on Labor and Employment. https://escholarship.org/content/qt9rp5w1jw/qt9rp5w1jw.pdf

77 M. Choobineh and M. Kalantar, "Introduction of a Novel Operational Framework for distribution network considering distributed generations and load elasticity effect," in *16th Electrical Power Distribution Conference*, IEEE, 2011, pp. 1–6.

78 J. Saebi, M. H. Javidi, and M. O. Buygi, "Toward mitigating wind-uncertainty costs in power system operation: A demand response exchange market framework," *Electric Power Systems Research*, vol. 119, pp. 157–167, 2015.

79 H. R. Arasteh, M. P. Moghaddam, and M. K. Sheikh-el-Eslami, "Bidding strategy in demand response exchange market," in *2012 Proceedings of 17th Conference on Electrical Power Distribution*, IEEE, 2012, pp. 1–5.

80 Q. Qdr (2006). Benefits of Demand Response in Electricity Markets and Recommendations for Achieving Them. U.S. Department Energy.

81 Y. Xiang, H. Cai, C. Gu, and X. Shen, "Cost-benefit analysis of integrated energy system planning considering demand response," *Energy*, vol. 192, p. 116632, 2020.

82 H. Arasteh, V. Vahidinasab, M. S. Sepasian, and J. Aghaei, "Stochastic system of systems architecture for adaptive expansion of smart distribution grids," *IEEE Transactions on Industrial Informatics*, vol. 15, no. 1, pp. 377–389, 2018.

83 H. R. Arasteh, M. P. Moghaddam, M. K. Sheikh-El-Eslami, and A. Abdollahi, "Integrating commercial demand response resources with unit commitment," *International Journal of Electrical Power & Energy Systems*, vol. 51, pp. 153–161, 2013.

84 S. Zhang, Y. Jiao, and W. Chen, "Demand-side management (DSM) in the context of China's on-going power sector reform," *Energy Policy*, vol. 100, pp. 1–8, 2017.

85 M. B. Anwar and M. O'Malley, "Strategic participation of residential thermal demand response in energy and capacity markets," *IEEE Transactions on Smart Grid*, vol. 12, no. 4, pp. 3070–3085, 2021.

86 A. S. M. Khan and L. J. de Vries (2017). The Impact of Demand Response and Electrical Energy Storage on (the Need for) a Capacity Market. Technische Universität Wien.

87 A. S. M. Khan, R. A. Verzijlbergh, O. C. Sakinci, and L. J. de Vries, "How do demand response and electrical energy storage affect (the need for) a capacity market?," *Applied Energy*, vol. 214, pp. 39–62, 2018.

88 L. Guo, T. Sriyakul, S. Nojavan, and K. Jermsittiparsert, "Risk-based traded demand response between consumers' aggregator and retailer using downside risk constraints technique," *IEEE Access*, vol. 8, pp. 90957–90968, 2020.

89 A. Tabandeh and M. J. Hossain, "Hybrid scenario-IGDT-based congestion management considering uncertain demand response firms and wind farms," *IEEE Systems Journal*, vol. 16, no. 2, pp. 3108–3119, 2021.

90 H. Ming, L. Xie, M. C. Campi, S. Garatti, and P. R. Kumar, "Scenario-based economic dispatch with uncertain demand response," *IEEE Transactions on Smart Grid*, vol. 10, no. 2, pp. 1858–1868, 2017.

91 J. G. Kirkerud, N. O. Nagel, and T. F. Bolkesjø, "The role of demand response in the future renewable northern European energy system," *Energy*, vol. 235, p. 121336, 2021.

92 M. Miri and M. McPherson, "Demand response programs: Comparing price signals and direct load control," *Energy*, vol. 288, p. 129673, 2024.

93 P. Mishra, V. Varadharajan, U. Tupakula, and E. S. Pilli, "A detailed investigation and analysis of using machine learning techniques for intrusion detection," *IEEE Communications Surveys and Tutorials*, vol. 21, no. 1, pp. 686–728, 2018.

94 S. Annala et al., "Regulation as an enabler of demand response in electricity markets and power systems," *Journal of Cleaner Production*, vol. 195, pp. 1139–1148, 2018.

95 I. Lamprinos, N. D. Hatziargyriou, I. Kokos, and A. D. Dimeas, "Making demand response a reality in Europe: Policy, regulations, and deployment status," *IEEE Communications Magazine*, vol. 54, no. 12, pp. 108–113, 2016.

96 X. Lu, K. Li, H. Xu, F. Wang, Z. Zhou, and Y. Zhang, "Fundamentals and business model for resource aggregator of demand response in electricity markets," *Energy*, vol. 204, p. 117885, 2020.

97 M. Hamwi, I. Lizarralde, and J. Legardeur, "Demand response business model canvas: A tool for flexibility creation in the electricity markets," *Journal of Cleaner Production*, vol. 282, p. 124539, 2021.

98 M. Barbero, L. C. Casals, A. Colet-Subirachs, J. Salom, and C. Corchero, "Demand response approaches in a research project versus a real business," *Sustainable Energy, Grids and Networks*, vol. 35, p. 101090, 2023.

99 M. J. Berry and G. S. Linoff, *Data Mining Techniques: For Marketing, Sales, and Customer Relationship Management*. John Wiley & Sons, 2004.

100 J. Xie and C.-C. Liu, "Multi-agent systems and their applications," *Journal of International Council on Electrical Engineering*, vol. 7, no. 1, pp. 188–197, 2017.

101 Federal Energy Regulatory Commission (2006). Federal Energy Regulatory Commission Survey on Demand Response and Time Based Rate Programs/ Tariffs. August.

102 Z. Ma, K. Kuusinen, and M. B. Kjærgaard, "A survey of demand response adoption in retail stores DR control preferences, stakeholder engagement, and cross-national differences," *Energy Informatics*, vol. 2, no. 1, pp. 1–20, 2019.

103 C. Vivekananthan, Y. Mishra, G. Ledwich, and F. Li, "Demand response for residential appliances via customer reward scheme," *IEEE Transactions on Smart Grid*, vol. 5, no. 2, pp. 809–820, 2014.

104 S. M. Ghorashi, M. Rastegar, S. Senemmar, and A. R. Seifi, "Optimal design of reward-penalty demand response programs in smart power grids," *Sustainable Cities and Society*, vol. 60, p. 102150, 2020.

105 Y. Liu, L. Xiao, G. Yao, and S. Bu, "Pricing-based demand response for a smart home with various types of household appliances considering customer satisfaction," *IEEE Access*, vol. 7, pp. 86463–86472, 2019.

106 Y. Wang et al., "The optimal emergency demand response (EDR) mechanism for rural power grid considering consumers' satisfaction," *Energy Reports*, vol. 7, pp. 118–125, 2021.

107 S. M. Hashemi, V. Vahidinasab, M. Ghazizadeh, and J. Aghaei, "Load control mechanism for operation of microgrids in contingency state," *IET Generation, Transmission & Distribution*, vol. 14, no. 23, pp. 5407–5417, 2020.

108 S. M. Hashemi, V. Vahidinasab, M. S. Ghazizadeh, and J. Aghaei, "Security constrained operation of radial micro grids based on the loads' vulnerabilities and flexibilities," *TABRIZ Journal of Electrical Engineering*, vol. 50, no. 3, pp. 1441–1453, 2020.

109 S. M. Hashemi, V. Vahidinasab, M. S. Ghazizadeh, and J. Aghaei, "Reliability-oriented DG allocation in radial microgrids equipped with smart consumer switching capability," in *2019 International Conference on Smart Energy Systems and Technologies (SEST)*, IEEE, 2019, pp. 1–6.

110 P. Palensky and D. Dietrich, "Demand side management: Demand response, intelligent energy systems, and smart loads," *IEEE Transactions on Industrial Informatics*, vol. 7, no. 3, pp. 381–388, 2011.

111 G. Strbac, "Demand side management: Benefits and challenges," *Energy Policy*, vol. 36, no. 12, pp. 4419–4426, 2008.

112 H. A. Aalami and S. Nojavan, "Energy storage system and demand response program effects on stochastic energy procurement of large consumers considering renewable generation," *IET Generation, Transmission & Distribution*, vol. 10, no. 1, pp. 107–114, 2016.

113 E. Hirst et al., *Handbook of Evaluation of Utility DSM Programs*. Oak Ridge National Laboratory, 1991.

114 D. Yu et al., "Modeling and prioritizing dynamic demand response programs in the electricity markets," *Sustainable Cities and Society*, vol. 53, p. 101921, 2020.
115 S. Mohseni, A. C. Brent, and D. Burmester, "A demand response-centred approach to the long-term equipment capacity planning of grid-independent micro-grids optimized by the moth-flame optimization algorithm," *Energy Conversion and Management*, vol. 200, p. 112105, 2019.
116 S. Ren, X. Dou, Z. Wang, J. Wang, and X. Wang, "Medium-and long-term integrated demand response of integrated energy system based on system dynamics," *Energies*, vol. 13, no. 3, p. 710, 2020.
117 M. Barani, S. Backe, R. O'Reilly, and P. C. del Granado, "Residential demand response in the European power system: No significant impact on capacity expansion and cost savings," *Sustainable Energy, Grids and Networks*, vol. 38, p. 101198, 2023.
118 G. G. Dranka and P. Ferreira, "Review and assessment of the different categories of demand response potentials," *Energy*, vol. 179, pp. 280–294, 2019.
119 M. Hu, F. Xiao, and L. Wang, "Investigation of demand response potentials of residential air conditioners in smart grids using grey-box room thermal model," *Applied Energy*, vol. 207, pp. 324–335, 2017.
120 K. Wohlfarth, M. Klobasa, and R. Gutknecht, "Demand response in the service sector–Theoretical, technical and practical potentials," *Applied Energy*, vol. 258, p. 114089, 2020.
121 J. Blanke, C. Beder, and M. Klepal, "An integrated behavioural model towards evaluating and influencing energy behaviour—The role of motivation in behaviour demand response," *Buildings*, vol. 7, no. 4, p. 119, 2017.
122 O. Ivanov, S. Chattopadhyay, S. Banerjee, B.-C. Neagu, G. Grigoras, and M. Gavrilas, "A novel algorithm with multiple consumer demand response priorities in residential unbalanced LV electricity distribution networks," *Mathematics*, vol. 8, no. 8, p. 1220, 2020.
123 A. S. Deese and J. Daum, "Application of ZigBee-based Internet of Things technology to demand response in smart grids," *IFAC-PapersOnLine*, vol. 51, no. 28, pp. 43–48, 2018.
124 A. Saha, M. Kuzlu, M. Pipattanasomporn, and S. Rahman, "Enabling residential demand response applications with a ZigBee-based load controller system," *Intelligent Industrial Systems*, vol. 2, no. 4, pp. 303–318, 2016.
125 M. Tostado-Véliz, H. M. Hasanien, A. R. Jordehi, R. A. Turky, and F. Jurado, "Risk-averse optimal participation of a DR-intensive microgrid in competitive clusters considering response fatigue," *Applied Energy*, vol. 339, p. 120960, 2023.
126 M. Z. B. M. Zahid and H. Aki, "Analyses on the impact of consumers' participation by demand response for flexibility procurement," *Sustainable Energy, Grids and Networks*, vol. 38, p. 101237, 2024.

127 E. McKenna, I. Richardson, and M. Thomson, "Smart meter data: Balancing consumer privacy concerns with legitimate applications," *Energy Policy*, vol. 41, pp. 807–814, 2012.

128 M. R. Asghar, G. Dán, D. Miorandi, and I. Chlamtac, "Smart meter data privacy: A survey," *IEEE Communications Surveys and Tutorials*, vol. 19, no. 4, pp. 2820–2835, 2017.

129 J. Wang, C. N. Bloyd, Z. Hu, and Z. Tan, "Demand response in China," *Energy*, vol. 35, no. 4, pp. 1592–1597, 2010.

130 N. Venkatesan, J. Solanki, and S. K. Solanki, "Residential demand response model and impact on voltage profile and losses of an electric distribution network," *Applied Energy*, vol. 96, pp. 84–91, 2012.

131 Q. Xie et al., "Use of demand response for voltage regulation in power distribution systems with flexible resources," *IET Generation, Transmission & Distribution*, vol. 14, no. 5, pp. 883–892, 2020.

132 A. Faruqui and S. Sergici (2010). Household Response to Dynamic Pricing of Electricity-A Survey of the Empirical Evidence. SSRN. https://papers.ssrn.com/sol3/papers.cfm?abstract_id=2471056

133 P. Cappers, C. Goldman, and D. Kathan, "Demand response in US electricity markets: Empirical evidence," *Energy*, vol. 35, no. 4, pp. 1526–1535, 2010.

134 J. W. Zarnikau, "Demand participation in the restructured Electric Reliability Council of Texas market," *Energy*, vol. 35, no. 4, pp. 1536–1543, 2010.

135 N. Gerami, A. Ghasemi, A. Lotfi, L. G. Kaigutha, and M. Marzband, "Energy consumption modeling of production process for industrial factories in a day ahead scheduling with demand response," *Sustainable Energy, Grids and Networks*, vol. 25, p. 100420, 2021.

136 A. Faruqui and S. Sergici, "Household response to dynamic pricing of electricity: A survey of 15 experiments," *Journal of Regulatory Economics*, vol. 38, no. 2, pp. 193–225, 2010.

137 N. Good, K. A. Ellis, and P. Mancarella, "Review and classification of barriers and enablers of demand response in the smart grid," *Renewable and Sustainable Energy Reviews*, vol. 72, pp. 57–72, 2017.

138 D. T. Nguyen, M. Negnevitsky, and M. de Groot, "Pool-based demand response exchange—concept and modeling," *IEEE Transactions on Power Systems*, vol. 26, no. 3, pp. 1677–1685, 2010.

139 H. A. Aalami and A. Khatibzadeh, "Regulation of market clearing price based on nonlinear models of demand bidding and emergency demand response programs," *International Transactions on Electrical Energy Systems*, vol. 26, no. 11, pp. 2463–2478, 2016.

140 H. A. Aalami, H. Pashaei-Didani, and S. Nojavan, "Deriving nonlinear models for incentive-based demand response programs," *International Journal of Electrical Power & Energy Systems*, vol. 106, pp. 223–231, 2019.

141 F. Alfaverh, M. Denai, and Y. Sun, "A dynamic peer-to-peer electricity market model for a community microgrid with price-based demand response," *IEEE Transactions on Smart Grid*, vol. 14, no. 5, pp. 3976–3991, 2023. https://doi.org/10.1109/TSG.2023.3246083

142 S. Aghniaey, T. M. Lawrence, T. N. Sharpton, S. P. Douglass, T. Oliver, and M. Sutter, "Thermal comfort evaluation in campus classrooms during room temperature adjustment corresponding to demand response," *Building and Environment*, vol. 148, pp. 488–497, 2019.

143 M. Waseem, Z. Lin, S. Liu, I. A. Sajjad, and T. Aziz, "Optimal GWCSO-based home appliances scheduling for demand response considering end-users comfort," *Electric Power Systems Research*, vol. 187, p. 106477, 2020.

144 L. Malehmirchegini and H. Farzaneh, "Incentive-based demand response modeling in a day-ahead wholesale electricity market in Japan, considering the impact of customer satisfaction on social welfare and profitability," *Sustainable Energy, Grids and Networks*, vol. 34, p. 101044, 2023.

145 C. E. Kement, H. Gultekin, and B. Tavli, "A holistic analysis of privacy-aware smart grid demand response," *IEEE Transactions on Industrial Electronics*, vol. 68, no. 8, pp. 7631–7641, 2020.

146 Y. Chen et al., "Privacy-preserving baseline load reconstruction for residential demand response considering distributed energy resources," *IEEE Transactions on Industrial Informatics*, vol. 18, no. 5, pp. 3541–3550, 2022. https://doi.org/10.1109/TII.2021.3107400

147 H. Cheng, T. Lu, R. Hao, J. Li, and Q. Ai, "Incentive-based demand response optimization method based on federated learning with a focus on user privacy protection," *Applied Energy*, vol. 358, p. 122570, 2024.

148 F. A. Diawuo, M. Sakah, S. de la Rue du Can, P. C. Baptista, and C. A. Silva, "Assessment of multiple-based demand response actions for peak residential electricity reduction in Ghana," *Sustainable Cities and Society*, vol. 59, p. 102235, 2020.

149 T. H. Christensen et al., "The role of competences, engagement, and devices in configuring the impact of prices in energy demand response: Findings from three smart energy pilots with households," *Energy Policy*, vol. 137, p. 111142, 2020.

150 C. Ribeiro, T. Pinto, Z. Vale, and J. Baptista, "Data mining for remuneration of consumers demand response participation," in *International Conference on Practical Applications of Agents and Multi-Agent Systems*, Springer, 2020, pp. 326–338.

151 K. Li et al., "Impact factors analysis on the probability characterized effects of time of use demand response tariffs using association rule mining method," *Energy Conversion and Management*, vol. 197, p. 111891, 2019.

152 S. Chen, B. Guo, Y. Ye, B. Li, B. Qi, and H. Yan, "Research on integrated demand response system interconnection framework and its key technologies in reliable communication mode," in *2020 IEEE 4th Information Technology, Networking, Electronic and Automation Control Conference (ITNEC)*, IEEE, 2020, pp. 2047–2054.

153 M. Shakeri et al., "An overview of the building energy management system considering the demand response programs, smart strategies and smart grid," *Energies*, vol. 13, no. 13, p. 3299, 2020.

154 S. Ahmadzadeh, G. Parr, and W. Zhao, "A review on communication aspects of demand response management for future 5G IoT-based smart grids," *IEEE Access*, vol. 9, pp. 77555–77571, 2021. https://doi.org/10.1109/ACCESS.2021.3082430

155 V. Durvasulu and T. M. Hansen, "Benefits of a demand response exchange participating in existing bulk-power markets," *Energies*, vol. 11, no. 12, p. 3361, 2018.

156 M. Zhang, F. Eliassen, A. Taherkordi, H.-A. Jacobsen, H.-M. Chung, and Y. Zhang, "Energy trading with demand response in a community-based P2P energy market," in *2019 IEEE International Conference on Communications, Control, and Computing Technologies for Smart Grids (SmartGridComm)*, IEEE, 2019, pp. 1–6.

157 D. Kanakadhurga and N. Prabaharan, "Demand response-based peer-to-peer energy trading among the prosumers and consumers," *Energy Reports*, vol. 7, pp. 7825–7834, 2021.

158 K. Wohlfarth, E. Worrell, and W. Eichhammer, "Energy efficiency and demand response–two sides of the same coin?," *Energy Policy*, vol. 137, p. 111070, 2020.

159 P. Warren, "Demand-side policy: Global evidence base and implementation patterns," *Energy Environment*, vol. 29, no. 5, pp. 706–731, 2018.

160 E. McKenna, S. Higginson, P. Grunewald, and S. J. Darby, "Simulating residential demand response: Improving socio-technical assumptions in activity-based models of energy demand," *Energy Efficiency*, vol. 11, no. 7, pp. 1583–1597, 2018.

161 T. Hargreaves, "Beyond energy feedback," *Building Research & Information*, vol. 46, no. 3, pp. 332–342, 2018.

162 B. W. Pratt and J. D. Erickson, "Defeat the Peak: Behavioral insights for electricity demand response program design," *Energy Research & Social Science*, vol. 61, p. 101352, 2020.

163 C. Wang, H. Sun, and Y. Yang, "Research on demand response model of electric power interruptible load based on big data analysis," in *International Conference on Advanced Hybrid Information Processing*, Springer, 2019, pp. 336–345.

164 A. Tascikaraoglu, "On data-driven approaches for demand response," in *Big Data Application in Power Systems*, Elsevier, 2018, pp. 243–259.

165 I.-H. Lu, M.-Y. Cho, H.-C. Hsu, K.-H. Yang, and W.-C. Tung, "Research and development of air-conditioning chiller IoT controller and experiment of automatic demand response unloading strategy," *International Journal of Organizational Innovation*, vol. 13, no. 3, pp. 60–76, 2021.

166 B. E. Sedhom, M. M. El-Saadawi, M. S. El Moursi, M. A. Hassan, and A. A. Eladl, "IoT-based optimal demand side management and control scheme for smart microgrid," *International Journal of Electrical Power & Energy Systems*, vol. 127, p. 106674, 2021.

167 A. Jindal, G. S. Aujla, N. Kumar, and M. Villari, "GUARDIAN: Blockchain-based secure demand response management in smart grid system," *IEEE Transactions on Services Computing*, vol. 13, no. 4, pp. 613–624, 2019.

168 S. Wen, W. Xiong, J. Tan, S. Chen, and Q. Li, "Blockchain enhanced price incentive demand response for building user energy network in sustainable society," *Sustainable Cities and Society*, vol. 68, p. 102748, 2021.

169 A. Lucas, D. Geneiatakis, Y. Soupionis, I. Nai-Fovino, and E. Kotsakis, "Blockchain technology applied to energy demand response service tracking and data sharing," *Energies*, vol. 14, no. 7, p. 1881, 2021.

170 N. Voulis, M. J. van Etten, É. J. Chappin, M. Warnier, and F. M. Brazier, "Rethinking European energy taxation to incentivise consumer demand response participation," *Energy Policy*, vol. 124, pp. 156–168, 2019.

171 Z. Wang, R. Paranjape, Z. Chen, and K. Zeng, "Layered stochastic approach for residential demand response based on real-time pricing and incentive mechanism," *IET Generation, Transmission & Distribution*, vol. 14, no. 3, pp. 423–431, 2019.

172 H. Karimi, S. Jadid, and H. Saboori, "Multi-objective bi-level optimisation to design real-time pricing for demand response programs in retail markets," *IET Generation, Transmission & Distribution*, vol. 13, no. 8, pp. 1287–1296, 2019.

173 L. Tao and Y. Gao, "Real-time pricing for smart grid with distributed energy and storage: A noncooperative game method considering spatially and temporally coupled constraints," *International Journal of Electrical Power & Energy Systems*, vol. 115, p. 105487, 2020.

174 T. H. Christensen, S. P. A. K. Larsen, and H. N. Knudsen, "How to engage households in energy demand response solutions?," in *ECEEE 2019 Summer Study on Energy Efficiency: Is Efficient Sufficient?*, European Council for an Energy Efficient Economy, 2019.

175 A. S. Hopkins and M. Whited, *Best Practices in Utility Demand Response Programs*. Synapse Energy Economics Inc., 2017.

176 U.S. Department of Energy (2014). Strategic Plan, 2014-2018.

177 International Energy Agency Demand Side Management Programme, "Strategic Plan for the IEA Demand-Side Management Programme 2004-2009," 2004. [Online]. Available: http://dsm.iea.org

178 Department of Energy (2006). The Department of Energy's Strategic Plan, Mission Discovering the Solutions to Power and Secure America's Future.

179 M. T. Devine and V. Bertsch, "The role of demand response in mitigating market power: A quantitative analysis using a stochastic market equilibrium model," *OR Spectrum*, vol. 45, no. 2, pp. 555–597, 2023. https://doi.org/10.1007/s00291-022-00700-0.

180 J. Huang, M. Yuan, Z. Zou, and Y. Sun, "Low-carbon economic optimization scheduling in distributed energy systems considering carbon emission responsibility and demand response," *International Journal of Electrical Power & Energy Systems*, vol. 169, p. 110768, 2025.

181 S. Fawzy, E. E. Abd-Raboh, and A. A. Eladl, "Interactive demand response and dynamic thermal line rating for minimizing the wind power spillage and carbon emissions," *Electric Power Systems Research*, vol. 232, p. 110399, 2024.

182 J. Yang, Y. Xiang, Z. Tang, S. Zhang, and Z. Hu, "Closed-loop corrective EV demand response with composite customer directrix load for renewable energy accommodation in distribution networks," *Renewable Energy*, vol. 256, p. 124403, 2026.

183 B. Du et al., "Evaluating the carbon-emission reduction potential of employing low-carbon demand response to guide electric-vehicle charging: A Chinese case study," *Applied Energy*, vol. 397, p. 126196, 2025. https://doi.org/10.1016/j.apenergy.2025.126196

184 H. Tlili, "Blockchain and tokenized carbon markets: Empirical evidence on market efficiency and transaction dynamics," *Sustainable Futures*, vol. 10, p. 101109, 2025.

185 H. Arasteh, N. Moslemi, P. Siano, and J. M. Guerrero, *Data-Driven Energy Management and Tariff Optimization in Power Systems: Shaping the Future of Electricity Distribution Through Analytics*. Wiley-IEEE Press, 2025.

186 M. D. de Souza Dutra and N. Alguacil, "Fairness of prosumers' incentives in residential demand response: A practical decentralized optimization approach," *International Journal of Electrical Power & Energy Systems*, vol. 148, p. 109015, 2023.

187 L. Li, S. Fan, J. Xiao, H. Zhou, Y. Shen, and G. He, "Fair trading strategy in multi-energy systems considering design optimization and demand response based on consumer psychology," *Energy*, vol. 306, p. 132393, 2024.

188 P. Calver and N. Simcock, "Demand response and energy justice: A critical overview of ethical risks and opportunities within digital, decentralised, and decarbonised futures," *Energy Policy*, vol. 151, p. 112198, 2021.

189 L. Hou, H. Lin, X. Yang, H. Ren, and D. Shao, "Low-carbon demand response strategy of buildings considering load rebound," *Energy Reports*, vol. 10, pp. 3599–3607, 2023.

190 P. Musilek and A. Hussain, “Equity-oriented power sharing and demand response for enhancing green energy access in mixed communities,” *Energy Buildings*, vol. 308, p. 114029, 2024.

191 R. Anto, R. Singh, D. Ray, S. H. Imam, and A. Arora, “Leveraging enhanced consumer engagement to achieve demand response and flexibility in an Indian energy community,” *Utilities Policy*, vol. 98, p. 102077, 2026.

3

Impact of Deep Decarbonization on Digitalization and Industrialization of Modern Energy Societies

Asma Nasiri[1]*, *Alireza Zare Ansaroudi*[1], *Sina Nabighadim*[1], *Nima Nasiri*[2], *and Mehdi Abapour*[1]

[1] *Energy System Research Institute (ESRI), Faculty of Electrical and Computer Engineering, University of Tabriz, Tabriz, Iran*

[2] *Resilient Smart Grids Research Lab, Electrical Engineering Department, Azarbaijan Shahid Madani University, Tabriz, Iran*

*Corresponding author: n.asma1380@gmail.com

List of Acronyms

BEV	Battery-electric vehicle
CCUS	Carbon capture, utilization, and storage
CO_2-e	Carbon dioxide equivalent
EV	Electric vehicle
FCEV	Fuel cell electric vehicle
GHG	Greenhouse gas
Gt	Gigatons
IEA	International Energy Agency
PEV	Plug-in electric vehicle
PHEV	Plug-in hybrid electric vehicle
PV	Photovoltaic
RES	Renewable energy source
TWh	Terawatt-hour
V2G	Vehicle-to-grid
WT	Wind turbine

3.1 Introduction

Climate change and global warming are the major challenges that the whole world is facing today. In the Paris Agreement, adopted at the 2015 Paris Climate Change Conference, nations committed to limit the increase in global temperatures to below 2°C and, ideally, under 1.5°C compared to pre-industrial levels [1]. Four years later, a study was released that reports that global emissions are rising by 1.5% each year. Figure 3.1 illustrates worldwide carbon dioxide (CO_2) emissions from 1995 to 2024, showing a rapid growth in CO_2 emissions over the past three decades [2]. At the same time, the top four biggest carbon emissions generators, represented by China, the United States, India, and Russia, respectively, show CO_2 equivalent (CO_2-e) emissions of 12,604, 5,130, 3,122, and 2,176 million tons, respectively (Figure 3.2) [3]. Therefore, deep decarbonization is a vital strategy to address global warming challenges, reducing greenhouse gas (GHG) emissions and providing a pathway to achieving goals under the Paris Agreement [1].

The rising global warming will drive the world toward deep decarbonization through the use of low-carbon technologies and lead to restructuring in modern energy systems. Achieving near-zero emissions or deep decarbonization will need systemic changes in several sectors, including power, transport, agriculture, and industry [4]. At the first stage, deep decarbonization will change the manufacturing industry, energy generation, energy consumption, transportation, and even the way food is produced, which leads to industrialization in these sectors [5]. At the second stage, digital technologies can play a crucial role in carbon emission reduction as it enhances the efficiency. Digital technologies like

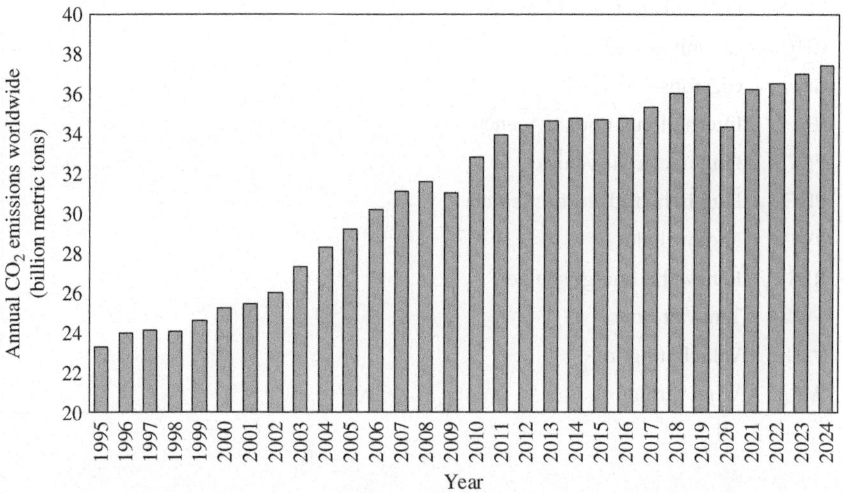

Figure 3.1 Annual CO_2 emissions worldwide (billion metric tons). *Source*: Adapted from [2].

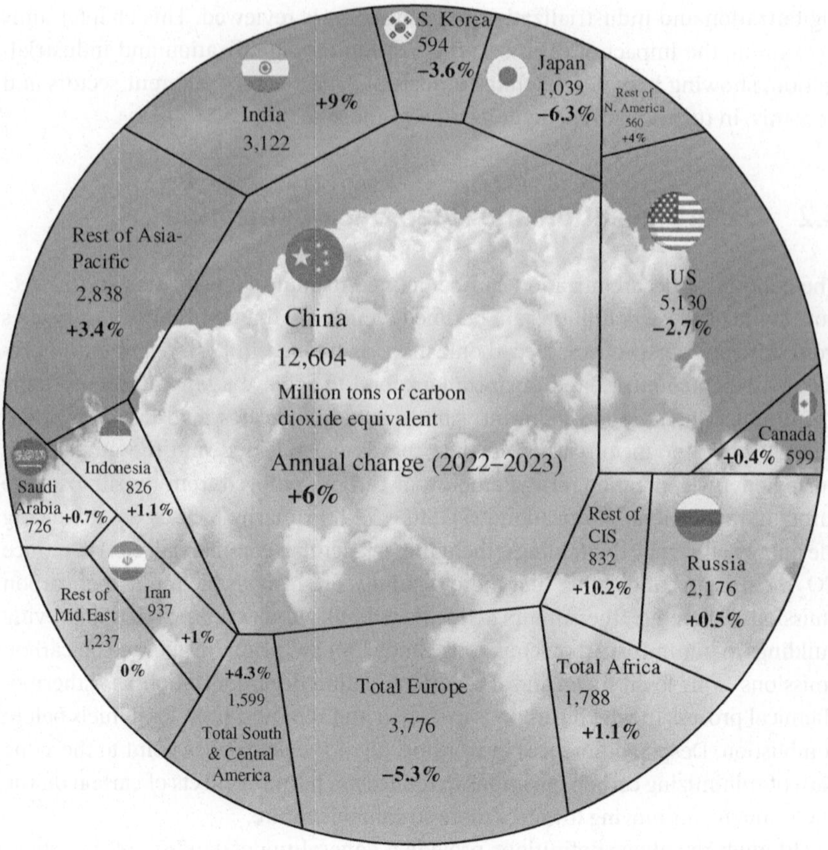

Figure 3.2 CO_2-e emissions of countries (million tons) and annual change in emissions in 2022–2023 (%). *Source*: Adapted from [3].

networked devices, data analytics, and sensors are transferring the way energy is used [4]. As digitalization advances, it creates new opportunities for optimizing energy consumption and reducing GHG emissions. Therefore, deep decarbonization will bring industrialization and digitalization in modern energy societies and smart cities [6].

Many studies have examined the interaction between deep decarbonization and industrial transformation, showing the role of decarbonization in industrialization [4, 5, 7–16]. Meanwhile, some research has focused on the relation between deep decarbonization and digital technologies, exploring how digital technologies can reduce carbon emissions [17–26]. However, none of the research studied the impact of deep decarbonization on both digitalization and industrialization. In this study, the four sectors are examined, including power, transportation, industry, and agriculture sectors, and the impact of deep decarbonization on digitalization and industrialization in each sector is reviewed. This chapter aims to examine the impact of deep decarbonization on digitalization and industrialization, showing how it accelerates industrialization across different sectors and how this, in turn, can drive digitalization in those sectors.

3.2 Decarbonization and Deep Decarbonization

The concept of decarbonization can be defined in multiple ways, and in this section, some of these definitions are outlined, which are drawn from various studies from different perspectives. Decarbonization, as defined in [27], refers to the process by which countries and organizations tend to achieve a low-carbon economy or minimize their carbon footprint. Another study indicates that decarbonization refers to the amount of carbon dioxide that is not released into the atmosphere during energy generation, or the amount of carbon removed from the atmosphere through atmospheric purification [28]. Study [29] primarily focuses on enhancing the energy efficiency of buildings (including residential, commercial, etc.) to reduce CO_2-e emissions, indicating that these buildings are responsible for 28% of carbon emissions. Therefore, the authors define decarbonization as a process of improving buildings to minimize carbon emissions. Study [30] has primarily focused on carbon emissions from fossil fuels, and the authors define decarbonization as a thermochemical process in which carbon is extracted and removed from fossil fuels before combustion. Despite differences in wording, all of the definitions point to the same path of minimizing carbon emissions to reduce the harmful effects of carbon on the environment, for moving toward a more sustainable future.

Although the above definitions provide a general understanding of decarbonization, recent studies have introduced the concept of deep decarbonization, which entails a comprehensive and nearly total reduction of GHG emissions. Deep decarbonization can be defined as a process of minimizing all CO_2-es to nearly zero, aiming to limit the global warming temperature to 1.5 °C above the levels observed before the onset of industrialization. Deep decarbonization, electrification, renewable energy sources (RESs), hydrogen-based fuels, increasing energy efficiency, energy storage, and flexible grids [7, 9]. However, the adoption of these methods requires progress in industrialization and digitalization.

3.3 Impact of Deep Decarbonization on Industrialization

Before today's wave of industrialization and digitalization, humanity had gone through three different industrial revolutions. The first industrialization took place in the 18th century, with the introduction of steam power [31]. The First Industrial Revolution led to significant economic growth, created better living conditions, and improved productivity. The Second Industrial Revolution occurred in the late 18th to early 19th century, and it started with the introduction of the internal combustion engine and linear assembly lines fueled by oil and natural gas energy [31]. The Third Industrial Revolution was driven by power and industrial electronics, semiconductors, computers, and communication technologies, such as the internet, and it opened up new investment opportunities [31]. The Fourth Industrial Revolution, which is known as Industry 4.0, emerged with the rise of disruptive intelligence and information technologies. Since the industrialization and energy sectors are closely linked, the Fourth Industrial Revolution focuses on the digitalization of the manufacturing industry and the energy sector to build a pathway to modern energy societies [10].

As is known, the three earlier industrial revolutions primarily relied on fossil fuel-based resources, resulting in remarkable carbon emissions. However, the Fourth Industrial Revolution is more likely based on RESs, hydrogen fuels, electric vehicles (EVs), and other green technologies that will help us to reach our goals in the deep decarbonization process and build a low-carbon society [10]. As shown in Figure 3.3, the industrial process and energy sectors, which include power, industry, transportation, building, and fuel production, are responsible for 9 and 69% of GHG emissions in 2023, respectively [32]. The statistics on GHGs show an urgent need for green industrialization in the industrial process and energy sectors. In other words, for achieving deep decarbonization goals, acceleration of industrialization is needed in an environmentally friendly way, such as integrating RESs, hydrogen, and other green technologies in the industry and energy

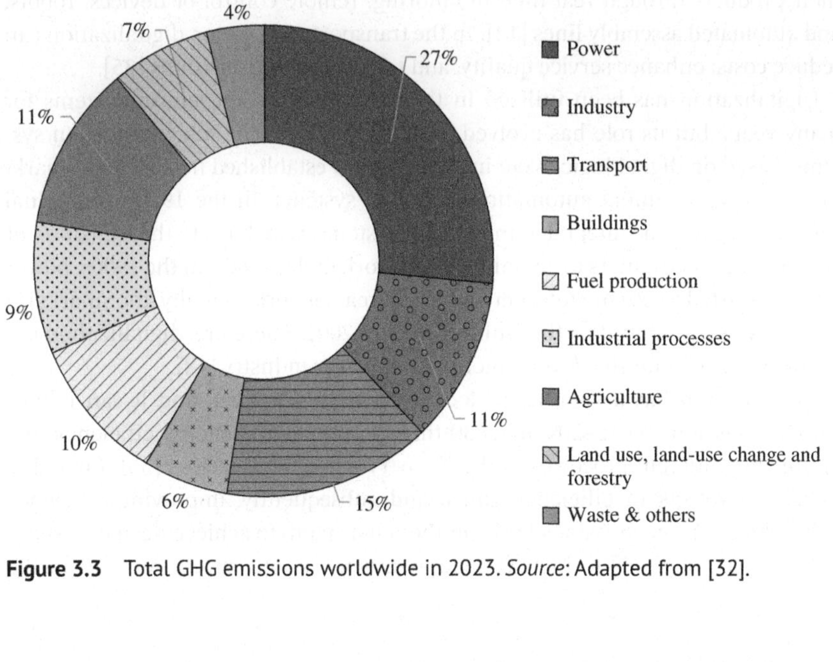

Figure 3.3 Total GHG emissions worldwide in 2023. *Source*: Adapted from [32].

sectors. Therefore, it's essential to examine the impact of deep decarbonization on industrialization in modern energy societies.

Deep decarbonization challenges the industry and forces it to use low-carbon or carbon-neutral processes. In another way, deep decarbonization is the driving force behind the industrial transformation, which is reshaping industrialization into a new model that can be called green industrialization. The green industrialization refers to a process that leads the industry to create "green goods" that are environmentally friendly and that consume less energy [15]. Clean energy generation like RESs, the use of alternative fuels, hydrogen, carbon capture, promoting recycling, and using other green technologies all fall under the umbrella of green industrialization, which plays a vital role in achieving a low-carbon or carbon-free society [33]. In brief, deep decarbonization policies are progressively structuring industrial innovation and affecting the development of green technologies [11].

3.4 Impact of Deep Decarbonization on Digitalization

Digitalization is currently one of the main trends in energy societies. As mentioned earlier, the world is currently experiencing the Fourth Industrial Revolution, which is reshaping lives through digital technologies such as artificial intelligence, robotics, mobile connectivity, and the Internet of Things [34]. Digitalization can be defined as the process of integrating digital technologies into new or existing structures. In other words, digitalization connects the physical world to the software. Digitalization helps industries to have better operational control over their products through real-time monitoring, remote control of devices, robots, and automated assembly lines [12]. In the transportation sector, digitalization can reduce costs, enhance service quality, and reduce carbon emissions [35].

Digitalization has been utilized in the energy sector and power systems for many years, but its role has evolved over the decades. The relay protection systems, based on digital logic, were introduced and established in the 1960s, marking the onset of initial automation in power systems. In the 1970s, individual software began to be integrated into power systems, which led to the transition of power system computers to an online framework in the 1980s. In the 1990s, power systems started to use the internet and local area networks. Finally, the smart grid concept was introduced by the United States in 2001. Therefore, digitalization is a significant driver for the development of the power industry [18].

Digitalization has a significant decarbonization potential, and it can reduce carbon emissions by 15–20% by 2030 through enhancing energy efficiency and optimizing different processes in the industry and energy sectors [36]. First, digitalization results in efficiency gains, and subsequently, improving efficiency will reduce carbon emissions [18]. On the other hand, to achieve deep decarbonization goals, it needs to increase energy efficiency and have a revision in the existing industry structure, or the way energy is generated or consumed [22]. Therefore, digital technologies can be used to achieve deep decarbonization. For deep decarbonization, the way of energy generation has to be changed and start to use environmentally friendly sources like RESs and distributed generations, and transportation systems must be changed to likely cleaner systems, such as EVs [22]. As a result, for moving toward deep decarbonization, it is crucial to use all of these technologies, and all of these will make systems more complicated, bringing up new issues. Therefore, it will be essential to control these complicated systems, and this is the point at which digitalization can help us. Digitalization will help control these systems designed to achieve deep decarbonization. In summary, the deep decarbonization process will create the need for digitalization in modern energy societies.

3.5 The Role of Deep Decarbonization in Revolution of Digitalization and Industrialization in Different Sectors

Deep decarbonization is reshaping the operation of industries and societies, forcing different sectors to adopt low-carbon technologies and processes that underpin a modern energy society. Deep decarbonization goes beyond emission reduction, leading to industrial and digital transformation. This section investigates how deep decarbonization impacts these four sectors, including power, transportation, industry, and agriculture. In addition to this, it will be demonstrated how deep decarbonization drives technological innovation and efficiency, and contributes to digital solutions integrations. Figure 3.4 demonstrates the pathway toward deep decarbonization by highlighting that these goals can be achieved by industrializing and digitalizing the power, transportation, industry, and agriculture sectors. Looking into these sectors reveals that deep decarbonization accelerates both industrialization and digitalization, opening a path toward a sustainable and modern energy society.

3.5.1 Power Sector

Decarbonization of the power sector is a core part of energy sector decarbonization, affecting both the economy and carbon emissions with electrification and the use of electricity-based fuels [14]. In 2024, power generation experienced a slight increase of 1% in carbon emissions, following a 1.4% decrease in 2023. This growth was driven by a 1.3% year-on-year increase in fossil-fuel-based generations, alongside global electricity growth of 4.3%. With emissions reaching approximately 13,800 million tons of carbon, the power sector remains the largest source of carbon emissions. In 2023, the power sector was responsible for 27% of global GHG emissions [32]. Therefore, there is an urgent need to accelerate

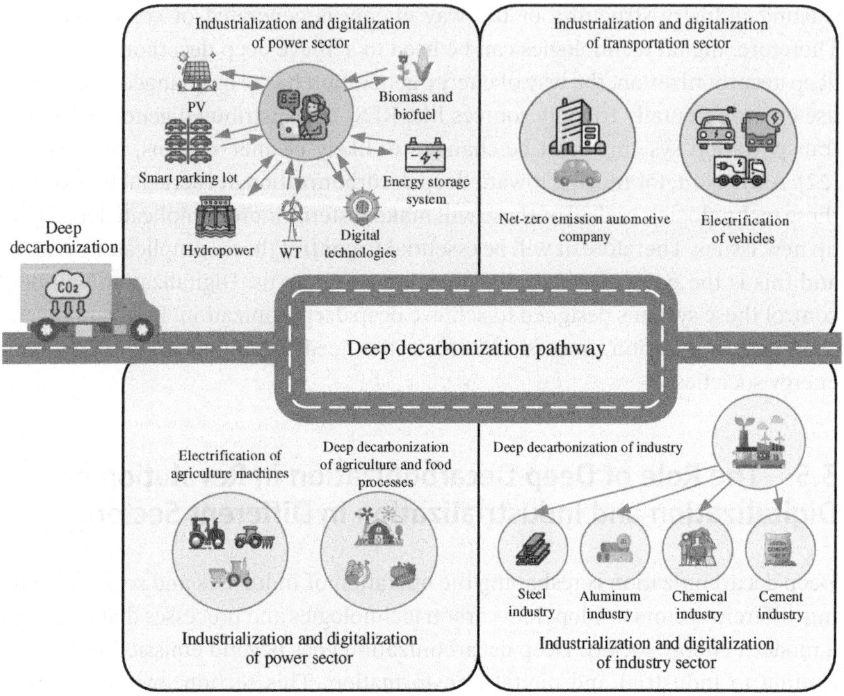

Figure 3.4 Deep decarbonization pathway.

the green industrialization and digitalization of the power sector to expedite deep decarbonization in this sector [37].

As is known, the traditional fossil-fuel-based generations lead to environmental pollution, climate change, and a remarkable amount of carbon emissions. Therefore, a green and fundamental industrialization of the power sector is essential to shift the power generation toward cleaner sources. Clean power sources refer to clean, efficient, and sustainable power generation technologies, including biofuels, wind turbines (WTs), photovoltaics (PVs), hydropower, geothermal, and biomass [38]. These RESs have been used for electricity generation since the mid-20th century. However, they have become cost-efficient for widespread development in the power sector since 2000. For instance, the cost of PVs has decreased 89% since 2010, while the cost of WTs has fallen by 50% since 2008. RESs and energy storage have become cost-effective that several utilities are considering substituting gas-fired peak plants, which operate for only a few hours each year and incur high maintenance costs, with RESs and energy storages [20]. On the other hand, using RESs is one of the most effective strategies of deep decarbonization in the power and energy sector [39]. RESs play a crucial

Table 3.1 Share of each RES in CO_2 reduction.

RESs type (1 TWh generation)	Reduction in CO_2 per capita (metric tons)	Effectiveness of each technology in CO_2 reduction	Key considerations
Biofuel	0.02	Low	It would not be advisable to use plants as a fuel, as they play a crucial role in sequestering CO_2 in the atmosphere
Geothermal and biomass	0.12	High	It's the most effective one in carbon reduction, and since it is available worldwide, it is not affected by weather factors. However, it requires a high investment cost in installation and electricity generation
Hydropower	0.09	High	It has a relatively low cost of electricity production. However, it depends on the potential of each region.
Solar	0.08	Moderate to high	PVs have the lowest installation cost among all of the RESs and they have a low cost of electricity generation
Wind	0.02	Low	The reason for the less impactful role of wind power in terms of CO_2 reduction is that the number of processes involved in WT construction involves different processes like extracting and processing raw materials, transportation, and maintenance, which release a remarkable amount of CO_2 to the atmosphere. However, the electricity generated by WTs is cost-efficient and it has the lowest cost among RESs, so it's economically sustainable

Source: Data from [40].

role in mitigating human environmental impact, with a 1% increase in overall RES energy consumption or production leading to a 0.1–0.6% reduction in CO_2 emissions.

All of these RESs play different roles and have varying effects on CO_2 reduction. A study conducted on 27 OECD[1] countries from 1960 to 2020 measured the contribution of each RES to reducing CO_2 emissions. As shown in Table 3.1, generating 1 TWh (terawatt-hour) from geothermal and biomass energy results in a CO_2

1 Organization for Economic Co-operation and Development.

reduction of 0.12 metric tons. Hydropower contributes to a reduction of 0.09 metric tons, while solar energy and wind energy each lead to a reduction of 0.08 metric tons. Biofuels result in a CO_2 reduction of 0.02 metric tons [40]. Therefore, global expansion of RESs is a key way to achieve deep decarbonization, a process that started many years ago and continues to advance. As shown in Figure 3.5, most of the countries have already started to increase the share of RESs in their energy and power sector, in which solar and wind are leading the way, while geothermal and hydropower have a gradual growth [41]. In summary, the deep decarbonization of the power sector pulls this sector into a new industrialization.

As mentioned above, achieving deep decarbonization in the power sector requires a new and green industrialization in this sector, such as integrating the RESs in the power sector [42]. However, in contrast to fossil fuels, RESs are not dispatchable, as they are not available at all times and depend on daily and seasonal conditions. Therefore, this issue can cause challenges in the reliability of the power system. Nevertheless, implementing digital technologies can help the power system overcome the challenges associated with RESs, offering significant expansion opportunities. On the other hand, the digital transformation of the power sector has shown the potential in reducing the carbon intensity associated with electricity generation. The studies show that a one-unit increase in digital input will result in a 5.6-unit reduction in the carbon intensity of energy production [18]. The mentioned result increases the importance of digitalization in the decarbonization process of the power sector. The efforts of such decarbonized power systems have resulted in huge investments by power firms in digitalization

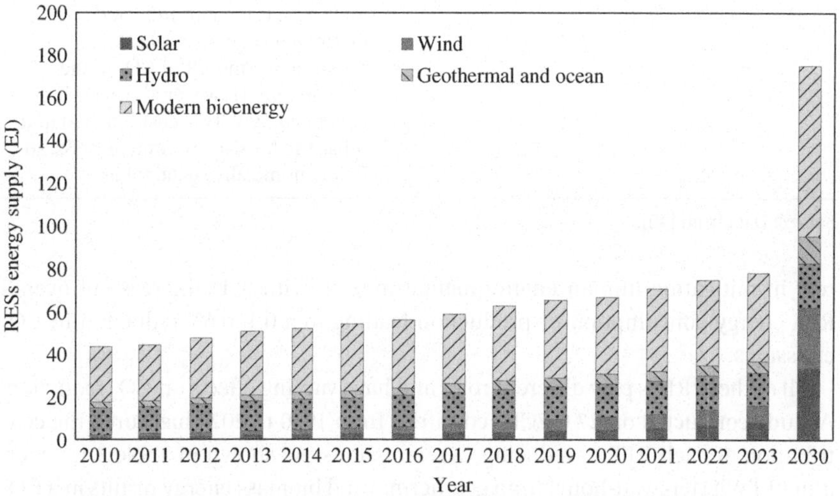

Figure 3.5 RESs supplied by technologies in the net-zero scenario (2015–2022). *Source*: Adapted from [41].

techniques to improve efficiency and reliability [24]. The International Energy Agency (IEA) reports that investment in digital infrastructure in transmission and distribution of electricity grids in 2022 has grown more than 40% compared to 2015 (Figure 3.6) [43]. Therefore, deep decarbonization is the driver of digitalization in the power and energy sector.

As mentioned above, for deep decarbonization of the power sector, new technologies such as RESs and energy storage will be a driver of digitalization in this sector. Theoretically, more use of RESs in the power generation sector will lead to more carbon reduction in the electricity-generating process. However, integration of RESs in the power system will face challenges due to the instability of RESs [18]. These challenges increase the need for new digital technologies such as substation automation, energy management systems, distribution automation, RESs forecasting systems, energy storage management technologies, and grid monitoring systems [4]. Alongside the technologies mentioned, big data and artificial intelligence contribute to balancing supply and demand in power systems with data collection, large-scale statistical data analysis, and providing visual reports. Furthermore, digital twin technology facilitates the integration of RESs. These digital technologies help the power system operators to forecast more accurately and control the variable RESs generation to manage the real-time demand of

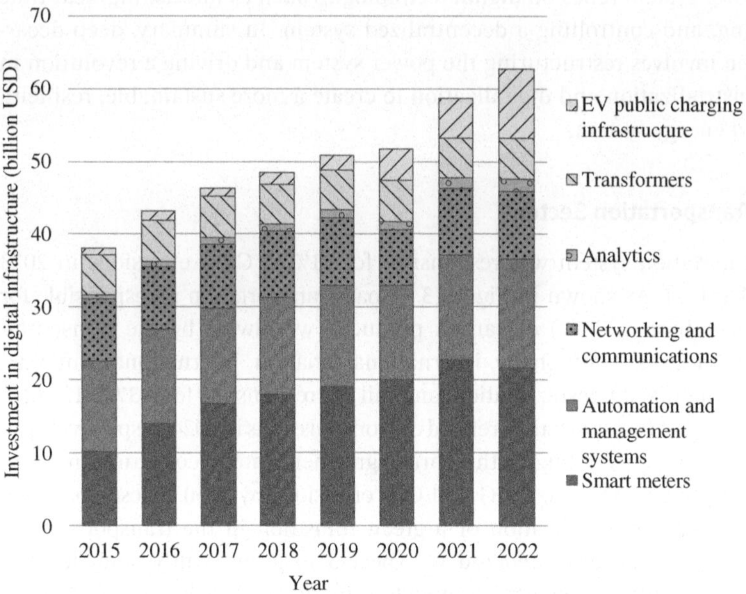

Figure 3.6 Investment in digital infrastructure in transmission and distribution electricity grids, 2015–2022. *Source*: Adapted from [43].

energy and increase grid flexibility and reliability to ensure the energy security of the power system [18].

The traditional power system was structured for a one-way flow of power in which electricity is transmitted from centralized power plants to consumers over long distances through high-voltage transmission lines. Deep decarbonization will force the power systems to use distributed resources like RESs, energy storage, and EVs [44]. The studies report that electrification of the transport and heating sectors will lead to integration of billions of distributed energy resources into the existing power system by 2030 [21]. All of these will lead to a two-way power flow, which requires a structure upgrade to support prosumers (i.e., individuals or entities that both produce and consume electricity) [4]. Only the use of advanced digitalization will allow the integration and management of these decentralized and complicated systems in modern energy societies.

Deep decarbonization of the power sector requires a fundamental change that goes beyond increasing the share of RESs and energy storage. While RESs contribute to reducing carbon emissions, achieving deep decarbonization will require a new wave of industrialization in the power sector, including the development of clean technologies within a decentralized framework and the use of advanced infrastructures. This industrial transformation drives the digitalization, since effective integration of RESs without threatening the reliability of the power system relies on digital technologies such as forecasting, real-time monitoring, and controlling a decentralized system. In summary, deep decarbonization involves restructuring the power system and driving a revolution in both industrialization and digitalization to create a more sustainable, resilient, and smart energy system.

3.5.2 Transportation Sector

The transportation system was responsible for 11% of GHG emissions in 2023 (Figure 3.3) [32]. As shown in Figure 3.7, road transportation is responsible for about three-quarters (74%) of carbon produced worldwide by the transportation sector in 2024. Additionally, international aviation, international shipping, domestic shipping, domestic aviation, and rail are responsible for 6.37, 6.12, 4.55, 3.53, and 1.05% of transportation-related carbon emissions in 2024, respectively [2]. At the same time, according to the Paris Agreement, many countries are committed to achieving net-zero GHG and CO_2 emissions by 2050. These goals will require a green industrialization or a green transition in the transport sector. Meanwhile, transportation demand is expected to grow swiftly, and demand for passenger travel is forecasted to double through 2020–2050 worldwide [45]. Therefore, there is an urgent need to switch to low-carbon transportation, and a fundamental industrial revolution is needed in the transport sector.

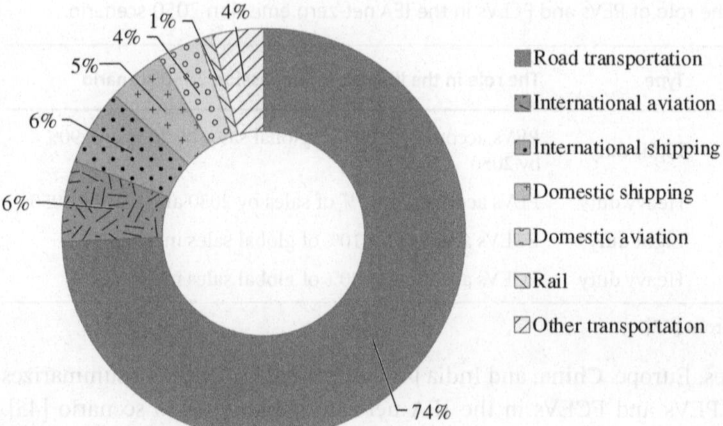

Figure 3.7 Share of each subsector in the transportation sector in carbon dioxide emissions in 2024. *Source*: Adapted from [2].

Table 3.2 Carbon reduction in the life cycle compared to conventional vehicles for different countries.

	Carbon reduction in the life cycle compared to conventional vehicles (estimated in 2021)			
Vehicle type	**Europe (%)**	**United States (%)**	**China (%)**	**India (%)**
BEVs	63–69	57–68	34–46	19–49
PHEVs	25–31	35–46	10	—
FCEVs	76–79	—	33–34	—

Source: Data from [47].

Efforts toward deep decarbonization are reshaping the vehicles and pushing the transportation sector toward zero-emission vehicles. Zero-emission vehicles typically refer to those that operate on grid electricity or hydrogen. Electric fueled vehicles include the battery-electric vehicles (BEVs), which only run on an electric motor, and plug-in hybrid electric vehicles (PHEVs), which can operate through plug-in charging or an internal combustion engine. Both BEVs and PHEVs are included as plug-in electric vehicles (PEVs). The PEVs can reduce emissions by 60–95% compared to a traditional internal combustion engine [46]. Similarly, fuel cell electric vehicles (FCEVs) utilize a fuel cell to generate electricity by converting hydrogen into electricity for powering the vehicle's electric motor [45]. Table 3.2 states the carbon reduction in life cycle compared to conventional vehicles in the

Table 3.3 The role of PEVs and FCEVs in the IEA net-zero emission 2050 scenario.

Technology	Type	The role in the IEA net-zero emission 2050 scenario
PEV	Light duty	PEVs account for 60% of global sales by 2030 and 90% by 2050
	Heavy duty	PEVs account for 17% of sales by 2030 and 68% by 2050
FCEVs	Light duty	FCEVs account for 10% of global sales in 2050
	Heavy duty	FCEVs account for 30% of global sales in 2050

Source: Data from [45].

United States, Europe, China, and India [47]. Additionally, Table 3.3 summarizes the role of PEVs and FCEVs in the IEA net-zero emission 2050 scenario [45]. Therefore, the push for deep decarbonization has transformed vehicles, leading to a shift from conventional vehicles to PEVs and FCEVs. Deep decarbonization not only changes the vehicles' structures but also affects the automotive industry.

Deep decarbonization and associated regulations are prompting many global automakers to set their plans on the carbon neutrality or zero-carbon emission path. On March 15, 2022, the European Union adopted the Carbon Border Adjustment Mechanism, establishing the world's first official carbon tariff, which applies to all products regulated under the European Union Emissions Trading System [48]. These directives enforce the automotive sector to consider carbon emission regulation across the full life cycle, including development, production, utilization, and recycling, rather than focusing on emissions during utilization. Table 3.4 provides a summary of several automakers' strategies for the deep decarbonization process. Despite the different plans each has, all companies have specified a year for achieving net-zero carbon emissions, which shows that deep decarbonization is making a new industrialization in the transport industry [48].

Deep decarbonization is not only reshaping the energy base of transportation but also accelerating its digital transformation. Smart traffic management systems, route optimization systems, and fleet management are examples of digitalization that are necessary in the transport sector [4]. In addition to what was mentioned, there is a trending concept of integrating EVs with the power grid in modern energy societies. As mentioned earlier, the development of EVs will aid in the deep decarbonization process, but this large-scale development will create challenges for the power system and lead to environmental impacts [49]. As is known, EVs are usually charged at night or in the evenings, when the residential demand is at its peak. This peak-time charging of EVs will lead to an increase in carbon emissions, as often thermal power plants supply the majority of these demands. Additionally, this uncoordinated charging will lead to instability in the power system. Vehicle-to-grid (V2G) technology is the solution to this problem, as this

Table 3.4 Plans of different automakers for the deep decarbonization of the transport sector.

Automotive company	Plans for achieving deep decarbonization
Volkswagen	Reducing total carbon emissions in the entire life cycle of vehicles by 30% by 2025 compared with 2015. Additionally, they plan to provide 22 million EVs globally by 2028 and achieve carbon neutrality for the entire organization by 2050
BMW	By 2025, 25% of vehicles sold in China are expected to be electric. Additionally, they aim to reduce carbon emissions per vehicle to one-third by 2030 relative to 2019. Furthermore, they are planning to phase out internal combustion engine passenger vehicles by 2039
Daimler	By 2025, they plan to achieve full electrification, with 50% of their sales consisting of EVs, and achieve carbon neutrality over all stages of the vehicle value chain
Volvo	By 2030, aiming to achieve net-zero carbon emissions over the entire value chain
Porsche	By 2030, they intend to achieve net-zero carbon emissions over the entire value chain and use RESs in the production process
Nissan	Achieving net-zero carbon emissions across all operations and the product life cycle by 2050
General Motors	Planning to stop the sale of gasoline-powered vehicles and shift to net-zero emission structures by 2030
Kia	Achieving net-zero emissions by 2040
BYD	Completed China's first zero-carbon industrial park in 2022, aiming to achieve net-zero emissions by 2050
FAW Group	Achieving net-zero carbon emissions by 2053, with emphasis on using hydrogen fuel cells for their commercial vehicles

Note: The value chain includes the production, the use, and the end-of-life stages of vehicles.
Source: Data from [48].

technology uses EVs as an energy storage system and enables two-way interaction with the power system through charging and discharging. The V2G manages the charging and discharging of EVs more flexibly. EVs can be charged during periods of low demand and times of low-carbon power generation. Moreover, EVs can be discharged during peak demand or when power generation will probably be more carbon-intensive [49]. According to [49], V2G reduces uncontrolled EV charging emissions by 20.6% and reduces the power system's carbon footprint by replacing thermal power with discharged electricity. Many countries, such as Denmark, the Netherlands, and Japan, have initiated pilot projects to explore V2G applications. As a result, implementing V2G technology and the other technologies will require fundamental digitalization in the transportation sector.

In summary, deep decarbonization is the driver of both digital and industrial transformations. Deep decarbonization accelerates the process of shifting from conventional vehicles to EVs and FCEVs, and it forces the automotive companies to redesign their production process, supply chains, and use innovative strategies. At the same time, for maximum utilization of EVs, a digital transformation is needed. Therefore, for achieving deep decarbonization, both an industrial and a digital revolution are needed in the transportation sector.

3.5.3 Industry Sector

The industry sector is responsible for 20% of GHG emissions in 2023 (Figure 3.3) [32]. According to the IEA, industry sectors produced approximately 8.98 Gt (gigatons) of CO_2 in 2022 [50]. Reducing carbon emissions in the industry sector is more difficult than in other sectors, often referred to as hard to abate due to a lack of decarbonization technologies and their strong dependence on fossil fuels for both energy and raw materials [8]. However, it's essential to focus on the decarbonization of the industry sector due to the remarkable contribution to carbon emissions and increasing demand for these industrial products [51]. Deep decarbonization is forcing structural changes in industrial processes like steel, cement, chemicals, and aluminum industries. This industrial revolution in manufacturing will require digitalization in industry sectors. Therefore, the impact of deep decarbonization on industrialization and digitalization of steel, cement, chemicals, and aluminum sectors will be examined.

The steel and iron sector is responsible for a large share of global industrial emissions, about 29.2% (Figure 3.8), and it annually emits about 2.6 Gt CO_2, which is the highest among the heavy industry [52]. Figure 3.9 shows a 2% increase in carbon emissions from 2019 to 2023 [51]. Deep decarbonization of

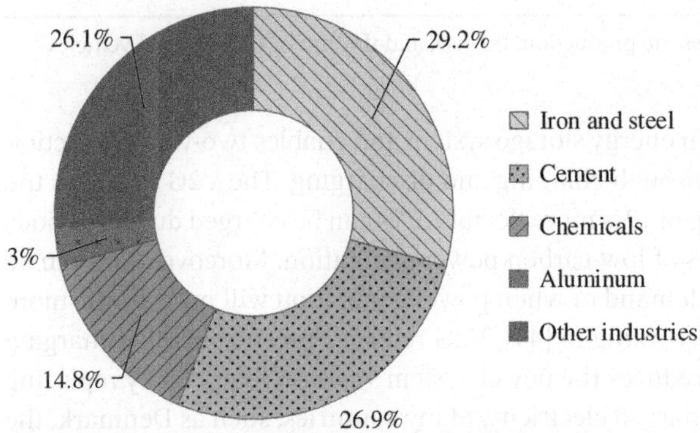

Figure 3.8 Industrial CO_2 emission by subsectors (%). *Source*: Adapted from [50].

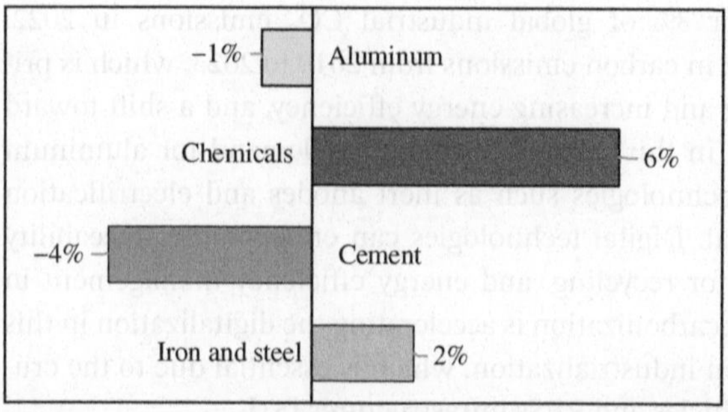

Figure 3.9 Industrial CO_2 emissions changes by subsectors during 2019–2023. *Source*: Adapted from [51].

this sector relies on electrification, hydrogen-based direct reduced iron, and carbon capture, utilization, and storage (CCUS) [13, 51]. This sector requires high temperatures, which makes electrification challenging, so innovation and alternative fuels can play a crucial role in the deep decarbonization process. Smart sensors and cyber-physical systems allow precise monitoring and process control, which leads to optimized energy consumption and reduced wear of equipment, leading to lower carbon emissions. Additionally, condition monitoring and predictive maintenance minimize unplanned outages. Therefore, digitalization becomes a key enabler for achieving deep decarbonization to produce greener steel [23].

The cement industry is responsible for the share of 26.9% of global industrial emissions. As shown in Figure 3.8, cement emission fell around 4% over 2019–2023, but the intensity remained flat because of the CO_2 emission from the clinker. The CCUS technology can remove the emissions of this sector by up to 60%, while biomass and waste-derived fuels substitute for coal [51]. Digitalization in the cement industry focuses on creating an interconnected ecosystem where smart equipment, real-time data analytics, and integrated value chains increase efficacy, minimize cost, and enhance sustainability, which lead to lowering the carbon emission [25].

Chemicals account for 14.8% of global industrial emissions of CO_2 in 2022, and there has been 6% increase from 2019 to 2023 (Figure 3.9), with sharp demand growth driven by plastics and fertilizer. This sector highly relies on fossil-based feedstocks like naphtha and natural gas, which make it carbon-intensive [51]. The technologies that can be used for deep decarbonization include shifting to green hydrogen feedstocks, adopting CCUS technology, and developing recycling clusters. Deep decarbonization requires digital chemical plants that are equipped with blockchain-based carbon accounting, efficiency optimization, platform-based data integration, and smart equipment [26].

Aluminum accounted for 3% of global industrial CO_2 emissions in 2022. There has been 1% decrease in carbon emissions from 2019 to 2023, which is primarily because of recycling and increasing energy efficiency, and a shift toward RESs electricity generation in this sector. However, the demand for aluminum is growing, and utilizing technologies such as inert anodes and electrification of smelting will be essential. Digital technologies can enhance the traceability of supply chains, logistics for recycling, and energy efficiency management in smelters. Therefore, deep decarbonization is accelerating the digitalization in this sector and leading to a green industrialization, which is essential due to the crucial role of aluminum in vehicles and RESs infrastructures [51].

The four industrial sectors that have a major role in carbon emissions are reviewed and examined to assess the role of deep decarbonization in these industries. In summary, deep decarbonization is not only reshaping the industry and manufacturing sector but also accelerating the adoption of digital technologies. Through the integration of smart systems, advanced data analytics, and innovative traceability tools, the industry sector can cut emissions, enhance efficacy, and achieve sustainable growth. This transformation shows how decarbonization is driving digitalization and industrialization, shaping a new, green industrial paradigm for the future.

3.5.4 Agriculture Sector

Since the 19th century, GHGs from fossil fuels and land use have steadily risen, driven by the widespread adoption of machinery following the Industrial Revolution. The mechanization of the agricultural sector and the replacement of manual labor or animals with machines lead to carbon and GHG emissions in the agricultural industry. In 2023, the agriculture sector and the land-use and forestry sector were responsible for 11 and 7% of GHG emissions, respectively (Figure 3.3) [32]. Moreover, increasing global population growth in recent years is expected to drive a substantial increase in food demand over the next 50years. This surge in food demand will increase carbon emissions from the agricultural sector, leading to the intensification of global climate change [53]. Therefore, GHG emissions from agriculture will rise significantly unless some changes are made in the current food supply. The agricultural sector is divided into four stages, and the source and level of carbon emissions in each stage are reviewed.

Agricultural production is the first stage of the agricultural sector, where food, meat, fruits, and other agricultural products are produced. At this stage, agricultural machinery such as tractors, cultivators, plows, loaders, and cultipackers that run on fossil fuels is a source of carbon emissions. Moreover, spraying, irrigation, and fertilization systems are another source of carbon emissions in this stage. The second stage is the manufacturing and processing of final products, in which the product is transported to the food manufacturing centers. The carbon emissions of the second stage primarily come from food transportation, which contributes to pollution and energy consumption in processing, storage, and packaging, as these processes often rely on electricity generated from fossil fuels. The third stage is the food distribution, in which carbon emissions arise from transportation. Finally, the fourth stage is related to food consumption, in which carbon emissions result from transportation, cooking, and food waste [53]. As shown in Figure 3.10, agriculture production, manufacturing and processing of final products, food distribution, and food consumption account for 71.4, 8.8, 8.3, and 11.5% of total GHG emissions from the food and agricultural sectors in 2021 [54].

Based on previous observations, the agricultural sector needs to be decarbonized to meet the goals of the Paris Agreement. Deep decarbonization of agriculture and the food sector will lead to an Industrial Revolution in the technologies of these sectors. The deep decarbonization of the second and third stages, which are primarily dependent on transportation, can be achieved by electrifying the food transportation system [53]. In the fourth stage, deep decarbonization can be reached by electrifying and integrating RESs in the cooking process, and about the food waste, changing diets, and better distribution, like smaller portion sizes, can be helpful in the deep decarbonization process [55]. The primary concern is regarding the first stage, which accounts for 71.4% of total agricultural GHG emissions. Therefore, for the deep decarbonization of the first stage, a fundamental industrialization is required, and needs to be examined in more detail. Three subsectors of the agricultural production stage have been chosen to study, including on-farm energy use, fertilizer, and rice cultivation, which are among the significant sources of carbon emissions in the first stage. The amount of GHG emissions for each is stated in Table 3.5 [16].

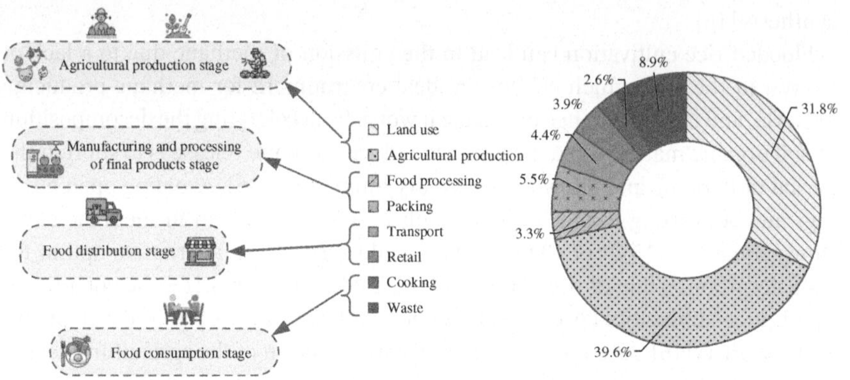

Figure 3.10 Share of each stage in the total carbon emission of the agricultural sector in 2021. *Source*: Adapted from [54].

Table 3.5 GHG emissions of each subsector in the agriculture production stage (for the year 2020).

Subsector	Amount of GHG emissions (Gt)	Share in global GHG emissions (%)
On-farm energy use	1.03	2
Fertilizer	1.01	2
Rice cultivation	0.69	1.2

Source: Data from [16].

Deep decarbonization of on-farm energy use can be achieved with electrification. One of the significant sources of carbon emissions in this subsector is the use of fossil-fueled machinery systems. Switching from fossil fuels to clean electricity in the machinery area can lead to a 70% reduction of carbon footprint, considering the life cycle assessment of this machinery [44]. For example, electric pumping and irrigation, powered by low-carbon electricity, can result in a 95% reduction in carbon emissions and energy use in the agricultural sector [16]. Moreover, farms have great potential in generating their own electricity with RESs, such as installing WTs and PVs in farm areas to generate clean energy for their energy demands [16].

Fertilizer production contributes significantly to carbon emissions. The production of synthetic nitrogen fertilizers accounts for 85% of fertilizer-related GHG emissions, and ammonia production alone consumes 2% of global energy, primarily from coal and natural gas. Deep decarbonization of this subsector can be achieved through low-carbon energy sources, carbon capture and storage, low-carbon powered heat sources such as hydrogen boiler, electrification, or novel processes like electrochemical, plasma-activated, and solar-driven ammonia synthesis [16].

Flooded rice cultivation can lead to the emission of methane due to a lack of oxygen in the soil, which creates an ideal environment for methane-producing bacteria, and climate change can make it worse by accelerating the decomposition of soil organic matter. A practical way to reduce methane release is to shorten the period of flooding and allow the soil to dry more often [16]. Avoiding open burnings and converting to biochar will result in a 30% reduction in methane emissions, equal to 0.21 Gt of CO_2-e, which could serve as an alternative option for deep decarbonization of this subsector [56]. Table 3.6 summarizes the potential of these three subsectors for carbon reduction and shows the current development in these areas [16]. As a result, for deep decarbonization of the agricultural sector, technological changes will be necessary, and all work in the food and agriculture sector must be reconsidered. Therefore, deep decarbonization of the agricultural sector will require fundamental industrialization in the agriculture sector.

Table 3.6 Summarization of CO_2-e emission reduction in subsectors of the agriculture production stage and the development level of each subsector.

Subsector	CO_2-e emission reduction potential (% reduction/Gt CO_2-e per year)	Development level
On-farm energy use	70/0.72	Currently, 28% of on-farm energy use is electrified, while 40% of electricity is generated by low-carbon sources
Fertilizer	95/0.37	Only 0.01% of ammonia is produced by low-carbon electricity
Rice cultivation	Up to 30/0.21	0%

Source: Data from [16].

Deep decarbonization not only drives industrial changes in the agricultural sector but also accelerates the adoption of digital technologies in this sector. Technologies such as remote sensing, robotics, cloud computing, AI, and decision-support systems can be used to improve yields and efficacy and reduce carbon emissions [17]. Through the application of real-time feedback and automated adjustments, digital technologies in the agricultural sector can reduce the overuse of resources such as fertilizers, pesticides, and water, thereby resulting in lower emissions of GHG and increased productivity. Some research works indicate that digital technology can be used to reduce N_2O emissions in soil, lower life cycle emissions from fertilizer manufacturing and distribution, and boost energy efficiency in agricultural processes [57]. Furthermore, technology such as irrigation systems with automated controllers and soil moisture sensors can boost efficiency and lower energy use, and thereby lower emissions [16].

Research on rice, wheat, sugarcane, dairy, and cotton production indicates that input management can reduce emissions while improving resource-use efficiency and profitability. In paddy fields, precision fertilizer management improved nitrogen-use efficiency and rice yields while reducing N_2O emissions compared to conventional systems in fields [58]. In Brazil, applying digital technologies in sugarcane production can bring down the global warming effect by 13% per unit of product output [59]. Using optical crop sensors to apply nitrogen at variable rates in Austrian wheat production led to a lower global warming by 8.6% per unit product output. In China's dairy industry, precision feeding and manure management have improved input-use efficiency and resulted in a 12% reduction of carbon emissions in comparison to the traditional dairy industry [60]. Similarly, in India's cotton sector, the low-cost nitrogen management method reduced N_2O

emissions by about 67% without affecting yields [61]. As a result, these studies show how the digitalization of the agricultural system can lead to a reduction in GHG emissions while enhancing efficacy and productivity.

In summary, the agricultural sector is analyzed through four stages, including agricultural production, manufacturing and processing of final products, food distribution, and food consumption. Afterward, the carbon emissions of the first stage are discussed in more detail. As a result, for deep decarbonization of the agricultural sector, existing structures must be transitioned to modern, green ones, leading to green industrialization. Additionally, implementing a new structure requires the use of digital technologies, leading to digitalization in the agricultural sector.

3.6 Deep Decarbonization as Enabler of Modernization in Energy Societies

Deep decarbonization strategies and plans not only aim to reduce GHG emissions but also provide a strategic approach to building a modern, digitalized energy society. On the path to achieving deep decarbonization goals, energy systems are being redesigned with low-carbon or net-zero emission technologies, which have equipped them with digital infrastructure such as smart meters, automated control systems, real-time data analysis, integrated energy storage, and RESs [62]. Therefore, these strategies and technologies create opportunities for modernizing energy systems [63]. From a digitalization perspective, deep decarbonization brings the integration of RESs, energy storage, electrified systems, and flexible demand and energy management, which brings the need for digital infrastructure [64]. All of these upgrade the technological capacity of the energy sector and lead to a digitalized modern energy society. Simultaneously, deep decarbonization drives industrial transformation, including shifting energy sources from fossil fuels to clean energy, an electrified, digitally monitored process [65]. These transformations aligned with the modernization goals of energy societies. Therefore, deep decarbonization plans can facilitate modern, digitalized, and industrial energy societies.

3.7 Social Visions of Digitalization and Industrialization for Deep Decarbonization

As mentioned earlier, digitalization is a crucial component of deep decarbonization, offering opportunities to improve existing infrastructure [66]. However, if this digitalization doesn't happen under equitable policies and in a controlled framework, it leads to uneven social impacts. Digitalization can help sectors like energy networks manage their systems more fairly, through fairer supply and demand management. However, it can also lead to social inequalities and create new vulnerabilities [67]. These inequalities can be included as unequal access to digital technologies, disparities in digital literacy, uneven distribution of costs and benefits, affordability of digital services, data ownership, and privacy concerns [67]. Therefore, depending on related policies, digitalization can both lead to social inequalities or a socially just deep decarbonization pathway [68].

Similarly, industrialization is an essential component of the deep decarbonization pathway, which enables low-carbon industries [69]. However, if industrialization is poorly managed, it can have social impacts. The structural shift from fossil fuels to clean energy will lead to job losses in fossil-fuel-based sectors and companies. Therefore, it requires support for workers through new skills development, social protection, and job placement [70]. These analyses show that countries should have a strong and fair industrial policy, and without it, they will face social inequalities and welfare losses. Therefore, even if digitalization and industrialization are crucial components of a deep decarbonization path, if they're not pursued under fair policies, they can lead to social inequality.

3.8 Conclusion

Deep decarbonization is not only a pathway for emissions reduction but also acts as a transformative factor that reshapes the digital and industrial infrastructure. By examining the interaction between deep decarbonization, industrialization, and digitalization, this study confirms that GHG emissions reduction cannot happen without systemic transformation and changes. This chapter achieved several key contributions, which are as follows:

1. A comprehensive review of deep decarbonization, industrialization, and digitalization is provided, showing that achieving deep decarbonization goals will require a framework that integrates the three dimensions.
2. The sectoral impacts of deep decarbonization across four sectors, including power, transportation, industry, and agriculture, are investigated. The carbon emissions of each sector are discussed, showing how each sector should undergo industrial and digital transitions to achieve the carbon emission goals.
3. This chapter showed that digital and industrial transformation of energy systems and sectors are essential to achieving net-zero emission goals. These findings link digitalization and industrialization with the achievement of sustainable and net-zero emission societies.
4. Finally, this chapter shows that deep decarbonization is not only a climate strategy but also a driver of modernization in industrial and digital societies. Deep decarbonization plans act as a roadmap for realizing modern energy societies in which technological innovations, digital infrastructures, and sustainability objectives are aligned.

In conclusion, deep decarbonization plays a role as a driver of systemic transformation, pushing sectors and economies to rethink traditional processes and transform them into new processes. Deep decarbonization promotes industrialization and digitalization, which build a developed and flexible system that not only addresses environmental matters, but also ensures development in the economic growth in modern energy societies.

References

1 M. Kueppers et al., "Decarbonization pathways of worldwide energy systems–Definition and modeling of archetypes," *Applied Energy*, vol. 285, p. 116438, 2021.

2 I. Tiseo, "Distribution of carbon dioxide emissions produced by the transportation sector worldwide in 2024, by subsector," 2024. [Online]. Available: https://www.statista.com/statistics/1185535/transport-carbon-dioxide-emissions-breakdown/

3 Energy Institute, "Carbon emissions," 2024. [Online]. Available: https://assets.kpmg.com/content/dam/kpmg/az/pdf/2024/Statistical-Review-of-World-Energy.pdf

4 D. Vine (2021). Power Infrastructure Needs for Economywide Decarbonization. Center for Climate and Energy Solutions.

5 C. Bataille et al., "A review of technology and policy deep decarbonization pathway options for making energy-intensive industry production consistent with the Paris Agreement," *Journal of Cleaner Production*, vol. 187, pp. 960–973, 2018.

6 N. Nasiri, A. Nasiri, S. N. Ravadanegh, B. Mohammadi-Ivatloo, and M. Abapour, "Multi-energy networked microgrid scheduling in smart cities under transactive energy concept," in *Sustainable Energy Resources in Smart Cities*, Springer, 2025, pp. 127–149.

7 A. Nurdiawati and F. Urban, "Towards deep decarbonisation of energy-intensive industries: A review of current status, technologies and policies," *Energies*, vol. 14, no. 9, p. 2408, 2021.

8 J. Meng et al., "Technologies and gaps in deep decarbonization of hard-to-abate industrial sectors," *Nature Reviews Clean Technology*, vol. 1, pp. 578–595, 2025.

9 B. Kazmi, S. A. A. Taqvi, and D. Juchelková, "State-of-the-art review on the steel decarbonization technologies based on process system engineering perspective," *Fuel*, vol. 347, p. 128459, 2023.

10 F. Siekmann, H. Schlör, and S. Venghaus, "Linking sustainability and the Fourth Industrial Revolution: A monitoring framework accounting for technological development," *Energy, Sustainability and Society*, vol. 13, no. 1, p. 26, 2023.

11 C. Bai, P. Dallasega, G. Orzes, and J. Sarkis, "Industry 4.0 technologies assessment: A sustainability perspective," *International Journal of Production Economics*, vol. 229, p. 107776, 2020.

12 D. T. Matt, G. Pedrini, A. Bonfanti, and G. Orzes, "Industrial digitalization: A systematic literature review and research agenda," *European Management Journal*, vol. 41, no. 1, pp. 47–78, 2023.

13 A. Boldrini, D. Koolen, W. Crijns-Graus, and M. van den Broek, "The impact of decarbonising the iron and steel industry on European power and hydrogen systems," *Applied Energy*, vol. 361, p. 122902, 2024.

14 J. E. Bistline and G. J. Blanford, "Impact of carbon dioxide removal technologies on deep decarbonization of the electric power sector," *Nature Communications*, vol. 12, no. 1, p. 3732, 2021.

15 Z. Fang, "Assessing the impact of renewable energy investment, green technology innovation, and industrialization on sustainable development: A case study of China," *Renewable Energy*, vol. 205, pp. 772–782, 2023.

16 L. Rosa and P. Gabrielli, "Achieving net-zero emissions in agriculture: A review," *Environmental Research Letters*, vol. 18, no. 6, p. 063002, 2023.

17 A. G. Green et al., "A scoping review of the digital agricultural revolution and ecosystem services: Implications for Canadian policy and research agendas," *Facets*, vol. 6, no. 1, pp. 1955–1985, 2021.

18 C. Huang and B. Lin, "Promoting decarbonization in the power sector: How important is digital transformation?," *Energy Policy*, vol. 182, p. 113735, 2023.

19 A. Nasiri, N. Nasiri, E. Babaei, and S. N. Ravadanegh, "Optimal operation of integrated electricity and natural gas distribution," in *Sustainable Energy Resources in Smart Cities*, Springer, 2025, pp. 97–126.

20 U. Cali, M. Kuzlu, M. Pipattanasomporn, J. Kempf, and L. Bai, "Introduction to the digitalization of power systems and markets," in *Digitalization of Power Markets and Systems Using Energy Informatics*, Springer, 2021, pp. 1–16.

21 J. Świątowiec-Szczepańska and B. Stępień, "Drivers of digitalization in the energy sector—The managerial perspective from the catching-up economy," *Energies*, vol. 15, no. 4, p. 1437, 2022.

22 A. de Hauteclocque, "Digitalization of the electricity sector and digitalization of society: Effects on the energy transition," in *A Force of Energy: Essays in Energy Law in Honour of Professor Martha Roggenkamp*, University of Groningen Press, 2022, pp. 161–172.

23 K. Herzog et al., "The digitalization of steel production," *BHM Berg- und Hüttenmännische Monatshefte*, vol. 162, no. 11, pp. 504–513, 2017.

24 M. El Zein and G. Gebresenbet, "Digitalization in the renewable energy sector," *Energies*, vol. 17, no. 9, p. 1985, 2024.

25 T. Walther, "Digital transformation of the global cement industry," in *2018 IEEE-IAS/PCA Cement Industry Conference (IAS/PCA)*, IEEE, 2018, pp. 1–8.

26 N. Kockmann, "Digital methods and tools for chemical equipment and plants," *Reaction Chemistry & Engineering*, vol. 4, no. 9, pp. 1522–1529, 2019.

27 O. D. Doleski, T. Kaiser, M. Metzger, S. Niessen, and S. Thiem, "Decarbonization as a strategic issue," in *Digital Decarbonization: Achieving Climate Targets with a Technology-Neutral Approach*, Springer, 2022, pp. 1–19.

28 J. Survilo, "Decarbonization, what it is," in *2024 IEEE 65th International Scientific Conference on Power and Electrical Engineering of Riga Technical University (RTUCON)*, IEEE, 2024, pp. 1–6.

29 M. A. Nazari et al., "An updated review on integration of solar photovoltaic modules and heat pumps towards decarbonization of buildings," *Energy for Sustainable Development*, vol. 72, pp. 230–242, 2023.

30 D. Hirsch, M. Epstein, and A. Steinfeld, "The solar thermal decarbonization of natural gas," *International Journal of Hydrogen Energy*, vol. 26, no. 10, pp. 1023–1033, 2001.

31 O. Apata, A. Adebayo, and P. Ainah, "Renewable energy systems and the fourth industrial revolution," in *2021 IEEE PES/IAS PowerAfrica Conference*, IEEE, 2021, pp. 1–5.

32 United Nations Environment Programme (2024). Emissions Gap Report 2024: No More Hot Air... Please! With a Massive Gap Between Rhetoric and Reality, Countries Draft New Climate Commitments. https://www.unep.org/resources/emissions-gap-report-2024

33 S. A. R. Khan, Z. Yu, A. Belhadi, and A. Mardani, "Investigating the effects of renewable energy on international trade and environmental quality," *Journal of Environmental Management*, vol. 272, p. 111089, 2020.

34 P. Bello, "The role of digitalization in decarbonizing the oil and gas industry," in *SPE Nigeria Annual International Conference and Exhibition*, SPE, 2021, p. D021S009R004.

35 A. Kwilinski, O. Lyulyov, and T. Pimonenko, "Environmental sustainability within attaining sustainable development goals: The role of digitalization and the transport sector," *Sustainability*, vol. 15, no. 14, p. 11282, 2023.

36 І. Максимова and В. Куриляк, "Digitalization and decarbonization: Aspects of synergy in the EU industries," *Економіка та* суспільство, no. 67, pp. 1–9, 2024.

37 IEA, "Emissions," 2025. [Online]. Available: https://www.iea.org/reports/electricity-2025/emissions?

38 Y. Geng, Q. Xiang, J. Gao, Y. Yan, and J. Li, "Progress and framework of clean energy production: Bibliometric analysis from 2002 to 2022," *Energy Strategy Reviews*, vol. 52, p. 101270, 2024.

39 N. Nasiri, S. Zeynali, S. N. Ravadanegh, S. Kubler, and Y. Le Traon, "Decentralized privacy-preserving distributionally robust restoration of electricity/natural-gas systems considering coordination of pump storage hydropower and wind farms," *IEEE Access*, vol. 12, pp. 13747–13762, 2024.

40 G. Perone, "The relationship between renewable energy production and CO_2 emissions in 27 OECD countries: A panel cointegration and Granger non-causality approach," *Journal of Cleaner Production*, vol. 434, p. 139655, 2024.
41 IEA, "Renewable energy supply by technology in the net zero scenario 2010–2030," 2025. [Online]. Available: https://www.iea.org/data-and-statistics/charts/renewable-energy-supply-by-technology-in-the-net-zero-scenario-2010-2030
42 N. Nasiri, S. N. Ravadanegh, and N. T. Kalantari, "A distributionally robust optimization approach for local electrical market interaction with wholesale electrical market and prosumers within the framework of transactive energy concepts," *IEEE Transactions on Industrial Informatics*, vol. 20, no. 4, pp. 5714–5724, 2023.
43 IEA, "Investment change in 2023–2024, and additional average annual change in investment in the Net Zero Scenario, 2023–2030," 2024. [Online]. Available: https://www.iea.org/data-and-statistics/charts/investment-change-in-2023-2024-and-additional-average-annual-change-in-investment-in-the-net-zero-scenario-2023-2030
44 S. Zeynali, N. Nasiri, S. N. Ravadanegh, S. Kubler, and Y. Le Traon, "Tri-level scheduling of integrated electric vehicle parking lots and solar distributed generations in local and wholesale electricity markets," *International Journal of Electrical Power & Energy Systems*, vol. 158, p. 109916, 2024.
45 United Nations Climate Change (2022). Deep Decarbonization Technologies for Sustainable Road Mobility. https://unfccc.int/ttclear/misc_/StaticFiles/gnwoerk_static/tec_transport/39230e72e40f4417a9dfaca1151f24b2/9823fd844afd4fceadc929a3555cb4a0.pdf
46 H. Ambrose, A. Kendall, M. Lozano, S. Wachche, and L. Fulton, "Trends in life cycle greenhouse gas emissions of future light duty electric vehicles," *Transportation Research Part D: Transport and Environment*, vol. 81, p. 102287, 2020.
47 Z. Mera and G. Bieker (2023). Comparison of the Life-Cycle Greenhouse Gas Emissions of Combustion Engine and Electric Passenger Cars and Two-Wheelers in Indonesia. The International Council on Clean Transportation.
48 Z. Gao et al., "Electric vehicle lifecycle carbon emission reduction: A review," *Carbon Neutralization*, vol. 2, no. 5, pp. 528–550, 2023.
49 J. Xu et al., "Emissions reduction potential and feasibility of vehicle-to-grid for Beijing's future electric vehicles," *Cell Reports Physical Science*, vol. 6, no. 6, p. 102616, 2025.
50 IEA, "Direct CO_2 emissions from industry in the Net Zero Scenario, 2000–2030," 2022. [Online]. Available: https://www.iea.org/data-and-statistics/charts/direct-co2-emissions-from-industry-in-the-net-zero-scenario-2000-2030

51 World Economic Forum (2024). Cross-Sector Findings. https://reports.weforum.org/docs/WEF_Net_Zero_Industry_Tracker_2024_Cross_Sector_Findings.pdf
52 J. Kim et al., "Decarbonizing the iron and steel industry: A systematic review of sociotechnical systems, technological innovations, and policy options," *Energy Research & Social Science*, vol. 89, p. 102565, 2022.
53 A. F. Soofi, S. D. Manshadi, and A. Saucedo, "Farm electrification: A road-map to decarbonize the agriculture sector," *The Electricity Journal*, vol. 35, no. 2, p. 107076, 2022.
54 H. Ritchie, "How much of global greenhouse gas emissions come from food?," 2021. [Online]. Available: https://ourworldindata.org/greenhouse-gas-emissions-food
55 H. Ritchie, "Food production is responsible for one-quarter of the world's greenhouse gas emissions," 2019. [Online]. Available: https://ourworldindata.org/food-ghg-emissions
56 J. Wang et al., "The role of rice cultivation in changes in atmospheric methane concentration and the Global Methane Pledge," *Global Change Biology*, vol. 29, no. 10, pp. 2776–2789, 2023.
57 K. Kazimierczuk, S. E. Barrows, M. V. Olarte, and N. P. Qafoku, "Decarbonization of agriculture: The greenhouse gas impacts and economics of existing and emerging climate-smart practices," *ACS Engineering Au*, vol. 3, no. 6, pp. 426–442, 2023.
58 K. Zhang et al., "Unveiling the environmental and socioeconomic benefits of precision nitrogen management for paddy fields in subtropical China," *European Journal of Agronomy*, vol. 142, p. 126663, 2023.
59 G. M. Sanches et al., "Towards greater sustainability of sugarcane production by precision agriculture to meet ethanol demands in south-central Brazil based on a life cycle assessment," *Biosystems Engineering*, vol. 229, pp. 57–68, 2023.
60 C. Liu, X. Wang, Z. Bai, H. Wang, and C. Li, "Does digital technology application promote carbon emission efficiency in dairy farms? Evidence from China," *Agriculture*, vol. 13, no. 4, p. 904, 2023.
61 M. R. Shaheb, A. Sarker, and S. A. Shearer, "Precision agriculture for sustainable soil and crop," in *Soil Science: Emerging Technologies, Global Perspectives and Applications*, IntechOpen, 2022, pp. 49–73.
62 M. Zheng and X. Zhang, "Digitalization and renewable energy development: Analysis based on cross-country panel data," *Energy*, vol. 319, p. 135077, 2025.
63 Z. Li, L. Wu, J. Li, and Y. Zhang, "The impact of digital industrialization on the level of green development," *International Review of Economics & Finance*, vol. 101, p. 104193, 2025.
64 Z. Wang, Y. Sun, H. Kong, and C. Xia-Bauer, "An in-depth review of key technologies and pathways to carbon neutrality: Classification and assessment of decarbonization technologies," *Carbon Neutrality*, vol. 4, no. 1, p. 15, 2025.

65 M. Kravchenko, K. Kopishynska, O. Trofymenko, I. Pyshnograiev, and K. Boiarynova, "A framework for modeling the decarbonization of the economy based on energy innovations in the context of Industry 5.0 and sustainable development: International perspective," *Problemy Ekorozwoju*, vol. 20, no. 1, pp. 207–220, 2025.

66 N. Nasiri, A. Nasiri, S. Najafi-Ravadanegeh, and K. Zare, "Strategic participation of multi-energy systems in local electricity markets under uncertainty within a decentralized optimization framework," *Sustainable Energy, Grids and Networks*, vol. 44, p. 102066, 2025.

67 S. Gantioler, J. Balest, S. Tomasi, F. Voltolini, and N. DellaValle, "Transformative disruptiveness or transition? Revealing digitalization and deep decarbonization pathways in the Italian smart electricity meter roll-out," *Energy Research & Social Science*, vol. 106, p. 103309, 2023.

68 S. Sareen et al., "Social implications of energy infrastructure digitalisation and decarbonisation," *Buildings and Cities*, vol. 4, no. 1, pp. 612–628, 2023.

69 A. Nasiri, N. Nasiri, B. Mohammadi-Ivatloo, and S. N. Ravadanegh, "Hybrid stochastic-robust optimization for smart parking lot trading with local electricity markets under a decentralized framework with renewable energy integration," *Renewable Energy Focus*, vol. 57, p. 100794, 2025.

70 International Labour Organization, "The social, economic and employment impacts of decarbonization and green industrial growth scenarios for the Middle East and North Africa region," 2023. [Online]. Available: https://www.ilo.org/publications/social-economic-and-employment-impacts-decarbonization-and-green-industrial

4

Emerging Trends, Paradigms, and Technologies for Developing Climate–Energy–Sustainability Nexus in Net-zero Emission Modern Energy Networks

Reza Gharibi[1], *Ali Darvishi*[1], *Rahman Dashti*[1], *and Behrooz Vahidi*[2]*

[1] *Clinical-Laboratory Center of Power System & Protection, Faculty of Intelligent Systems Engineering and Data Science, Persian Gulf University, Bushehr, Iran*
[2] *Electrical Engineering Department, Amirkabir University of Technology, Tehran, Iran*

*Corresponding author: vahidi@aut.ac.ir

List of Acronyms	
A-CAES	Adiabatic CAES
AI	Artificial intelligence
AMIs	Advanced metering infrastructures
AWES	Airborne wind energy systems
CAES	Compressed air energy storage
CCUS	Carbon capture, utilization, and storage
CESN	Climate–energy–sustainability nexus
DAC	Direct carbon dioxide capture
D-CAES	Diabatic CAES
DERs	Distributed energy resources
EGS	Enhanced geothermal systems
FOW	Floating offshore wind turbines
I-CAES	Isothermal CAES
IoT	Internet of Things
LAES	Liquid air energy storage
LDES	Long-duration energy storage
MENs	Modern energy networks
ML	Machine learning

List of Acronyms	
OPVs	Organic photovoltaics
OTEC	Ocean thermal energy conversion
OWC	Oscillating water columns
P2P	Peer-to-peer
PHES	Pumped hydroelectric energy storage
PMUs	Phasor measurement units
PtX	Power-to-X
PV	Photovoltaic
QD	Quantum dot
RERs	Renewable energy resources
SMRs	Small modular reactors
TES	Thermal energy storage
WEC	Wave energy converters

4.1 Introduction

4.1.1 Motivation and Problem Description

Population growth, urbanization, and industrial expansion make the rise in energy demand unavoidable. How this demand is met and the resources utilized to satisfy that are key determinants of the pollution levels associated with energy generation and consumption. The rise of environmental concerns and the onset of climate change have made it imperative to employ optimal methods in the energy supply chain and adopt clean energy sources. Therefore, effective energy management, the reduction of greenhouse gas emissions, and the utilization of renewable energy resources (RERs) and clean technologies have become particularly important as key solutions for carbon reduction [1]. However, merely integrating RERs is not enough to realize substantial emissions reductions; without curbing fossil fuel consumption, RERs will primarily supplement fossil fuels, increasing total energy output rather than reducing carbon emissions. To transition to carbon-free energy and achieve carbon neutrality, it is important to minimize dependency on fossil fuels while integrating RERs, preventing environmental burden-shifting between sectors, and tackling concerns linked to resilience, social justice, and energy security. This transition demands a combination of advanced technologies, innovative paradigms, and well-designed policies [2]. These pathways encompass the development and widespread deployment of RERs, such as solar and wind, as well as emerging energy technologies; improving energy efficiency across sectors; applying advanced storage solutions and smart energy management; and integrating electric transportation and modernizing infrastructure. When combined with effective regulations and targeted investments, these actions create a complete framework for reaching carbon neutrality while also maintaining energy system resilience and stability [3].

Modern energy networks (MENs) play a pivotal role in achieving the goal of net-zero carbon emissions and the transition to carbon-free energy sources. MENs help to optimize energy systems, monitor them in real time, make better decisions, and increase grid stability by utilizing sophisticated technology and artificial intelligence (AI). They also help to integrate RERs by managing changes in renewable production and incorporating them into optimal planning. These improvements need major updates to the current infrastructure, such as adding more energy transmission lines, installing smart grids, and improving energy storage systems to make sure they last. Concurrently, new technologies like high-efficiency solar cells, innovative wind turbine designs, and next-generation battery technologies are very important for making energy networks perform better [4]. Additionally, using new and underused energy sources, like hydrogen, is a key way to cut carbon emissions and make the energy portfolio more diverse. The development and large-scale implementation of these technologies will generate new opportunities in energy and emissions reduction [5]. Utilizing emerging technology resources is considered a crucial approach to reducing carbon dioxide (CO_2) emissions and achieving carbon neutrality. Expanding these systems can enhance energy security while decreasing reliance on fossil fuels. Integrating these resources and technologies within a robust, intelligent framework will accelerate the transition to a sustainable, low-carbon energy system [6].

4.1.2 Literature Review

Existing literature highlights how emerging trends and technologies drive the climate–energy-sustainability nexus (CESN) in net-zero systems. In [4], innovations and advances in RERs are discussed about, such as solar, hydro, geothermal, and other new sources. The study highlights the need to invest in infrastructure and using technologies that are beneficial for the environment. Study [7] emphasizes the necessity of advancing new solar technologies designed to enhance efficiency and decrease costs. It reviews the different generations of solar cells and investigates the methods, technologies, and specific features. In another study [8], solar energy systems based on spectral separation technology are examined. The study demonstrates that this approach can overcome the limitations of conventional solar systems and enhance their efficiency.

Another potential technique for solar energy usage is perovskite solar cells, which have grown in popularity due to their high efficiency and low production costs. In [9], the structural and fundamental properties of this technology are discussed, as well as its progress, problems, and future potential. Wind energy is also an important RER that is being widely utilized. Advancing wind energy technology and investigating new strategies to increase efficiency remain serious issues. Study [10] emphasizes the importance of wind energy in promoting carbon-free, sustainable development and addressing climate change. This study illustrates that wind energy benefits both the environment and the economy while significantly lowering greenhouse gas emissions. Paper [11] gives a thorough examination of modeling techniques for offshore floating turbines. It reviews recent advancements in the discipline and addresses current problems. Airborne wind systems are another new technology that is more efficient than traditional wind turbines while also providing better scalability and flexibility. In [12], these systems are reviewed, their applications categorized, and their mechanical design, practical examples, and associated challenges discussed.

While some clean energy sources have been underutilized to date, they are expected to make a more significant contribution in the future. These include geothermal energy and hydrogen. In [13], sustainable hydrogen generation technologies are investigated, with an emphasis on overcoming the associated technological and economic barriers. Integrating hydrogen into a variety of sectors, including transportation and energy generation, can help to hasten the transition to carbon neutrality. Researchers in [14] look at innovative technologies for using geothermal energy. The study suggests that integrating geothermal energy into smart systems and applying AI methods can lead to long-term reductions in both costs and carbon emissions. Heat ocean thermal energy conversion (OTEC) technology is a promising alternative to fossil fuels in coastal regions. Study [15] addresses this technology from technical, economic, and environmental perspectives.

The shift to sustainable, zero-carbon energy systems necessitates the acceptance of new concepts and the use of advanced generations of technology. In [16], the development of integrated energy systems and the power-to-X (PtX) paradigm are examined, as well as their role in improving grid flexibility and enhancing decarbonization. Study [17] describes electrification as a critical strategy for decarbonizing and upgrading energy networks. It promotes the environmental benefits of using electricity instead of gas and fossil fuels. In [18], the necessity of digital governance and policy frameworks for advancing low-carbon energy transitions and nuclear energy investment is highlighted. The study also states that contemporary energy infrastructures necessitate the use of digital technologies by governments for better administration, monitoring, and policy formation. The paper [19] introduces the notion of energy justice and equality as a basic prerequisite for contemporary energy networks. It highlights that attaining energy justice requires participatory initiatives and the formation of peer-to-peer (P2P) energy communities. Study [20] looks at P2P digital ecosystems, emphasizing their potential to promote transparent and secure energy exchange while also examining challenges and opportunities.

According to a review of the literature, a dedicated commitment to cutting-edge technologies and novel paradigms is required to achieve net-zero carbon emissions and the shift to MENs. The sustainable growth of renewable energies, together with the deployment of innovative technology, will increase their efficiency and lower related costs. This will facilitate the integration of these resources into the energy networks and provide their economic and technical justification. The development of RERs must be complemented by consideration of other promising energies. With continued technological progress, geothermal, ocean, airborne, and hydrogen energy could become pivotal contributors to MENs and the decarbonization process. The factors mentioned above must be combined within a coherent and intelligent framework to support the achievement of net-zero by utilizing existing potential. Electrification, digitalization, transparent and facilitated energy trade, and the incorporation of energy justice are key dimensions of this paradigm, each carrying particular importance. This chapter addresses these aspects along with emerging technologies and energy sources. It discusses the challenges, opportunities, and trends in the development of MENs aimed at achieving zero carbon emissions. Main objectives of this chapter are:

- To provide an overview of the net-zero imperative and highlight the role of MENs within CESN.
- To examine emerging technologies, including advanced renewables, long-duration storage, hydrogen, small modular reactors (SMRs), fusion, and novel carbon capture methods, as enablers of deep decarbonization.
- To explain how digitalization, including smart grids and blockchain, enhances system flexibility, transparency, and resilience.
- To present conceptual shifts such as sector coupling (SC), circular economy principles, and the rise of prosumerism and energy democracy, which reshape energy system design and governance.
- To highlight the importance of behavioral and social innovation, including demand-side flexibility and public engagement, for successful energy transitions.
- To introduce holistic paradigms linking climate, energy, and sustainability, emphasizing integrated approaches for resilient, equitable, and adaptive net-zero energy networks.
- To identify key challenges, including scalability, infrastructure demands, and intermittency, while recognizing opportunities for economic advancement, energy security, and improved public health.

4.1.3 Chapter Organization

This chapter is divided into five major sections. It begins with an introduction that explains the net-zero imperative and the role of MEN in CESN. Section 4.2 examines developing technologies, including next-generation renewable systems, enhanced storage, hydrogen, nuclear advancements, carbon capture, and digitization. Section 4.3 explores new paradigms and conceptual shifts, such as SC, circular economy principles, prosumerism, and behavioral and social innovation. Section 4.4 addresses the challenges and opportunities that accompany this transition. Finally, Section 4.5 introduces holistic paradigms linking climate, energy, and sustainability, emphasizing integrated approaches to achieving resilient and equitable net-zero energy networks, and concludes with a summary and a call to action.

4.2 Emerging Technologies for Net-zero MENs

4.2.1 Advanced Renewable Energy Systems

The global transition toward net-zero carbon systems is dependent not only on the development of renewable energies, but also on the development and improvement of technology associated with these systems. Solar, wind, hydro, and geothermal sources of energy are regarded as fundamental sources of clean and renewable energy, but they also face several challenges. These challenges include varying resource availability, land-use limitations, and the need for energy not only during daylight hours. To overcome these challenges and limitations, next-generation solutions that are adaptable to various environments and situations for these resources are required [4].

At the forefront of this shift and transition are innovative renewable energy systems. The primary goals of these technologies are to increase energy absorption, conversion, and integration into new energy systems. For instance, it is promised that the efficiency of photovoltaic (PV) panels, which is currently regarded as one of the system's limitations, will be improved in new solar panels by altering their architecture and component materials. High-altitude offshore winds can improve efficiency and supply more reliable and robust wind resources for wind energy systems. Such ideas, research, and potential in the fields of geothermal and ocean energy can also be a way forward [4].

The efficiency, variety, and coordination of clean energy can be significantly enhanced by the use of new renewable energy system technologies, which can also help to support pollution-free electrical grids and reduce dependence on traditional fuels. The next subsections address some of the most important changes in solar, wind, geothermal, and ocean energy.

4.2.1.1 Next-generation Solar

Solar energy has emerged as one of the most significant energy sources in recent decades, particularly due to its renewable nature. One of the main reasons to pay attention to solar energy is that it can absorb radiation and turn it into electrical power. This is an important way to address climate change and get clean energy. Traditionally, PV technologies, particularly crystalline silicon and thin-film panels, have been the basis of solar energy systems. These technologies have enabled some residences, workplaces, and even large industrial hubs to generate a percentage of their energy, which reduces the amount of energy that must be purchased from the power grid and, ultimately, lowers the overall cost. Although this technology is widely accepted in both industry and society, traditional solar technologies face limitations that prevent them from being sustainable, scalable, and adaptable. Despite increasing efficiency and lowering production costs, these constraints have prompted an effort to conduct a more thorough investigation of novel solar energy absorption systems. Important factors, such as material and environmental constraints, are taken into account in these studies [8].

First and second generations are the two main categories into which traditional solar technology can be divided. The first generation of these technologies, which nowadays make up around 90% of the solar panel market, includes crystalline silicon PV solar panels. These panels use highly pure silicon wafers to convert sunlight into electrical energy. With a 25-year lifespan and efficiencies of 15–22%, these cells offer acceptable efficiency and reliability [21]. On the other hand, thin-film solar cells, such as copper indium gallium selenide (CIGS) and cadmium telluride (CdTe), were employed to define the second generation of this technology. The second generation's flexibility for lightweight and portable applications, along with its reduced material needs, were its advantages [7].

Despite their evolution and demonstrated performance, traditional solar systems still face challenges. For example, making crystalline silicon panels needs a lot of silicon and has high temperature and energy needs, which makes them more expensive and uses more resources. They can also be harmful to the environment because they need rare and often hazardous elements like cadmium and indium. As the entire world's need for electricity grows, these concerns grow bigger, especially in places where the infrastructure cannot be good enough. and the conditions for installation are tough. Because of these limits, traditional solar panels might not be sufficiently affordable to supply the world's energy needs in the future [22].

Next-generation solar technologies will be formed by addressing these challenges and increasing the efficiency and reach of these technologies. Multiple challenges are being addressed simultaneously by emerging technologies. One of the numerous aspects of these challenges is the technological one, which, in the most recent versions, seeks to improve panel efficiency while reducing harmful substances and increasing panel flexibility. Emerging technologies are seeking circular designs and simpler recyclability from an environmental perspective. Additionally, energy justice is a pressing challenge in the energy sector. By improving solar energy efficiency with the next generation of panels, solar energy can be made more accessible in places lacking energy and industrial capacity, therefore meeting the social goals of energy justice [12].

The next generation of solar panels was developed in response to the difficulties and issues encountered by the first and second generations. Perovskite solar cell technology is one such effort. Named after the mineral perovskite, this PV technology is recognized for its similarity to the crystal structure of the mineral. The main advantages of focusing on this type of panel are its impressive performance in testing environments and its potential for low-cost production. The organo-mineral metal halide compounds in these panels give them excellent light absorption capabilities. One significant advantage of perovskites over silicon cells, which need complex fabrication and high purity, is their ability to be processed at low temperatures using solution-based techniques. These aspects decrease energy consumption and enhance the capacity of these cells for production [7].

The performance of perovskite solar cells has improved significantly over a relatively short period. In laboratory conditions, these cells have demonstrated single-junction efficiencies of 25% and combined perovskite and silicon layers to achieve efficiencies of roughly 27%. These efficiencies can rival or even exceed the performance of traditional cells. These cells are flexible and lightweight, which makes them good for curving windows and roofs. They also work better than other types of cells. Because of their excellent light-absorbing features, these cells are especially good for low-light areas and bad weather conditions, such as cloudy days. Even with these benefits, perovskite panels still have a lot of problems because they are still in the early stages of research. One of the most serious issues is their short lifespan. These panels have a two- to three-year lifespan, which is much less than that of silicon systems. Because perovskite panels can be harmed by factors such as oxygen, moisture, and UV light, it is necessary to protect them using advanced methods. Despite these developments, specific environmental issues persist with these cells, including the presence of chemicals like lead in some perovskite formulations. Maintaining laboratory production quality at the factory level presents another challenge. Perovskites are growing rapidly despite these challenges [7].

Organic photovoltaics (OPVs) are another promising segment of next-generation solar technologies. Due to specific characteristics, OPVs are suitable for indoor and low-light environments. The organic semiconductor materials used to make OPVs offer several advantages over silicon-based solar cells, including mechanical flexibility, low weight, and reduced production costs. The features that enable this technology to be used indoors make it suitable for Internet of Things (IoT) applications, where small, low-power devices such as wearable electronics, smart meters, and sensors require reliable, distributed energy sources. This technology can provide that. OPVs can be powered by harvesting light from indoor light sources due to their high absorption coefficient and tunable bandgap [23].

Recent developments have enhanced the efficiency of OPVs exposed to indoor lighting, with some achieving efficiencies of over 30% at light intensities ranging from 500 to 1,000 lux. This ability is the result of improvements in material design, heterogeneous bulk architectures, and the optimization of lighting-specific active layer thickness. AI and machine learning (ML) have also played a significant role in accelerating device optimization and material screening, leading to optimal OPV configurations. Stability, scalability, and a limited exciton diffusion length are some of the early challenges facing the technology. However, due to their advantages, OPVs are becoming increasingly acknowledged as a crucial component in achieving net-zero emissions and off-grid energy supply [23].

Other emerging advances in solar technology include quantum dot (QD) PV cells. The semiconductor particles in QD PV cells are nanometers in size and have a tunable bandgap. This characteristic enables QDs to absorb a wide range of light, including infrared light, making it possible to create flexible and inexpensive solar panels. However, QDs still face a significant energy conversion barrier, which is less than that of other established technologies. Furthermore, the dependence of QD cells on specific toxic substances presents environmental issues. To be commercialized, they also need to overcome challenges such as poor stability in real-world environments, particularly when air and moisture are present. Multi-junction PV cells, on the other hand, were created to overcome the efficiency limitation by stacking several semiconductor layers, each of which is responsible for absorbing a specific portion of sunlight. Despite its excellent efficiency, this design is still very complex and expensive. The primary cause of this is the use of costly and uncommon materials, along with sophisticated production techniques that are appropriate for space applications. Due to this, it is less appealing than more workable and scalable alternatives, such as OPVs or perovskites [7].

Perovskites, OPVs, QDs, and multi-junction cells are some of the next-generation solar technologies that, when combined, represent a significant advancement in capturing and utilizing solar energy. Each of these technologies presents unique development challenges and has its advantages and disadvantages. These technologies are very important for delivering low-carbon, distributed electricity in MENs. Figure 4.1 presents a classification of the different generations of solar cell technologies.

4.2.1.2 Advanced Wind

Wind energy, one of the most important types of renewable energy, plays a crucial role in the global transition to carbon-free energy systems. Wind power has become

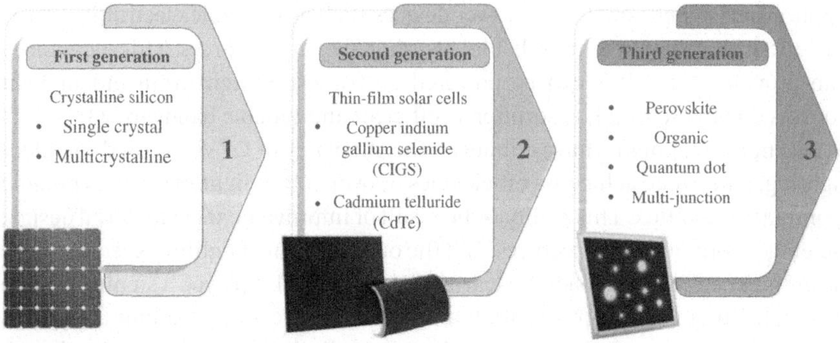

Figure 4.1 Different generations of solar cell technology.

the primary source of non-hydro renewable energy due to its high electricity production capacity, which was approximately 800 GW in 2021. Traditional and conventional wind systems have proven to be highly effective both on land and at sea [10]. By changing the architecture of turbines, including making the rotor bigger and taller, traditional systems have greatly enhanced energy. Even with these improvements, traditional systems still have physical limits. For example, onshore wind has problems with land use, not having the right physical environments, and getting people to accept it. On the other hand, offshore turbines cannot be used in waters deeper than 60 m [24]. These limitations, combined with the growing demand for clean energy, have necessitated the need to overcome them. Emerging approaches to wind energy have been considered.

Next-generation technologies include floating offshore wind turbines (FOW) and airborne wind energy systems (AWES). FOW are an exciting new way to generate clean electricity from wind in waters deeper than 60 m, which account for 80% of offshore wind potential and cannot be reached with traditional fixed turbines. These turbines are not directly attached to the seabed, but are held in place by floating platforms connected by mooring lines [11, 12].

This method allows access to stronger winds located further away. Hywind Scotland (30 MW), Kincardine (50 MW), and Windfloat Atlantic (25 MW) are only a few of the farms that are now in operation. There are also plans for more farms to be built or are already being built. There are three main types of floating wind systems: spar-buoy platforms, semisubmersibles, and tension leg platforms. Each of these systems has its own way of keeping the sea stable. This technology still isn't close to being as good as stationary wind turbines, even though it has a lot of advantages and more wind power capacity. But by 2028, its price is predicted to drop a lot and make it more competitive. As technology advances and larger farms are constructed, floating wind has the potential to significantly alter the approach to renewable energy [11].

Another emerging approach in this field is AWES, which are viewed as a promising alternative to traditional ground-based wind turbines. These systems take advantage of the fact that wind power grows as the cube of wind speed. This lets them use stronger and more consistent winds that are usually found between 200 m and 10 km above ground level. These systems use flying structures, such as kites or gliders, to catch wind energy at high altitudes and turn it into electricity on the ground (Ground-Gen systems) or in planes (Fly-Gen systems). The technology has a number of benefits, including needing less material, being more flexible in where it may be used, and being able to tap good wind resources in places where traditional wind turbines aren't practicable. There has been a lot of interest in AWES since the early 2000s. Many academic institutes and businesses have been working on and testing new prototypes. As this industry grows, AWES might greatly increase the world's supply of renewable energy and make wind energy easier to get at in new and difficult places [12]. As discussed, various types of wind power systems have been introduced; Figure 4.2 illustrates their configurations, including grounded, offshore, and airborne systems, in a schematic diagram.

4.2.1.3 Enhanced Geothermal Systems

Enhanced geothermal systems (EGS) represent a significant step forward in utilizing geothermal energy beyond the geographic boundaries of hydrothermal resources. Traditional systems rely on natural heat convection, permeability, and other factors, whereas EGS actively design subsurface reservoirs in hot, dry, and impermeable rock formations of the shallow continental crust. This engineering technology uses hydraulic fracturing, chemical stimulation, and thermal shock, among other things, to construct and improve interconnected fracture networks. This makes it possible to extract heat more efficiently where it would not occur

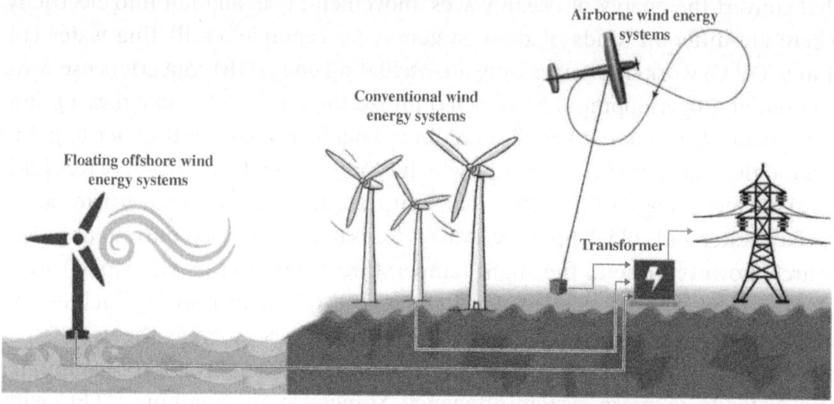

Figure 4.2 Wind power systems: grounded, offshore, and airborne.

naturally. This method converts geothermal energy from a regionally specific energy source to a widespread, baseload competitor, making it accessible in all regions where the initial conditions are met [25].

The advancement of tools is also critical in enhancing the goals of this technology. The development of instruments and technologies like plasma drilling, impact drilling, and rotary drilling has made it easier to get through hard rock and made it possible to work in tough rock situations. These tools provide a lot of advantages, such as lowering costs and making operations more efficient. In addition to physical instruments, the development of software tools like AI and ML can help with real-time maintenance and forecasting of geothermal systems, which can help with energy management. Additionally, the use of software-based optimization tools enables the integration of geothermal systems with other RERs, thereby ensuring grid stability by complementing the periodicity of other RERs [25]. Furthermore, the emergence and advancement of thermal energy storage methods such as sensible heat storage and the use of phase change materials have made this technology more efficient than before, storing and releasing excess heat as needed [26].

4.2.1.4 Ocean Energy

Ocean energy is a component of advanced and modern energy systems that harnesses the ocean's vast power and force using cutting-edge methods and technologies such as wave motion, tidal height changes, and OTEC. Tidal energy takes advantage of predictable fluctuations in sea level, either by extracting potential energy from the difference in water height using dams or by utilizing the kinetic energy of water currents through tidal stream technologies. Meanwhile, horizontal-axis turbines in tidal technology have made significant advances and are nearing maturity, making them superior to tidal dams [27]. Another type of ocean energy technology is wave energy converters (WEC). WEC are devices that convert the energy of ocean waves' movement, rise, and fall into electricity. There are different kinds of these systems. For example, oscillating water columns (OWC) work by compressing air, oscillating body (OB) converters use wave motion, and overtopping devices (OD) utilize the energy of water flowing into a reservoir. Wave energy supplies are more widely available and easier to get to around the world than tidal energy, which is only available in certain places [28].

OTEC technology utilizes the temperature difference between warm ocean surface water and cold deep-sea water. OTEC is referred to as a continuous power source; however, due to the slight temperature difference, its efficiency varies, ranging from 3 to 5%. Recent developments in OTEC technology include the investigation of new methods to improve performance, such as the use of special cycles known as Kalina and O'Hara, which use a combination of ammonia and water to increase system efficiency. Some systems combine OTEC with other technologies, such as solar heating or desalination, to improve efficiency.

Although these plants have a high initial operating cost due to the special equipment, such as heat exchangers and chilled water systems, once operational, their maintenance costs are extremely low because they do not require fuel. Figure 4.3 presents a schematic diagram of a closed-cycle OTEC system. Current research focuses on enhancing the exchanger design, improving turbine efficiency, and utilizing environmentally friendly fluids further to improve the efficiency and environmental sustainability of this technology [15]. Table 4.1 presents the projected roles of advanced renewable technologies in net-zero energy systems by 2050, including their approximate generation shares and capacity factors. Advanced wind and next-generation solar are expected to dominate bulk electricity supply, while enhanced geothermal and ocean energy provide complementary baseload and coastal contributions to support the low-carbon transition [29–32].

4.2.2 Breakthrough Energy Storage Solutions

To transition to low-carbon energy systems and grow advanced renewable energy systems, it is important to be able to store and deliver these energies when and where they are needed. The increased use of solar and wind energy in the energy grid, combined with their intermittent nature, presents significant challenges to the

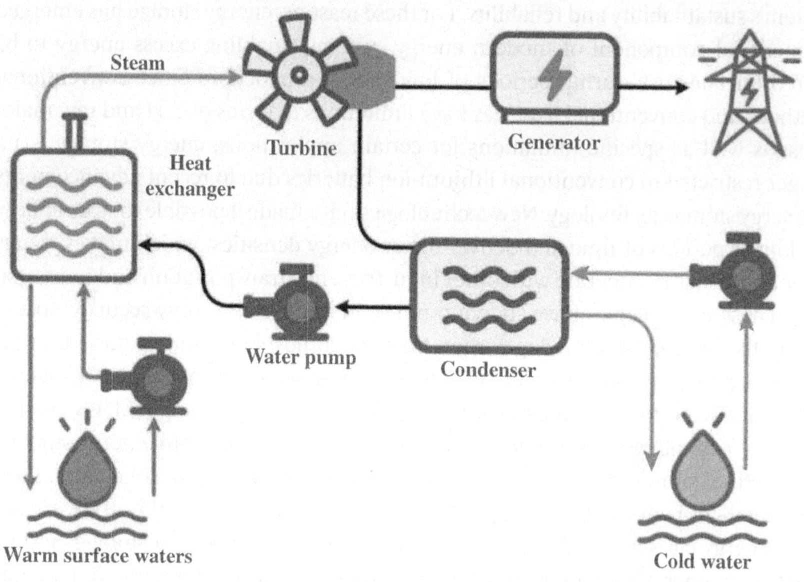

Figure 4.3 An example diagram of an OTEC system (closed cycle type).

Table 4.1 Contributions of emerging renewable energy technologies in net-zero pathways (global outlook).

Technology	Current/recent status	Projected role in net-zero scenarios	Notes
Advanced wind	~2,100 TWh generated globally in 2022	Requires ~17% annual growth to 2030; combined with solar, ~70% of global electricity by 2050	Includes offshore and next-generation turbine technologies
Next-generation solar	~70% of new renewable capacity additions in 2022	~20% of global electricity generation by 2030	Covers PV efficiency improvements and CSP
EGS	Limited deployment; pilot projects ongoing	Technical potential up to 4,600 GW_e; realistic ~256 GW_e by 2050	Still high cost, but scalable with technology advances
Ocean energy	Negligible contribution today	Expected to remain marginal by 2050 in net-zero scenarios	Includes tidal, wave, and ocean thermal technologies

system's sustainability and reliability. For these reasons, energy storage has emerged as a crucial component of modern energy systems, enabling excess energy to be stored for later use during periods of low energy production. Since conventional methods and conventional batteries have limitations in terms of cost and raw materials, as well as specific limitations for certain applications, energy storage is no longer restricted to conventional lithium-ion batteries due to recent advancements in energy storage technology. New technologies have made it possible to store energy for longer periods of time and deliver better energy densities, which makes things safer and easier to integrate with other industries, like transportation and industrial [33]. These technologies have shown promise in improving energy security, speeding up the process of removing carbon from the atmosphere, and opening up new business prospects. An overview of new energy storage technologies is provided below, divided into three categories: (i) long-duration energy storage (LDES), which includes technologies like liquid air, compressed air, pumped hydro energy storage, and thermal storage; (ii) next-generation batteries, which involve solid-state, flow, and metal–air batteries with chemistries other than lithium-ion; and (iii) hydrogen as an energy carrier, which covers advanced electrolysis, hydrogen storage, and its function in energy-to-energy applications. All of these technologies work together to create a more adaptable, durable, and sustainable energy future.

4.2.2.1 Long-duration Energy Storage

Storage technologies that can continuously discharge energy for extended periods, up to several days or weeks, are referred to as LDES. From this perspective, LDES technology is crucial because it can bridge the gap between the demand for electricity and the production of renewable energy, resulting in a robust and resilient energy system even when extreme weather disrupts renewable energy production. Because LDES may address the imbalance between daily and seasonal supply and demand, it is more important than ever that it be used in future networks that mostly rely on intermittent RERs [34]. The following divisions discuss four different types of LDES: compressed air energy storage, liquid air energy storage, pumped hydro energy storage, and thermal energy storage.

- **Compressed air energy storage (CAES):** CAES utilizes extremely high pressures to compress air, then releases the compressed air and mixes it with fuel to power turbines similar to conventional gas turbines. Air compressors, air storage tanks, high- and low-pressure turbines, combustion turbines, and an electric generator are some of the parts that CAES systems often contain. Because this approach is less expensive than artificial air storage, it is frequently utilized to store air in subterranean places such as caverns, abandoned mines, and specific aquifers. CAES is a critical component of LDES due to its low construction costs and ability to combine with intermittent RERs. Diabatic (D-CAES), adiabatic (A-CAES), and isothermal (I-CAES) technologies are the three types of CAES technologies that are frequently developed. D-CAES generates a significant amount of heat during air compression, which must be cooled to prevent damage to the equipment, resulting in heat loss and decreased efficiency. In addition to wasting heat, this kind uses fossil fuels, which contribute to carbon emissions and interfere with modern grid aspects. A-CAES increases efficiency by approximately 55–75% by absorbing the heat produced during the compression process, but it also lengthens the operating time. By maintaining a constant compression temperature, I-CAES, on the other hand, minimizes heat loss and thermal stress, ultimately achieving a potential efficiency of approximately 90%. I-CAES still experiences technical issues with true isothermal compression, despite these benefits. To achieve the target of zero emissions, research on these technologies has primarily focused on enhancing responsiveness, efficiency, and integration with renewable energy grids [33]. Table 4.2 lists some of the key characteristics of CAES technologies.

Table 4.2 Important characteristics and comparison of CAES technologies.

Technology	Efficiency (%)	Heat management	Carbon emissions	Power output	Fuel requirement
D-CAES	40–50	External cooling	Higher	High	Yes
A-CAES	55–75	Heat storage	Moderate	Moderate to high	Reduced
I-CAES	> 90	Constant temperature	Minimal	Lower	Minimal or none
TES	> 90 (heat), 50–70(electric)	Insulated heat storage	Very low	Medium to high	No fuel, needs electricity/heat
LAES	50–60	Cryogenic tanks, heat recovery	Low	Medium to high	No fuel, uses air + electricity
PHES	70–80	Minimal thermal issues	Very low	Very high (MW–GW)	No fuel, needs electricity

- **Liquid air energy storage (LAES):** the energy storage technology known as LAES stores energy in the form of liquid air, utilizing cryogenics. Conventional industrial liquefaction processes are used to compress and cool ambient air to cryogenic temperatures (about −196 °C) during times of excess or inexpensive electricity generation. The resultant liquid air is stored in insulated, low-pressure tanks because it has a volume that is roughly 700 times smaller than that of its gaseous state. When electrical energy is needed, the liquid air is turned into a high-pressure gas using particular methods. Then, a turbine expands this high-pressure gas to make electricity. The recovery of "cold exergy" and stored heat from the liquefaction process is crucial to enhancing the cycle's efficiency in this technology, which can be further improved under various configurations. Geographic flexibility is one of LAES's main advantages over CAES. LAES is suitable for urban and industrial locations, as it is not dependent on any specific geological features or conditions. To significantly improve performance, the technology can be combined with other procedures, such as liquefied natural gas (LNG) regasification, while also considering the geographic circumstances. LAES technology is a promising way to support grid balancing and renewable energy integration, and it is currently being developed at commercial scales with power outputs ranging from 10 to 100 MW [35].

- **Pumped hydroelectric energy storage (PHES):** the most developed and popular type of LDES is PHES, which accounts for approximately 96% of the global storage capacity. It is based on the simple principle that during periods of low energy demand or excess renewable energy generation, electrical energy is used to pump water to elevated reservoirs, which are then used to power turbines when required. PHES systems have demonstrated significant advantages due to their high storage capacity, long life of over 40 years, and high round-trip efficiency of 80–90%. They also have low operating costs and high reliability due to their structure. In addition to their technical characteristics, PHES reservoirs can serve other flood control purposes in regions with excessive rainfall. Despite these advantages, the installation of PHES is geographically constrained since it requires suitable land with large elevation changes and access to reliable water supplies, which limits and complicates its usage in flat, sandy areas. PHES development, with a focus on energy's future, strives to improve cycle efficiency, broaden operational flexibility, and reduce environmental impacts through new designs and complex management systems. In order to help accelerate the global shift to low-carbon energy systems, PHES is expected to be a crucial technology for the integration of RERs and future grid sustainability [33].
- **Thermal energy storage (TES):** heat is the energy that is absorbed by this storage and released when required. This feature enables the flexible integration of RERs into industrial processes and district heating applications. There are three primary types of TES: sensible heat storage, which gets efficiencies of 45–87% using common materials like water, sand, rock, or molten salts; latent heat storage, which uses phase-change materials to achieve high energy density and stable temperatures but still has problems with thermal conductivity and stability; the third type of storage is thermochemical storage, which offers up to 10 times the energy density of sensible heat and minimal long-term losses by storing heat in reversible chemical reactions. Also, some innovative designs, like self-insulating seasonal systems, aim to increase storage time and lower costs in order to meet long-term storage objectives and integrate into zero-carbon energy systems [33].

These LDES technologies, which include CAES, LAES, PHES, and TES, show how different techniques can address the intermittent nature of renewable energy generation while simultaneously promoting robust and sustainable energy systems. Finally, they help the switch to energy systems that reduce carbon, make the

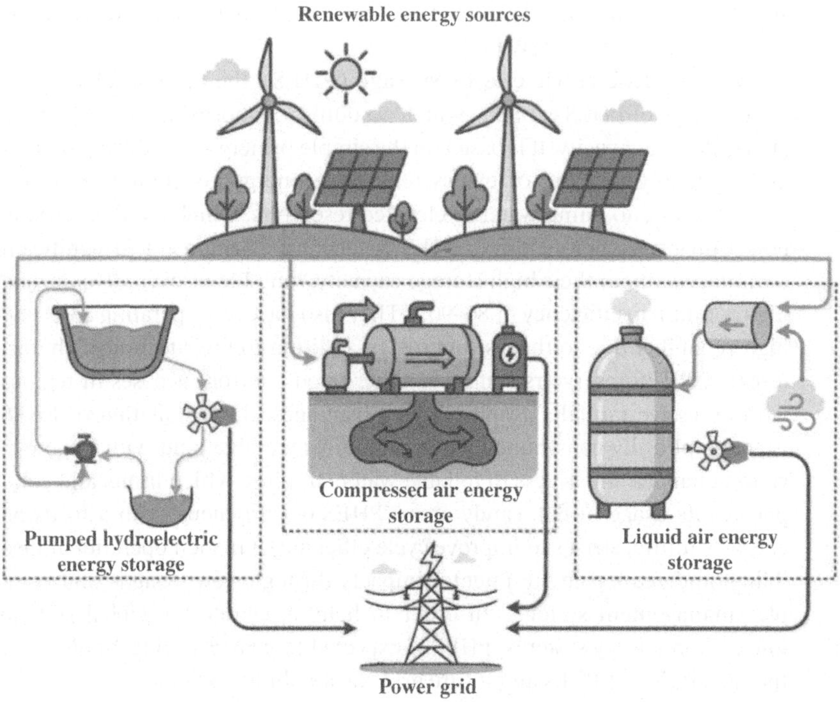

Figure 4.4 Some long-duration energy storage and renewable energy sources.

grid better, cut down on the need for fossil fuels, and encourage long-term energy transmission. Figure 4.4 provides an overview of long-duration energy storage technologies and renewable energy sources

4.2.2.2 Next-generation Batteries

The transition to zero-carbon energy grids is based on long-lasting batteries that power RERs, electric vehicles, and decentralized energy systems. Improved batteries minimize the reliance on combustion engines in modern grids, resulting in lower carbon emissions. However, traditional lithium-ion batteries have limitations, such as cost, safety, and energy density, which make next-generation batteries necessary.

One of the most promising developments is solid-state batteries. These batteries utilize solid materials instead of flammable liquid electrolytes, thereby enhancing safety and increasing energy density. By eliminating the risk of leakage and enabling the use of lithium metal anodes, solid-state designs can achieve theoretical energy densities exceeding 500 Wh/kg. This increases the driving range of electric vehicles and lessens their overall carbon footprint. Their long cycle life also reduces material use, which in turn lowers greenhouse gas emissions associated with production and recycling. Despite these benefits, production complexity remains a challenge, prompting accelerated research into their commercialization [36]. Alongside solid-state batteries are flow batteries, which store energy in external liquid electrolytes. This design increases the scalability of energy and power. Flow batteries are usually used in stationary applications. They help to integrate renewable energy into the power grid. These batteries can hold extra wind or solar energy for hours or even days. They release this energy when output is low, reducing dependency on fossil fuel-based power plants during these periods. Their long cycle life and simplicity of recycling make them ideal for cleaning up grid infrastructure. Additionally, emerging chemistries beyond conventional vanadium are aligning these technologies with carbon reduction goals, aiming to improve affordability and sustainability [14].

Metal–air batteries have gained significant attention. They use oxygen from the air as a reactant at the cathode, giving them a high energy potential that can sometimes rival that of gasoline. For electric vehicles, this capability can match or even exceed the performance of traditional gasoline cars without increasing the battery's weight or size. Among the various technologies in this field, aluminum–air and zinc–air batteries are seen as safer and more sustainable options [36]. However, these batteries still have problems such as short lifespan and energy loss, so research is underway to develop better materials and designs to make them more reliable. If successful, they could be used in heavy vehicles, long-distance transport, contributing to a reduction in greenhouse gas emissions where conventional batteries struggle [37]. Beyond lithium, innovation is focusing on common elements such as sodium, magnesium, and zinc. Sodium-ion batteries are in their early phases of commercial application. They provide a less expensive alternative to lithium while also increasing efficiency to 200 Wh/kg, making them feasible for grid storage and short-range vehicles. High-temperature sodium-sulfur designs and next-generation solid-state sodium-ion batteries promise lifetimes of over 12,000cycles. This increases their sustainability and impact on carbon reduction. Lithium–sulfur batteries also show potential, with theoretical energy densities of 2,600 Wh/kg and abundant, low-cost sulfur. But problems like losing materials and structural damage are currently being worked in [36].

These next-generation batteries represent a significant shift in how society stores and utilizes energy. Their primary objectives are to promote the use of clean energy and to address the entire energy storage cycle, encompassing raw material sourcing, production, and recycling at the end of their life. As these technologies get better, they will help build a global economy that does not generate any carbon. This will help to meet a variety of demands, including personal transportation and grid stability, as well as accelerate the transition to a sustainable, low-carbon future.

4.2.2.3 Hydrogen Energy Carrier: Green Production, Storage, and PtX

Hydrogen is acknowledged as a flexible energy source capable of bridging the gap between economical renewable energy generation and energy storage as we transition to a carbon-free energy future. Green hydrogen, also known as hydrogen generated via electrolysis with RERs, has sparked attention due to its potential for deep decarbonization, which is a substantial advantage over traditional hydrogen production using fossil fuels. The major way of manufacturing green hydrogen is electrolysis. This process has undergone significant improvements in recent years. It addresses concerns about cost, reducing emissions, and working with other types of energy carriers. The most commonly utilized electrolysis methods and technologies are alkaline electrolysis (AE), proton exchange membrane electrolysis (PEM), and high-temperature solid oxide electrolysis cells (SOEC). AE systems are mature and cost-effective, but they operate at lower pressures and current densities, whereas PEM electrolyzers produce hydrogen with higher purity and density, but they require more capital due to the use of precious metal catalysts. In contrast, SOEC technology is still evolving and offers greater efficiency by operating at high temperatures; however, material degradation remains a significant barrier to large-scale commercialization. Recent innovations include hybrid approaches that combine the low cost and high durability of alkaline units with the rapid response capability of PEM systems to capture variable solar and wind energy. These integrated solutions, supported by ML-based predictive control and optimization strategies, have yielded higher hydrogen yields, more than 13% longer electrolyzer lifetimes, and over 6% annual improvements in economic performance. These advancements are expected to reduce hydrogen production costs from approximately $7.26 per kg to less than $2 within the next decade, making green hydrogen competitive with conventional fuels in various markets [13].

Although hydrogen production is crucial, storage is also necessary for utilizing hydrogen as an energy carrier in various seasons and sectors. Hydrogen can be stored in compressed gas, cryogenic liquid, or solid chemical form using advanced carriers, such as metal hydrides and liquid organic hydrogen carriers. On a large scale, geological storage in salt caves is a promising option due to its high capacity and low risk of leakage. For example, a salt cave in Spain with a working volume of 515,000 m^3 can store up to 5.8 million kg of hydrogen, allowing for seasonal storage of approximately 23 GWh of injected energy and 84 GWh of recoverable energy per year [13]. Such storage infrastructures complement the design of multi-energy systems, in which hydrogen is combined with electricity, heating, and transportation systems to improve grid stability and reduce reliance on battery storage alone [38]. In distributed or mobile applications, compressed hydrogen tanks remain the dominant option, particularly for fuel cell vehicles, despite ongoing research into advanced materials to improve volumetric energy density and safety of smaller systems [13]. These developments establish a coherent hydrogen value chain that can balance the variable supply of renewable energy, decouple production from consumption, and facilitate sector integration.

Hydrogen plays a crucial role in PtX technologies, which convert excess renewable electricity into hydrogen (electricity to H_2) and then into energy carriers and chemical products, such as synthetic methane, methanol, ammonia, and liquid hydrocarbons. This approach enables applications in sectors that are challenging to directly supply with electricity, such as aviation, shipping, heavy industry, and chemical production, by converting electricity into storable and transportable chemical energy carriers [13]. The use of PtX technologies as part of an integrated energy hub in modern energy systems allows for flexible load management, reduced constraints, and active participation in electricity markets, particularly when combined with demand response schemes that shift energy consumption to periods of excess supply. A coordinated approach to using hydrogen as a storage and energy source can reduce operating costs, CO_2 emissions, and increase system flexibility [1]. Furthermore, combined cycle energy systems that combine hydrogen production with cogeneration of electricity and heat, carbon capture and utilization, and waste heat recovery from electrolyzers or fuel cells are more efficient and have lower life cycle costs than single-purpose renewable power plants [13].

With these features, the combination of advanced electrolysis, scalable hydrogen storage, and PtX conversion pathways enables a shift from linear, fossil-fuel-dependent models to flexible, cyclical, and low-carbon energy systems. Hydrogen enables renewable energy storage on timescales ranging from hours to entire seasons, aids in the decarbonization of sectors that have previously been resistant to electrification, and creates new opportunities for the production of synthetic fuels and chemicals. As electrolyzer performance improves, operations get better, and integrated energy hub designs are made, PtX hydrogen-based systems are ready to play a big role in the global move to green energy.

4.2.3 Advanced Nuclear Technologies

The world's efforts to reduce carbon dioxide emissions depend on nuclear power, particularly in energy systems that aim to meet the climate targets outlined in the Paris Agreement. Whereas fossil fuels such as coal and natural gas are used to generate electricity, nuclear power plants do not emit carbon into the atmosphere. Additionally, this technology produces stable and dependable electricity, which helps reduce the fluctuation of RERs like solar and wind [18].

International energy agencies and intergovernmental organizations view nuclear power as a crucial component of the transition to clean energy systems. Spending more on energy storage, increasing demand-side flexibility, and continuing to develop RERs are all necessary if we wish to make this transition without or with less nuclear power. High-temperature process heat for applications such as district heating, seawater desalination, hydrogen production, and other industrial processes that typically rely on fossil fuels can also be produced by next-generation nuclear technologies [39]. New nuclear technologies, such as fusion energy systems and SMRs, are viewed as significant advancements in this regard. Because SMRs are quick to set up and provide a range of energy options, they can quickly meet short-term needs. However, nuclear fusion technology has a longer timeline and is revolutionizing the energy production process [40].

4.2.3.1 Small Modular Reactors

SMRs are designed to make between 35 and 300MW of electricity, which is a lot less than what standard gigawatt-scale power plants can do. The small size makes it easier for factories to make reactor modules, which makes it easy to transportation them to the installation site. This method reduces the time required for on-site project implementation and mitigates risks associated with investment expenses [40].

A considerable number of SMR designs utilize the integrated pressurized water reactor (iPWR) configuration. This architecture includes important components such as the reactor core, steam generators, pressure vessel, and coolant pumps, all contained in a single pressurized vessel. This integrated design reduces the substantial primary circuit plumbing that may lead to loss-of-coolant accidents (LOCA) in traditional reactors. The reactor's smaller size improves passive safety measures, allowing waste heat evacuation by natural convection in a variety of emergency situations, removing the need for direct operator involvement [40]. The working concepts of the SMR are outlined below:

- **Operation of the reactor core:** low-enriched uranium fuel, usually with less than 5% U-235 enrichment, is used in the SMR reactor core. This method ensures complete compatibility with the current nuclear fuel cycle infrastructure and is similar to the technology used in traditional light-water reactors.
- **Coolant circulation:** the majority of designs utilize a coolant, typically light water, to absorb the heat generated by nuclear fission. In numerous SMR designs, fluid circulation primarily occurs through natural convection, thereby significantly reducing the need for continuous operation of electrically powered pumps.
- **Heat transfer:** through the use of steam generators situated within the pressurized vessel, the heat produced in the reactor core is moved from the primary circuit to the secondary chamber.

- **Power generation:** waste heat is eventually eliminated by cooling towers or natural water sources, while secondary circuit steam powers turbines to produce electricity.
- **Passive safety:** many SMR reactors feature passive safety mechanisms, including air-cooled containment vessels that operate independently of external power sources, gravity-fed emergency cooling systems, and waste heat removal through natural circulation.

Due to their high design flexibility, SMRs can be deployed in areas with weak or isolated power grids, industrial complexes, and remote locations such as islands and mining operations. New models, such as floating nuclear power plants, have been developed to facilitate offshore oil platforms and other operations, as well as to provide energy to people in polar regions. SMR applications extend beyond generating electricity; they also encompass district heating, seawater desalination, and hydrogen production. These kinds of applications significantly help reduce carbon emissions by providing people with alternative energy options [39]. Table 4.3 shows a variety of SMR approaches and how SMR nuclear technology may meet a wide range of demands in regions, industries, and the environment, as well as improve safety and deployment flexibility [40].

Table 4.3 Representative SMR designs and key characteristics.

Design name	Power output (MW)	Unique features	Intended applications
NuScale (United States)	60 per module	Below-grade installation, natural circulation cooling, modular scalability	Grid power, load-following, remote deployment
SMART (South Korea)	100	Integral design for cogeneration, optimized for desalination and heating	Electricity, seawater desalination, district heating
SMR-160 (United States)	160	Passive safety systems, long refueling intervals (> 3years), compact site footprint	Electricity, industrial use
Linglong One (China)	125	Multipurpose applications, modular iPWR design	Electricity, district heating, hydrogen production
ACPR50S (China)	60	Offshore floating reactor platform	Remote grids, maritime industry
CAREM (Argentina)	32–120	Simplified core and natural circulation	Grid and off-grid electricity, desalination

4.2.3.2 Fusion Energy

Fusion energy is seen as a secure and low-carbon resource capable of revolutionizing future energy systems. Unlike nuclear fission, which creates energy by dividing heavy nuclei, fusion involves the merger of two light nuclei, commonly deuterium and tritium, to produce helium while also generating high-energy neutrons and a significant amount of energy. An important component of the method is the absence of chain reactions, which lessens the likelihood of runaway reactions and the production of long-lived radioactive waste [41]. Also, the fuels needed for fusion, including deuterium from seawater and tritium from lithium, are easy to find and numerous all over the world. This means that there may be an almost infinite source of energy [42].

To reach this aim, nuclear fusion power plants need to heat hydrogen isotopes to very high temperatures. This creates a plasma state in which electrons and nuclei are separated and may flow around freely. For fusion reactions to keep happening, this plasma needs to be stable, which means that improved confinement devices need to be put in place. The magnetic confinement method (MCF), which uses devices like tokamaks, is the most sophisticated method currently known. Tokamaks are toroidal chambers that use annular and poloidal magnetic fields to hold and stabilize plasma. Some people have suggested stellarators as an alternative since they are more stable and less sensitive to magnetohydrodynamic instabilities. However, they are harder to design and build [41]. In addition to the containment systems, the peripheral components of the device play a crucial role in the fuel cycle. The breeder blanket modules, positioned around the plasma chamber, absorb high-energy neutrons from the reactions and convert their kinetic energy into thermal energy. When neutrons hit lithium, they make tritium, which helps the fuel regenerate. Coolants, such as helium or molten salts, carry the heat from this process to power generation systems, which then turn the turbines. Vacuum tanks keep the pressure low, and radiation barriers protect people and equipment from neutron flux, which makes sure that everything is safe to use [42].

This technology has moved past theoretical frameworks and made a lot of progress in experiments. The Joint European Torus (JET) successfully showed that stable deuterium–tritium fusion is possible, producing 59 MJ of energy in five seconds. Lawrence Livermore National Laboratory (LLNL) reached a major milestone when the energy output from the fusion reaction was more than the energy input to the target fuel, with 3.15 MJ compared to 2.05 MJ. There are now 143 fusion systems in operation, under development, or in the planning stage throughout the world. This shows how much potential fusion technology has. The complex comprises 98 experimental facilities and 12 DEMO concepts aimed at demonstrating grid-connected fusion power [41]. Many active designs have opted for the tokamak configuration due to its established reliability; however, alternative concepts, such as magnetic target fusion and inertial confinement, are also being concurrently explored.

Fusion energy presents considerable environmental and economic potential. It produces no greenhouse gas emissions during operation and can deliver a stable baseload power supply to the grid, thereby supporting the integration of variable RERs. The applications extend beyond electricity generation to encompass hydrogen production, high-temperature industrial processes, medical isotope production, and space propulsion [41, 42]. Nonetheless, realizing this vision presents several challenges. Plasma stability, material resistance to intense neutron flux, tritium regeneration, and reduced engineering costs are significant challenges. Even if there are certain problems, international cooperation, new technology like high-temperature superconducting magnets, and a lot of money from the business sector are speeding up progress. This makes it more likely that commercial fusion power plants will be built this century [42].

4.2.4 Enhanced Carbon Capture

Carbon capture, utilization, and storage (CCUS) technology is crucial to net-zero emissions in modern energy grids. It maximizes energy recovery and reduces human-caused carbon dioxide emissions. CCUS is crucial for power generation, steel, cement, and petrochemical industries, where emission reduction is difficult or expensive. The need for scalable CCUS technologies is growing as global CO_2 emissions reached 33.1 Gt in 2018, accounting for over 67% of total greenhouse gas emissions [43].

The process of extracting CO_2 from industrial point sources or straight from ambient air is the core of this technology. Chemical absorption is the most popular and technologically advanced technique available because it can be easily integrated into existing infrastructure. Although problems such as high energy consumption in the regeneration step, solvent chemical degradation, and equipment corrosion persist, solvents like mono-ethanolamine (MEA) have been used for many years [43]. Efforts have been undertaken to develop a new generation of solvents to address these limitations, including biphasic, nonaqueous solvents and ionic liquids. Biphasic solvents facilitate the regeneration process by creating two immiscible phases: one enriched in CO_2 and the other depleted of it, thereby minimizing the thermal energy required for the process. However, despite issues such as high cost, high viscosity, and scalability concerns, ionic liquids are promising alternatives due to their tunable molecular structure and extremely low vapor pressure [43, 44]. Figure 4.5 illustrates the typical chemical absorption process for CO_2 separation.

Researchers are looking into physical and membrane-based technologies as possible ways to improve the efficiency of CO_2 absorption. Advanced membranes,

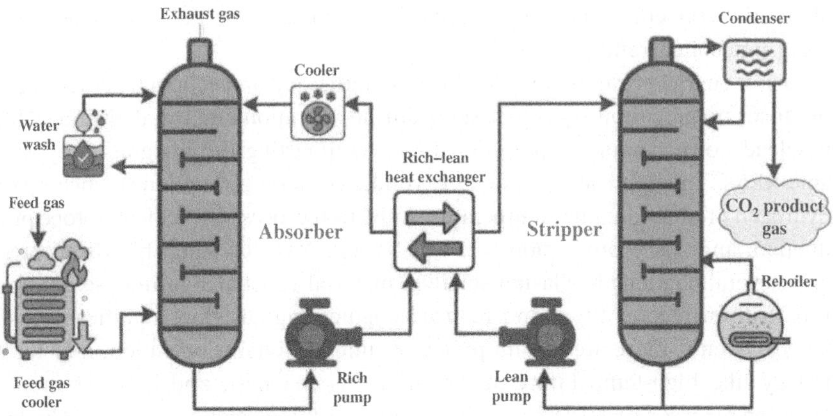

Figure 4.5 Example diagram of a CO_2 separation technique utilizing conventional chemical absorption.

especially mixed matrix membranes that combine polymers with nanoporous materials like metal–organic frameworks (MOFs), operate better when it comes to selectivity and permeability. There are a lot of problems with this technique, namely surface flaws that happen during manufacturing and inadequate long-term stability. Solid adsorbents, including activated carbon and zeolites, exhibit a significant capacity for CO_2 absorption under high-pressure conditions, which is attributable to their elevated surface area. One big problem with these materials is that they do not function as well in humid environments, especially in post-combustion systems [43].

Direct carbon dioxide capture (DAC) methods, on the other hand, are seen as a way to improve point-source capture, even if they are currently expensive. The present cost of capturing CO_2 using DAC is between $94 and $232 per ton. However, as the technology improves, it is expected to drop to less than $60 per ton by 2040 [43]. The success of this technique hinges on progress in making energy-efficient solvents and solid adsorbents that work well even when there isn't much CO_2 in the air.

It is possible to use captured carbon dioxide as both a target for geological storage and a feedstock for the chemical industry. CO_2 utilization methods include using CO_2 directly in industries like making food and drinks, and turning chemicals into fuels and other useful items. One of them is the electrocatalytic reduction of CO_2 using renewable power, which might lead to the production of carbon-neutral molecules such as syngas and methanol. At present, two-electron transfer products, including carbon monoxide and formic acid, have demonstrated faradaic efficiencies exceeding 95%. However, there are still many technical obstacles in the way of producing compounds with a few carbon atoms in a selective manner. The global utilization of captured CO_2 currently represents less than 0.5% of total emissions. However, integrating these technologies with high-purity CO_2 sources or within a circular carbon economy framework may provide substantial economic advantages [43, 44].

Geological storage represents an important strategy for mitigating substantial greenhouse gas emissions. Depleted oil and gas reserves, unmined coal seams, and deep saline aquifers are also good places to store this type of energy. In these places, CO_2 is stored in a number of ways, such as through structural, dissolution, mineralization, and capillary processes. Saline aquifers have a lot of capacity, but there are still worries about long-term leaks and what they mean for regulations [43].

CCUS can be improved through the integration of advanced hydrocarbon recovery techniques, including CO_2-EGR and CO_2-ECBM. Supercritical CO_2 in shale formations can effectively displace absorbed methane, resulting in a recovery increase of up to 24%. Alternative methods have demonstrated enhancements ranging from 6 to 15%. Swelling resulting from CO_2 absorption may decrease permeability and interfere with the process [45].

It is still very hard to go forward with optimized CCUS since it is not clear how it will work in the real world. The technology's viability is affected by the high cost of capturing CO_2, which is expected to be between $60 and $110 per ton, and the fact that geological reactivity changes with time. Enhanced hydrocarbon recovery may lower certain expenses, but how well it works depends on where it is done and how it is injected in the first place. Combining CCUS with policy tools like carbon pricing, emissions markets, and targeted subsidies can make it more financially stable. For CCUS to reach its full potential in reaching net-zero emissions, researchers, policymakers, and businesses need to work together. Standardizing rules, making capture technologies that are cheap, improving reservoir modeling, and adding CCUS to broad decarbonization plans are all very important. Through ongoing innovation and policy alignment, CCUS could emerge as a fundamental component of global climate change strategies [44].

4.2.5 Digitalization and Grid Modernization

Grid modernization is the process of changing old power grids into new ones that are more flexible, efficient, and sustainable. These new grids can handle and connect to large-scale RERs while still being very reliable and resilient. Digitalization and the use of cutting-edge technology for sensing, communication, automation, and data analysis are key to this change. These technologies make it possible to monitor, regulate, and optimize the flow of power in real time at every step of generation, transmission, distribution, and consumption. This adjustment is not merely a technical one; it is a strategic move toward reaching net-zero carbon, which means using fewer fossil fuels by combining different energy sources [19]. Modern smart grids combine operational technology and digital intelligence to provide two-way power flow, dynamic demand management, and seamless coordination between centralized and distributed equipment. These systems make operations run more smoothly by gathering accurate, up-to-date information and using advanced control methods. This enables utilities to optimize asset utilization, minimize technical losses, and reduce outage durations. This capability is critical for decarbonization because it maximizes the use of renewable energy, reduces grid constraints, and promotes the electrification of sectors such as transportation and heating [19, 46].

Digitalization also helps with energy justice by making sure that everyone has fair access to clean, cheap power and by getting more people involved in making energy decisions. Smart metering, smart analytics, and decentralized market systems make it possible for customers to become "prosumers," which means they may actively produce, store, and trade renewable energy. These features are what make sophisticated smart grid technologies and distributed ledger-based energy markets possible. Both of these things are important for decentralizing, cutting down on carbon emissions, and boosting democracy.

4.2.5.1 Advanced Smart Grid Technologies

Advanced smart grid technologies enhance the fundamental capabilities of digital networks by incorporating real-time grid optimization, intelligent fault detection, and self-healing features. These tasks rely on sensor networks that work together, fast communication, and advanced computational intelligence to monitor the system's status and take the appropriate actions automatically [46].

- **Real-time grid optimization:** advanced smart grids utilize hierarchical control systems and adaptive optimization algorithms to balance power generation and consumption dynamically. These systems employ spatio-temporal data from phasor measurement units (PMUs), advanced metering infrastructures (AMIs), and distributed sensors to find changes in demand, control voltage profiles, and work together with distributed energy resources (DERs). Methods based on AI and ML also make it easier to predict grid problems, which helps operators avoid and deal with congestion, cut losses, and better integrate the changing output of RERs [46].
- **Fault detection:** new analytical technologies, such as high-resolution waveform analysis, wavelet transform, and adaptive Kalman filter, are being used more frequently to detect grid parameter anomalies early. These methods enable the early detection of equipment failures, insulation issues, and even cyber-physical disturbances before they worsen. Advanced protection systems that comply with the IEC 61850 standard enable rapid data exchange

between intelligent electronic devices, ensuring that protective measures are implemented promptly under both normal and critical conditions [46].

- **Self-healing grids:** self-healing networks can automatically isolate faults and restore services without human intervention. When this feature is used with real-time monitoring and automated switching devices, it is possible to redirect power flows to regions that need it, making sure that power is always available. Even when RERs are widely used, multi-agent control frameworks and distributed optimization algorithms make networks recover faster and perform more reliably. Field testing has demonstrated that these solutions can lower the number of outages, keep the voltage steady within certain limits, and make the grid more resilient to both physical and cyber disturbances [47].

These improvements enhance the system's reliability and efficiency, and they also facilitate the wider adoption of renewable energy. This helps to reach carbon reduction goals by reducing the need for fossil fuel-based reserves and preventing RERs from becoming limited.

4.2.5.2 Distributed Ledger Technologies (Blockchain) in Energy Systems

Distributed ledger technologies (DLT), particularly blockchain, are emerging as key enablers of P2P energy trading and decentralized participation in modern grid markets. These systems let end prosumers exchange extra energy directly with other prosumers by giving them transaction records that are safe, clear, and unchangeable, so they don't need to use traditional intermediaries. One of the advantages of this technology is that it promotes the use of renewable energy while reducing the use of expensive fossil fuels, decreasing carbon emissions in the grid [20].

- **Applications for P2P energy trading:** blockchain-based P2P markets enable households or businesses with excess renewable energy output to sell it directly to nearby consumers. Blockchain platforms with smart contracts handle transactions automatically, such as checking that energy is delivered, making payments, and enforcing agreed-upon conditions without the need for third-party monitoring. This decentralization lowers the cost of transactions, promotes confidence, and lets local renewable energy markets work in real time [48].
- **Enhancing market transparency and security:** blockchain designs like Hyperledger Fabric let you regulate who may access what in a very detailed way. This keeps your data private while also allowing it to be audited. These systems can operate with IoT-based smart meters to keep track of verified consumption and generate data. This stops fraud and makes sure that payments are sent correctly. Alternative distributed ledger designs, such as IOTA's directed acyclic graph (DAG), address traditional blockchain's scalability and

energy efficiency restrictions by enabling high transaction throughput ideal for granular energy trades [20].

- **Integration with grid operations:** blockchain platforms can be integrated with distributed energy resource management systems (DERMS) to help align market activity with physical grid limits. For example, real-time network state data can help blockchain-based market algorithms ensure that transactions do not trigger voltage violations or overloads. This integration promotes transactive energy ecosystems, in which economic incentives coincide with technical needs, allowing for demand response, flexible load participation, and optimal renewable use [48].

DLT help to democratize energy access by enabling decentralized, automated, and trust-based energy transactions. They also stimulate renewable energy generation and lower the carbon footprint of electricity usage. To achieve zero carbon, these devices supplement advanced smart grid technologies by combining economic mechanisms with digital and current grid technologies. Figure 4.6 illustrates the advantages and achievements associated with the modernization and digitalization of electrical grids.

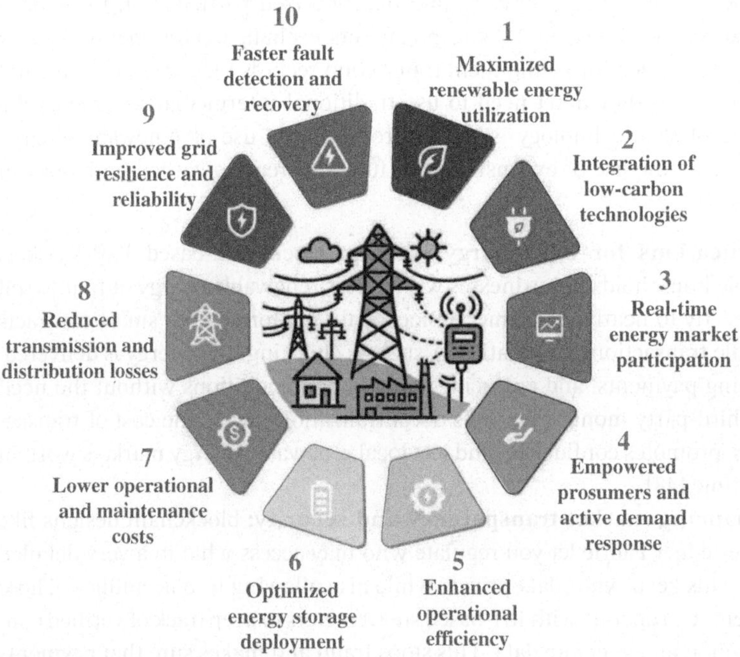

Figure 4.6 The benefits of modernization and digitalization of electrical grids.

4.3 Evolving Paradigms and Conceptual Shifts

Climate change, new technologies, new policies, and changing social values are all driving a big change in the global energy landscape with a zero-carbon goal. More decentralized, linked, and adaptable frameworks are replacing traditional, linear methods of energy production and consumption that were built on centralized fossil fuel systems. This change is not only about technology; it is also about ideas that change how energy systems are designed, operate, and used. New ideas like sector interconnection, circular economy principles, consumer participation, and community-based behavioral change are breaking down old ways of doing things, making it easier for different sectors to work together, and helping to create more sustainable and fair energy futures. Policymakers, business leaders, researchers, and communities all need to understand these changes because they impact how energy may help with environmental objectives, economic stability, and social well-being.

4.3.1 Sector Coupling and Integrated Energy Systems

As the need for deep decarbonization grows, networks including energy, heating and cooling, transportation, and industry are no longer working as separate networks. Sectoral connection and integrated energy systems are a big change from infrastructures that only serve one function and are separate from each other. Networks can connect and share resources, exchange energy in a variety of ways, and improve the overall system performance. By connecting these sectors with direct electrification and conversion technologies like PtX, it is feasible to balance the generation of fluctuating renewable energy, lower the total carbon intensity, and provide new methods to get flexible, reliable, and sustainable energy.

4.3.1.1 Electrification of End Uses: Transport, Heating, and Industry

SC, which is the systematic integration of power, heating/cooling, and transportation to make the most use of renewable energy and cut down on carbon emissions, is becoming more and more important as the energy sector becomes less carbon-intensive. A fundamental approach for SC is electrification of end uses, which means substituting burning fossil fuels directly with electricity. This is made possible by improvements in heat pump technology, electric boilers, and industrial electric furnaces [49, 50]. This method works well because the carbon intensity of grid power is still going down. In Europe, it dropped by about 16% between 2005 and 2014, and it is expected to drop much more by 2030 because of measures to phase out coal and the growth of renewable energy [17].

High-efficiency air and ground source heat pumps may get coefficients of performance (COP) of 2–3 in the building and heating sectors. This means that they can supply two or three times as much thermal energy per unit of electrical input as regular resistance systems [17]. Their use in space and water heating, especially by making the energy supply less carbon-intensive, greatly lowers CO_2 emissions throughout the course of their lives. Electric boilers and thermal systems that store heat also make it easier to move loads, sync heat production with times when renewable energy is being generated, and offer extra grid services [49].

There are both big opportunities and big problems in industrial applications. A thorough bottom-up look of EU industry indicates that up to 78% of industrial final energy consumption may be electrified with technologies that are already on the market. This number goes up to 99% when technologies that are still being developed are taken into account. Mature heat pumps, mechanical vapor recompression, and resistive heating can be used to electrify low- and medium-temperature processes (below 400 °C) in industries including food, paper, and textiles. Electric arc furnaces, induction, and plasma technologies may be used for high-temperature operations like metal smelting and ceramic manufacture. However, using them on a wide scale frequently involves a lot of money and changes to how things work [50].

Battery electric vehicles (BEVs) are the most direct way to electrification in transportation. This is because lithium-ion battery technology has improved, giving them high specific energy (150–190 Wh/kg for $LiCoO_2$ chemistry) and cycle lifetimes that last more than 1,000cycles. Their extensive usage cuts down on air pollution in cities and, when driven by low-carbon electricity, cuts down on CO_2 emissions from well to wheel by a lot [49]. Connecting electric car charging to smart grids and vehicle-to-grid (V2G) technology makes the transportation sector even more a part of an energy system that is flexible and based on RERs [17].

Overall, large-scale electrification for homes, businesses, and transportation is the main way to connect sectors that will help reduce carbon emissions while making the system more flexible and resilient.

4.3.1.2 PtX Concepts: Converting Renewable Electricity into Hydrogen, Synthetic Fuels, or Chemicals

PtX is the key component of integrated energy systems. It is a flexible platform that turns fluctuating renewable electricity into energy carriers and industrial feeds that may be stored. It connects power, heat, transport, and industry where direct electrification stops. PtX is not meant to be a single tool; instead, it is meant to be an infrastructure for integration. The technology absorbs intermittent renewables, recycles carbon, and provides fuels and chemicals that are compatible with hard-to-electrify applications. This is why PtX is a top priority in sectoral connectivity roadmaps for transport and industry, and why it is clearly included in EU plans and standards (RED II/RFNBOs, ReFuelEU/FuelEU) to open up e-fuel markets [16]. PtX doesn't only regard hydrogen as a fuel; it also sees it as an intermediate feedstock, which means it's a building block that can be used to make a lot of other things. Some uses for PtX are as follows:

- **Power-to-methane (PtM):** the Sabatier reaction, which gives out heat, combines hydrogen and CO_2 to make synthetic natural gas (SNG). SNG is chemically the same as fossil natural gas, thus it may be stored in gas caverns on a terawatt-hour scale and moved via the gas system without changing. This lets renewable energy generated in the summer be utilized to heat homes in the winter, something batteries can't do on the same scale [51].
- **Power-to-liquids (PtL):** this method turns renewable hydrogen and collected carbon into synthetic diesel, jet fuel, methanol, or dimethyl ether using Fischer–Tropsch synthesis or other catalytic processes. These droplet fuels are necessary for the decarbonization of long-haul aircraft, marine transport, and the manufacture of industrial raw materials. In these areas, direct electrification is not possible because of energy density, storage stability, or compatibility with older engines [51].
- **Power-to-chemicals (PtC):** makes ammonia for fertilizers, urea for polymers, or methanol for making chemicals. This takes the role of feedstocks made from fossil fuels and adds renewable energy directly to the chemical value chain [16].

PtX is different from "hydrogen production" since it combines multiple areas and employs hydrogen as one step in a conversion cycle. The real benefit of PtX is that it can turn renewable power into a wide range of energy carriers, such as synthetic gases, liquids, and chemical feedstocks, that may be distributed throughout the economy based on need. PtX can link and balance the power, heating, transportation, and industrial sectors. It can also transform renewable energy generation that generate sometimes into a dependable supply of energy and materials, and it can connect electricity generation that changes all the time with a wide range of end-use applications [16, 51]. PtX is an infrastructure enabler from the point of view of system planning. One can use existing gas grids, port facilities, and fuel distribution networks to handle PtX goods, which means they don't need to build new logistical chains. This means that PtX is a bridging technology that connects different energy sectors and helps the shift from fossil-based to renewable supply chains [16].

In summary, PtX should be seen as the molecular arm of the renewable energy system rather than as a single technology. It takes the variable and location-dependent output of wind and solar power plants and turns it into energy carriers that can be moved about and used in many different ways. Not simply making hydrogen, but also being able to convert and link things is what makes PtX so important in future energy networks that don't need carbon. Figure 4.7 summarizes several Power-to-X (PTX) production pathways and associated energy conversion processes.

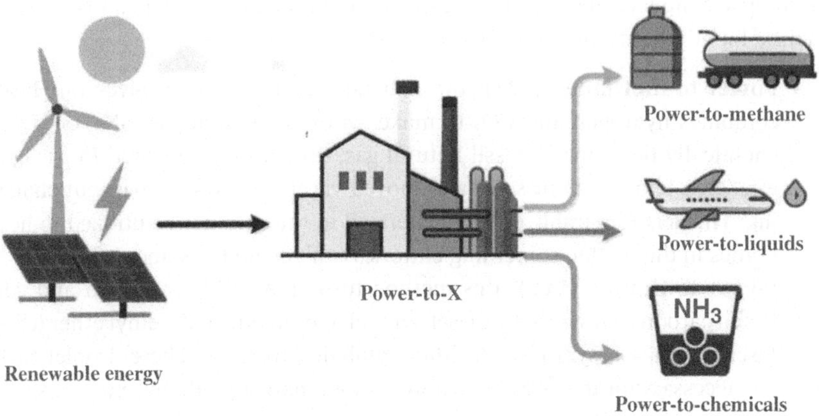

Figure 4.7 A few of the several PTX production and energy conversions.

4.3.2 Circular Economy Principles in Energy

In the energy industry, a circular economy starts with using fewer raw materials and then making sure that materials are utilized as long as possible once they are consumed. Traditional energy systems work in a straight line: they gather resources, consume them, and then throw them away. This generates waste and puts strain on limited resources. A circular strategy, on the other hand, focuses on building infrastructure that does more with less resources and makes sure that parts can be reused and recycled when they are no longer needed.

This is seen by the increased emphasis on dematerialization and modular design in renewable energy. Thin-film solar panels, for example, have demonstrated that CdTe layers may be lowered to approximately one micron without losing efficiency, reducing reliance on limited tellurium and lowering prices [52]. Wind turbine makers are also employing lightweight composites to reduce material consumption while improving performance. Aside from design changes, resource efficiency is achieved by expanding device operational lifespans. Predictive maintenance and digital monitoring help turbines, panels, and batteries survive longer, reducing the need for new materials [53]. These techniques are consistent with a larger academic trend that emphasizes resource efficiency and life cycle thinking as critical components of the renewable energy transition [54]. The circular approach goes beyond decreasing input; it considers what happens when systems reach the end of their useful life. Batteries, solar panels, and turbines all show how recycling and reuse solutions close material loops. Lithium-ion batteries, for example, frequently maintain 70–80% of their capacity after usage in electric vehicles and may be reused in stationary energy storage before being recycled for cobalt, nickel, and lithium [52]. Solar panels, which are predicted to generate huge end-of-life waste streams in the next decades, are becoming more susceptible to extended producer responsibility programs that require collection and recycling. These techniques recover glass, aluminum, silicon, and even valuable metals like silver. Companies like First Solar are currently operating closed-loop systems that reintegrate cadmium and tellurium into new modules [4, 53].

Wind turbines are harder to recycle, especially their blades, which are built of composite materials that don't break down easily. However, circular solutions are emerging here as well. Some nations, like Denmark, are reusing outdated blades as construction materials [54]. While companies like Vestas are creating recyclable thermoplastic polymers with the goal of delivering zero-waste turbines by 2040 [53]. Towers and nacelles, which include metals such as steel, copper, and rare earth magnets, are already easier to recycle, providing an ongoing source of secondary materials. Despite these advancements, many important elements required for clean energy technologies, such as indium, gallium, and rare earths, continue to have near-zero recycling rates due to technological and economic constraints [52]. To close these gaps, monies must be directed into collection networks, improved separation procedures, and international recycling infrastructure. It also necessitates collaboration: governments, manufacturers, recyclers, and consumers must work together to implement legislation such as landfill bans, recycling objectives, and take-back programs, as well as advances in reverse logistics and digital material tracking [53].

The most important thing is that materials used to make energy should never be considered disposable. Circular strategies that minimize material consumption, lengthen product lives, and recover resources at the end of life preserve finite resources while also lowering carbon emissions associated with mining and processing. Every reused or recycled battery, panel, or turbine component represents a reduction in extraction and greenhouse gas emissions. In this approach, the circular economy helps make materials safer while also helping to reach climate and decarbonization targets. Based on the above discussion, the principles of the circular economy in energy can be summarized as follows:

- Minimize material use: reduce consumption of raw materials through lightweight, modular, and efficient design.
- Extend lifespan and efficiency: enhance durability and performance via predictive maintenance and digital monitoring.
- Reuse and repurpose components: implement second-life applications, such as reusing batteries in stationary storage systems.
- Recycle and recover resources: recover valuable materials from end-of-life components, including metals, glass, and polymers.
- Collaborative and systemic management: engage stakeholders through policies, take-back programs, and digital tracking to close material loops and reduce environmental impact.

4.3.3 Prosumerism and Energy Democracy

Prosumerism and energy democracy are essential for reimagining the role of citizens in the energy transition, particularly in the context of promoting sustainable practices and reducing carbon emissions. These notions emphasize residents as active players in energy production, decision-making, and government, rather than just consumers. By producing and using renewable energy, households and communities become "energy citizens," influencing new cultural and political processes around the transition away from fossil fuels [55–57].

Consumer empowerment is based on their capacity to access distributed generating technologies such as solar panels, wind microturbines, and home batteries, and integrate them into smart homes and community energy initiatives. These behaviors illustrate energy citizenship, in which people adopt new routines like flexible demand response, energy efficiency measures, or local collaborative self-consumption programs [56, 57]. Aside from cutting home energy costs, such measures boost resilience, generate potential for local value, and directly reduce greenhouse gas emissions. According to studies conducted throughout Europe, prosumer collectives frequently function as cooperatives or energy communities, providing social advantages such as job creation, energy literacy, and the relief of energy poverty [55]. These programs increase public acceptability of renewable technology while also making carbon reduction a shared responsibility by encouraging local ownership and transparency.

Decentralization is critical in this transition. Traditional centralized systems, dominated by fossil and nuclear power and regulated by technical decision-making, restrict citizen participation and support entrenched interests. Decentralized solutions, on the other hand, provide people and communities more control over local energy resources [55, 57]. Energy democracy describes this transition as a greater transfer of political and economic power, in which citizens become stakeholders and account holders in the energy industry [57]. However, decentralization must be balanced against system-wide stability: integrating large amounts of variable renewables necessitates coordination via smart grids, demand response, and aggregation methods [56]. The issue is to gain autonomy while maintaining the reliability of the electrical supply.

Finally, prosumerism and energy democracy change more than just technological structures; they modify the social fabric of energy systems. Furthermore, they promote diversity, communal responsibility, and procedural fairness, as well as affordability and equity in the context of decarbonization [55, 57]. These initiatives make the change more human by letting people take part both individually and as a group. This makes sure that reducing carbon is a democratic effort based on participation and shared ownership, not just a technology aim. The key factors influencing prosumerism and energy democracy are summarized below:

- access to distributed generation (solar, wind, batteries);
- smart homes and digital tools;
- supportive policies and regulations;
- social empowerment and education;
- equity, inclusiveness, and energy justice;
- environmental sustainability and grid stability.

4.3.4 Behavioral and Social Innovation

To attain climate goals, it is necessary to make big changes to how people act and how society is organized, not just changes to technology. Behavioral and social innovations support technological initiatives by promoting flexible energy use, community involvement, and broad acceptance of low-carbon transitions. They aim not only to optimize energy systems, but also to develop new social behaviors and governance structures that enable individuals and communities to participate in carbon reduction initiatives [58, 59].

A key factor is demand-side flexibility, which refers to the capacity of consumers to change when and how they use energy in response to system demands or incentives. Demand-side innovations can reduce, shift, or improve energy usage, resulting in system-wide flexibility that reduces peak demand and allows for greater integration of RERs [60]. Smart meters, automated demand response, and dynamic tariffs are examples of methods that put this promise into action, allowing consumers and companies to modify usage in real time [56, 58]. These techniques not only improve grid stability, but also allow individuals to actively contribute to carbon reduction by timing their behavior to periods of renewable energy supply. When demand-side flexibility appears as chances for civic involvement, it becomes more than a technical adjustment; it becomes an exercise in energy citizenship, in which collective small-scale actions result in considerable reductions in greenhouse gas emissions [58, 59].

However, the effective application of flexibility measures necessitates social involvement and trust. Research indicates that public support for renewable energy infrastructure and policy is shaped by perceptions of fairness, transparency, inclusion, and economic incentives [58]. Community energy projects, living labs, and participatory governance platforms are examples of social innovations that have come up as ways to close these gaps. For example, regional governance efforts in the Netherlands show how local "living laboratories" may experiment with energy-neutral housing schemes, with social networks proving more important than technical conditions in deciding success [59]. Engagement approaches also include "green nudges," which are subtle behavioral interventions that encourage sustainable choices. Examples include default energy-efficient settings or serious gaming tools that make flexible consumption tangible and rewarding

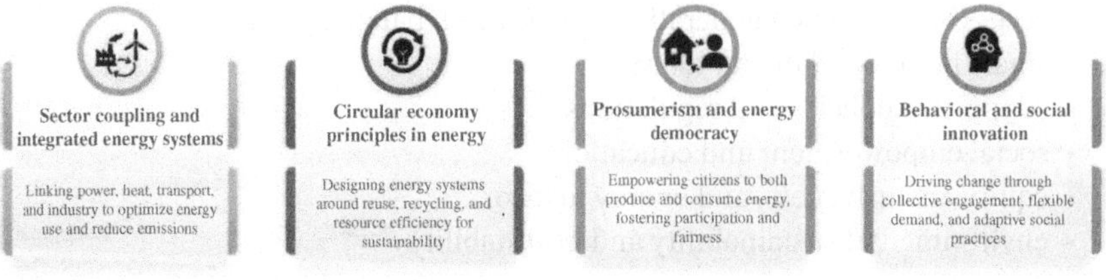

Figure 4.8 Some evolving paradigms in sustainable energy systems.

[58]. These innovations demonstrate that behavioral change is most effective when it is matched with values, emotions, and local identities rather than being imposed purely by external control [58, 59].

Social and behavioral innovation is important because it links the use of technology to bigger goals like justice and empowerment. People are more inclined to support low-carbon infrastructure like wind farms, smart grids, or hydrogen pipelines when they perceive other benefits, like job growth, lower energy costs, or stronger communities. European experiences with community energy cooperatives, such as SOM Energia in Spain, demonstrate that supporting legislative frameworks are critical for maintaining citizen-led projects, but policy uncertainties or rollbacks could decrease confidence and participation [59]. Governments and civil society may turn possible resistance into active support by including energy projects in social contexts. This will make sure that attempts to decarbonize are not stopped by popular opposition [58, 59].

In short, behavioral and social innovation shows that decarbonization isn't only about putting in new technology; it's also about changing how individuals, communities, and households use energy. Citizens are redefined as agents of change through adaptable consumer habits, participatory government, and societal involvement. This connection between social innovation and low-carbon targets makes sure that the energy transition is not only possible from a technical point of view, but also fair, strong, and socially acceptable [58, 59]. As discussed, an overview of some evolving paradigms in sustainable energy systems is presented in Figure 4.8.

4.4 Challenges and Opportunities

The shift to net-zero contemporary energy networks is one of the most ambitious changes in society and technology in the 21st century. It is not just a matter of replacing fossil fuels with RERs; it is a major overhaul of infrastructures, markets, and behaviors in the larger CESN. To reach carbon neutrality by the middle of the century, new technologies, new policies, new ways to pay for things, and societal acceptability all need to work together. However, this change is happening in a situation where there are competing pressures: on the one side, there is an urgent need to decrease emissions, and on the other side, there are structural, institutional, and cultural hurdles that hinder development [60, 61]. Looking at the problems and possibilities at the same time gives us a better idea of both the problems that need to be solved and the many benefits that may be gained by working toward a sustainable and resilient energy future.

4.4.1 Challenges

Scalability and cost reduction are two of the most significant challenges on the path to achieving net-zero MENs. While renewable technologies like solar PVs and wind have already seen significant cost reductions, large-scale deployment of enabling solutions like advanced storage systems, hydrogen production, and carbon capture technologies still requires significant upfront investment and has uncertain cost trajectories [12, 61]. In developing nations, limited financing and unstable policies hinder investment in clean energy, resulting in slower progress toward decarbonization [53]. Achieving competitiveness at scale requires not only ongoing research and innovation but also supporting financial methods and targeted subsidies during the early deployment period.

Infrastructure development is closely related to costs. To incorporate a significant fraction of variable RERs, net-zero energy networks require significant expansion and upgrading of transmission, distribution, and storage systems [62]. Conventional grid designs were not intended for decentralized, multidirectional energy flows, and hence risk creating barriers. Furthermore, the integration of electric vehicles, distributed solar energy, and electrified heating systems places additional demand on existing networks [56]. Delays in permission, extended construction lead times, and the necessity for cross-border collaboration all complicate the process [49]. Even cost-effective clean technologies will struggle to attain their full potential unless significant infrastructural investments are made.

Intermittency and grid stability remain key challenges. High wind and solar penetration cause fluctuation in generation and demand patterns, necessitating complex flexibility solutions [33, 62]. Integrated energy networks (IENs) that include electricity, heat, gas, and storage might be able to handle these changes, but they are technically difficult to run and are affected by changes in supply and demand [62]. Digitalization and demand-response systems make demand-side flexibility possible. This might be benefit, but it needs strong market processes, client engagement, and safe data frameworks [17]. Without these circumstances, uncertainty might hurt both the reliability of the energy transition and the public's faith in it.

Social and governance challenges complicate the shift beyond technical issues. Energy justice and fairness are key to net-zero [19]. Large-scale change may be resisted due to affordability, energy poverty, and geographical disparities. New infrastructure may displace communities and workers in carbon-intensive industries. The public acceptability of renewables, transmission corridors, and storage facilities is variable and requires involvement, participation, and compensation [52]. Governance issues, including fragmented regulatory frameworks, inadequate enforcement of climate commitments, and uneven carbon-pricing systems, impede development [61].

Another issue is supply-chain dependence on essential minerals and components. Batteries, solar modules, and wind turbines utilize lithium, cobalt, rare earth elements, and other vital materials. The geographic concentration of these supply chains makes them vulnerable to geopolitical conflicts and market instability [60]. Extraction's environmental and social repercussions also raise sustainability problems. Critical material dependence might slow and unfairly transition the energy transition without recycling, replacement, and a diverse supply.

Finally, climate resilience in energy infrastructure is becoming increasingly important. Heatwaves, wildfires, storms, and flooding are straining networks and compromising operational reliability [62]. As renewable energy penetration increases, system managers must consider compounding risks, such as droughts reducing hydropower supplies or prolonged periods of low-wind conditions. To keep low-carbon systems safe and sustainable as climate change gets worse, it is crucial to build adaptive capacity through strong system planning, redundancy, and smart monitoring [52].

4.4.2 Opportunities

The transition to net-zero energy networks presents several challenges, but also offers a significant opportunity. Economic growth and job creation are top priorities. One of the fastest-growing sectors worldwide, solar PV and wind provide millions of jobs and new industrial pathways. The local manufacture of clean-energy components and the provision of supplemental services, including installation, maintenance, and digital solutions, offer jobs for all skill levels. This economic growth, combined with retraining and equitable transition policies, can help balance the losses of the fossil fuel industry and boost regional development [12, 61].

Improving energy security and independence is another potential. Countries can avoid unpredictable commodity markets and geopolitical concerns by decreasing fossil fuel imports [49]. Domestic RERs, storage, and innovative grid technologies boost self-sufficiency and resilience. Power-to-gas, hydrogen storage, and SC technologies make the country and regions more flexible, which is good for both the economy and strategy [62].

Public health and the environment will gain significantly from the energy transition. Renewable electricity and electrified end uses reduce particulate matter, sulfur oxides, and nitrogen oxides, improving air quality [17]. Lower rates of respiratory and cardiovascular disease result in reduced public health costs and improved quality of life [19]. Decarbonization immediately tackles climate change, reducing ecological disturbance, agricultural loss, and biodiversity loss [61].

Digitalization and smart energy management provide further chances for optimization. The IoT, AI, and blockchain make it possible to monitor things in real time, use predictive analytics, and take part in a decentralized market [62]. These digital enablers facilitate the cost-effective integration of distributed energy resources, demand response, and grid resilience. Digital technologies, dynamic pricing, and consumer engagement may democratize the energy system, empowering prosumers and communities [19].

Ultimately, net-zero systems drive innovation in technology and system design. Extensive deployment of hydrogen, long-term storage, and carbon capture technologies is improving efficiency and performance [56]. IENs combine power, heating, mobility, and industry to optimize resource utilization and reduce emissions [62]. These innovations enable technical leadership, new governance models, and market structures that favor sustainability, adaptability, and inclusion [53]. Figure 4.9 depicts the major challenges and opportunities associated with net-zero MENs in a schematic form.

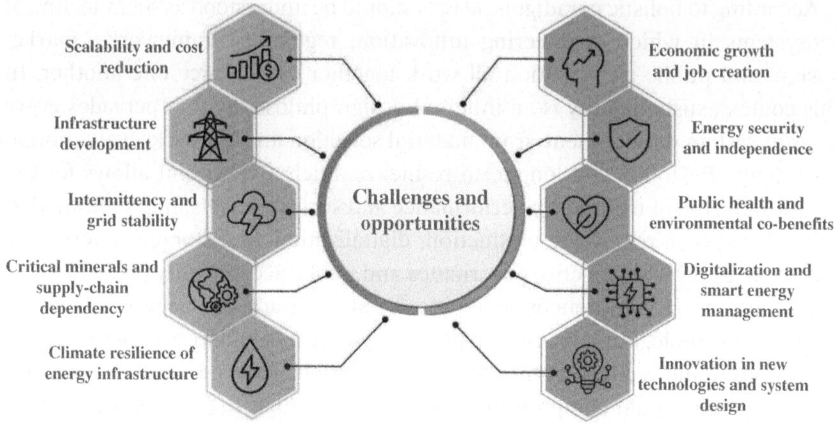

Figure 4.9 Some of the challenges and opportunities in net-zero MENs.

4.5 Holistic Paradigms Linking Climate, Energy, and Sustainability

To get to net-zero emissions in today's energy networks requires more than just small improvements in technology or separate regulatory efforts. It calls for a comprehensive approach that takes into account how climate change, energy systems, and sustainability all work together. The CESN, which is also known as this integrated approach, gives a complete framework for bringing together broken solutions on the way to long-term decarbonization and the well-being of society [60, 61]. In this context, independent actions such as using renewable energy, putting in place carbon pricing systems, or making things more efficient are useful but not enough on their own. They need to be part of a larger reform that takes into account how technical, economic, and social areas are all connected.

The core of this nexus is the recognition that climate goals cannot be met without a major change to the energy infrastructure. The continued use of fossil fuels is directly linked to rising global temperatures, extreme weather events, and damage to the environment. On the other hand, energy transitions need to be carefully planned so that they not only lower emissions but also follow broader ideas of sustainability, like using resources efficiently, protecting the environment, and being fair to future generations. This requires long-term planning, a lot of policy coordination, and an understanding of how routes depend on each other, which often keeps systems on paths that are harmful for the environment. Additionally, fair transition principles must be part of reaching climate goals. These principles protect vulnerable groups from the negative effects of changes in the way energy is produced and used [19].

According to holistic paradigms, MENs should be understood as socio-technical ecosystems in which engineering innovation, regulatory frameworks, market design, and public participation all work together to reinforce one another. In this context, sustainability is an inherent design philosophy that pervades every layer of system development, from material selection and life cycle evaluation to operational flexibility and long-term resilience. Such a viewpoint allows for the co-optimization of technology performance and societal results, highlighting that advancements in renewable production, digitalization, and storage systems can only be realized with effective governance and public acceptability [19, 63].

Furthermore, a comprehensive approach reduces trade-offs between opposing aims. For example, initiatives to decarbonize power must address issues of accessibility, cost, and security at the same time. Without such a balance, expansion in one industry could compromise outcomes in another. In practice, this means that large-scale electrification projects must be accompanied by investments in grid flexibility, energy justice, and demand-side engagement to avoid deepening

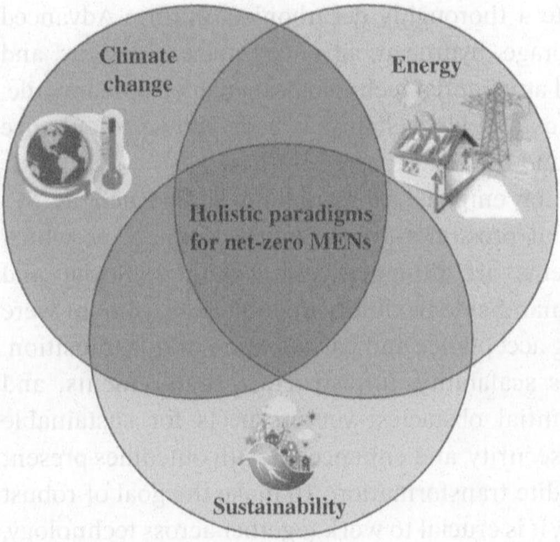

Figure 4.10 Integrated framework for net-zero energy transitions.

disparities or creating new vulnerabilities. Net-zero energy transitions that incorporate sustainability as a guiding paradigm can result in enormous economic, environmental, and social synergies. As a result, this synergy encourages innovation spillovers, long-term economic growth, and increased resilience to geopolitical and climate shocks [19, 64].

Finally, these comprehensive paradigms establish important and long-term connections between action on climate change, energy innovation, and sustainability goals. They serve as the foundation for the creation of MENs that are not only carbon-neutral but also resilient, equitable, and capable of sustaining long-term human and environmental well-being. In doing so, they emphasize the importance of an adaptive governance model, one that can adjust dynamically to uncertainty while maintaining sustainability as a core design tenet. Such a strategy would ensure that the transition to net-zero is not only technically feasible but also socially acceptable and environmentally beneficial, strengthening the idea of a sustainable energy future. Figure 4.10 presents an integrated framework for net-zero energy transitions.

4.6 Conclusion

This chapter has examined how emerging technologies, evolving paradigms, and social innovation can converge to guide MENs toward net-zero emissions within CESN. The discussion emphasized that technology advancement alone is insufficient; instead, daring innovation, systemic integration, and societal participation must all work together to create a thoroughly decarbonized future. Advanced renewable energy, long-term storage, hydrogen, advanced nuclear designs, and carbon capture have all emerged as essential technological enablers. Meanwhile, digitalization through smart grids and blockchain has been shown to enhance system flexibility, transparency, and resilience.

Equally important, the chapter emphasized paradigm shifts such as SC, circular economy principles, and prosumer-driven energy democracy, which reframe the way energy systems are designed and governed. Social and behavioral dimensions from demand-side flexibility to public engagement were underscored as vital for securing acceptance and participation in this transition. Meanwhile, challenges such as scalability, infrastructure requirements, and intermittency persist as substantial obstacles; yet prospects for sustainable economic development, energy security, and enhanced health outcomes present compelling motivations to expedite transformation. To make the goal of robust and fair net-zero MENs a reality, it is crucial to work together across technology, policy, and society. The energy transition can provide not only sustainability but also justice and resilience to future generations by combining innovation and inclusive methods. The main achievements of this chapter can be highlighted as follows:

- Establishing the net-zero imperative within the CESN framework and positioning MENs as pivotal actors linking climate, energy, and sustainability goals.
- Identifying key technological enablers, including advanced renewables, long-duration storage, hydrogen systems, nuclear innovations, carbon capture, and digitalization tools that collectively underpin the decarbonization process.
- Highlighting conceptual and systemic paradigms, such as SC, circular economy principles, prosumerism, and energy democracy, that redefine how energy systems are structured and governed.
- Demonstrating the importance of holistic paradigms that integrate climate, energy, and sustainability dimensions, providing an adaptive and equitable framework for net-zero MENs.
- Underscoring the role of social and behavioral innovation, including demand-side flexibility, public engagement, and inclusive governance, as critical components of a just and participatory transition.
- Outlining the key challenges (scalability, infrastructure development, intermittency) alongside the opportunities (sustainable economic growth, energy independence, improved health, and environmental quality) that characterize the pathway to sustainable MENs.

References

1 A. Darvishi, B. Ranjbar, R. Gharibi, R. Khalili, and R. Dashti, "Multi-objective optimization of a socio-economic energy hub with demand response program and considering customer satisfaction," *Journal of Energy Storage*, vol. 100, p. 113624, 2024.

2 S. Önnered and I. Bravić, "Systemic innovation in the European energy sector: A collective outlook," *Journal of Innovation Management*, vol. 12, no. 2, pp. 50–73, 2024.

3 Y. Zhou, "Low-carbon urban–rural modern energy systems with energy resilience under climate change and extreme events in China—A state-of-the-art review," *Energy and Buildings*, vol. 321, p. 114661, 2024.

4 R. Jayabal, "Towards a carbon-free society: Innovations in green energy for a sustainable future," *Results in Engineering*, vol. 24, p. 103121, 2024.

5 K. Wei, P. Shi, H. Zong, and Y. Qin, "Research on the development of key technologies in the hydrogen energy industry," in *2025 4th International Conference on Smart Grid and Green Energy (ICSGGE)*, IEEE, 2025, pp. 44–49.

6 G. M. Idroes et al., "Exploring the role of geothermal energy consumption in achieving carbon neutrality and environmental sustainability," *Heliyon*, vol. 10, no. 23, p. e40709, 2024.

7 J. Pastuszak and P. Węgierek, "Photovoltaic cell generations and current research directions for their development," *Materials*, vol. 15, no. 16, p. 5542, 2022.

8 J. Guan, J. Sun, X. Shi, and J. Wei, "Review of full-spectrum solar energy systems based on spectral splitting technology," *International Journal of Hydrogen Energy*, vol. 81, pp. 1235–1255, 2024.

9 T. Seyisi et al., "Major challenges for commercialization of perovskite solar cells: A critical review," *Energy Reports*, vol. 13, pp. 1400–1415, 2025.

10 A. G. Olabi et al., "Wind energy contribution to the sustainable development goals: Case study on London array," *Sustainability*, vol. 15, no. 5, p. 4641, 2023.

11 X. Chen et al., "A review of experiment methods, simulation approaches and wake characteristics of floating offshore wind turbines," *Journal of Marine Science and Engineering*, vol. 13, no. 2, p. 208, 2025.

12 R. Joshi, D. Von Terzi, and R. Schmehl, "System design and scaling trends in airborne wind energy demonstrated for a ground-generation concept," *Wind Energy Science*, vol. 10, no. 4, pp. 695–718, 2025.

13 S. Shanmugasundaram, J. Thangaraja, S. Rajkumar, S. D. Ashok, A. Sivaramakrishna, and T. Shamim, "A review on green hydrogen production pathways and optimization techniques," *Process Safety and Environmental Protection*, vol. 197, p. 107070, 2025.

14 Z. Hou et al., “Towards a high efficiency and low-cost aqueous redox flow battery: A short review,” *Journal of Power Sources*, vol. 601, p. 234242, 2024.

15 T. H. Nakib, M. Hasanuzzaman, N. A. Rahim, M. A. Habib, N. N. Adzman, and N. Amin, “Global challenges of ocean thermal energy conversion and its prospects: A review,” *Journal of Ocean Engineering and Marine Energy*, vol. 11, no. 1, pp. 197–231, 2025.

16 R. Gharibi, B. Vahidi, R. Dashti, and R. Khalili, “From electrons to solutions: Power-to-X strategies for energy systems integration,” in *Power-to-X in Regional Energy Systems*, CRC Press, 2025, pp. 201–227.

17 K. Dennis, “Environmentally beneficial electrification: Electricity as the end-use option,” *The Electricity Journal*, vol. 28, no. 9, pp. 100–112, 2015.

18 N. Alsagr, I. Ozturk, and A. Usman, “Digital government, political stability, and nuclear energy investment: The role of the Paris agreement,” *Nuclear Engineering and Technology*, vol. 57, no. 5, p. 103350, 2025.

19 R. Gharibi, B. Vahidi, and R. Dashti, “Importance of energy justice and equity in grid modernization,” in *Communities for Clean Energy Justice and Equity in Grid Modernization*, John Wiley & Sons, 2025, pp. 11–50.

20 O. Jogunola, M. Hammoudeh, K. Anoh, and B. Adebisi, “Distributed ledger technologies for peer-to-peer energy trading,” in *2020 IEEE Electric Power and Energy Conference (EPEC)*, IEEE, 2020, pp. 1–6.

21 G. Bevan, “What is next-generation solar and how can it contribute to a just energy transition?,” *Sunrise Network*, 2025. Accessed: Aug. 19, 2025. [Online]. Available: https://www.sunrisenetwork.org/news/what-is-next-generation-solar-and-how-can-it-contribute-to-a-just-energy-transition/

22 G. Yandem and M. Jabłońska-Czapla, “Review of indium, gallium, and germanium as emerging contaminants: Occurrence, speciation and evaluation of the potential environmental impact,” *Archives of Environmental Protection*, vol. 50, no. 3, pp. 84–99, 2024.

23 S. A. Abubaker and M. Z. Pakhuruddin, “Progress and development of organic photovoltaic cells for indoor applications,” *Renewable and Sustainable Energy Reviews*, vol. 203, p. 114738, 2024.

24 W. Huang, “Research directions in synthetic fiber ropes applied as mooring lines for floating offshore wind turbines,” *Renewable and Sustainable Energy Reviews*, vol. 225, p. 116183, 2026.

25 V. H. U. Eze, E. C. Eze, G. U. Alaneme, and P. E. Bubu, “Recent progress and emerging technologies in geothermal energy utilization for sustainable building heating and cooling: A focus on smart system integration and enhanced efficiency solutions,” *Frontiers in Built Environment*, vol. 11, p. 1594355, 2025.

26 T. Sharmin, N. R. Khan, M. S. Akram, and M. M. Ehsan, “A state-of-the-art review on geothermal energy extraction, utilization, and improvement strategies: Conventional, hybridized, and enhanced geothermal systems,” *International Journal of Thermofluids*, vol. 18, p. 100323, 2023.

27 T. Thennakoon et al., "Harnessing the power of ocean energy: A comprehensive review of power generation technologies and future perspectives," *Journal of Research in Technology and Engineering*, vol. 4, no. 3, pp. 73–102, 2023.

28 M. Z. A. Khan, H. A. Khan, and M. Aziz, "Harvesting energy from ocean: Technologies and perspectives," *Energies*, vol. 15, no. 9, p. 3456, 2022.

29 International Energy Agency (IEA), "Wind has significant potential to increase countries' renewable capacity growth," 2025. Accessed: Sep. 12, 2025. [Online]. Available: https://www.iea.org/energy-system/renewables/wind

30 REN21 (2025). Renewables 2023 Global Status Report. https://www.ren21.net/gsr-2023/modules/global_overview/

31 G. Axelsson, "The future of geothermal energy," in *Living with Climate Change*, Elsevier, 2024, pp. 397–422.

32 International Energy Agency (IEA) (2021). Net Zero by 2050. https://www.iea.org/reports/net-zero-by-2050

33 Y. Zeng, T. Zhou, T. Wang, M. Zhang, S. Zhang, and H. Yang, "Long-duration energy storage: A critical enabler for renewable integration and decarbonization," *Energies*, vol. 18, no. 3, p. 466, 2025.

34 J. Twitchell, K. DeSomber, and D. Bhatnagar, "Defining long duration energy storage," *Journal of Energy Storage*, vol. 60, p. 105787, 2023.

35 X. Ding et al., "A systematic review on liquid air energy storage system," *Renewable and Sustainable Energy Reviews*, vol. 210, p. 115164, 2025.

36 K. Itani and A. De Bernardinis, "Review on new-generation batteries technologies: Trends and future directions," *Energies*, vol. 16, no. 22, p. 7530, 2023.

37 B. Ramasubramanian, G. K. Dalapati, M. V. V. Reddy, K. Zaghib, V. Chellappan, and S. Ramakrishna, "Progress and complexities in metal–air battery technology," *Energy Technology*, vol. 12, no. 5, p. 2301375, 2024.

38 R. Gharibi, R. Khalili, B. Vahidi, A. F. Nematollahi, R. Dashti, and M. Marzband, "Enhancing energy hub performance: A comprehensive model for efficient integration of hydrogen energy and renewable sources with advanced uncertainty management strategies," *Journal of Energy Storage*, vol. 107, p. 114948, 2025.

39 L. M. Asif, B. Solomon, and C. Adulugba, "Prospects of nuclear power in a sustainable energy transition," *Arabian Journal for Science and Engineering*, vol. 50, no. 5, pp. 3467–3477, 2025.

40 A. C. Kadak, *A Comparison of Advanced Nuclear Technologies*. Columbia University in the City of New York, 2017.

41 M. Mohamed, N. D. Zakuan, T. N. A. Tengku Hassan, S. S. M. Lock, and A. Mohd Shariff, "Global development and readiness of nuclear fusion technology as the alternative source for clean energy supply," *Sustainability*, vol. 16, no. 10, p. 4089, 2024.

42 S. Meschini et al., “Review of commercial nuclear fusion projects,” *Frontiers in Energy Research*, vol. 11, p. 1157394, 2023.

43 Q. Lin, X. Zhang, T. Wang, C. Zheng, and X. Gao, “Technical perspective of carbon capture, utilization, and storage,” *Engineering*, vol. 14, pp. 27–32, 2022.

44 E. Liu, X. Lu, and D. Wang, “A systematic review of carbon capture, utilization and storage: Status, progress and challenges,” *Energies*, vol. 16, no. 6, p. 2865, 2023.

45 G. Al-khulaidi, Y. Sun, A. G. Alareqi, A.-W. Ibrahim, A. Magaji, and X. Zhang, “Review on carbon capture, utilization, and storage for enhancing gas recovery,” *Energy & Fuels*, vol. 38, no. 10, pp. 8355–8384, 2024.

46 Q. Gao et al., “Transformer-enhanced intelligent microgrid self-healing: Integrating large language models and adaptive optimization for real-time fault detection and recovery,” *Energy Engineering: Journal of the Association of Energy Engineers*, vol. 122, no. 7, p. 2767, 2025.

47 I. Hamdeen, E. A. Badran, M. F. Kotb, and M. Elgamal, “Self-healing multi-agent techniques in electric power distribution systems: A review,” *Renewable and Sustainable Energy Reviews*, vol. 224, p. 116132, 2025.

48 I. Nazir et al., “Blockchain enabled reserved pricing and privacy preserving model for peer-to-peer energy trading in smart grid,” *IEEE Transactions on Consumer Electronics*, vol. 71, no. 3, pp. 8716–8726. https://doi.org/10.1109/TCE.2025.3596907

49 Q. Wang et al., “Enhancing energy transition through sector coupling: A review of technologies and models,” *Energies*, vol. 16, no. 13, p. 5226, 2023.

50 S. Madeddu et al., “The CO_2 reduction potential for the European industry via direct electrification of heat supply (power-to-heat),” *Environmental Research Letters*, vol. 15, no. 12, p. 124004, 2020.

51 M. Sterner and M. Specht, “Power-to-gas and power-to-X—The history and results of developing a new storage concept,” *Energies*, vol. 14, no. 20, p. 6594, 2021.

52 G. Gaustad, M. Krystofik, M. Bustamante, and K. Badami, “Circular economy strategies for mitigating critical material supply issues,” *Resources, Conservation and Recycling*, vol. 135, pp. 24–33, 2018.

53 E. C. Onukwulu, M. O. Agho, and N. L. Eyo-Udo, “Circular economy models for sustainable resource management in energy supply chains,” *World Journal of Advanced Science and Technology*, vol. 2, no. 2, pp. 34–57, 2022.

54 A. A. Abdirahman, M. Asif, and O. Mohsen, “Circular economy in the renewable energy sector: A review of growth trends, gaps and future directions,” *Energy Nexus*, vol. 17 p. 100395, 2025.

55 I. Campos and E. Marín-González, “People in transitions: Energy citizenship, prosumerism and social movements in Europe,” *Energy Research & Social Science*, vol. 69, p. 101718, 2020.

56 I. Ruostetsaari, “From consumers to energy citizens: Finns’ readiness for demand response and prosumerism in energy policy making,” *International Journal of Energy Sector Management*, vol. 14, no. 6, pp. 1157–1175, 2020.

57 M. Wahlund and J. Palm, “The role of energy democracy and energy citizenship for participatory energy transitions: A comprehensive review,” *Energy Research & Social Science*, vol. 87, p. 102482, 2022.

58 T. Hoppe and G. De Vries, *Social Innovation and the Energy Transition*. MDPI, 2018.

59 C. Popescu, E. Hysa, A. Kruja, and E. Mansi, “Social innovation, circularity and energy transition for environmental, social and governance (ESG) practices—A comprehensive review,” *Energies*, vol. 15, no. 23, p. 9028, 2022.

60 N. Z. Khanna, J. Zhang, H. Lu, W. Feng, M. Johnson-Wang, and N. Zhou, “Conceptualizing demand-side technological and social innovations in modeling pathways to carbon neutrality,” *Energy Research & Social Science*, vol. 100, p. 103115, 2023.

61 D. S. Renné, “Progress, opportunities and challenges of achieving net-zero emissions and 100% renewables,” *Solar Compass*, vol. 1, p. 100007, 2022.

62 S. Aziz, I. Ahmed, K. Khan, and M. Khalid, “Emerging trends and approaches for designing net-zero low-carbon integrated energy networks: A review of current practices,” *Arabian Journal for Science and Engineering*, vol. 49, no. 5, pp. 6163–6185, 2024.

63 P. K. Kurella and S. Mikkili, “A comprehensive review of DERMS for smart and secure energy networks: Standards, protocols, and security considerations,” *International Journal of Ambient Energy*, vol. 46, no. 1, p. 2506703, 2025.

64 M. Daneshvar, B. Mohammadi-Ivatloo, and A. Anvari-Moghaddam, *Communities for Clean Energy Justice and Equity in Grid Modernization*. John Wiley & Sons, 2025.

5

Energy Security and Justice in Climate–Energy–Sustainability Nexus Frameworks for Realizing Low-carbon Societies

Alireza Zare Ansaroudi, Asma Nasiri, Sina Nabighadim, Hamed Tajalli, Mehdi Abapour, and Heresh Seyedi*

Energy System Research Institute (ESRI), Faculty of Electrical and Computer Engineering, University of Tabriz, Tabriz, Iran

*Corresponding author: zare.alireza.2000@gmail.com

List of Acronyms	
AI	Artificial intelligence
CESN	Climate–energy–sustainability nexus
EV	Electric vehicle
IEA	International Energy Agency
OPEC	Organization of the Petroleum Exporting Countries
TES	Total energy supply
V2G	Vehicle-to-grid

5.1 Introduction

Currently, one of the main threats facing the world today is climate change. The main causes of climate change include urbanization, agriculture, industrial production, and greenhouse gases [1]. The problems linked to climate change, including the increase in the temperature levels of the globe, have contributed to the melting of the polar ice caps, rising sea levels, reduction in the levels of freshwater resources, reduction in the different species of plants and animals, and the increase in the rate of natural disasters, including floods, storms, and others.

Among the known factors contributing to these problems are the high consumption of fossil fuels and greenhouse gas emissions. As shown in Figure 5.1, from 1970 to 2024, more than 60% of the world's energy needs were met by fossil fuels and traditional biomass, indicating a strong dependence on fossil fuels and a resulting sharp increase in greenhouse gas emissions, such as carbon dioxide. According to the International Energy Agency (IEA), the amount of carbon dioxide emitted from energy consumption increased by 0.8% in 2024, reaching a record high of 37.8 Gt [2]. If this trend continues, life will be tough for future generations. Therefore, minimizing greenhouse gas emissions and moving toward clean and renewable energies is no longer an option, but a necessity to improve the future of humans and the Earth in terms of climate and geographical conditions.

The mentioned issues have high importance; therefore, many countries have decided to move toward low-carbon infrastructure. In this approach, important measures have been taken, the most important of which are the use of renewable energy sources, especially solar and wind energy, the use of smart homes, investment in new technologies such as electric cars, and the participation of countries in international agreements. In 2015, one of the most important agreements, called the Paris Agreement, was created to combat global warming and climate change [4]. Under the agreement, countries committed to keeping global temperature increases below 2 °C above pre-industrial levels. However, there are still many challenges to achieving a low-carbon society.

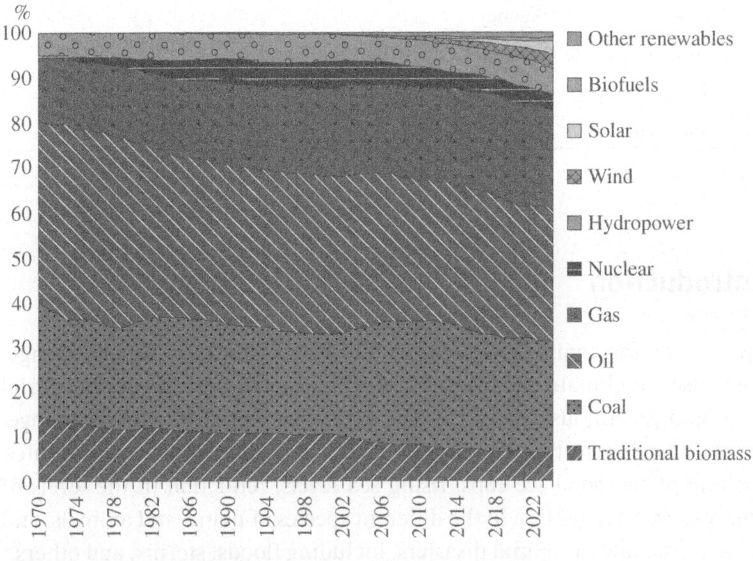

Figure 5.1 Global primary energy consumption by sources from 1970 to 2022. *Source*: Adapted from [3].

Energy demand is increasing with population growth and urbanization. The global energy consumption from 1990 to 2022 is shown in Figure 5.2, which clearly shows the global growth in energy consumption. However, energy consumption cannot be limited to reduce carbon emissions, because energy plays an important role in economic development and social well-being. Energy consumption drives growth and improves the quality of life, and sustainable access to it is a fundamental factor. On the other hand, the most important and largest source of greenhouse gas emissions is energy consumption. For this reason, there is a complex and sensitive relationship between climate, sustainability, and energy. In this regard, the concept of the climate–energy–sustainability nexus (CESN) has been proposed as a new framework to show that decision-making in the energy sector cannot be done separately from climate considerations and sustainability requirements. Following the aforementioned method, energy policies can consider three specific objectives at the same time; this approach will result in lowering carbon emissions to mitigate climate change, ensuring energy reliability, and environmental and social sustainability for a better and greener future. To lower carbon emissions and mitigate the effects of climate change, there should be a balance among these three objectives. Furthermore, energy security and energy justice principles should be addressed in this framework. In a sustainable and equitable outcome, energy justice and energy security should be highlighted. Therefore, a promising carbon reduction framework should be a comprehensive strategy for reducing greenhouse gas emissions that also utilizes energy justice and energy security as guiding principles, leading societies to a sustainable future.

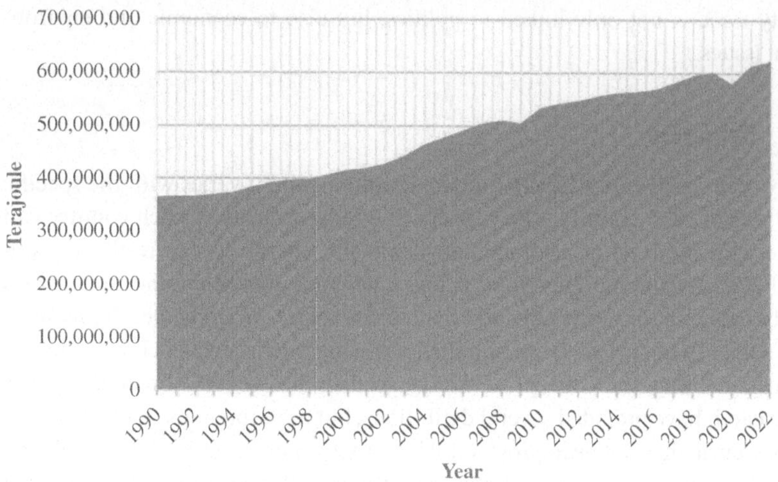

Figure 5.2 Total energy supply (TES), world, 1990–2022. *Source*: Adapted from [5].

Given that energy security and justice play a fundamental role in the ways to realize a low-carbon society, this chapter seeks to show how new policies and technologies in the energy sector—including the development of renewable sources, the spread of electric vehicles (EVs), and smart homes—can simultaneously affect the two key components of energy security and energy justice. In Sections 5.2 and 5.3, respectively, the fundamental concepts of energy security and energy justice, along with their dimensions and indicators, are discussed to provide the necessary conceptual framework for a better understanding of the subject. Then, in Section 5.4, the opportunities, challenges, and social, economic, and environmental consequences of each of these three technologies will be examined in relation to the two components of energy security and justice.

5.2 Energy Security

Since the beginning of recorded history, energy has played a crucial role in the lives and development of societies. Today, our lives are much more dependent on energy than ever before. If there is not enough energy available, the work and life of people and the entire society will face serious problems. For this reason, having access to energy can be considered a fundamental aspect of security. When we talk about security, most people and even researchers usually understand it as the absence of danger and risk [6]. In recent years, due to the global crises and fluctuations in the energy field, countries and international organizations have paid special attention to the issue of energy security. Also, these days, the energy security topic is not only linked to politics, but also to economic and environmental issues.

5.2.1 Definition

Energy security does not have a unique definition, and it varies with the specific conditions of each country. The concept of energy security of each country can change with the level of economic development, energy system resilience, and geopolitical conditions. Therefore, it has a dynamic definition and dimensions that change with developments and new technologies. With the development of technologies in energy societies and the increasing importance of climate change and sustainability, the concept of energy security has become broader. Studies and research have provided several definitions, but a comprehensive definition of energy security has not been provided. This variety of perspectives has introduced various indicators to measure energy security for different countries, which can be used for comparative purposes between countries.

5.2.2 Dimensions

Many definitions of energy security provided by national institutions, the IEA, Organization of the Petroleum Exporting Countries (OPEC), the European Commission, and the International Atomic Energy Agency, are rooted in the traditional approach, often known in the literature as the four As, which is summarized in four main areas: availability, affordability, acceptability, and accountability. In this approach, energy security is considered a comprehensive issue that should not be addressed solely in terms of access to energy. This definition requires attention to many social, economic, environmental, and geopolitical aspects. In this regard, new definitions have been made to include the aspects above. From these definitions and studies, seven main dimensions of energy security can be defined as shown in Figure 5.3 [7]. In other studies, conducted on energy security, not all of these dimensions are examined, and usually only a few specific dimensions are focused on. The following provides the definitions of these dimensions:

Energy availability: one of the important factors in energy security is the possibility of sustainable access to different sources. Studies have shown that the diversity of energy sources can have numerous benefits for countries. From a political perspective, if a country relies on a single foreign source, there is always the possibility of pressure and abuse from that supplier. However, when energy is supplied from multiple countries or diverse sources, this dependence is reduced, and political independence is strengthened. From an economic perspective, a country with access to a variety of sources is less vulnerable to potential disruptions in energy supply, such as a reduction or cessation of exports from a particular supplier. It has several sources to support sustainable energy security while stimulating innovation, investment, and economic growth. On the environmental front, renewable sources eliminate energy shortfalls and outages. It also reduces environmental impacts compared to fossil fuels [8]. For example, in 2017, Germany had a more stable energy supply than France. In that year, one of France's nuclear power plants was shut down due to safety issues, and as a result, France faced an electricity shortage and increased

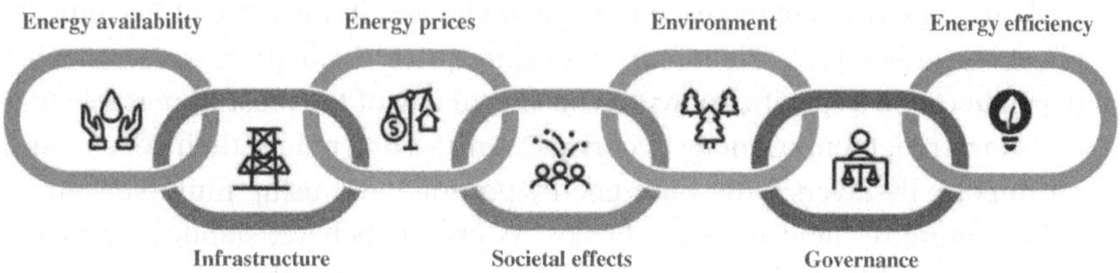

Figure 5.3 Dimensions of energy security.

electricity imports from its neighbors because its energy supply was not diverse enough. However, Germany, with its significant share of renewable energy, was able to maintain its energy supply stability well and also use coal and gas power plants to compensate for its needs [9]. Therefore, the more diverse the mix of energy sources and the higher the share of renewable energies, the less likely the system will be to experience outages and instability, and the greater the energy security.

Energy infrastructure: infrastructure is the foundation of a reliable and sustainable energy supply. This infrastructure includes refineries, power plants, pipelines, transmission networks, substations, and energy storage facilities. Proper investment in infrastructure can be considered a guarantee of a stable energy supply in the short and long terms. Failure or malfunction of infrastructure can even cause widespread blackouts, so the reliability of infrastructure is of great importance. The digitalization of the process of controlling, managing, and operating energy networks provides the ground for cyberattacks, and the infrastructure may be damaged by these attacks. Disruption attacks, data manipulation attacks, or information attacks may destabilize the power system or even trigger massive blackouts. For instance, in 2015, there was a cyberattack on the Ukrainian power grid, which resulted in massive blackouts [10]. To promote energy security, there is a need to appropriately secure the infrastructure of the communication network resources. To achieve the foregoing, one needs to apply strategies or guidelines that will prevent attacks or even sabotage. To support the stability of the network from attacks, there is also a need for the network's infrastructure to be secured.

Energy prices: price stability is one of the key components of energy security. Price spikes in the global fossil fuel market create problems for long-range planning for countries as well as corporations. Price instability affects investment in the construction of new energy facilities, as well as budgeting for middle-class citizens or enterprises. The main reason for these instabilities is the buying and selling of crude oil involving US dollars. Given this, exchange rate fluctuations can change the cost of energy imports for countries, even if the global price of oil remains constant. Research shows that the prices of fossil fuels (oil, gas, and coal) are closely related to each other, and changes in the price of one quickly affect the other [11]. Therefore, managing price fluctuations and reducing dependence on a specific source is considered one of the most important strategies for strengthening energy security. Countries that reduce their dependence on imports by diversifying their energy portfolio and using multiple sources will be more resilient to price shocks. As experts believe, stability in energy prices is one of the main foundations of energy security [7].

Societal effects: energy is one of the basic needs of life, and for this reason, in some studies, social welfare is part of energy security [7]. Concerns in this regard mostly include energy poverty. Energy poverty does not mean not having electricity and gas, but goes beyond this; a widespread deprivation of access to sufficient energy services for a decent life and human development is called energy poverty. Energy poverty means the lack of sufficient choice in access to appropriate, affordable, reliable, high-quality, safe, and environmentally friendly energy services to support economic and human development [12]. This definition shows that energy poverty is not only about the amount of energy consumed, but also about the services that are provided through it. Essential services such as cooking, heating, cooling, lighting, and access to communication and health technologies all require energy. The lack of energy and the resulting lack of these services reduce comfort and can even limit access to education, health, and active participation in society.

Environment: there is a direct relationship between energy security and environmental issues. In addition to the fact that greenhouse gas and pollutant emissions cause climate change and air pollution, activities such as the construction of hydroelectric dams and oil spills during extraction and transportation contribute to environmental degradation. For this reason, the European Commission and many researchers emphasize that special attention should be paid to environmental sustainability in the design of energy policies [7]. Regarding the relationship between the environment and energy security, a study was conducted on 45 Asian countries, and based on these studies, it was concluded that the greater the environmental degradation, the lower the energy security, because the excessive exploitation of natural resources and the resulting pollution weaken the ability of countries to provide sustainable energy [13]. The opposite is also true: the more energy sustainable a country is, the less pressure it places on the environment. Therefore, in addition to being important for nature, protecting the environment also ensures energy security and sustainability.

Governance: governments have a crucial role to play in managing and mitigating short-term and long-term energy disruptions. Forward-looking governments can strengthen long-term energy security by planning infrastructure. Tax and subsidy policies play a key role in this regard. In addition to the above, energy diplomacy is also an important tool to ensure sustainable imports. Governments should also act as the main source of information, as reliable and accurate data are the foundation of energy policy.

Energy efficiency: increasing efficiency through new technologies and methods of optimizing consumption reduces overall energy demand and enhances energy security. A related concept in this context is energy intensity, which is the amount of energy required to produce a unit of output. Reducing energy intensity allows countries to work with fewer resources and be less vulnerable to energy shocks. For example, industries such as steel are highly vulnerable to changes in energy prices or shortages, but less energy-intensive industries are more resilient [7]. Many researchers consider improving efficiency and reducing energy intensity to be key policies for enhancing energy security.

5.2.3 Energy Security as a Member of the Energy Trilemma – Problems and Contradictions

At the national or transnational level, it is insufficient to focus solely on a single aspect, such as security, to address energy issues. A more comprehensive approach is needed that addresses several competing objectives. This approach can be summarized in the concept of the energy triad, which considers the three main energy objectives, shown in Figure 5.4, simultaneously [7].

These three objectives are sometimes in conflict with each other. For example, if a country seeks to enhance its energy security by relying on inexpensive domestic sources such as coal, this approach may compromise environmental sustainability. For example, Estonia is in a better position in terms of energy security than the other two Baltic countries, because by relying on domestic oil shale resources, it has been able to reduce its dependence on imports and strengthen the security of energy supply [14]. This same reliance on oil shale has led Estonia to become one of the most carbon-intensive economies in the European Union, whereby the carbon dioxide emissions are reported to be over double the average of the European Union. This illustrates the extent to which energy security might have negative consequences in the absence of sustainability considerations. Therefore, energy policymaking must consider all dimensions of security, justice, and sustainability to achieve sustainable development.

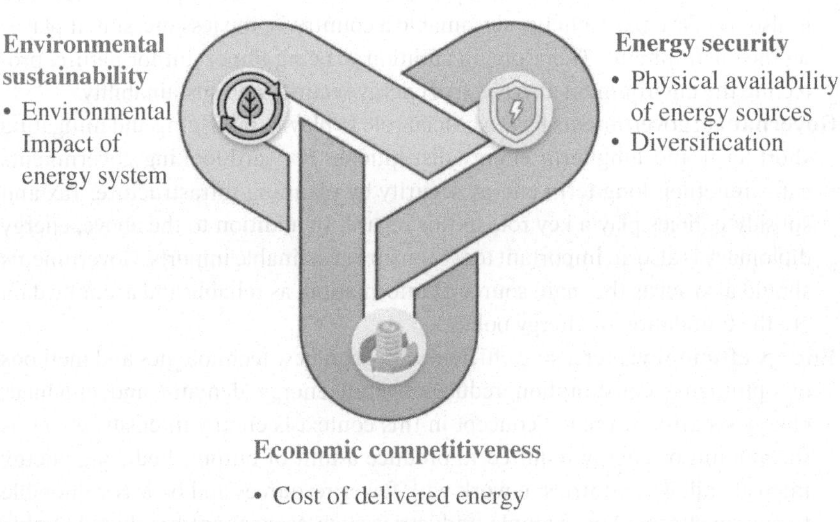

Figure 5.4 Energy trilemma goals.

5.3 Energy Justice

5.3.1 Definition

Energy justice means that decisions made about energy production, distribution, and consumption must pay special attention to social, environmental, and ethical consequences, in addition to technical and economic issues [15]. According to researchers, energy justice must be taken seriously because failure to do so will render projects aimed at developing or transitioning to clean energy ineffective and may even increase social inequality, energy poverty, and pressure on vulnerable groups [16, 17].

5.3.2 Dimensions

Energy justice has three main dimensions, each addressing an aspect of fairness in the energy sector. Each of these dimensions is discussed below:

Distributive justice: this dimension addresses how the benefits and costs of energy projects are distributed among those who use clean, cheap energy and those affected by pollution, environmental degradation, or even migration [15, 17]. This issue is of great importance in the transition to low-carbon energy. An example is the hydropower projects in Vietnam, which, although reducing carbon emissions, have caused the displacement of people, loss of agricultural land, and reduced access to natural resources. At the same time, a large part of the electricity generated is transmitted to urban and industrial areas [17]. It is therefore clear that clean energy can also cause injustice without considering the fairness of the distribution of its benefits and costs.

Procedural justice: this dimension addresses the decision-making process in the energy sector and issues such as transparency, access to information, and genuine public participation [15, 16]. This dimension of energy justice emphasizes that full and clear information should be made available to local communities and vulnerable groups and that no decisions should be made without their participation [17]. It also emphasizes the existence of legal and administrative avenues for protest and redress [15].

Recognition justice: recognition and respect for minorities and vulnerable people are considered in this dimension. Decisions made about energy should take into account the needs of these groups. Failure to pay attention to the needs of these groups exacerbates social inequalities.

Energy justice is neither merely theoretical nor merely practical, but also serves the role of an analytical tool for the estimation of energy problems, taking into consideration their economic, engineering, moral, and social aspects [15, 16]. It is because of this approach that the policymakers are able to design their decisions just, efficient, sustainable, and humane. In order to provide an ethical and practical framework for policymakers and planners, eight key principles have been introduced and discussed below:

Availability: everyone has access to sufficient and sustainable energy to meet their needs. This principle also considers issues such as the availability of reserves, resilient infrastructure, production technology, transmission, storage, and necessary investments.

Affordability: energy should be affordable for all people, especially low-income groups, and energy bills should not burden people. Access to energy is meaningful when people can afford it.

Due process: all decisions in the energy sector should be fair and transparent. Stakeholders should be involved in the decision-making and policy-making process. Methods for compensation should also be considered.

Good governance: energy policies must be transparent, accountable, and responsive. The more transparent and accessible information is to the public, and the more financial accountability is done, the less corruption there will be, and the more democracy and social trust there will be.

Sustainability: energy resources should be used to last for future generations. This principle highlights the government's responsibility in the sustainable use of natural resources, to avoid compromising ecosystems, and to remedy any impacts resulting from resource extraction.

Intragenerational equity: people of the present generation have the right to access energy services that are sufficient to enable them to experience a minimum level of well-being.

Intergenerational equity: future generations have the same right to enjoy a healthy and good life as today's generation, and today's decisions should not destroy their environment and resources.

Responsibility: countries, organizations, and even present generations should be held accountable for the environmental and social consequences of energy.

5.3.3 Operationalizing Energy Justice Given Effective Factors

In moving toward a clean energy society, the principles mentioned in the previous subsection must be followed to succeed. Therefore, genuine public participation in decision-making is essential. This means that people should not only be the addressees of policies, but also have a role in their design and implementation. Also, the management of projects must be fully transparent. Public trust is increased by this transparency, and corruption is prevented, and ultimately, the benefits and costs of these projects are distributed fairly among all, so that new injustices do not arise. The principles mentioned are the key dimensions of energy justice. Energy justice should not remain just a slogan, but should be present in all energy policies as a practical and ethical tool. If these principles are observed, the transition to clean energy will be accompanied by reduced pollution and environmental protection, improved social welfare, and a just society will be created. In other words, sustainable development can only be achieved when all its dimensions, of which justice is one, are considered simultaneously.

Since energy security and energy justice are both important concepts that relate to numerous dimensions, it is evident that any new policy or technology in the energy sector should be assessed in terms of its potential economic and environmental efficiencies, and also in terms of its potential implications for energy security and energy justice. Low-carbon solutions such as renewable energy development, EVs, and smart homes represent the principal avenues for a transition to a low-carbon society. Beyond their ability to reduce greenhouse gas emissions, these low-carbon solutions can affect energy security and energy justice. This will be explored in the following section.

5.4 Strategic Implications of Carbon Reduction for Energy Security and Justice

5.4.1 Consequences of Using Renewable Energy Sources

Renewable energy is considered one of the main strategies for reducing greenhouse gas emissions to create a low-carbon society. As evidenced by international statistics, the installed capacity of renewables is growing rapidly; furthermore, the share of renewable electricity will increase significantly by 2030. Figure 5.5 further illustrates the imminent global increase in the use of renewable energies. By 2025, nearly 34.5% of the world's electricity was generated from renewable sources, and this level is expected to rise to 45.6% in the near future, by 2030. The differences across countries are considerable—Iceland generates nearly 100% of its electricity needs from renewables, while other high-ranking countries, including Brazil, Austria, and New Zealand, generate more than 80% of energy from renewables. Northern European countries, including Sweden, Germany, and Finland, also rank high, indicating that serious policies have been put in place in this area. It should be noted, however, that the United States and India, with 24.10 and 19.80% of their electricity supplies from renewable sources, lag behind in this regard.

In preceding sections, it was stated that as countries increase their use of renewable energy, the topic involves more than simply reducing reliance on fossil fuels and carbon emissions; energy security and justice, too, are directly impacted and should be examined and considered.

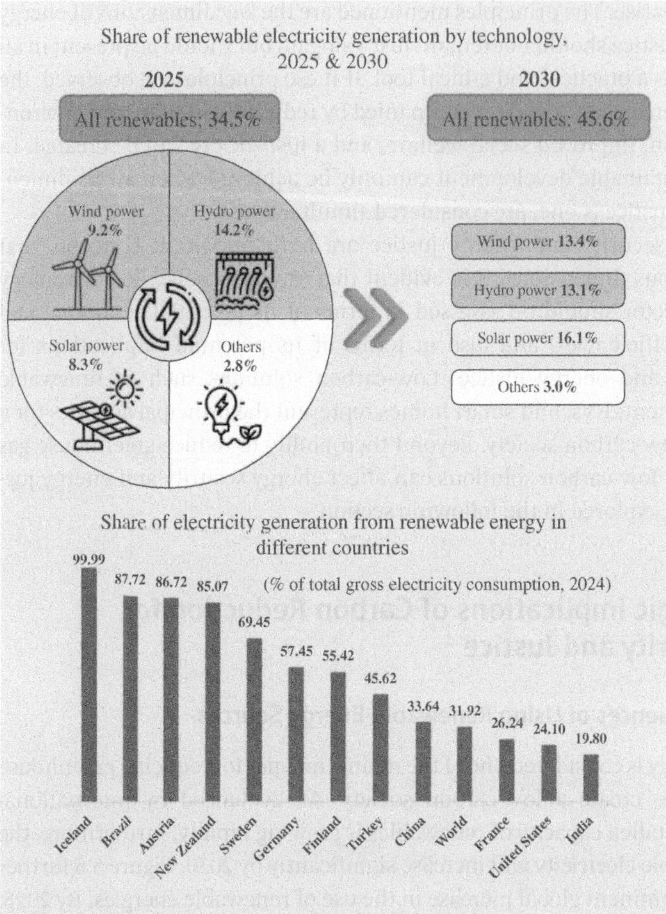

Figure 5.5 Share of renewable electricity generation. *Source*: Adapted from [18].

5.4.1.1 Renewables Implications for Energy Security

National energy independence: with a transition to renewable energy, nations increasingly free themselves from the constraints of traditional sources such as oil and gas, which are primarily sourced from politically sensitive regions of the world. Reduced dependence lessens the risk of price shocks, supply disruptions, or political sanctions arising from fluctuations in the supply or demand for fossil fuels [19].

Increased long-term supply stability: renewable energy sources, unlike fossil fuels, are inexhaustible. For this reason, they have a long-term stable energy supply. When a country utilizes solar and wind energy, it no longer has to worry about when the oil and gas supply will run out or about unstable oil and gas prices in the future. Stability allows for greater energy security and greater national independence in terms of energy supply.

The challenge of production instability: as more renewable energy sources, such as solar and wind, are used, the challenges of instability in electrical energy production will increase. Instead of producing a constant supply of electricity, the output from switching to these energy sources depends on available sunlight and wind. For this reason, we cannot merely assume these sources will allow us to completely move away from the fossil fuels utilized today. To address the challenges previously mentioned, grids can be configured to sufficiently handle variations in electricity generation. Additionally, traditional power plants, such as thermal plants, should still operate and be a backup for electric shortages that will occur during times of low renewable energy production [20]. In addition to the two solutions previously mentioned, the use of some type of storage device, such as batteries, could be utilized to act whenever needed and assist in reducing the instability.

Need for flexible grid infrastructure and investment: the implementation of renewable energy sources relies on a strong and intelligent infrastructure. The higher the proportion of renewable energy sources used, the greater the need for flexibility in the electrical grid, because power plants may be located in remote areas, and new transmission lines need to be constructed to get the electricity produced in those areas to the cities and centers of consumption [21]. In addition, the grid itself must be intelligent and flexible enough to foresee and respond to fluctuations in the quantities of electricity generated from these variable sources.

Creating new dependencies: producing equipment such as solar panels, wind turbines, and batteries requires critical minerals (lithium, cobalt, and rare earth elements), rather than fossil fuels. The extraction and processing of these critical minerals is concentrated in a limited number of countries; therefore, while renewable energy will reduce dependency on fossil fuels, it creates dependencies on other countries, directly impacting energy security in times of geopolitical crises or changes in global markets [22]. Mitigating the vulnerability from critical dependence on foreign mineral sources will require diversifying supply (multiple sources) and developing recycling technologies to recover used materials.

Vulnerability to climate change: river flows and water reserves affect the amount of hydroelectric power generation. Therefore, changing precipitation patterns, increasing temperatures, and accelerating snowmelt, all part of climate change, affect production levels. If precipitation decreases and drought occurs, reservoir water levels will drop, reducing the capacity of hydroelectric power plants. On the other hand, if rainfall is sudden and heavy, flooding will occur, causing problems for infrastructure. For this reason, extensive research has been conducted on this topic, and it has been found that climate change reduces the reliability of hydroelectric power and creates challenges for water resource management and energy security planning [23].

5.4.1.2 Renewables Implications for Energy Justice

Increasing access to energy: the availability of energy is increasing due to the emergence of renewable energy sources, particularly in marginalized areas. For instance, off-grid solar energy resources are efficient strategies that help create fair access to electricity in these areas. The off-grid system provides sustainable support for the production of localized energy, prevents the development of the electric grid, and ensures fast, affordable electrification. When analyzing the energy access projects initiated for the rural population, for instance, in Bangladeshi rural areas, it becomes obvious that almost all households are connected to home-based solar energy, which has contributed substantially to the betterment of living conditions, social protection, disaster preparedness, and microeconomic development, among others [24]. Renewable energy, including off-grid solar, can create jobs, promote economic development, improve access to energy, ensure greater energy justice, and improve quality of life if facilitated with proper policies, financial services, and community involvement, particularly from the disadvantaged communities.

Cost and investment requirements: low-income households face challenges in using renewable technologies. These technologies have a high initial cost, and the low-income group cannot afford this cost. To cover this cost, they need loans and facilities, which require guarantees, which hinder access to these facilities and, as a result, prevent investment in this field and force them to use traditional methods. If the aforementioned barriers are overcome, the lack of appropriate infrastructure, the lack of technical services in deprived areas, and the need for installation and maintenance will hinder their use. All of this shows that for equal access to clean energy, in addition to providing adequate equipment, financial support, training, and infrastructure must also be provided [25].

Employment creation and local development: the use of renewable resources in remote areas not only improves access to energy but also provides a platform for job creation and local development. The experience of home solar systems in Bangladesh shows that after having access to sustainable electricity, households and small businesses have been able to start new economic activities such as tailoring, poultry farming, or providing local small services and earn part of their income from this route [24]. According to a comparison made in remote communities in Congo, Australia, and Canada, renewable energy projects, if designed with the financial and managerial participation of local people, create more job opportunities in installation, operation, and maintenance, and ensure that economic benefits remain in the same area [26]. This model can also be applied to wind, biomass, hydro, and even marine energy projects; that is, by using indigenous resources and local manpower, in addition to producing clean electricity, skills training capacities, social ownership, and economic development are formed in disadvantaged communities. These communities are transformed from the role of consumers to active players in energy production and management.

Social and environmental damage: the war in Russia and Ukraine has put pressure on global food and energy markets, making biofuels more difficult for low-income communities to use. Because biofuels are produced from food crops or agricultural land, competition for land and resources increases, and some land and water are diverted from food production to fuel production [27]. This both reduces food production and increases food prices, with low-income households taking the hardest hit. The environmental impacts of biofuel production may lead to deforestation, loss of biodiversity, and increased fertilizer and water use. These pressures could then lead to migration and social instability; as such, biofuel development must be taken with food security and social justice in mind, because it could also disrupt the balance between food, energy, and justice in crisis situations. In addition to the benefits of clean energy production, the social and environmental impacts of wind farm development are substantial, such as impacts from the wind turbines, when placed in grasslands, on plant functional diversity, which may subsequently negatively affect ecosystem productivity and health [28]. In one region of India, wind farms have had a negative impact on the population of birds of prey, and this change has led to an increase in the population of lizards and changes in their behavior and physiology, which is an example of the chain effects of wind farms on the ecosystem [29]. Additionally, if wind projects are located in sensitive or protected areas, they can lead to competition for land and pressure on natural habitats. Therefore, it is recommended that they be built in more degraded areas [30]. From a social perspective, research also shows that residents near wind turbines have reported noise pollution from the sound of the blades, which negatively impacts their quality of life and mental well-being [31]. Additionally, the presence of turbines in natural landscapes can also negatively impact property values.

Geography as a factor of inequity: geothermal energy is a stable and reliable source, but its use is highly dependent on geological conditions. Even in modern technologies such as "enhanced geothermal systems," the presence of an appropriate thermal gradient, rock permeability, and access to groundwater are essential; factors that are found only in limited areas of the world. Therefore, despite technological advances, geographical and technical limitations prevent the widespread exploitation of this resource, thus also affecting energy equity between countries [32].

To better understand the implications of renewables and to analyze them systematically, it is necessary to examine each of these implications within the framework of the dimensions of energy security and the principles of energy justice. This categorization helps to clarify which dimension of energy security or justice each implication relates to. Table 5.1 illustrates the relationship between the impact and the main dimensions of energy security and justice.

Table 5.1 Causes and effects in the context of energy security and justice within renewable energy systems.

Effects	**Causes in the case of energy security**
National energy independence	• Energy availability • Governance • Energy prices
Increased long-term supply stability	• Energy availability • Environment
The challenge of production instability	• Energy availability • Infrastructure • Energy efficiency
Need for flexible grid infrastructure and investment	• Infrastructure • Governance • Energy efficiency
Increasing job security and a resilient economy	• Energy availability • Societal effects
Effects	**Causes in the case of energy justice**
Increasing access to energy	• Availability • Affordability • Responsibility
Fair energy cost and investment	• Affordability • Due process • Good governance
Employment creation and local development	• Responsibility • Sustainability • Good governance
Low social and environmental damage	• Sustainability • Responsibility • Intragenerational equity

5.4.2 Consequences of Using Electric Cars

Transport has become the second-largest source of global carbon emissions [33], and decarbonizing the transport sector is essential to reducing greenhouse gas emissions. Private cars play a significant role in carbon emissions, with nearly 50% of carbon emissions in the transport sector coming from private cars and vans, as shown in Figure 5.6. EVs have been introduced as one of the most important tools to reduce carbon emissions in the transport sector.

The share of new cars sold in 2024 is shown in Figure 5.7. As the data shows, Norway ranks first in the world with a share of 92%. Almost all new cars sold in this country are electric. Norway is a leader among the Nordic countries in this regard. Sweden has a share of 58%, Denmark has a share of 56%, and Finland has a share of 50%. This statistic indicates that these countries have increased the adoption of electric cars through supportive policies, tax breaks, government subsidies, and extensive charging infrastructure.

The Netherlands and China, both with a 48% share, show that the trend toward EVs is accelerating even in large markets. Belgium (43%) and Iceland (42%) are also among the leading countries. The United Kingdom (28%) and France and Austria (24%) each have a share. On average, 22% of new cars worldwide are expected to be electric in 2024, meaning one in five new cars will be electric. However, the overall global trend toward electrification is accelerating, and the share of EVs in the global market is expected to increase in the coming years.

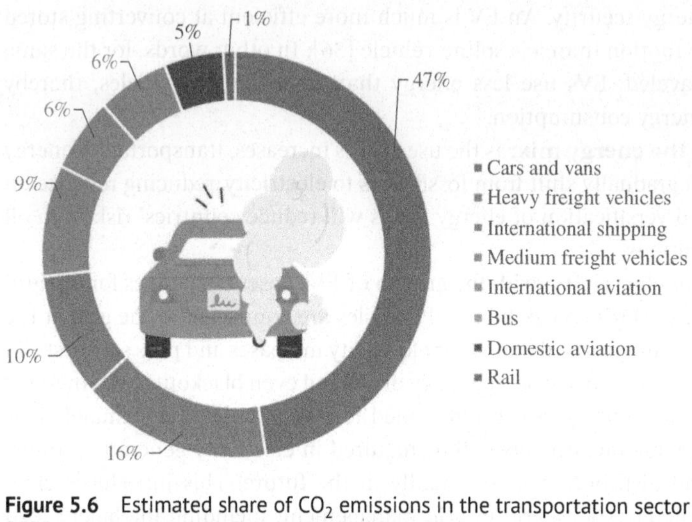

Figure 5.6 Estimated share of CO_2 emissions in the transportation sector worldwide in 2022, by transport type. *Source*: Adapted from [34].

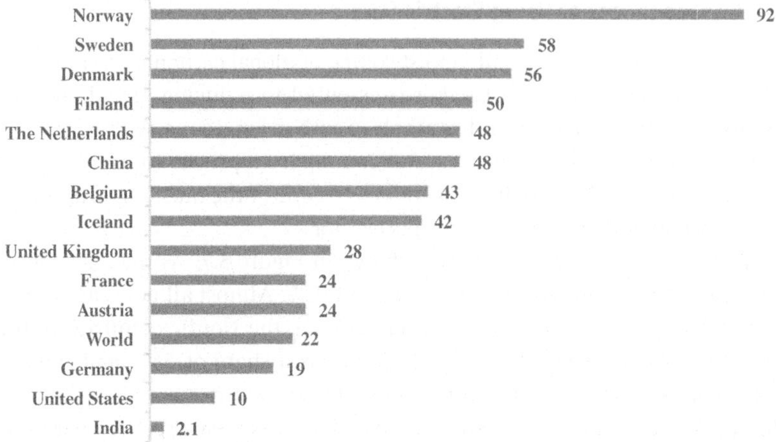

Figure 5.7 Share of new cars sold that are electric, 2024. *Source*: Adapted from [35].

5.4.2.1 EVs' Implications for Energy Security

The transition away from fossil fuels to EVs is significant and presents both challenges and benefits for energy security. The following is a review of the issues raised:

Reducing dependence on imported fossil fuels: in EVs, electricity has replaced gasoline and diesel, reducing countries' dependence on fossil fuel imports. The use of these vehicles in countries without oil resources can improve energy security. An EV is much more efficient at converting stored energy into motion than a gasoline vehicle [36]. In other words, for the same distance traveled, EVs use less energy than fossil-fueled vehicles, thereby reducing energy consumption.

Diversifying the energy mix: as the use of EVs increases, transportation energy sources will gradually shift from fossil fuels to electricity, reducing reliance on them. This diversification of energy needs will reduce countries' risk from oil price fluctuations.

Impact on the electricity grid: the growth of EVs poses challenges for the grid infrastructure. During peak hours, if vehicles are connected to the grid at the same time to charge, the demand for electricity increases and puts severe strain on the grid, which can cause frequency drops and even blackouts [37]. In many places, the electricity grid is not designed for the new level of demand. This means that large investments will be required in electricity generation, transmission, and distribution more broadly in the future. This introduces challenges for the electricity grid in grid management, including intelligent load management. Intelligent load management applies intelligent technologies to help manage the time and rate of electricity charging for vehicles. For instance, it can delay the charging to nonpeak demand periods, like midnight, in order to minimize grid stress.

Building dependence on key minerals: lithium is an essential component of EV batteries. As a result of lithium's high energy density and ability to be held for long periods of time, lithium-ion batteries are now the heart of EV technology. Because of these features, EVs can compete with gasoline-powered vehicles, and lithium-ion batteries play an important role in reducing greenhouse gas pollution and dependence on fossil fuels. With lithium's importance, though, come challenges. The lithium extraction process is environmentally damaging, and supplies are limited to a few countries. Recycling lithium from old batteries and reusing lithium and other valuable materials reduce pressure on new and natural lithium resources. Also, batteries that are no longer suitable for vehicles may still find use in home energy storage systems for low-power applications [38].

Vehicle-to-grid (V2G) opportunity: EVs are not the only consumers of electricity. Rather, with V2G technology, they can also store it. EVs connected to the grid can give electricity in their batteries back to the grid when electricity demand is at its peak and then recharge at off-peak times. This technology can help stabilize the grid, reduce frequency fluctuations, and contribute flexibility to the grid, particularly with increasing usage of intermittent renewable energy, such as solar and wind [39].

Cyber risks: the growth of EV charging locations has resulted in additional demand on the power grid. Since charging points draw on a high volume of electricity, smart charging has emerged to alleviate stress on the grid. Smart charging means that charging points are managed remotely to ensure that power delivery is in line with grid conditions. Unfortunately, this will create additional vulnerability for cyber and physical intrusion into power grid security [40]. If charging points aren't cyber-checked, there could be some pretty bad outcomes. An example could be a hacker simply shutting down a series of charging points, upsetting the balance of the power grid, and potentially leading to a regional or nationwide blackout, or they can simply disable their smart system, which would make their charging points run uncontrolled, which would stress power cables and stations, and potentially lead to a fire.

5.4.2.2 EVs' Implications for Energy Justice

Unequal access: EVs require charging, and charging stations are located primarily in uptown and wealthy neighborhoods. As a result, residents of lower-income neighborhoods are more restricted in their use of EVs [41].

High price: one of the most important problems with EVs is that they are expensive. Therefore, only those with good financial standing can buy them. Of course, the government provides financial assistance to buy these cars, but this assistance goes more to the wealthy, not to those who really need it [41].

Increased mining: with the spread of EVs, air pollution in large cities decreases, which benefits city residents. However, the production of batteries requires the extraction of minerals such as lithium and cobalt. These processes often occur in developing countries and can generate significant pollution in the local environment. In this way, pollution is being transported from cities to mining areas. There is evidence that the rapid increase in demand for lithium-ion batteries has led to a decline in educational attainment among people living in cobalt-rich areas of the Democratic Republic of the Congo. Children exposed to cobalt mining are also experiencing lower physical growth [42]. The concentration of critical minerals needed for electric transportation in countries with poor governance will speed up the pace of contracting, and the pressure for mining development by companies and countries is likely to exacerbate corruption in the extractive sector [43].

Job creation: such developments will create new jobs in battery manufacturing, EV manufacturing, and the installation of charging stations. However, jobs related to internal combustion engine components are gradually being eliminated or reduced. The shift in manufacturing from mechanical to electrochemical processes is changing the type and skills of the workforce, and some current workers may become unemployed if they do not update their skills [44].

5.4.2.3 Achieving Energy Security and Justice in the Field of EVs

Based on the mentioned factors in previous subsections, to ensure that everyone can benefit from EVs and the energy transition is equitable, several key and coordinated actions must be taken. One of these actions is that the infrastructure required for EVs, such as charging stations and smart grids, needs to be expanded nationwide, including in rural areas. In addition, the pricing of electricity for charging vehicles needs to be fair, so that it is both affordable for people and does not put unnecessary strain on the grid. Must ensure the raw materials for battery manufacturing are sustainably and ethically sourced from a reliable and surveyed supply. And finally, direct income support for low-income families can enable their transition to adopting this new technology, while also ensuring that the transition to EVs is equitable for all members of society.

To fully grasp the implications of the rise of EVs, it is important to analyze these effects through the lens of energy security and justice principles. In this context, Table 5.2 highlights how each implication relates to various aspects of energy security and justice, along with its potential impact on these principles.

Table 5.2 Causes and effects in the context of energy security and justice within EVs.

Effects	Causes in the case of energy security
Dependence reduction on imported fossil fuels	• Energy availability • Energy efficiency
The energy mix diversity	• Energy availability • Energy prices
Impact on the electricity grid	• Infrastructure • Governance • Energy efficiency
V2G opportunity	• Energy efficiency • Infrastructure • Environment
Cyber risk reduction	• Governance • Infrastructure
Effects	**Causes in the case of energy justice**
Equality in access	• Availability • Affordability
Fair pricing	• Affordability • Intragenerational equity
Extraction rate for raw material preparation	• Sustainability • Responsibility • Intragenerational equity
Job creation	• Intragenerational equity • Responsibility

5.4.3 Consequences of Using Smart Homes

Smart homes, which are part of the digital transformation, have great potential to improve the quality of life and reduce environmental impacts. However, these effects do not occur in the short term and may even lead to increased energy consumption in the early stages. Initially, households purchase a large number of new smart devices that are often on standby or active mode, and the total operating time of the equipment increases. At this stage, users utilize more convenience and entertainment facilities, and do not activate energy-saving features; additionally, the perception of low-power devices leads to longer use and increased consumption. The smartization process also electrifies previously nonelectric devices, adding a new load to the grid. Over time and with increasing familiarity, households gradually activate energy-saving features such as automatic on/off, scheduling, low-power mode, and intelligent energy management. These systems monitor the consumption of each device and shift or optimize it to off-peak hours, while also raising users' environmental awareness by providing consumption reports. At this stage, families switch to purchasing more energy-efficient equipment, and the combined effect of "energy-efficient equipment + smart control" is formed. Finally, artificial intelligence (AI) and the integration of different systems lower human errors, and smart homes truly help reduce energy consumption and lower carbon dioxide emissions [45].

As smart homes progress from energy consumption growth to efficiency and savings, their overall significance extends beyond regulating household consumption. By modifying consumption patterns, distributing load in the grid, and allowing households to engage more prominently in the energy market, smart homes build dimensions that extend past individual savings. The dissemination of smart homes may very well influence the structure of the energy supply and demand system at both national and regional scales, with consequences for energy security and energy justice. In the subsequent subsection, we can explore the implications of smart homes for two significant components—energy security and energy justice—to elucidate how smart homes, specifically, and energy technologies in general, can influence policy, infrastructure, and social disparities, in addition to optimizing energy consumption.

5.4.3.1 Smart Homes and Implications for Energy Security

Increased energy availability: smart homes lower grid stress and the possibility of outages by managing consumption and shifting loads to off-peak times. This increases energy availability for everyone during peak. It should be noted that peak load decrease occurs without diminishing user comfort and convenience [46]. Availability is lost if the internet or the electricity infrastructure is unstable, or if a device is not functioning, which could also create new vulnerabilities.

Need for reliable infrastructure: smart home expansion will require a reliable electricity grid, fast internet, data centers, and communication standards. Long-term, these developments can improve infrastructure and resiliency in energy systems.

Energy prices: smart homes can lower energy costs by managing consumption and using time-varying tariffs, reducing the cost of electricity for the consumer. Nevertheless, if regulations are not appropriate or consumers are forced to pay for smart devices, the total energy cost for low-income households may increase [46].

Environment: these homes reduce energy consumption and CO_2 emissions by consuming energy during off-peak hours and integrating renewable energy where possible. Homes equipped with smart technology can improve a building's energy performance rating by 37%. But the construction of the equipment and hidden energy consumption used to power servers and standby mode can counteract the positive environmental effects of smart technology [47].

Need for new regulations: smart homes will also need strict regulations for cybersecurity, privacy, and data transparency. Smart home design and implementation will also have to be adapted to the conditions present in each community, and local groups and environmental organizations must be involved in the decision-making. Otherwise, there is the risk of data abuse, monopolization of power in manufacturers, and inequality in access [48].

Increased energy efficiency: smart homes are inherently designed to be more energy-efficient. A smart home can use intelligent control of lighting, heating/cooling, etc., and load management to reduce the whole-house energy consumption, thereby significantly improving energy efficiency. However, if automation is designed to provide convenience or if the device's idle state is not well managed, actual productivity will be lower than expected [49].

5.4.3.2 Smart Homes Implications for Energy Justice

Social impacts: the new technologies offer improvements to social well-being and can lead to increased consumer agency by generating new jobs in installation, service, and manufacturing, as well as reducing their energy bills. Conversely, if access to these technologies is limited, social disparities will worsen. Further, the complexity of new technology can be burdensome for consumers with lower levels of digital literacy [49].

Improved accessibility and affordability: the efficient management of energy provided by the smart home system also provides sustainable access to energy for the home. With the smart home system, the energy used in the house is efficiently planned, and the heating and cooling system of the house is controlled, which helps in efficient consumption of energy, thereby reducing wastage of energy with minimal consumption but without lowering the comfort of the residents of the house [50]. The problem with the poor is that they are unable to enjoy the benefits of these gadgets due to the high cost involved in purchasing, installing, and sustaining the gadgets, along with the requirement of constant connectivity with the internet, the electric grid, etc.

Fair treatment: a major way in which smart homes are able to enhance the level of transparency on the part of consumers is if the system is developed within the framework of clear-cut policies, with the involvement of stakeholders. Policies are required regarding the quality, service provided, cybersecurity, and the aspects of consumers' privacy, so that the consumers feel the process is fair, with redressal available in the event of some discrepancy or error on the part of the system. The involvement of the local community, with their own interpretation of technology in the context of the smart grids project, has a significant bearing on the degree of acceptance on the part of the community itself [48].

However, if the policies are ambiguous or there is a lack of surveillance, the consumers will be left with the drawbacks of faulty or inaccurate readings on the part of the smart meters. According to a report in the United Kingdom, almost a third of smart meters do not work properly, which could erode public trust [51].

Recognizing differences and intergenerational fairness: smart technologies can help everyone if they are designed for different groups, such as the elderly, large families, or renters. This makes access fair and supports intergenerational justice. Research shows that people's lifestyles and incomes affect how they use technology [52]. But if designers ignore these differences, technologies will only be built for the average person, and many people will not be able to use them well. Furthermore, building many smart devices without good e-waste management harms the environment and future generations. Reducing CO_2 emissions from these technologies will help protect the planet for the future.

5.5 Conclusion

The findings from this chapter have brought to light the fact that becoming a low-carbon society has many dimensions aside from the technological viewpoint. Fair social, economic, and institutional choices play a significant role in becoming a low-carbon society. The findings from the results of the analysis include:

1. The concept of energy security and justice is distinct, yet complementary, and without them, it is not possible to achieve development in the energy sector. An approach solely concentrated on reducing levels of carbon emissions can result in disparities in the social context.
2. The development of renewable energy sources promotes energy independence, as it diminishes the use of fossil fuels, but it also faces challenges, such as the instability associated with the production of these sources, as well as creating a dependency on important minerals.
3. In view of the significance of EV development in reducing greenhouse gas emissions and balancing the energy mix, a lack of emphasis on supporting policies could lead to inequality in the use of the technology, strain on the electricity distribution grid, and challenges associated with mining materials, among others.
4. Smart homes can assist in utilizing energy well and even conserve it, but if people don't have access to the internet and funds, then smart homes can increase gaps in both social and technological areas.

In essence, to achieve and maintain a low-carbon society, it is necessary to consider the three main ideas of energy security, energy justice, and sustainability within the context of decision-making.

References

1 M. Kabir et al., “Climate change due to increasing concentration of carbon dioxide and its impacts on environment in 21st century: A mini review,” *Journal of King Saud University-Science*, vol. 35, no. 5, p. 102693, 2023.

2 IEA (2025). CO_2 Emissions—Global Energy Review 2025—Analysis. International Energy Agency. https://www.iea.org/reports/global-energy-review-2025/co2-emissions

3 H. Ritchie, “How have the world’s energy sources changed over the last two centuries?,” *Our World in Data*, 2021. [Online]. Available: https://ourworldindata.org/global-energy-200-years

4 G. P. Peters et al., “Key indicators to track current progress and future ambition of the Paris Agreement,” *Nature Climate Change*, vol. 7, no. 2, pp. 118–122, 2017.

5 IEA, “Data and statistics,” 2025. [Online]. Available: https://www.iea.org/data-and-statistics

6 J. Strojny, A. Krakowiak-Bal, J. Knaga, and P. Kacorzyk, “Energy security: A conceptual overview,” *Energies*, vol. 16, no. 13, p. 5042, 2023.

7 B. W. Ang, W. L. Choong, and T. S. Ng, “Energy security: Definitions, dimensions and indexes,” *Renewable and Sustainable Energy Reviews*, vol. 42, pp. 1077–1093, 2015.

8 L. Gitelman, M. Kozhevnikov, and Y. Visotskaya, “Diversification as a method of ensuring the sustainability of energy supply within the energy transition,” *Resources*, vol. 12, no. 2, p. 19, 2023.

9 A. Azzuni and C. Breyer, “Definitions and dimensions of energy security: A literature review,” *Wiley Interdisciplinary Reviews: Energy and Environment*, vol. 7, no. 1, p. e268, 2018.

10 C. K. G. Ang and N. P. Utomo, “Cyber security in the energy world,” in *2017 Asian Conference on Energy, Power and Transportation Electrification (ACEPT)*, IEEE, 2017, pp. 1–5.

11 M. Radovanović, S. Filipović, and D. Pavlović, “Energy prices and energy security in the European Union: Panel data analysis,” *Energy Sources, Part B: Economics, Planning, and Policy*, vol. 12, no. 5, pp. 415–419, 2017.

12 M. González-Eguino, “Energy poverty: An overview,” *Renewable and Sustainable Energy Reviews*, vol. 47, pp. 377–385, 2015.

13 W. Shittu, F. F. Adedoyin, M. I. Shah, and H. O. Musibau, “An investigation of the nexus between natural resources, environmental performance, energy security and environmental degradation: Evidence from Asia,” *Resources Policy*, vol. 73, p. 102227, 2021.

14 J. Augutis, R. Krikštolaitis, L. Martišauskas, S. Urbonienė, R. Urbonas, and A. B. Ušpurienė, “Analysis of energy security level in the Baltic States based on indicator approach,” *Energy*, vol. 199, p. 117427, 2020.

15 B. K. Sovacool and M. H. Dworkin, "Energy justice: Conceptual insights and practical applications," *Applied Energy*, vol. 142, pp. 435–444, 2015.

16 K. Jenkins, D. McCauley, R. Heffron, H. Stephan, and R. Rehner, "Energy justice: A conceptual review," *Energy Research & Social Science*, vol. 11, pp. 174–182, 2016.

17 G. Siciliano, L. Wallbott, F. Urban, A. N. Dang, and M. Lederer, "Low-carbon energy, sustainable development, and justice: Towards a just energy transition for the society and the environment," *Sustainable Development*, vol. 29, no. 6, pp. 1049–1061, 2021.

18 IEA, "Renewables," 2025. [Online]. Available: https://www.iea.org/energy-system/renewables

19 K. Khan, A. Khurshid, J. Cifuentes-Faura, and D. Xianjun, "Does renewable energy development enhance energy security?," *Utilities Policy*, vol. 87, p. 101725, 2024.

20 S. Chen, P. Liu, and Z. Li, "Low carbon transition pathway of power sector with high penetration of renewable energy," *Renewable and Sustainable Energy Reviews*, vol. 130, p. 109985, 2020.

21 V. Mladenov, V. Chobanov, and A. Georgiev, "Impact of renewable energy sources on power system flexibility requirements," *Energies*, vol. 14, no. 10, p. 2813, 2021.

22 IEA, "Clean energy supply chains vulnerabilities—Energy technology perspectives 2023," 2023. [Online]. Available: https://www.iea.org/reports/energy-technology-perspectives-2023/clean-energy-supply-chains-vulnerabilities

23 S.-e.-h. Soomro et al., "How does the climate change effect on hydropower potential, freshwater fisheries, and hydrological response of snow on water availability?," *Applied Water Science*, vol. 14, no. 4, p. 65, 2024.

24 K. J. Shapna, J. Li, and M. L. Hossain, "Solar energy integration in off-grid communities: Empowering remote areas in Bangladesh for sustainable development," *Discover Sustainability*, vol. 6, no. 1, p. 371, 2025.

25 S. Berkouwer and J. Dean, *Barriers to Energy Efficiency Adoption in Low-Income Communities*. Kleinman Center for Energy Policy, 2021.

26 J. Lemaire, R. Castro, and F. Montemor, "Empowering remote and off-grid renewable energy communities: Case studies in Congo, Australia, and Canada," *Energies*, vol. 17, no. 19, p. 4848, 2024.

27 A. Albatayneh, "The energy-food dilemma for utilizing biofuels in low-income communities amidst the Russian–Ukrainian conflict," *Energy Exploration & Exploitation*, vol. 41, no. 6, pp. 1942–1955, 2023.

28 X. Zhao et al., "Wind farms reduce grassland plant community diversity and lead to plant community convergence," *BMC Ecology and Evolution*, vol. 25, no. 1, p. 10, 2025.

29 M. Thaker, A. Zambre, and H. Bhosale, "Wind farms have cascading impacts on ecosystems across trophic levels," *Nature Ecology & Evolution*, vol. 2, no. 12, pp. 1854–1858, 2018.

30 V. Kati, C. Kassara, Z. Vrontisi, and A. Moustakas, "The biodiversity–wind energy–land use nexus in a global biodiversity hotspot," *Science of the Total Environment*, vol. 768, p. 144471, 2021.

31 J. Guan and H. Zepp, "Factors affecting the community acceptance of onshore wind farms: A case study of the Zhongying Wind Farm in Eastern China," *Sustainability*, vol. 12, no. 17, p. 6894, 2020.

32 A. Boretti, "Enhanced geothermal systems: Potential, challenges, and a realistic path to integration in a sustainable energy future," *Next Energy*, vol. 8, p. 100332, 2025.

33 Q.-L. Jing, H.-Z. Liu, W.-Q. Yu, and X. He, "The impact of public transportation on carbon emissions—from the perspective of energy consumption," *Sustainability*, vol. 14, no. 10, p. 6248, 2022.

34 A. Fleck, "Estimated share of CO_2 emissions in the transportation sector (cars cause biggest share of transportation CO_2 emissions)," *Statista*, 2024. [Online]. Available: https://www.statista.com/chart/30890/estimated-share-of-co2-emissions-in-the-transportation-sector/

35 International Energy Agency and H. Ritchie, "Share of new cars sold that are electric [dataset]," *Our World in Data*, 2025. [Online]. Available: https://ourworldindata.org/grapher/electric-car-sales-share

36 W. Achariyaviriya, W. Wongsapai, D. Rinchumphu, N. Tippayawong, K. Y. Tippayawong, and P. Suttakul, "A comparative study of vehicle powertrain efficiency: Data-driven analyzing energy consumption and environmental impact," *Transportation Engineering*, vol. 18, p. 100286, 2024.

37 D. Vásquez-Cardona, S. D. Saldarriaga-Zuluaga, S. Bustamante-Mesa, J. M. López-Lezama, and N. Muñoz-Galeano, "Impacts of electric vehicle penetration on the frequency stability of Curaçao's power network," *World Electric Vehicle Journal*, vol. 16, no. 5, p. 264, 2025.

38 V. O. Vega-Muratalla, C. Ramírez-Márquez, L. F. Lira-Barragán, and J. M. Ponce-Ortega, "Review of lithium as a strategic resource for electric vehicle battery production: Availability, extraction, and future prospects," *Resources*, vol. 13, no. 11, p. 148, 2024.

39 M. R. H. Mojumder, F. Ahmed Antara, M. Hasanuzzaman, B. Alamri, and M. Alsharef, "Electric vehicle-to-grid (V2G) technologies: Impact on the power grid and battery," *Sustainability*, vol. 14, no. 21, p. 13856, 2022.

40 H. van den Brink and P. Broos, "Cyber security challenges in the electric vehicle infrastructure," in *IET Conference Proceedings CP798*, vol. 2022, no. 3, IET, 2022, pp. 429–432.

41 S. J. Choi, J. Jiao, and T. Mendez, "Who owns electric vehicles (EVs)? The relationship between EV adoption and socio-demographic characteristics across different price segments and brands in the Texas triangle," *Research in Transportation Business & Management*, vol. 57, p. 101225, 2024.

42 M. Malpede (2025). The Dark Side of Batteries: Child Labor and Cobalt Mining in the Democratic Republic of Congo. GREEN Working papers 22, GREEN, Centre for Research on Geography, Resources, Environment, Energy & Networks, Universita' Bocconi, Milano, Italy.

43 F. G. Boateng and J. M. Klopp, "The electric vehicle transition: A blessing or a curse for improving extractive industries and mineral supply chains?," *Energy Research & Social Science*, vol. 113, p. 103541, 2024.

44 T. Cotterman, E. R. Fuchs, K. S. Whitefoot, and C. Combemale, "The transition to electrified vehicles: Evaluating the labor demand of manufacturing conventional versus battery electric vehicle powertrains," *Energy Policy*, vol. 188, p. 114064, 2024.

45 Y. Han, X. Du, H. Zhang, J. Ni, and F. Fan, "Does smart home adoption reduce household electricity-related CO_2 emissions? Evidence from Hangzhou city, China," *Energy*, vol. 289, p. 129890, 2024.

46 M. Zehnder, H. Wache, H.-F. Witschel, D. Zanatta, and M. Rodriguez, "Energy saving in smart homes based on consumer behavior: A case study," in *2015 IEEE First International Smart Cities Conference (ISC2)*, IEEE, 2015, pp. 1–6.

47 F. AlFaris, A. Juaidi, and F. Manzano-Agugliaro, "Intelligent homes' technologies to optimize the energy performance for the net zero energy home," *Energy and Buildings*, vol. 153, pp. 262–274, 2017.

48 T. M. Skjølsvold and M. Ryghaug, "Embedding smart energy technology in built environments: A comparative study of four smart grid demonstration projects," *Indoor and Built Environment*, vol. 24, no. 7, pp. 878–890, 2015.

49 U. U. Rehman, P. Faria, L. Gomes, and Z. Vale, "Future of energy management models in smart homes: A systematic literature review of research trends, gaps, and future directions," *Process Integration and Optimization for Sustainability*, vol. 9, pp. 1–30, 2025.

50 B. Lashkari, Y. Chen, and P. Musilek, "Energy management for smart homes—state of the art," *Applied Sciences*, vol. 9, no. 17, p. 3459, 2019.

51 R. Jones, "Almost a third of household smart meters not working properly, says Citizens Advice," *The Guardian*, 2024. [Online]. Available: https://www.theguardian.com/business/article/2024/may/23/almost-a-third-of-household-smart-meters-not-working-properly-says-citizens-advice

52 E. Allameh, M. Heidari, and B. de Vries, "Living preference modeling of smart homes for different target groups," *Journal of Ambient Intelligence and Smart Environments*, vol. 10, no. 2, pp. 103–125, 2018.

6

Exploring Energy Storage Pathways for the Climate–Energy–Sustainability Nexus in Deep Decarbonization of Modern Energy Networks

Prince Aduama[1], Vikash Kumar Saini[1], Ameena S. Al-Sumaiti[1], and Rajesh Kumar[2,3]*

[1] *Smart Operation Research (OR) Lab, Advanced Power and Energy Center, Department of Electrical Engineering, Khalifa University, Abu Dhabi, UAE*
[2] *RAMAN Lab, Department of Electrical Engineering, Malaviya National Institute of Technology, Jaipur, India*
[3] *Department of Human Anatomy and Physiology, Faculty of Health Sciences, University of Johannesburg, Johannesburg, South Africa*

*Corresponding author: 100060989@ku.ac.ae

Nomenclature ***Symbols*** ***Sets/indices***	
$t \in \mathscr{T}$	Time steps
$s \in \mathscr{S}$	Storage technologies
$r \in \mathscr{R}$	Renewable technologies
Parameters (given)	
D_t (kW)	Electrical demand at time t
$\widehat{G}_{r,t}$ (kW)	Available renewable from tech r
$c_{\text{grid},t}$ (\$/kWh)	Grid import tariff
$p_{\text{sell},t}$ (\$/kWh)	Export tariff
$\eta_{\text{ch},s}, \eta_{\text{dis},s} \in (0,1)$	Charge/discharge efficiencies
$P_{\text{max,ch},s}, P_{\text{max,dis},s}$ (kW)	Charge/discharge power limits
$\text{SoC}_{\min,s}, \text{SoC}_{\max,s} \in (0,1)$	Min/max state of charge fractions
EF_{grid} (kgCO_2/kWh)	Grid emissions factor

Nomenclature ***Symbols*** ***Sets/indices***	
π_{CO_2} (\$/tCO$_2$)	Carbon price
Δt (h)	Time-step length
c_s^{E} (\$/kWh)	Installed energy capacity cost for storage system, *s*
c_s^{P} (\$/kW)	Installed power capacity cost for storage system, *s*
ε	Emission cap/constraint
$\overline{P_{ch,s}^{max}}, \overline{P_{dis,s}^{max}}$ (kW)	Upper bounds for charging and discharge power capacity sizing for storage *s* respectively
$\overline{E_s^{max}}$ (kWh)	Upper bounds for energy capacity sizing for storage *s* respectively
Decision variables	
$P_{imp,t}, P_{exp,t}$ (kW)	Grid import/export
$G_{r,t}$ (kW)	Renewable utilized from tech *r*
$C_{r,t}$ (kW)	Renewable curtailment
$P_{ch,s,t}, P_{dis,s,t}$ (kW)	Storage charge/discharge at time *t*
$E_{s,t}$ (kWh)	Storage SoC at time *t*
$L_{shed,t}$ (kW)	Load shedding, penalized in objective
P_s^{max} (kW)	Sized power capacity for storage system, *s*
E_s^{max} (kWh)	Energy capacity for storage system, *s*
E_{CO_2} (tCO$_2$)	Amount of CO_2 emissions abated

6.1 Introduction

Due to the world's effort to reduce carbon emissions and combat climate change, modern energy systems are undergoing decarbonization through the replacement of traditional fossil fuel-based power plants. Fossil fuel-based plants, both short-term and long-term, degrade air quality and the environment [1]. Renewable energy (RE)-based power grids on the other hand, have a limited environmental impact, offering the capability to reduce carbon footprints, ultimately combating climate change.

As a result, fast expansion of RE in modern power networks, especially solar and wind penetration, has become key in the efforts to decarbonize modern energy systems [2]. In [3], a microgrid configuration that provides a cost-optimal configuration of the microgrid, which can lead to a 58–82% reduction in CO_2 emissions is proposed. Furthermore, in [4], a novel techno-economic model is developed for microgrid energy supply considering environmental issues. The model includes a transactive energy strategy that aims at co-optimizing significant conflicting services. Papageorgiou et al. [5] focus on the climate change impact assessment of a microgrid after integrating a solar PV system. In addition, the study assessed the life-cycle climate change impact of microgrids and the influence of the variation in the electricity supply. Moreover, in [6], a sustainable-based integrated microgrid configuration was developed for energy transformation. Additionally, an urban microgrid architecture is proposed in [7]. In the proposed architecture, regulatory policies are used to improve the resilience of the energy grid and mitigate the effects of climate change. In [8], a microgrid structure and hierarchical dispatch optimization are proposed to improve climate concerns. This study also focuses on RE and load uncertainty modeling.

In power systems, it is critical to ensure balanced power flow between generation and consumer nodes, power quality, and reliability within the system [9]. However, the increased penetration of RE systems comes with some problems for the power grid's reliability and stability due to the uncertain nature of RE resources.

Energy storage (ES) systems offer several benefits to the grid, including energy shifting, peak demand saving, and ancillary services, which mitigate the risk from variable generation of RE. As a result, power grids can achieve enhanced decarbonization through the effective deployment of ES technologies. As a result, ES systems are key components of the modern energy system. It stores the excess energy and releases the energy in peak load duration to minimize the energy cost [10]. They are important in matching increasing load demands and ensuring sustainable, cost-effective, and reliable power system operations.

Storage technologies such as Li-ion, redox flow battery, pump hydro storage, and other storage technologies have been explored in the literature for decarbonizing the power grid [11, 12]. In [13], six criteria, including cost, efficiency, power quality, load management, lifespan, and technical maturity, are discussed in the choice of the appropriate storage system. Additionally, this study focused on developing multi-decision criteria for sustainability through various storage technologies. The work presented in [14] focuses on the contribution of Li-ion and lead-acid batteries in sustainability analysis for grid-level ES systems utilizing a cradle-to-grave approach incorporating end-of-life, operational environment, and manufacturing specifications. The environmental impact assessments are particulate matter, climate change, minerals, resource use, and acidification potential. Furthermore, the study in [15] conducts an environmental impact assessment of Li-ion batteries for electric vehicle applications. Fouzia et al [16] conducted a comparative study of Li-ion and lead-acid batteries for the decarbonized energy grid power supply. In this system, the PV is directly integrated with the battery system, and an energy management strategy is implemented to optimize the power flow in ES. Furthermore, hybrid ES technology is implemented with a standalone microgrid to demonstrate the potential savings and optimized operations under different climate conditions [17]. In addition, the pumped hydro energy storage (PHS), a mechanical type of ES, is the largest form of ES system available today, accounting for more than 90% of global ES according to IEA [18]. In PHS systems, energy is stored as the potential energy of water at an elevation. This is achieved by pumping water from a lower elevation to a higher elevation during off-peak periods with excess electricity, generating carbon emission-free electricity. The efficiency of this type of plant depends on the construction status, life, and location [19]. Gravity storage technology is an emerging technology with immense potential for grid storage and, therefore, aids in decarbonization efforts. This technology employs a weight-based approach in which lifting a mass releases energy that has been stored. This approach is analogous to PHS systems, where the weight in PHS is water whereas the weights in gravity storage are materials such as rocks [20]. During operation, electric motors lift a heavy mass to a higher elevation, storing energy as gravitational potential energy. When energy is needed, the mass is lowered, and the potential energy is converted back into electrical energy through generators [21].

Recent advances in marginal abatement cost analysis have highlighted the complexity of achieving deep decarbonization in energy systems. Studies using energy system optimization models show that net-zero CO_2 emissions can be achieved at costs under \$400 per tCO_2-eq, but with exponentially increasing marginal costs for final emissions reductions [22]. As shown by Farbes et al. [23], traditional approaches fail to address the coordinated deployment needed for net-zero targets. As a result, new methodologies should emphasize capturing critical interactions between decarbonization measures and across sectors. Furthermore, analysis of storage technology deployment patterns indicates that multiple technologies work complementarily rather than competitively to improve system flexibility [24]. Johnson [25] demonstrates that no single mitigation strategy is sufficient for deep decarbonization, requiring portfolio approaches that account for technology interactions and regional characteristics.

Despite extensive research on ES integration in RE systems, several gaps remain in understanding storage-enabled deep decarbonization. Most studies focus on the benefits of storage deployment without quantifying the marginal costs of emissions reduction in high-renewable systems. Previous research typically evaluates storage technologies independently rather than analyzing their complementary deployment patterns for emissions reduction. Additionally, the literature lacks a comparative analysis of how the characteristics of renewable resources influence carbon price thresholds for storage deployment. Finally, limited research quantifies how marginal abatement costs escalate as systems approach near-zero operational emissions.

This chapter addresses these gaps by providing a comprehensive analysis of storage deployment patterns, technology complementarity, and regional differences in carbon pricing effectiveness for deep decarbonization of high-renewable power systems.

The main objectives and contributions of this work are:

1. Quantification of the operational emissions reduction potential of ES systems in grids with existing high renewable penetration, and examining whether storage deployment can meaningfully reduce emissions from remaining fossil generation.
2. This chapter identifies technology-specific deployment thresholds and regional factors that determine storage economics across different renewable resource mixes and system configurations.
3. In addition, the developed optimization model characterizes the complementary roles of multiple storage technologies and establishes the economic boundaries where marginal decarbonization costs become prohibitive. Through optimization of representative systems in Australia and Germany, this work details the capabilities and limitations of storage-enabled decarbonization strategies.

6.2 Problem Development

The developed system consists of a grid-connected prosumer system with two variable RE sources (wind and PV) and three ES systems (PHS, Li-ion, and gravity). The study adopts an energy balance approach that excludes explicit modeling of the transmission and distribution network constraints, such as line capacities and power flow equations. The rationale for this is outlined below:

1. The primary objective of the work is to analyze the role of ES systems in emission reduction through optimal dispatch rather than network planning.
2. In addition, the model represents aggregated regional systems (Germany and Australia) where network constraints are implicit in the solar, wind, and load profiles.

The schematic of the system is shown in Figure 6.1. The optimization model and framework are discussed in the next section.

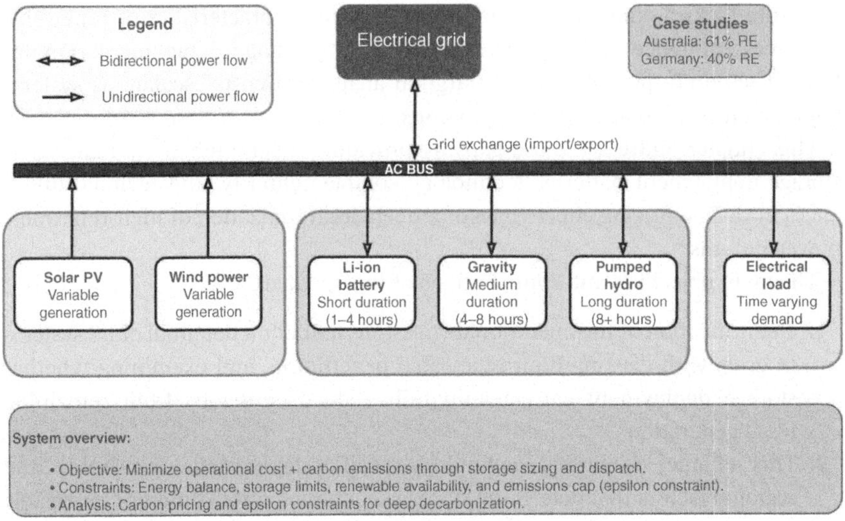

Figure 6.1 Schematic diagram of the grid-connected prosumer system with renewable generation, ES technologies, and load demand used in the optimization model.

6.3 Modeling Framework and Constraints

6.3.1 Objective Function

The objective of the optimization problem is to minimize total operating cost, while appropriately sizing relevant ES systems to ensure a reduction in CO_2 emissions. The operational role of the storage systems in the optimization problem aids in analyzing the roles they can play towards the deep decarbonization of high renewable penetration systems.

The objective function is given by Eq. (6.1),

$$\begin{aligned}\min J = & \underbrace{\sum_{t\in T}(c_{\text{grid},t}P_{\text{imp},t} - p_{\text{sell},t}P_{\text{exp},t})\Delta t}_{\text{Energy cost}} + \underbrace{\sum_{s\in S}\sum_{t\in T}C_{\text{cyc},s}(P_{\text{ch},s,t} + P_{\text{dis},s,t})\Delta t}_{\text{Storage cycling cost}} \\ & + \underbrace{\sum_{t\in T}c_{\text{shed}}L_{\text{shed},t}\Delta t}_{\text{Unserved energy penalty}} + \underbrace{\sum_{s\in S}(c_s^{\text{P,wk}}(P_{\text{ch},s}^{\max} + P_{\text{dis},s}^{\max}) + c_s^{\text{E,wk}}E_s^{\max})}_{\text{Capacity (power \& energy) cost, weekly}} \\ & + \underbrace{\sum_{s\in S}\sum_{t\in T}C_{\text{simul},s}P_{\text{simul},s,t}\Delta t}_{\text{Simultaneity charge/discharge penalty}} + \underbrace{\pi_{CO_2}\frac{E_{CO_2}}{1000}}_{\text{Carbon price (\$/tCO}_2)}\end{aligned} \tag{6.1}$$

In Eq. (6.1), the weekly cost of storage power and energy is given by $c_s^{P,\text{wk}} = \frac{c_s^P \text{CRF}(r, n_s)}{52}$, and $c_s^{E,\text{wk}} = \frac{c_s^E \text{CRF}(r, n_s)}{52}$, respectively, where $\text{CRF}(r, n) = \frac{r(1+r)^n}{(1+r)^n - 1}$ is the capital recovery factor and c_s^P and c_s^E are the installed capacity costs of the ES systems in \$/kW and \$/kWh, respectively. To ensure reliability of the system, a penalty is put on load shedding, represented by the third term of the objective function. Furthermore, a penalty is placed on the simultaneous charging and discharging of the ES systems. The carbon price term of the objective function penalizes emissions.

6.3.2 Constraints

Energy balance: The energy balance constraint is given by Eq. (6.2). This ensures that the power inflows are equivalent to the power outflows,

$$\sum_r G_{r,t} + \sum_s P_{\text{dis},s,t} + P_{\text{imp},t} = D_t - L_{\text{shed},t} + \sum_s P_{\text{ch},s,t} + P_{\text{exp},t} \tag{6.2}$$

Renewable contribution and curtailment: Equation (6.3) indicates that more than available RE cannot be used. In addition, Eq. (6.4) expresses the curtailment. This quantifies the "spilled" variable RE when supply exceeds what can be absorbed by load, storage, or exports,

$$0 \le G_{r,t} \le \widehat{G}_{r,t} \tag{6.3}$$

$$C_{r,t} = \widehat{G}_{r,t} - G_{r,t}, \quad \text{with } C_{r,t} \ge 0 \tag{6.4}$$

Sizing bounds and operating links: The sizing bounds of the optimization are shown in Eqs. (6.5)–(6.9),

$$0 \le P_{\text{ch},s}^{\text{max}} \le \overline{P}_{\text{ch},s}^{\text{max}}, \quad \forall s \in S \tag{6.5}$$

$$0 \le P_{\text{dis},s}^{\text{max}} \le \overline{P}_{\text{dis},s}^{\text{max}}, \quad \forall s \in S \tag{6.6}$$

$$0 \le E_s^{\text{max}} \le \overline{E}_s^{\text{max}}, \quad \forall s \in S \tag{6.7}$$

$$0 \le P_{\text{ch},s,t} \le P_{\text{ch},s}^{\text{max}}, \quad \forall s \in S, \ \forall t \in T \tag{6.8}$$

$$0 \le P_{\text{dis},s,t} \le P_{\text{dis},s}^{\text{max}}, \quad \forall s \in S, \ \forall t \in T \tag{6.9}$$

C-rate links: The C-rate couples the storage's power to its energy, allowing for realism, sizing, and a measure of degradation for the storage system. This prevents the optimizer from using small energy with huge power and vice versa, which is significant for battery health. In this work, the concept of C-rate is extended to other storage systems, as the power-to-energy ratio, related below as Eqs. (6.10) and (6.11),

$$P_{\text{ch},s}^{\max} \leq C_s^{\text{ch}} E_s^{\max}, \quad \forall s \in S \tag{6.10}$$

$$P_{\text{dis},s}^{\max} \leq C_s^{\text{dis}} E_s^{\max}, \quad \forall s \in S \tag{6.11}$$

Storage sizing constraints: The initial state of charge is given by Eq. (6.12). In addition, the initial and final state of charge is set cyclic and is represented by Eq. (6.13). Then the state of charge at any time is given by Eqs. (6.14) and (6.15),

$$E_{s,t_0} = \alpha_s^{\text{init}} E_s^{\max}, \quad \forall s \in S \tag{6.12}$$

$$E_{s,t_N} = E_{s,t_0}, \quad \forall s \in S \tag{6.13}$$

$$\text{SoC}_{\min,s}\, E_s^{\max} \leq E_{s,t} \leq \text{SoC}_{\max,s}\, E_s^{\max}, \quad \forall s \in S, \forall t \in T \tag{6.14}$$

$$E_{s,t} = E_{s,t-1} + \eta_{\text{ch},s}\, P_{\text{ch},s,t}\, \Delta t - \frac{1}{\eta_{\text{dis},s}} P_{\text{dis},s,t}\, \Delta t, \quad \forall s \in S, \forall t \in T \setminus \{t_0\} \tag{6.15}$$

Decarbonization: Equation (6.16) defines the decarbonization relation. When the carbon price approach is utilized, then Eq. (6.16) fits directly into the objective function. However, if the ε-cap approach is utilized, then Eq. (6.16) is modeled as a constraint as $E_{CO_2} \leq \varepsilon$. The epsilon-constraint method has been applied in recent studies such as Asim et al. [26],

$$\text{Emissions}\ \left(E_{CO_2}\right) = \sum_t \text{EF}_{\text{grid}} \cdot P_{\text{imp},t} \cdot \Delta t \tag{6.16}$$

To enhance computational efficiency, a representative week selection approach is utilized for this work. This involved selecting a week that most represents a typical day with respect to interactions between the PV, wind energy, and the load data. To construct the weekly representation data for each region, firstly, the features for the PV, wind, and load data are normalized using the *min–max scaler*. Then clusters of weeks are developed using *K*-means, with a single *representative week* being the one closest to the largest cluster's centroid. These are described in the five key steps below:

1. Let $x_t^{(r)} \in \{\text{PV}, \text{Wind}, \text{Load}\}$ denote the hourly power time series (kW) for region r at hour index $t = 0, 1, \ldots, T-1$. Each t is mapped to a day d and hour h via $t = 24d + h, h \in \{0, \ldots, 23\}$. Then the week index is defined as $w = \lfloor d/7 \rfloor$. Let W be the set of all week indices, then $H_w = \{t : \lfloor t/24 \rfloor / 7 = w\}$ is the 168 hourly indices in week w.
2. For each variable $x \in \{\text{PV}, \text{Wind}, \text{Load}\}$ and week $w \in W$, the mean, standard deviation, and maximum values are found using Eqs. (6.17)–(6.19), respectively:

$$\mu_{x,w} = \frac{1}{|H_w|} \sum_{t \in H_w} x_t \tag{6.17}$$

$$\sigma_{x,w} = \sqrt{\frac{1}{|H_w| - 1} \sum_{t \in H_w} (x_t - \mu_{x,w})^2} \tag{6.18}$$

$$m_{x,w} = \max_{t \in H_w} x_t \tag{6.19}$$

The results from Eqs. (6.17)–(6.19) could be represented as a vector as given below:

$$\mathbf{f}_w = [\mu_{\mathrm{PV},w}, \sigma_{\mathrm{PV},w}, m_{\mathrm{PV},w}, \mu_{\mathrm{Wind},w}, \sigma_{\mathrm{Wind},w}, m_{\mathrm{Wind},w}, \mu_{\mathrm{Load},w}, \sigma_{\mathrm{Load},w}, m_{\mathrm{Load},w}]^{\mathrm{T}} \in \mathbb{R}^9 \tag{6.20}$$

3. Let j index the components of $\mathbf{f}_w$. Each of the components across weeks is normalized using the min–max scaler as Eq. (6.21):

$$\tilde{f}_{w,j} = \frac{f_{w,j} - \min_{u \in W} f_{u,j}}{\max_{u \in W} f_{u,j} - \min_{u \in W} f_{u,j} + \epsilon}, \epsilon > 0 \tag{6.21}$$

where $\tilde{\mathbf{f}}_w$ denotes the normalized vector.

4. Then K-means clustering with $K = 5$ is done. Here, the weeks are partitioned into clusters $C_1, \ldots, C_K$ with assignments $z_w \in \{1, \ldots, K\}$ by minimizing the within-cluster sum of squares given by Eqs. (6.22) and (6.23):

$$\min_{\{z_w\},\{\mathbf{c}_k\}} \sum_{w \in W} \|\tilde{\mathbf{f}}_w - \mathbf{c}_{z_w}\|_2^2 \tag{6.22}$$

$$\mathbf{c}_k = \frac{1}{|C_k|} \sum_{w \in C_k} \tilde{\mathbf{f}}_w,\ C_k = \{w \in W : z_w = k\},\ k = 1, \ldots, K \tag{6.23}$$

5. Finally, the largest cluster and the week closest to its centroid are selected as the most representative week to the region's data for this application, given by Eqs. (6.24) and (6.25):

$$k^{\star} = \arg \max_{k \in \{1, \ldots, K\}} |C_k| \tag{6.24}$$

$$w^{\star} = \arg\min_{w \in C_{k^{\star}}} \|\tilde{\mathbf{f}}_w - \mathbf{c}_{k^{\star}}\|_2 \tag{6.25}$$

6.3.3 Climate Factor Integration and Quantification in Model

In dealing with the CES nexus, climate factors must be incorporated in the model. The model described in Sections 6.3.1 and 6.3.2 achieves this through several pathways:

1. The PV and wind energy profiles ($\widehat{G}_{r,t}$) inherently capture climate-driven variability, which includes the solar irradiance patterns affected by seasonal changes and weather conditions, as well as wind speed variations influenced

by meteorological patterns. In addition, regional climates also influence the renewable resource quality and predictability.

2. The *K*-means clustering approach (Eqs. 6.17–6.25) identifies typical operational weeks that represent climate-influenced generation and demand patterns. This captures seasonal RE generation cycles and climate-driven load demand variations.
3. Moreover, the utilization of data from two regions (Australia and Germany) incorporates distinct climate regimes, with Australia and Germany representing different climates.

6.4 Results and Discussion

PV, wind, and load data from two regions, Australia and Germany, are sourced from Huang [27] and ENTSOE [28], respectively. The Australian data, which has a data resolution of five minutes interval is converted to an equivalent one-hour interval. Similarly, Germany's energy data at 15-minute intervals is converted to an equivalent one-hour-interval PV, load, and wind data. Then the most representative week for these regions is found as described in the previous section. The daily profiles of the most representative weeks for the two regions are presented in Figures 6.2 and 6.3. Figures 6.2 and 6.3 illustrate the temporal correlation between PV generation (blue lines), wind generation (orange lines, with evening peaks), and electrical load demand (green lines).

The total load demands for the two regions are 432,266.97 and 8,309,199.41 MWh for Australia and Germany, respectively. Additionally, the RE penetrations for the same period are 262,546 and 3,347,123.10 MWh, representing 60.7 and 40.2% for Australia and Germany, respectively. These high penetration numbers inherently offset high CO_2 emissions in the original system. The high variability in renewable generation highlights the need for storage systems to manage supply–demand imbalances. The developed model tests how these ES systems could further aid in the deep decarbonization of high RE penetration systems.

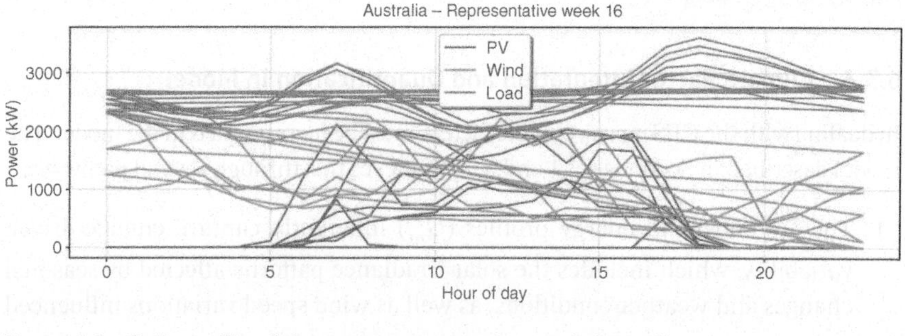

Figure 6.2 Daily profile of the most representative week, Australia.

Two scenarios of decarbonization are analyzed: using the carbon price approach and the ε-cap approach.

The three selected ES systems are based on the applications: battery (Li-ion) for short-duration applications (one to four hours), gravity for medium applications (four to eight hours), and the PHS for longer durations (eight+ hours). To analyze the role of the ES systems, they are optimally sized according to the operation and economics of the test country.

Tables 6.1–6.3 present the simulation parameters utilized in this work.

Global scalars used in the model: load-shedding penalty c_{shed} = \$2,000/MWh; discount rate $r = 0.08$.

6.4.1 Cost–Emissions Trade-offs in Deep Decarbonization

The sizing results for Australia show how ES decarbonizes highly variable renewable energy (VRE) systems. Table 6.4 presents the operational characteristics of the

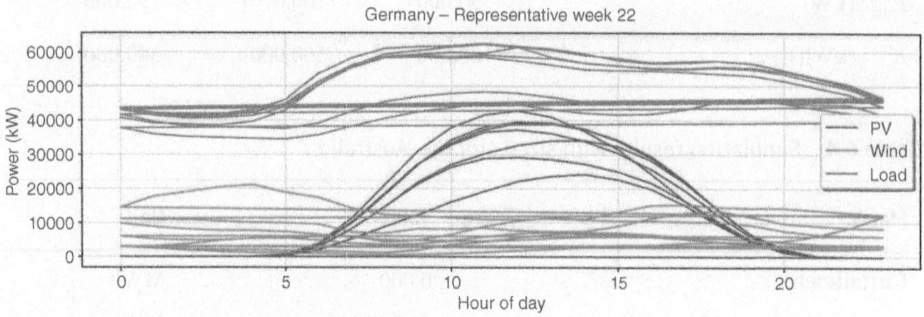

Figure 6.3 Daily profile of the most representative week, Germany.

Table 6.1 Storage operational characteristics.

Parameter	Battery	PHS	Gravity
Charge efficiency (η_{ch})	0.90	0.85	0.85
Discharge efficiency (η_{dis})	0.90	0.90	0.85
Min SoC (SoC_{min})	0.10	0.10	0.10
Max SoC (SoC_{max})	0.90	0.95	0.95
C-rate charge ($C_{rate,ch}$) (1/h)	0.50	0.125	0.25
C-rate discharge ($C_{rate,dis}$) (1/h)	0.50	0.125	0.25
Lifetime (years)	15	35	25

Table 6.2 Cost parameters and economic settings.

Parameter	Battery	PHS	Gravity
Cycling cost, c_{cyc} (\$/kWh)	0.010	0.002	0.003
Simul. ch./dis. penalty, c_{sim} (\$/kW)	1,000	500	750
Power capacity cost, c_{cap}^{P} (\$/kW)	400	1,500	800
Energy capacity cost, c_{cap}^{E} (\$/kWh)	140	50	100

Table 6.3 Upper bounds for sizing decisions.

Decision upper bound	Battery	PHS	Gravity
$\overline{P}_{ch}^{max}$ (kW)	50,000	100,000	75,000
$\overline{P}_{dis}^{max}$ (kW)	50,000	100,000	75,000
E_{s}^{max} (kWh)	100,000	500,000	300,000

Table 6.4 Simulation results with sized storage: Australia.

Metric	Value	Unit
Curtailment	0.000	MWh
Imports (peak 16–22 hours)	86,380.162	MWh
Imports (off-peak)	101,158.109	MWh
Total imports	187,538.271	MWh
Exports	16,332.882	MWh
Peak share of imports	46.06	%

optimally sized storage system for Australia. This table highlights zero curtailment, indicating full renewable utilization, substantial energy exports (16,333 MWh), and a peak-concentrated import pattern where 46% of grid imports occur during evening hours (16–22 hours) when renewable generation is reduced.

In this setting, the storage systems reduce import-weighted CO_2 by reshaping the period when energy is drawn from the grid rather than by soaking up spill.

In terms of cost, Table 6.5 indicates that the weekly system cost is largely driven by net energy purchases. The weekly CAPEX, although noteworthy, is modest relative to energy cost.

Furthermore, this table provides an economic breakdown showing that energy purchases dominate system costs ($40.7 million) with export revenues providing a partial offset ($0.85 million). Storage-related costs (cycling and CAPEX) represent less than 1% of total system costs, indicating that operational savings drive storage deployment rather than capital considerations. Table 6.6 details the optimal storage sizing results for the Australia case, showing no battery deployment under pure economic optimization, substantial PHS capacity (100 MW charge and 54 MW discharge), and moderate gravity storage deployment. The capacity factors (22–30%) indicate storage systems operate primarily during critical periods rather than continuous cycling.

The ε-cap frontier reveals the role that ES can play in reducing operational emissions from high-renewable power systems. As described, the base systems have approximated 61% renewable penetration in Australia and 40% in Germany. The simulation and analysis examine how ES deployment enables further emissions cuts through optimized system operation. Figure 6.4 illustrates the epsilon-constraint frontier for Australia, showing (1) the relationship between

Table 6.5 Cost decomposition for sizing simulation: Australia.

Component	Value
Energy purchases	$40,667,301.89
Export revenue	$849,309.86
Net energy cost	$39,817,992.03
Cycle wear cost	$25,145.93
Weekly CAPEX (sum)	$379,053.79

Table 6.6 Storage sizing and utilization in the sizing run: Australia.

Parameter	Battery	PHS	Gravity
P_{ch}^{max} (MW)	0.000	100.000	56.703
P_{dis}^{max} (MW)	0.000	54.000	38.250
E_{ch} (MWh)	0.000	3,695.490	2,306.080
E_{dis} (MWh)	0.000	2,714.753	1,802.401
CF_{ch} (%)	0.000	21.997	24.208
CF_{dis} (%)	0.000	29.925	28.049

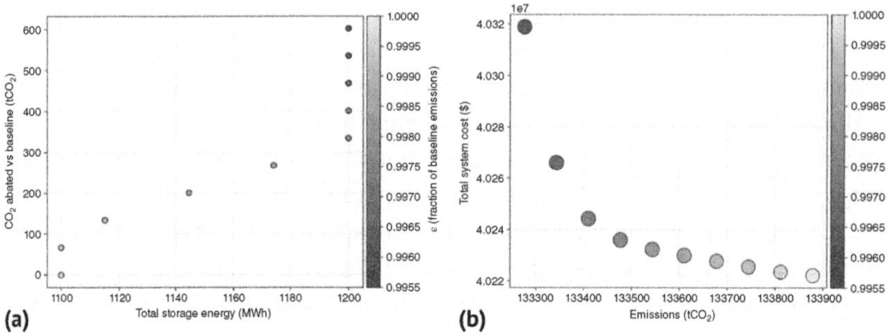

Figure 6.4 Carbon emission abatement with ε-cap approach for Australia: (a) abatement vs. storage; (b) abatement vs. total cost.

Table 6.7 Epsilon-constraint sweep results for Australia.

ε	Cost ($ million)	Emissions (tCO_2)	Curtail (MWh)	Battery (MWh)	PHS (MWh)	Gravity (MWh)
1.000000	1,902.931	1,793,207	0.0	0.0	0.0	0.0
0.999969	1,902.942	1,793,151	0.0	0.0	0.0	115.9
0.999938	1,902.952	1,793,096	0.0	0.0	0.0	231.8
0.999907	1,902.968	1,793,040	0.0	27.8	0.0	300.0
0.999876	1,902.991	1,792,985	0.0	95.2	0.0	300.0
0.999845	1,903.019	1,792,929	0.0	100.0	77.4	300.0
0.999814	1,903.049	1,792,874	0.0	100.0	297.6	300.0
0.999783	1,903.079	1,792,818	0.0	100.0	517.7	300.0
0.999752	1,903.109	1,792,763	0.0	100.0	737.9	300.0
0.999722	1,903.201	1,792,707	0.0	100.0	800.0	300.0

total storage energy capacity and emissions abatement, and (2) the cost–abatement trade-off. Furthermore, the results in Table 6.7 demonstrate that operational emissions can be reduced from 1,793,207 tCO_2 at the baseline to 1,792,707 tCO_2 through strategic storage deployment. It is realized that sequential storage deployment occurs as emissions constraints tighten with gravity storage appearing first, followed by Li-ion battery and PHS systems. The cost escalation accelerates as the epsilon constraint approaches 99.97% emissions reduction, demonstrating exponentially increasing marginal abatement costs. Figure 6.5 shows Germany's epsilon-constraint frontier with total system cost vs. emissions levels. The narrow emissions reduction range and steep cost escalation demonstrate the economic

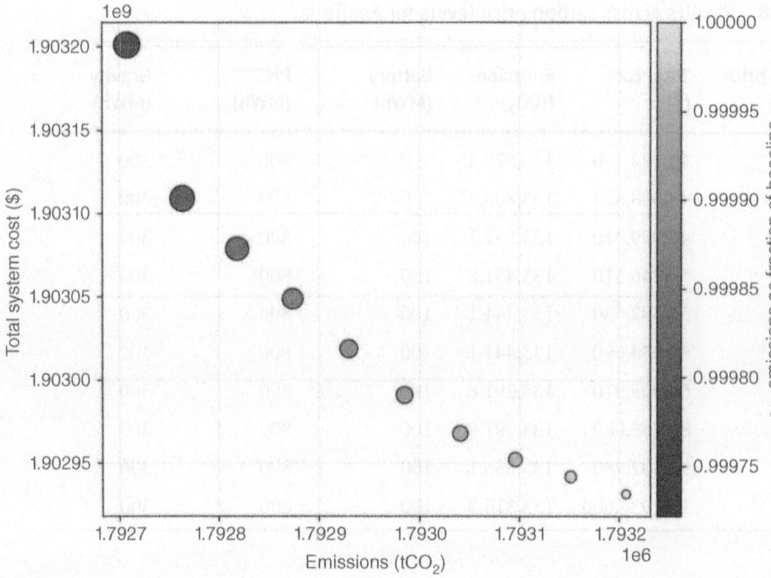

Figure 6.5 Carbon emission abatement with ε-cap approach: Germany.

challenge of storage-enabled decarbonization in Germany's system, with high marginal abatement costs for final emissions reductions.

6.4.2 Carbon Pricing and Storage-enabled Operational Optimization

Carbon pricing is the second approach analyzed for cutting emissions. The carbon pricing analysis reveals how different price levels incentivize storage deployment to optimize system operation and reduce emissions.

From Table 6.8 and Figure 6.6 for Australia's 61% renewable system, a carbon price of $\$50/tCO_2$ triggers additional storage (battery) deployment (100 MWh), enabling the system to reduce operational emissions from 133,800.4 to 133,552.1 tCO_2. This reduction is achieved through improved dispatch and storage-enabled load shifting. As carbon prices increase to $\$500/tCO_2$, operational emissions further decrease to 133,316.2 tCO_2, representing a total 0.42% reduction without load shedding. Notably, PHS capacity remains constant at 800 MWh and gravity at 300 MWh across all price levels, suggesting these technologies provide essential grid services based on the operating cycles regardless of carbon pricing. Moreover, from Table 6.9 and Figure 6.7 for Germany's 40% VRE system, substantially higher carbon prices are required to justify the deployment of ES systems due to the nature and profiles of the load, PV, and wind powers. Gravity storage first

Table 6.8 Results across carbon price levels for Australia.

Carbon price (tCO_2$)	Total cost ($)	Emissions (tCO_2)	Battery (MWh)	PHS (MWh)	Gravity (MWh)
0	40,222,190	133,878.3	0	800	300
25	43,568,820	133,800.4	0	800	300
50	46,909,410	133,552.1	100	800	300
75	50,246,510	133,451.8	100	800	300
100	53,582,630	133,441.1	100	800	300
150	60,254,690	133,441.1	100	800	300
200	66,925,910	133,391.8	100	800	300
300	80,265,090	133,391.8	100	800	300
400	93,602,780	133,356.1	100	800	300
500	106,936,600	133,316.2	100	800	300

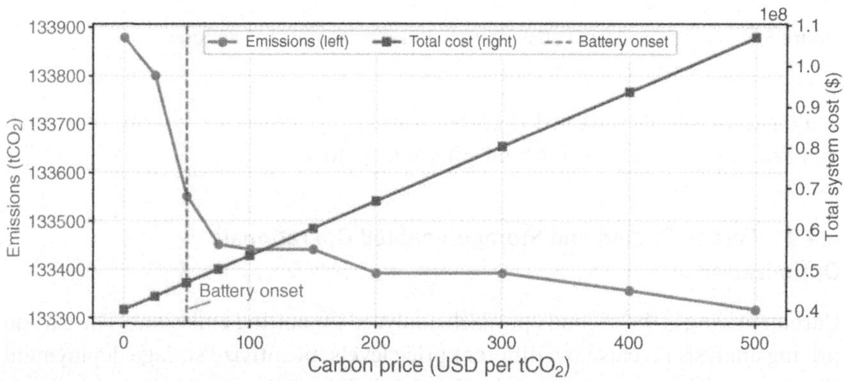

Figure 6.6 Carbon emission abatement with carbon cost approach: Australia.

deploys at $200/tCO_2$, enabling minor operational improvements, while battery storage requires $500/tCO_2$ to become economically viable for reducing dispatch emissions. Even at this high carbon price, operational emissions decrease by only 0.014% (from 1,793,207 to 1,792,949 tCO_2), while system costs increase by 47% to $2.799 billion. The optimization of ES utilization and management has been shown to improve grid reliability and system economy [29], though these benefits must be weighed against the substantial carbon pricing thresholds required for deployment in different regional contexts.

Table 6.9 Results across carbon price levels for Germany.

Carbon price (tCO_2$)	Total cost ($ million)	Emissions (tCO$_2$)	Battery (MWh)	PHS (MWh)	Gravity (MWh)
0	1,902.931	1,793,207	0	0	0
25	1,947.762	1,793,207	0	0	0
50	1,992.592	1,793,207	0	0	0
75	2,037.422	1,793,207	0	0	0
100	2,082.252	1,793,207	0	0	0
150	2,171.912	1,793,207	0	0	0
200	2,261.571	1,793,063	0	0	300
300	2,440.877	1,793,063	0	0	300
400	2,620.184	1,793,063	0	0	300
500	2,799.483	1,792,949	100	0	300

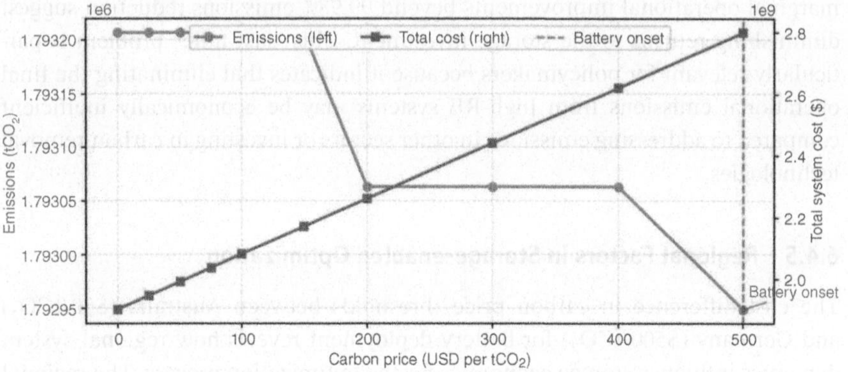

Figure 6.7 Carbon emission abatement with carbon cost approach: Germany.

6.4.3 Storage Utilization Patterns in High-renewable Operation

The results further reveal how different storage technologies enable operational flexibility in systems already dominated by variable renewable generation. In the case of Australia, PHS provides 3,695 MWh of energy capacity operating at 22% CF, while gravity storage offers 2,306 MWh at 24.2% CF. These moderate utilization rates indicate that storage systems primarily manage critical operational periods such as evening peaks after PV ceases or during periods of extended wind lulls, rather than continuous cycling operations. The operation of the overall system shows complementary behaviors before the storage systems: the PHS handles bulk energy shifting across relatively longer hours (eight+ hours), gravity storage provides intermediate-duration services (four to eight hours), and batteries (when deployed) manage short-duration (one to four hours) imbalances. This complementary operation is essential for managing the complex operational challenges of high-renewable systems where generation variability must be balanced against inflexible baseload constraints and demand patterns.

6.4.4 The Operational Emissions Challenge in High-renewable Systems

Despite Australia's 61% and Germany's 40% renewable generation, the remaining fossil generation creates substantial emissions during periods of renewable scarcity. The modest emissions reductions achieved through storage deployment (0.42% in Australia and 0.014% in Germany) highlight that operational optimization alone may struggle to achieve substantial deep decarbonization changes to eliminate the fossil capacity entirely. The exponential cost increases for marginal operational improvements beyond 99.98% emissions reduction suggest diminishing returns to the storage investment. This "last mile" problem is particularly relevant for policymakers because it indicates that eliminating the final operational emissions from high-RE systems may be economically inefficient compared to addressing emissions in other sectors or investing in carbon removal technologies.

6.4.5 Regional Factors in Storage-enabled Optimization

The cost difference in carbon price thresholds between Australia (\$50/tCO_2) and Germany (\$500/tCO_2) for battery deployment reveals how regional system dynamics influence storage economics for the optimization process. The regional differences have profound implications for global decarbonization strategies. Contrary to initial expectations, the predictability of renewable resources does not directly correlate with the economics of storage deployment. Germany, despite exhibiting more predictable PV generation patterns than Australia's highly variable solar output, requires 10 times higher carbon prices for battery deployment. This counterintuitive finding demonstrates that renewable penetration levels (Australia's 61% vs. Germany's 40%) and overall system flexibility needs, rather than individual resource characteristics, fundamentally determine storage economics. This may necessitate additional approaches to deep decarbonization, such as transmission expansion, demand flexibility, or maintaining some firm low-carbon generation.

6.4.6 Beyond Operational Optimization: Implications for System Planning

The limited operational emissions reductions achieved through storage deployment underscore that deep decarbonization requires fundamental system transformation rather than operational optimization alone. The results suggest a two-phase approach to power system decarbonization: first, increasing renewable and clean energy capacity to displace fossil generation (as already achieved with 61% and 40% penetration), and second, deploying ES systems to optimize the operation of remaining fossil units until they can be economically retired. The low CF observed (22–30%) indicates that ES systems provide value primarily through capacity and flexibility services in the operational optimization process. This suggests that energy-only markets may systematically undervalue storage, which may lead to underinvestment. Therefore, capacity markets, flexibility products, or direct procurement mechanisms may be necessary to ensure adequate ES deployment to manage highly VRE systems, even before considering deep decarbonization goals. Additionally, the deployment of multiple storage technologies at high decarbonization levels challenges traditional least-cost planning approaches that typically select a single dominant technology. The complementarity observed with PHS, batteries, and gravity storage systems serving distinct operational niches suggests that portfolio-based planning approaches may be more appropriate for high-renewable systems where temporal variability spans multiple timescales from sub-hourly to seasonal.

6.5 ES Pathways in the Climate–Energy–Sustainability Nexus

The results reveal key ES deployment pathways within the complex interactions of the CES nexus in deep decarbonization efforts.

6.5.1 Climate/System Configuration Pathways

Climate and system characteristics directly influence storage deployment patterns through renewable penetration levels and grid flexibility requirements. Australia's higher renewable penetration (61%) creates more frequent opportunities for storage arbitrage despite variable PV and wind generation patterns, enabling effective battery deployment at moderate carbon prices (\$50/tCO_2). On the other hand, Germany's relatively lower renewable penetration (40%) results in greater grid dependence and fewer storage arbitrage opportunities, requiring substantially higher carbon price thresholds (\$500/tCO_2) for economic storage deployment, demonstrating how renewable penetration levels rather than individual resource predictability fundamentally alter decarbonization pathways.

6.5.2 Energy System Pathways

Based on temporal energy management needs, three complementary storage pathways emerge: (1) short-duration batteries (one to four hours) address daily RE-load mismatches and grid balancing; (2) medium-duration gravity storage (four to eight hours) manages extended renewable lulls and peak shaving; and (3) long-duration pumped hydro (eight+ hours) provides bulk energy shifting across weather systems. The simultaneous deployment of all three technologies under stringent emissions constraints indicates pathway complementarity rather than competition. However, it is noteworthy that PHS systems are geography-specific. In geographies where PHS deployment is not an option, the gravity storage systems could also be deployed as long-duration storage. Other options include the redox flow batteries and hydrogen storage systems, although their economics are not as competitive as compared to PHS systems yet.

6.5.3 Sustainability Assessment Pathways

Economic sustainability constrains storage deployment through escalating marginal abatement costs. The epsilon-constraint analysis reveals that achieving beyond 99.98% emissions reduction creates unsustainable cost trajectories, suggesting storage-only pathways cannot achieve complete decarbonization. Additionally, environmental sustainability benefits remain modest (0.42% emissions reduction in Australia and 0.014% in Germany) despite optimal deployment, indicating storage pathways primarily enable better utilization of existing renewable capacity rather than completely displacing fossil generation.

6.5.4 Integrated CES Pathways

The analysis of the CES nexus further reveals that effective deep decarbonization requires coordinated rather than isolated pathways. Climate-informed storage sizing must account for regional renewable resource patterns, energy system flexibility needs, and economic sustainability constraints. The limited operational emissions reductions achieved through storage deployment indicate that nexus-based approaches should prioritize structural system transformation (increased renewable capacity, transmission expansion, and demand flexibility) over operational optimization alone for achieving net-zero objectives.

6.6 Conclusion

This chapter presents the role of ES systems in achieving deep decarbonization of power networks with existing high renewable penetration. The results reveal three major findings that challenge conventional approaches to power system decarbonization. First, operational emissions reductions through storage deployment were surprisingly limited in a high-renewable system. Despite optimal ES system deployment including batteries, PHS, and gravity storage, emissions decreased by only 0.42% in Australia and 0.014% in Germany. This demonstrates that although storage is essential for managing renewable variability, it cannot substitute for the structural transformation needed to eliminate fossil generation capacity. Second, regional characteristics influence the economics of storage-enabled decarbonization. Australia's system achieved operational improvements at carbon prices of \$50/tCO_2, while Germany's system required \$500/tCO_2 for comparable storage deployment. This massive difference underscores that universal carbon pricing mechanisms will produce highly uneven decarbonization outcomes across different power systems. Third, it was realized that achieving the final percentage points of emissions reduction incurs exponentially increasing costs. Furthermore, the deployment of multiple storage systems at these targets, each operating at modest capacity factors of 22–30%, suggests that deep decarbonization requires a portfolio approach rather than relying on a single storage technology solution.

These findings indicate that the CES nexus in modern power systems cannot be addressed through operational optimization of ES systems alone. While ES enables better utilization of existing renewable capacity, achieving net-zero emissions will require coordinated policies addressing both system planning and operation, as well as technology-specific support mechanisms beyond carbon pricing, and ε-cap. Furthermore, it should be recognized that regional differences necessitate tailored rather than uniform decarbonization pathways.

References

1 A. Altamimi and P. P. Kesuma, "Long-term reliability and degradation analysis of a microgrid with wind farms subjected to climate change conditions and age," *Arabian Journal for Science and Engineering*, vol. 48, no. 5, pp. 6577–6593, 2023.

2 K. Pang, J. Zhou, S. Tsianikas, D. W. Coit, and Y. Ma, "Long-term microgrid expansion planning with resilience and environmental benefits using deep reinforcement learning," *Renewable and Sustainable Energy Reviews*, vol. 191, p. 114068, 2024.

3 K. Anderson, A. Farthing, E. Elgqvist, and A. Warren, "Looking beyond bill savings to equity in renewable energy microgrid deployment," *Renewable Energy Focus*, vol. 41, pp. 15–32, 2022.

4 E. A. M. Ceseña, N. Good, A. L. Syrri, and P. Mancarella, "Techno-economic and business case assessment of multi-energy microgrids with co-optimization of energy, reserve and reliability services," *Applied Energy*, vol. 210, pp. 896–913, 2018.

5 A. Papageorgiou, A. Ashok, T. H. Farzad, and C. Sundberg, "Climate change impact of integrating a solar microgrid system into the Swedish electricity grid," *Applied Energy*, vol. 268, p. 114981, 2020.

6 S. S. Ottenburger et al., "Sustainable urban transformations based on integrated microgrid designs," *Nature Sustainability*, vol. 7, no. 8, pp. 1067–1079, 2024.

7 K. B. Jones, S. J. Bartell, D. Nugent, J. Hart, and A. Shrestha, "The urban microgrid: Smart legal and regulatory policies to support electric grid resiliency and climate mitigation," *Fordham Urban Law Journal*, vol. 41, p. 1695, 2013.

8 M. Macmillan, A. Zolan, M. Bazilian, and D. L. Villa, "Microgrid design and multi-year dispatch optimization under climate-informed load and renewable resource uncertainty," *Applied Energy*, vol. 368, p. 123355, 2024.

9 N. T. Mbungu, R. C. Bansal, R. M. Naidoo, and M. J.-P. Bazolana, "Discriminatory protection analysis of three-phase asynchronous motors during power disturbances," *Electric Power Components and Systems*, vol. 47, nos. 4–5, pp. 431–443, 2019.

10 C. Ghenai and M. Bettayeb, "Modelling and performance analysis of a stand-alone hybrid solar PV/fuel cell/diesel generator power system for university building," *Energy*, vol. 171, pp. 180–189, 2019.

11 M. Jafari, A. Botterud, and A. Sakti, "Decarbonizing power systems: A critical review of the role of energy storage," *Renewable and Sustainable Energy Reviews*, vol. 158, p. 112077, 2022.

12 T. Chen et al., "Applications of lithium-ion batteries in grid-scale energy storage systems," *Transactions of Tianjin University*, vol. 26, no. 3, pp. 208–217, 2020.

13 J. Ren and X. Ren, "Sustainability ranking of energy storage technologies under uncertainties," *Journal of Cleaner Production*, vol. 170, pp. 1387–1398, 2018.

14 R. Yudhistira, D. Khatiwada, and F. Sanchez, "A comparative life cycle assessment of lithium-ion and lead-acid batteries for grid energy storage," *Journal of Cleaner Production*, vol. 358, p. 131999, 2022.

15 M. A. Cusenza, S. Bobba, F. Ardente, M. Cellura, and F. Di Persio, "Energy and environmental assessment of a traction lithium-ion battery pack for plug-in hybrid electric vehicles," *Journal of Cleaner Production*, vol. 215, pp. 634–649, 2019.

16 B. Fouzia, "Technico-economical efficient multiyear comparative analysis of temperature and cycling effect on Li-ion and lead-acid batteries—A case study," *Journal of Energy Storage*, vol. 74, p. 109310, 2023.

17 D. Wang and M. Grimmelt, "Climate influence on the optimal stand-alone microgrid system with hybrid storage – A comparative study," *Renewable Energy*, vol. 208, pp. 657–664, 2023.

18 IEA, "Grid-scale storage," 2024. Accessed: Aug. 28, 2024. [Online]. Available: https://www.iea.org/energy-system/electricity/grid-scale-storage

19 C. Acar, "A comprehensive evaluation of energy storage options for better sustainability," *International Journal of Energy Research*, vol. 42, no. 12, pp. 3732–3746, 2018.

20 W. Tong et al., "Solid gravity energy storage: A review," *Journal of Energy Storage*, vol. 53, p. 105226, 2022.

21 C. Botha and M. Kamper, "Capability study of dry gravity energy storage," *Journal of Energy Storage*, vol. 23, pp. 159–174, 2019.

22 M. Blackhurst et al., "Marginal abatement costs for greenhouse gas emissions in the United States using an energy systems approach," *Environmental Research: Energy*, vol. 2, no. 1, p. 015012, 2025.

23 J. Farbes, B. Haley, and R. Jones, "Marginal abatement cost curves for U.S. net-zero energy systems," 2021. [Online]. Available: https://www.edf.org/sites/default/files/documents/MACC_2.0%20report_Evolved_EDF.pdf

24 P. Liqun, L. Jiang, and H. Gang, "China needs to expand both pumped hydro and battery storage," 2024. Accessed: Oct. 22, 2025. [Online]. Available: https://www.eco-business.com/news/china-needs-to-expand-both-pumped-hydro-and-battery-storage/

25 J. X. Johnson, "Oeo blog: You can't just do one thing (when it comes to decarbonization)," 2025. Accessed: Oct. 22, 2025. [Online]. Available: https://energy.cmu.edu/news/2025/02/25-oeo-decarbonization.html

26 A. M. Asim, A. S. Awad, and M. A. Attia, "Integrated optimization of energy storage and green hydrogen systems for resilient and sustainable future power grids," *Scientific Reports*, vol. 15, no. 1, p. 25656, 2025.

27 K. Huang, "Prediction-free coordinated dispatch of microgrid: A data-driven online optimization approach," Mendeley Data, V2, 2024.

28 ENTSOE, "Electricity generation, transportation and consumption for the European market," 2025.

29 S. Kumaresh, R. Devarapalli, F. P. García Márquez, and H. Acaroğlu, "A comprehensive review of optimization, market strategies, and AI applications in energy storage systems," *Evolutionary Intelligence*, vol. 18, no. 4, pp. 1–23, 2025.

7

Carbon Dioxide Removal Practical Strategies in Deep Decarbonization of MENs

Mohammad Mehdi Amiri[1,*], *Mehrdad Setayesh Nazar*[1], *Hamdi Abdi*[2], *and Pierluigi Siano*[3]

[1] *Faculty of Electrical Engineering, Shahid Beheshti University, Tehran, Iran*
[2] *Electrical Engineering Department, Razi University, Kermanshah, Iran*
[3] *Department of Management & Innovation Systems, University of Salerno, Fisciano, Salerno, Italy*

*Corresponding author: moh_amiri@sbu.ac.ir

List of Acronyms

BECCS	Bioenergy with carbon capture and storage
CAPEX	Capital expenditures
CCUS	Carbon capture, utilization, and storage
CDR	Carbon dioxide removal
CLC	Chemical looping combustion
DAC	Direct air capture
DER	Distributed energy resource
DERMS	Distributed energy resource management systems
IPCC	Intergovernmental Panel on Climate Change
MCFCs	Molten carbonate fuel cells
MENs	Modern energy networks
MRV	Monitoring, reporting, and verification
RERs	Renewable energy resources
SPECCA	Specific primary energy consumption per CO_2 avoided
TEA	Techno-economic assessment
TSEA	Techno-socio-economic assessment

7.1 Introduction

In a world in which ambitious global climate-change targets, like net-zero emissions by 2050, are needed, carbon dioxide removal (CDR) can play a critical role in supplementing emissions-reduction efforts, with its focus on mitigating residual emissions in those sectors that are tough to decarbonize, like heavy industry and transport. Included in CDR is a range of both technological and natural approaches that remove CO_2 directly from the air and store it forever, effectively helping to mitigate the climate crisis, as well as enabling the integration of variable energy sources and renewable energy into current power grid networks [1].

It is in this regard that CDR, which is predicted by some as the intentional removal and long-term sequestration of atmospheric CO_2, has been given substantial priority as a transitional means toward deep decarbonization and the solution of residual emissions in the hard-to-decarbonize sectors. In this chapter, after a general definition of the concept of deep decarbonization, concepts related to carbon removal and then carbon removal strategies will be presented. Deep decarbonization of modern energy networks (MENs) cannot rely on emissions abatement alone. Multiple net-zero pathways indicate that CDR must scale from 2$GtCO_2$/yr in 2024 to 7–9$GtCO_2$/yr by 2050. CDR is therefore a complement to, not a substitute for, rapid emissions cuts, with durable removal and robust verification pivotal to climate integrity [2].

7.1.1 Modern Energy Networks

MENs represent the ongoing transformation of conventional power systems into multi-vector, digitalized, and intelligent infrastructures designed to integrate high shares of renewable energy resources (RERs) while ensuring reliability, resilience, and market accessibility. In the past decade, fast improvements in distributed generation, power electronics, digital control systems, and intelligent edge devices have drastically altered the physical and working design of energy systems. These modifications are prompted not only by the need to decarbonize in order to address climate change as rapidly as possible but also by the increasing multi-energy interaction throughout the electricity, gas, heating, cooling, and transport sectors [3]. These features are illustrated in Figure 7.1:

- Full inverter-based penetration (wind, solar, and storage) with high RERs has been facilitated by advanced grid-forming and stability-enhancing controls.
- Cyber-physical coordination provisional of sophisticated distribution management systems, distributed energy resource management systems (DERMS), and extended area-based monitoring/protecting platforms. Multi-vector integration combines electricity and hydrogen, thermal networks, and e-fuels to maximize system-wide flexibility.

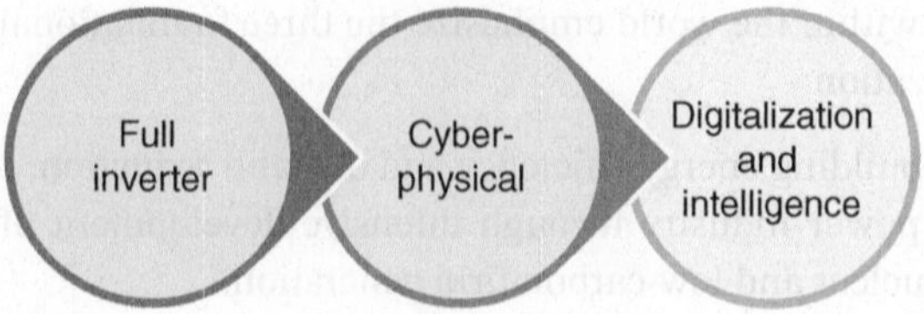

Figure 7.1 Main characteristics of MENs.

- Digitalization and intelligence, to balance distributed resources on an instantaneous basis, synchrophasor networks, smart meters, and AI-driven optimization.

Involving a functional need to actively participate in the market, which includes not only end users but prosumers and aggregators developing ancillary services, variability in demand, and carbon-removal opportunities with changing regulatory frameworks. Thus, energy networks modernizations are a concomitant response to both the environmental needs—mitigate emissions and facilitate deep decarbonization—and the societal requirements, safe, affordable, and participatory energy services. From this perspective, MENs are no longer grids with a higher RER share but larger ecosystems combining technological innovation, market designs, and policy-making to offer sustainability and resilience at scale. This holistic view of MENs enables integrating CDR plans for them, ensuring that the networks themselves actively intervene in net-negative emissions to ensure deep decarbonization [4].

7.1.2 Deep Decarbonization Concept

Deep decarbonization is "the idea, action, and policy of asserting a moral obligation to raise prices on carbon dioxide, especially where there is no political consensus to do so, across the economy" [5]. In other words, from this point forward, emissions of greenhouse gases, especially CO_2, must be radically reduced. This is not about some modest lifestyle adjustment: it is about changing foundations. This is our energy networks and our ways of life that have developed over the past few centuries based on fossil fuels. This implies a future in which energy resources are used without leaving any trace in the atmosphere as we power our progress. The scientific community, through bodies like the Intergovernmental Panel on Climate Change (IPCC), clearly states that achieving net-zero emissions by the mid-century is essential to limit global warming to a manageable 1.5 or 2 °C, averting the most severe consequences of a changing climate. Deep decarbonization means industrial-scale innovation and commercialization in

MENs [6]. Current syntheses within the world emphasize the three foundational pillars of profound decarbonization:

- Transport, industry, and building energy efficiency and demand reduction.
- The decarbonization of power industry through intensive development of RERs, with the help of nuclear and low-carbon firm generation.
- Electrification and fuel switching: the replacement of fossil fuels with clean electricity and renewable hydrogen in end-use areas.

However, despite the most forceful mitigation scenarios, stragglers of the hard-to-decarbonize sectors (aviation, cement, steel, and agriculture) remain. Due to this fact, CDR is nowadays being viewed as the pillar number 4 of deep decarbonization. The so-called technological enablers to deep decarbonization comprise the extreme price reductions in solar PV, wind, and batteries, enhanced control functions by renewable inverters to form a grid, innovations in green hydrogen and synthetic fuels, and using MENs with digitized control, connection to other sectors, and resiliency [7]. MENs are imperative as they furnish the working base to consolidate large percentages of renewables with bioenergy with carbon capture and storage (BECCS), direct air capture (DAC), and power-to-X (PtX) centers placed as carbon-unobtrusive investments. Simultaneously, profound decarbonization is a governance and equity dilemma. Strong policy solutions (e.g., EU Carbon Removals Certification Framework 2024; US Inflation Reduction Act, including 45Q tax credits), big financial mobilization (trillions of dollars annually), and just transition plans are necessary. The formulation of monitoring, reporting, and verification (MRV) protocols, which ensure permanence, additionality, and integrity of CDR, is equally important [8, 9].

7.2 Carbon Dioxide Removal

CDR intersects with MENs along three axes: dispatchable negative-emissions generation; electricity-intensive engineered removals; and carbon-value chains, which transform captured CO_2 into storable products or geologically mineralized forms. MEN attributes high-share variable renewables, HVDC interties, advanced distribution management, and flexibility markets, create operational niches where CDR can absorb surplus generation, provide controllable load, and in some cases, co-produce energy services. Scaling trajectories for carbon capture utilization and storage (CCUS) and CDR remain uncertain but have accelerated. Announced capture and storage capacities for 2030 have risen sharply, and about 45 commercial CCUS facilities operate today across industry and power [10].

The coupling between the MEN and CDR is illustrated by two families of distinctive CDRs. Recent syntheses compare pathway costs, life cycle performance, and systems impacts of BECCS, while cautioning about land, water, and biodiversity constraints. Dominant critics notwithstanding, when sustainable biomass supply, capture rates, and storage integrity are met, BECCS can provide firm, dispatchable power with net removals. DAC relies heavily on electricity, and it is location-independent; it can and may well locate wherever low-carbon power is or is dispatched to, with high-capacity storage, including geological mineralization. Quality, accounting, and market rules for carbon removals are being defined in emerging metrological and trading frameworks [11].

MENs enable system-level co-optimization to manage seasonal and diurnal variability in renewable generation while facilitating carbon-value chain amplification and geographic dispersion. The integration of MEN infrastructure with CDR pathways represents a bidirectional coupling opportunity, with California's regulatory and market frameworks offering a critical testbed for validating this approach at scale. Flexible DAC/carbon capture and storage (CCS) can mitigate renewable curtailment, provide a demand-side absorptive capacity during periods of excess generation, and stand down during scarcity, thereby reducing integration costs. Deposition characteristics, sensibilities, and decisions play a central role in MENs because they are not fuels; they must find occupancy [12, 13].

MEN–CDR paradigms must address MRV, additionality, leakage, and automotive sweep across capture, transport, utilization, and storage. Energy-system dispatch and carbon accounting ought to re-align through near-term investments in MRV protocols while encouraging distributed CDR market access by creating flexible resource grids by lowering their marginal opportunity costs. Policies like FERC Order 2222 allow distributed energy resource (DER) aggregation in wholesale markets. Market access to distributed flexibility can indirectly support CDR because it creates the flexible resources that make grid-coupled removals environmentally and economically feasible [14].

Finally, it is important to note that the economic landscape for removals is rapidly changing. It has been suggested that if certification, financing, and demand-side policies mature in the 2030–2035 window, the global CDR credit market may grow toward \$100 billion/yr. However, scholars are warning against over-relying on future removals at the expense of immediate abatement in the interim. Planners for the MEN sector are therefore required to consider portfolio-level trade-offs, such as allocating renewable and low-carbon energy between electrification, firming resources, and CDR duties, balancing land and water footprints, and selecting transport and storage options that will ensure social acceptance and permanence [15].

7.3 CO_2 Removal Modalities in MENs with Techno-economic Signatures

7.3.1 BECCS

By combining bioenergy conversion (combustion, gasification, and fermentation) with post-combustion capture and geological storage of CO_2, BECCS provides net CO_2 removals when biomass is sustainably sourced, and full value-chain emissions are bound. In MENs, BECCS offers dispatchable, firm capacity and negative emissions, reducing residuals associated with hard-to-abate sectors and supporting adequacy and inertia. The potential for negative emissions as well as the constraints on system/land use that limit scalable deployment have been highlighted in recent syntheses and policy reviews [1, 16]. BECCS is in the early commercialization phase, with operational capacity capturing approximately 2 $MtCO_2$ per year globally as of mid-2025, of which less than 1 $MtCO_2$ is stored in dedicated geological sites. Around 90% of current capture occurs at bioethanol facilities, where high CO_2 concentrations enable low-cost integration. Key operational plants include the Illinois Industrial CCS Project in the United States (capturing 1 $MtCO_2$/yr since 2018), Red Trail Energy (North Dakota, operational since 2022), and Blue Flint (North Dakota, since 2023), all targeting permanent storage. Solely as of 2025, the Mikawa Power Plant with a 50 MW capacity stands as Japan's uniquely fully functional biomass-fired BECCS facility, demonstrating the viability of small-scale deployment. Upcoming projects also signify the regional pace: Stockholm Exergi's BECCS plant in Sweden is expected to yield 0.8 $MtCO_2$/yr, while the AtmosClear venture in Louisiana, United States, cooperates with ExxonMobil on CO_2 transportation and storage [17–19].

Policy strides are also impactful. For instance, the US Department of Energy's $35 million CDR pilot prize has stimulated BECCS plant offtake agreements. Furthermore, corporate initiatives, such as Microsoft's Q2 2025 BECCS purchase, underscore growing market demand. Despite sporadic delays in CO_2 transportation infrastructure and limited biomass supply chains, BECCS continues to demonstrate steady progress [10, 17–19]. BECCS scales are classified by facility size:

- pilot (< 1 $MtCO_2$/yr)
- demonstration (1–5 $MtCO_2$/yr)
- commercial (5–10 $MtCO_2$/yr)
- large-scale (> 10 $MtCO_2$/yr)

Today, current operations cover pilot to demonstration; bioethanol is the most pronounced technology due to its maturity. Referring to early- and advanced-stage projects, it is possible to reach up to 60 $MtCO_2$/yr by 2030. These projections include 70 more bioethanol facilities (almost 20 $MtCO_2$/yr in capacity, for instance,

Midwest Carbon Express in the United States) and 30 $MtCO_2$/yr from heat and power plants—more than 80% will relate to dedicated biopower. Some large-scale plans are Drax's UK power plant, which can reach up to 8 $MtCO_2$/yr with the help of government funding. Similarly, two US facilities can cumulate 6 $MtCO_2$/yr by 2030. Other sectors also follow: seven cement plants fostering biomass with CCS, three pulp and paper mills deciding to retrofit, and five hydrogen facilities investing in both biomass and CCS [18].

7.3.1.1 Cost Structures and Economic Viability

BECCS is expected to cost \$80–200/tCO_2 in 2025, depending on feedstock, technology, scale, and location. Bioethanol routes are expected to cost the least with the simplest process. Traditional cost assessment via techno-economic assessment (TEA) shows negative profitability when carbon credits are not provided, with an example of NPV equal to −\$460 million for a wheat-straw CHP plant, requiring \$240/tCO_2 for parity with renewables. The largest cost drivers are capital expenditures (CAPEX) for capture units (30–40%), biomass supply (20–30%), and operational energy penalties [18, 20].

With the use of newer frameworks such as techno-socio-economic assessment (TSEA), the viability is significantly improved by monetizing societal benefits. NPV up to \$2.28 billion was achieved in optimized scenarios. Costs could drop to \$50–100/tCO_2 at gigaton scales with learning curves and subsidies, but deployment barriers remain without policy support. Life cycle assessments (LCAs) note their net-negative potential, though infrastructure costs continue to pose challenges [20–22].

7.3.1.2 BECCS Role in MENs

Finally, besides anchoring resource adequacy portfolios, BECCS supplies ramping support and, along with proper plant design, grid services as a firm, synchronous-compatible resource with net-negative potential. However, MEN planners should also explicitly price land and water externalities as well as transport and storage availabilities in portfolio models [16].

7.3.1.3 Recent Advancements in BECCS

These advancements underscore BECCS's role in 1.5 °C pathways, with calls for interdisciplinary efforts to address land competition and economic barriers [1]:

- Technology comparisons and performance. Finally, idealized real-world comparisons across representative scenarios or average lifetimes (e.g., 53 BECCS configurations based on seven technologies reviewed in [23] using exergy efficiency and specific primary energy consumption per CO_2 avoided [SPECCA] present molten carbonate fuel cells [MCFCs] with negative energy gains, followed by chemical looping combustion [CLC] at 0.1–1 MJ/kg CO_2, while calcium looping incurs modest penalties but significant capture costs [mean SPECCA 6.85 MJ/kg CO_2]). A more generic and revised SPECCA formulation ensures multi-output systems like CHP remain valid within the comparison.

- Economic and socio-economic frameworks: Almena et al. [20] present TSEA for BECCS, which converts societal benefits into monetizable gains to demonstrate profitability ($2.28 billion NPV in electricity-maximizing modes), emphasizing the influence of social carbon pricing and supportive policy frameworks. Reference [24] examines methane removal synergies with BECCS, estimating associated costs for complementary CDR pathways.
- Regional and sectoral applications: Dufour and Möllersten [25] evaluate Nordic BECCS potentials for net-zero, describe capture from point sources (e.g., pulp mills) as 10–20 $MtCO_2$/yr, and highlight residual emission counterbalancing. In Brazil, Pereira et al. [26] analyze BECCS relative to biochar, suggesting negative emissions of 20–50% of national GHGs by 2050.
- Finally, sustainability and integration. Reference [18] assesses the BECCS challenges and concludes that optimized biomass conversion and DAC integration would reduce the land use by 0.1–0.4 ha/tCO_2 and water demands, with costs of $80–200/$tCO_2$. Future research in the field should focus on low-penalty technologies, such as CLC, and infrastructure modeling. Reference [27] explores the aspect of biomass blending for power-BECCS that would support enhanced fuel flexibility and efficiency.

7.3.2 DAC with Mineralization

DAC separates CO_2 from ambient air (solvent/sorbent systems), after which mineralization injects CO_2 into reactive rocks (e.g., basalts) where it converts to stable carbonates on short geological timescales. This pairing offers high-durability storage and location flexibility, enabling siting near low-carbon electricity and heat in MENs [1]. As of September 2025, DAC is transitioning from hype to realism, marked by policy advancements, pilot-scale demonstrations, and regional innovations aimed at scaling up operations [28]. Three main resource impacts of DAC are shown in Figure 7.2.

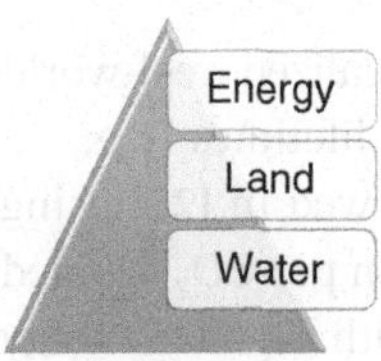

Figure 7.2 Main resource impacts of DAC.

7.3.2.1 Latest Status and Scales

As of mid-2025, DAC technology remains in the early commercialization phase, with approximately 130 facilities in various planning stages worldwide, though only about 15 are in advanced development or under construction. Global operational capacity has seen significant growth, projected to increase by 873% from 59 kilotons of CO_2 per year ($ktCO_2$/yr) in 2024 to 569 $ktCO_2$/yr in 2025. This expansion is driven by a mix of solid- and liquid-sorbent technologies, with leading companies like Climeworks, Carbon Engineering (now part of Occidental Petroleum), and emerging players such as Mission Zero advancing modular designs [29, 30].

Key operational facilities highlight this progress. The Mammoth plant in Iceland, operated by Climeworks, is the largest DAC facility as of August 2025, with a capacity of 36 $ktCO_2$/yr, utilizing geothermal energy for low-carbon operations, as shown in Figure 7.3 [31, 32].

In the United States, the STRATOS project in Texas, developed by 1PointFive (a subsidiary of Occidental), is slated for completion by late 2025, aiming to capture 500 $ktCO_2$/yr using liquid-sorbent technology, as shown in Figure 7.4 [33]. Internationally, HIF Global's Haru Oni facility in Chile is advancing DAC integration with e-fuel production, with commissioning expected in Q4 2025, marking the first such deployment outside Europe. Additionally, Saudi Arabia's inaugural DAC unit in Riyadh, a collaboration between KAPSARC and Climeworks, was inaugurated in July 2025, demonstrating modular solid-sorbent capture in arid environments [30, 32].

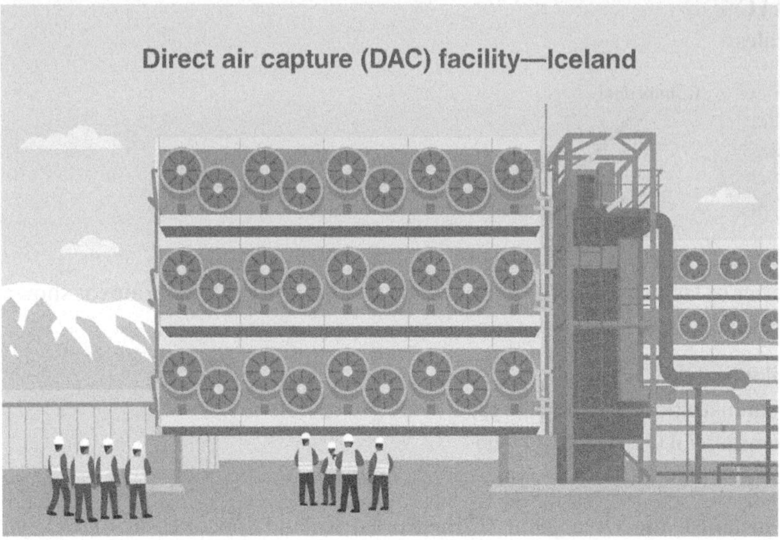

Figure 7.3 Climeworks' Mammoth plant in Iceland. *Source*: Adapted from [32].

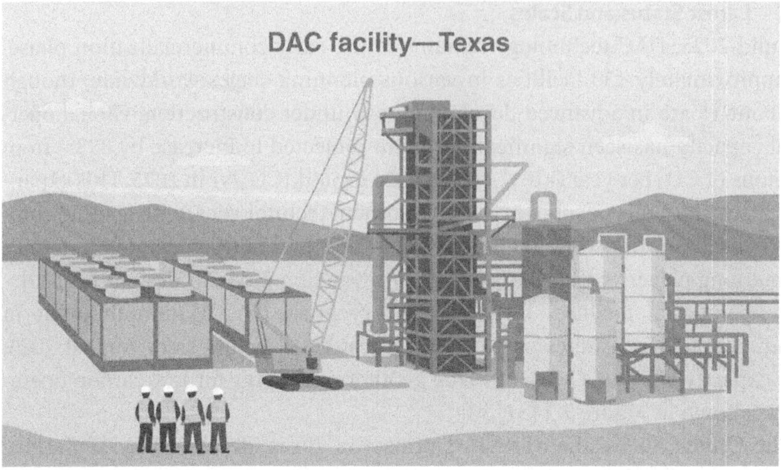

Figure 7.4 Project STRATOS under construction in Ector County, Texas. *Source*: Adapted from [33].

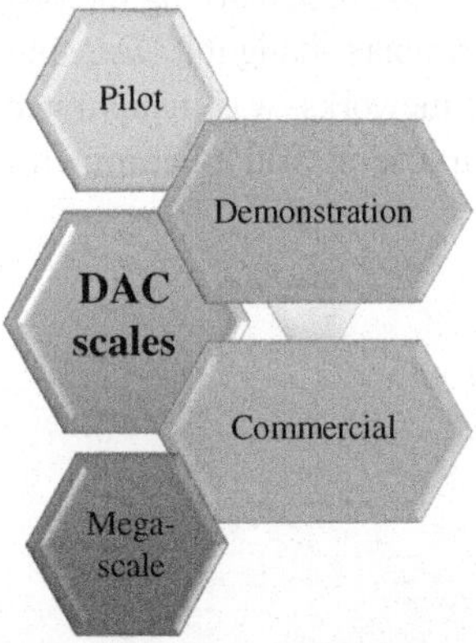

Figure 7.5 Groups of DAC in terms of facility size.

In terms of facility size, DAC scales can be divided into the four groups as shown in Figure 7.5:

- pilot (< 1 ktCO_2/yr)
- demonstration (1–10 ktCO_2/yr)
- commercial (10–100 ktCO_2/yr)
- mega-scale (> 100 ktCO_2/yr)

For instance, the Orca plant (Climeworks, Iceland) operates at 4 ktCO_2/yr, while upcoming projects like Mammoth and STRATOS represent a shift toward mega-scale.

7.3.2.2 Cost Structures and Economic Viability

DAC costs have decreased but remain high, ranging from \$200–600/tCO_2 in 2025, depending on energy sources, sorbent efficiency, and scale. At billion-ton scales, estimates project \$385–530/tCO_2, though pathways to < \$100/tCO_2 are feasible with technological advancements. Key cost drivers include CAPEX for contactors (40–50% of total), operational energy (20–30%), and regeneration processes [34–36].

A component-based learning curve analysis conducted from 2025 projects cost reductions of approximately 50–70% in solid-sorbent DAC by 2030, driven by modularization and further sorbent innovations; however, economy-wide modeling indicates deployment barriers at \$180–\$1,000/tCO_2 in the absence of subsidies [36, 37].

7.3.2.3 DAC Role in MENs

DAC plants should be considered as flexible, controllable loads modeled after absorption of surplus VRE, avoidance of curtailment in full, and modulation under scarcity, functionality perfectly suited to MEN flexibility markets and DERMS/ADMS-based dispatch, especially when sited near renewable clusters and storage infrastructure, reducing transport and grid impacts and increasing carbon integrity.

7.3.2.4 Recent Advancements in DAC

Both studies highlight the importance of cross-disciplinary progress in materials science, engineering, and economics in breaking these barriers. In particular, Kotowicz et al. [30] survey energy intensity and location strategies that would enable reducing costs to \$100–200/tCO_2, with co-location with renewables:

- Process innovations: A distributed DAC process using synergistic water-lean solvents was demonstrated, achieving a working capacity of up to 1.5 mol/kg at a minimum desorption energy of 50–70 kJ/mol CO_2 [38, 39]. At the same time, solid adsorbent alternatives for DAC were evaluated, and amine-functionalized adsorbents were identified as a promising option because of their high CO_2/N_2 selectivity (> 1,000) and specific energy consumption of < 1.5 GJ/tCO_2.
- Modeling and optimization: review of optimization models for goals relating to sustainability to reach low-carbon goals [40]. An open-source dynamic model of performance for DAC in [41] creates a model of the real-world capture system and provides predictions for capture efficiencies of 80–90% at varying humidity. DAC is also combined with sustainable fuels for models of aviation neutrality. Brazzola et al. [42] found CO_2 removal needs of 0.5–1 GtCO_2/yr by 2050, utilizing these models as the basis for the prediction.
- Economic and LCA analyses: the most recent detailed LCA review in [43] considered the DACCS impact and concluded that when powered by renewables,

their net removal is 0.7–0.9 tCO_2/t captured. Reference [44] quantified cost trajectories for four DAC types within a rationalized framework and made a projection conditional on learning rates of 15–20%: in 2040, < \$150/tCO_2. An evaluation based on industrialization potentials ranked adsorption in electro-swing mode ideal in terms of gigaton scalability because of their low energy demand of 1–1.5 MWh/tCO_2 to facilitate a gigaton [45].
- Integration and applications: Kar et al. [46] report a gas-phase DAC reactor for solar fuel production, which achieves 0.5–1 kg/h of syngas at 70% faradaic efficiency. Reference [47] proposes cost-allocation mechanisms for DAC in power systems, which decrease shared costs by 20–30% by keeping the power plant capacity stable. The development of membrane-based DAC by the authors in [48] leads to capture rates of 100–200 $gCO_2/m^2 \cdot h$.

7.3.3 CO_2-to-fuels (PtX) with MEN Coupling

CO_2-to-fuels pathways synthesize energy carriers by combining captured CO_2 and renewable H_2. These are technology examples closing carbon loops and providing sector coupling across electricity, hydrogen, and fuels. "Important: CCU alone is not CDR—permanent storage or long-lived products are necessary to achieve net removals" [49].

7.3.3.1 Techno-economic Signature

Overall electricity-to-fuel conversion efficiencies are estimated at ~40–60% in recent reviews [50, 51], with competitiveness being extremely sensitive to the cost of renewable electricity at scale and electrolyzer evolution: system-level assessments imply significant deployment this decade almost regardless, with little matching of scenarios to current policies: and large sensitivity analysis assessments to electricity price and capacity factor, and CO_2 source purity and logistics.

7.3.3.2 System Role in MENs

As multi-energy balancing hubs, PtX facilities absorb VRE surpluses (via electrolysis), offer long-duration/seasonal storage in chemical form, and support cross-sector decarbonization (aviation/shipping). PtX can only be considered for removal in this case if it is associated with permanent sequestration; therefore, it must have a split stream to storage or produce a durable carbon-containing hazardous material rather than temporarily displace decarbonized short-cycle fuels [49].

7.3.4 Cross-cutting Signatures Relevant to MEN Planning

- Permanence and MRV: mineralization offers long-term as well as highly durable storage; a rigorous MRV framework would be required throughout

Table 7.1 Classification of technologies for deep decarbonization in MENs.

CDR modality	Capture/ efficiency	Cost range (USD/tCO_2)	Permanence	System role in MENs
BECCS	≥ 90% capture	≈$100–200	Geological storage, multi-millennial	Firm, dispatchable negative-emission power; synchronous support
DAC + mineralization	Energy intensity high; FOAK capture ~1–4 ktCO_2/yr per module	$400–600+ (falling to $100–300)	Very high (mineral carbonation)	Flexible controllable load; absorbs surplus VRE; location-flexible
CO_2-to-fuels (PtX)	40–60% elec → fuel conversion efficiency	$250–600 (utilization, not strict CDR)	Low unless paired with storage/durable products	Sector coupling; long-duration/ seasonal storage; e-fuels for transport

capture, transportation, injection, and monitoring to prevent double counting—highlighted in 2024 State of CDR and Oxford Energy briefings [49, 52].

- System integration value: DAC's demand flexibility and BECCS's firm output create complementary operational niches in MENs; global CCUS momentum (project pipeline growth, storage capacity announcements) reduces infrastructure risk.
- Scaling outlook: 2024 State of CDR estimates that sustainable CDR must scale to multi-gigaton levels by mid-century; technology portfolios should be region-specific and infrastructure-aware (biomass basins, storage plays, and renewable hubs) [52, 53].

Table 7.1 shows the BECCS, DAC + mineralization, and CO_2-to-fuels (PtX) with their capture/efficiency, cost ranges, permanence, and system roles in MENs.

7.4 Operational Integration via ADMS/DERMS, Markets, and Grid-support Functions

7.4.1 Control Architecture in MENs: ADMS–DERMS "Two-layer" Stack

MENs orchestrate heterogeneous assets (IBR renewables, storage, flexible loads, DAC/CCS plants, BECCS units, and PtX hubs) through a two-layer control stack: a distribution system operator layer (ADMS) and a DER orchestration layer (DERMS), linked to bulk-system/market interfaces. ADMS delivers distribution-side observability and control—state estimation, Volt/VAR optimization (VVO/CVR), FLISR/FDIR, switching, and outage management—validated increasingly through hardware-in-the-loop (HIL) ADMS test beds. DERMS forecasts, aggregates, constraints, and dispatches behind- and front-of-the-meter DERs (including flexible CDR loads like DAC), coordinating telemetry/commands (e.g., IEEE 2030.x/2030.5) and exchanging set-points with ADMS and markets. Recent NREL platforms (ADMS test bed; ARIES) and FAST–DERMS demonstrate scalable co-simulation/control and ADMS integration for feeder-level PV and DER fleets [54].

Design implications for CDR: DAC modules behave as controllable, interruptible loads; BECCS units act as firm, dispatchable generators. DERMS exposes these as resources with operational envelopes (power, ramp, and thermal limits) to ADMS for feeder security and to markets for monetization [55].

7.4.2 Markets and Participation Pathways (Retail ↔ Wholesale)

MENs translate operational flexibility into value through retail programs and wholesale markets. In the United States, FERC Order requires RTOs/ISOs to enable DER aggregations to participate in energy, capacity, and ancillary-service markets—codifying telemetry, metering, and coordination with distribution utilities. Recent updates specify compliance milestones (e.g., ISO-NE filings through 2024; capacity and E&AS participation schedules) and market-readiness timelines (PJM milestones for capacity and E&AS) [56–58]. Policy trackers and regulator briefs show accelerating implementation across regions. Parallel federal work and state-level frameworks outline scale-up pathways for aggregated DERs [14].

CDR-specific angle: market access allows DAC load shifting and BECCS energy/ancillary services to be scheduled economically: DAC ramps up during low prices/VRE surplus (absorbing curtailment), stands down at scarcity, while BECCS bids firm output and potentially inertia/fast reserves if enabled by interconnection rules and telemetry standards [59].

7.4.3 Grid-support Functions and Interoperability

In distribution systems with high DERs and CDR loads, smart inverter functions (per the IEEE 1547 family) provide autonomous grid support (volt-to-volt, volt-to-watt, frequency-to-watt, and ride-through) [60]. A new guide and white paper synthesize the implementation challenges, certification timelines, and utility adoption timelines. IEEE 2030 series (and 2030.5 profiles) provide interoperable communication frameworks for DER, DR, EV, and microgrid/VPP coordination, critical for multi-vendor DERMS and standardized device control at MEN scale [61].

Operationalization in ADMS/DERMS: recent NREL studies demonstrate real-time optimal power flow (OPF)-based DER coordination and co-optimization with ADMS, including HIL validation. Such stacks allow feeder-aware dispatch of DERs and flexible industrial loads (e.g., DAC fans/solvent regeneration heaters) while maintaining nodal voltage, thermal limits, and protection margins [62].

7.4.4 From Pilots to Portfolio Operations: VPPs and Flexibility Products

Contemporary VPP programs (utility/state pilots to multi-GW roadmaps) aggregate DERs to deliver peak shaving, frequency/voltage support, and reliability services. 2024–2025 syntheses track regulatory design (equity/participation rules), telemetry, and performance metrics; NREL's "Status of Power System Transformation: Leading Topics of 2024" flags VPPs, autonomous controls, and dynamic DER pricing as leading flexibility topics for high-VRE grids. For MENs targeting deep decarbonization with CDR, VPPs provide the market/operational wrapper to scale flexible DAC loads and storage alongside traditional DERs [63].

7.5 Summary of Findings

7.5.1 Current Status of CDR Strategies in MENs

In 2025, CDR strategies are characterized by accelerating policy momentum and market growth, but deployment lags behind ambitions. The voluntary carbon market for CDR has expanded; Q1 2025 contracted credits increased by 122%, primarily from biomass and marine methods, spurred by investments and off-take agreements from the likes of Microsoft and Amazon [64]. Engineered CDR with DAC at the fore has 27 funded facilities globally, operational, with each capturing ~0.01 $MtCO_2$/yr and > 130 plants funded, but not under construction, potentially reaching 65 $MtCO_2$/yr by 2030 if realized. While nature-based solutions dominate current removals, novel methods such as BECCS and enhanced weathering (EW) are growing in scale. The United States' DAC hubs (e.g., STRATOS at 500 $ktCO_2$/yr) and enhanced basalt application in agriculture emphasize the opportunities for integration. New policy frameworks such as the United States' Inflation Reduction Act's enhanced 45Q credits ($180/$tCO_2$) and the EU's Carbon Removals Certification Framework (finalized in February 2024) are catalyzing progress, alongside international efforts in Canada, Japan, and the United Kingdom. However, only ~2 $MtCO_2$/yr storage from bioethanol-integrated BECCS is currently sequestered geologically, highlighting the gap to gigaton-scale requirements [65, 66].

7.5.2 Key Challenges in CDR Strategies

Increasing the scale of CDR is fraught with multiple and concomitant challenges. Technological and economic obstacles are still highly significant: the cost of novel CDR options like DAC or BECCS often exceeds \$100/tCO_2 and can amount to \$200–1,000/tCO_2 due to energy intensity and the low level of scalability. High costs and land/water pressures, against the backdrop of social constraints and uncertain dynamics in the use of these technologies, result in significant risks of reaching a temperature overshoot by 0.4 °C by the end of the 21st century if the use of CDR under high-CDR scenarios fails, demanding 58% emissions decrease after 2045 and contributing 38% of total existing assets as stranded [67]. Environmental trade-offs, for example, land-use conflicts and damage to biodiversity, further exacerbate food security issues: the cost of land use for afforestation and biochar is 0.1–0.4 ha/tCO_2, and many respondents in studies doubt the economic viability of these techniques and the level of carbon durability. The moral hazard is also visible, since reliance on CDR increases dependence on fossil fuels, and in high-CDR measures, minimal reductions prior to 2050 are expected. Equity is another challenge: for Canada, emitters face carbon debts of 6.7–16.4 GtCO_2 based on per capita principles. High-emission countries will have to amplify CDR to levels as high as 20.3 GtCO_2 cumulative by 2100, creating additional challenges for the resources of countries in the Global South. The political and legislative matters include a lack of supply, unstable regulatory frameworks, and insufficient site development, while some options like fertilization face limits due to strong public skepticism [68–70].

7.5.3 Emerging Trends in CDR Strategies in MENs

Trends in 2025 emphasize diversification, innovation, and market integration to overcome barriers. Policy reforms are gaining speed: the United States has devoted \$3.5 billion to DAC hubs, Canada grants 60% tax credits, and the EU aims for 50 MtCO_2/yr storage by 2030 with CDR certification. Voluntary markets are expanding as biochar and DAC credits reshape dynamics, priced at \$600–1,000/tCO_2, with aggregators like Frontier facilitating purchases. New low-energy DAC variants—such as electro-swing adsorption and zeolite-based systems—plus renewable integrations like geothermal-DAC are increasing CO_2 utilization in synthetic fuels. Equity-informed approaches aim to increase CDR obligations for high emitters like Canada (308 MtCO_2/yr from DAC under per capita scenarios) to address global inequalities. Optimism on social media is growing (median growth factor 1.46 for positive tweets), with conventional methods like afforestation (37% positive) being favored, while BECCS and ocean fertilization remain polarizing. An "early leaders" strategy is emerging, where less-affected regions pioneer CDR to build knowledge and mitigate risks for broader implementation. Hybrid portfolios that combine CDR with industrial processes and agriculture—such as EW in the US Corn Belt projected for 0.16–0.49 $GtCO_2$/yr by 2070—offer co-benefits like soil enrichment and emissions reduction [64, 65, 70, 71].

7.5.4 Recent Advancements in CDR Strategies

These developments lay a foundation based on robust policies, diversified portfolios, and equity integration needed to effectively scale CDR.

- Risk and policy implications: Bindl et al. [67] model the risks of uncertain CDR dependence, showing that high-CDR planning under 2 °C targets could result in a 0.4 °C overshoot or necessitate 58% emission cuts if infeasible—raising stranded assets by 38% and disproportionately affecting the Global South. It recommends early learning (by 2030) to avoid up to $1 trillion in stranded assets. Reference [72] explores separating emission and removal targets, finding limited benefits (such as modest fossil phase-out acceleration) but stresses the importance of diversified CDR to prevent over-reliance.
- Equity and national obligations: Motlaghzadeh et al. [69] apply equity principles (per capita convergence), which increase Canada's cumulative CDR requirement to 20.3 $GtCO_2$ by 2100, expanding DAC and EW deployment and highlighting infrastructure needs for equitable global participation. Reference [73] examines national strategies, revealing disparities where high-income nations underutilize novel CDR methods, advocating for diversified portfolios to enhance both quality and fairness.
- Public perceptions and adoption: Repke et al. [70] analyze Twitter data (2010–2022), revealing exponential growth in CDR-related discussions (32% annually) and rising positivity for most methods—afforestation (37% positive)—while BECCS and ocean fertilization remain contentious. It recommends tailoring policies to context and leveraging social media for real-time sentiment tracking. Reference [68] proposes an "early leaders" strategy for biological CDR, mitigating uncertainties and viability concerns through phased implementation in resilient regions.
- Technological innovations: Beerling et al. [65] assess EW in US agriculture, projecting 0.16–0.49 $GtCO_2$/yr removal by 2070, offering co-benefits such as soil pH improvement and N_2O mitigation (90–120 $MtCO_2e$/yr avoided), with costs declining to \$100–130/tCO_2. It emphasizes monitoring trace metals and ensuring equity-focused engagement. Reference [74] reviews the CDR literature, noting wide deployment variability and urging interdisciplinary efforts to reduce uncertainties.

7.5.5 Comparative Conclusion

BECCS can provide long-lasting, underground carbon storage when it's paired with reliable CO_2 sequestration. Models consistently show that BECCS is one of the few biologically based carbon-removal options capable of reaching multi-gigaton-per-year scales by the middle of the century. Still, BECCS competes heavily for land and biomass and can create indirect emissions throughout its supply chain. Because of this, its truly sustainable potential depends on strict rules for sourcing biomass, accurate life cycle accounting, and regional limits on land use. Recent policy and technical analyses highlight that BECCS works best when it relies on waste or residue biomass and when storage sites are proven to be secure. In contrast, large-scale use of dedicated energy crops could harm biodiversity and create tensions with food production [75, 76].

DAC provides very high permanence because the captured CO_2 can be stored deep underground or turned into stable minerals, and it avoids most land-use conflicts. DAC technology is advancing quickly, though progress varies across projects. Existing plants show that the technology works, but energy use is still high and today's capture costs remain expensive. Industry reports and agency assessments indicate that current plant-level costs generally range from a few hundred dollars to more than a thousand dollars per ton. While learning-curve projections point to potential long-term cost reductions, DAC is unlikely to match the near-term costs of cheaper carbon-removal approaches. Since DAC requires large amounts of electricity and/or heat, both the carbon footprint and the marginal cost of that energy play a crucial role in determining whether the process is truly carbon-negative and how much it costs overall.

Land-based biological methods, such as planting new forests, restoring degraded ones, and improving how existing forests are managed, can remove carbon at relatively low cost in many regions. They also deliver quick, near-term carbon uptake and offer important co-benefits for ecosystems, wildlife, and water systems. Their biggest drawback is that the carbon they store can be lost: forests are vulnerable to fires, pests, and land-use changes. Ensuring long-term permanence, therefore, depends on strong governance, careful monitoring, and conservative carbon accounting. Recent reviews and meta-analyses highlight that while these approaches can provide significant near-term carbon sinks, they cannot replace deep emissions cuts, and any credited removals must properly account for the risk of future reversals [77, 78].

Enhanced rock weathering and ocean alkalinity enhancement—whether through spreading crushed minerals along coastlines or using electrochemical methods—aim to lock CO_2 into stable carbonate minerals or boost the ocean's natural ability to absorb carbon. In theory, these methods could offer long-lasting storage without needing large areas of land. But they are still in the early stages of development and face major practical hurdles, including the need for significant mining, grinding, and transportation of materials. Their potential ecological effects and true large-scale costs are also not yet well understood. Recent technical reviews stress the need for pilot and demonstration projects, paired with thorough environmental assessments, before any widespread deployment [79, 80]. Soil carbon sequestration, adding biochar to soils, and restoring coastal "blue carbon" ecosystems like mangroves and seagrasses can deliver meaningful near-term carbon removal while also improving ecosystem health—often at relatively low cost. However, their long-term permanence and ability to scale globally are limited, and accurately measuring and verifying the carbon they store is challenging. Because of these constraints, carbon markets and policy frameworks need to use conservative permanence assumptions, ensure strong MRV, and provide incentives for long-term stewardship so that any claimed removals are truly credible [80].

When weighing different carbon-removal strategies for policy or project decisions, two practical themes consistently show up in the research. First, it makes sense to prioritize low-cost, low-risk natural solutions, such as preventing deforestation, restoring ecosystems, and improving soil carbon, when they deliver strong co-benefits and can be reliably protected over time. Second, durable engineered options like DACCS, sustainably sourced BECCS, and mineralization need to be developed alongside these natural approaches, since they provide the long-term, high-certainty removals that climate models indicate will be essential later this century. While today's cost projections show that engineered solutions will stay more expensive in the near term, they are still necessary for achieving net-negative emissions because they offer permanence and can scale when paired with secure storage [81]. A strong national or corporate carbon-removal strategy needs to include full LCA, clear accounting of co-benefits and trade-offs, such as impacts on land, biodiversity, and water, along with reliable permanence guarantees and transparent MRV and governance systems that can handle any future carbon reversals. This means investment should balance fast deployment of scalable biological sinks with focused R&D, demonstration projects, and supportive policies for durable engineered removals. Institutions must also enforce rigorous accounting so that any reported removals are truly additional, verifiable, and maintained over the long term. These recommendations are broadly aligned across IPCC assessments, recent life cycle analysis studies, and independent market trackers, all of which highlight both the technical progress achieved so far and the substantial gaps that still exist in cost, scalability, and governance.

Acknowledgments

This chapter is based on real-world projects and operational data and inspired by them. It also provides valuable insights for the applied work. The authors would also like to acknowledge the use of artificial intelligence tools to enhance the clarity, consistency, and overall quality of the manuscript during the writing and editing process.

References

1 M. M. Amiri, M. T. Ameli, H. Ameli, M. R. Aghamohammadi, and G. Strbac, "Cutting-edge technologies in sustainable energy planning and decarbonization," in *Sustainable Supply Chain of Renewable Energy Networks*, Springer Nature, 2025, pp. 69–99.

2 Reuters, "Carbon removal needs to quadruple to meet climate goals, researchers say," June 5, 2024. Accessed: Jan. 21, 2025. [Online]. Available: https://www.reuters.com/business/environment/carbon-removal-needs-quadruple-meet-climate-goals-researchers-say-2024-06-05

3 M. Daneshvar, B. Mohammadi-Ivatloo, and K. Zare, "Overview of modern energy networks," in *Coordinated Operation and Planning of Modern Heat and Electricity Incorporated Networks*, John Wiley & Sons, 2022, pp. 1–11.

4 M. Daneshvar, B. Mohammadi-Ivatloo, A. Anvari-Moghaddam, and R. Razzaghi (Eds.), *Interconnected Modern Multi-energy Networks and Intelligent Transportation Systems: Towards a Green Economy and Sustainable Development*. John Wiley & Sons, 2024.

5 Q. Ma, Y. Ye, Z. Liu, X. Liu, and G. Strbac, "Carbon cap based multi-energy sharing among heterogeneous microgrids using multi-agent safe reinforcement learning method with credit assignment and sequential update," *Applied Energy*, vol. 393, p. 126018, 2025.

6 IPCC, "The synthesis report of the sixth assessment report," 2023. Accessed: Jan. 21, 2024. [Online]. Available: https://www.ipcc.ch/synthesis-report/

7 IRENA, "World energy transitions outlook 2023: 1.5 °C pathway," June 2023. Accessed: Jan. 21, 2025. [Online]. Available: https://www.irena.org/-/media/Files/IRENA/Agency/Publication/2023/Jun/IRENA_World_energy_transitions_outlook_2023.pdf

8 C. Bataille et al., "The deep decarbonization pathways project (DDPP): Insights and emerging issues," *Climate Policy*, vol. 16, no. sup1, pp. S1–S6, 2016.

9 L. Mercer, J. Burke, and S. Rodway-Dyer, "Towards improved cost estimates for monitoring, reporting and verification of carbon dioxide removal," London

School of Economics and Political Science, 2024. Accessed: Apr. 21, 2025. [Online]. Available: www.lse.ac.uk/granthaminstitute

10 IEA, "Carbon capture utilisation and storage," 2025. Accessed: April. 21, 2025. [Online]. Available: https://www.iea.org/energy-system/carbon-capture-utilisation-and-storage

11 M. A. Hayat, K. Alhadhrami, and A. M. Elshurafa, "Which bioenergy with carbon capture and storage (BECCS) pathways can provide net-negative emissions?," *International Journal of Greenhouse Gas Control*, vol. 135, p. 104164, 2024.

12 Y. Liu et al., "Addressing solar power curtailment by integrating flexible direct air capture," *Carbon Capture Science & Technology*, vol. 14, p. 100304, 2025.

13 E. J. Graham et al., "Optimization of a combined power plant CO_2 capture and direct air capture concept for flexible power plant operation," *Energy & Environmental Science*, vol. 17, no. 12, pp. 4157–4174, 2024.

14 Federal Energy Regulatory Commission, "FERC Order No. 2222 explainer: Facilitating participation in electricity markets by distributed energy resources," 2020.

15 S. Twidale, "Global carbon removal market could reach $100 billion/yr from 2030-35, report says," *Reuters*, June 27, 2024. Accessed: Jan. 21, 2024. [Online]. Available: https://www.reuters.com/sustainability/climate-energy/global-carbon-removal-market-could-reach-100-billionyr-2030-35-report-says-2024-06-27/

16 S. Oh, J. Greene, M. Honegger, and A. Michaelowa, "Review of economics and policies of carbon dioxide removal," *Current Sustainable/Renewable Energy Reports*, vol. 12, no. 1, p. 6, 2025.

17 J. Murdoch, "Greenhouse gas removals: Call for evidence," *Ember*, June 20, 2025. Accessed: September. 21, 2025. [Online]. Available: https://ember-energy.org/app/uploads/2025/07/UK-Greenhouse-gas-removals-call-for-evidence-PDF.pdf

18 H. Ishaq and C. Crawford, "Negative emission technologies: A way forward?," *RSC Sustainability*, vol. 3, no. 9, pp. 3652–3680, 2025.

19 S. Ranevska, "ExxonMobil to transport & store CO_2 from the AtmosClear BECCS project in Louisiana," *Carbon Herald*, September 23, 2025. Accessed: Jan. 21, 2024. [Online]. Available: https://carbonherald.com/exxonmobil-to-transport-store-co2-from-the-atmosclear-beccs-project-in-louisiana/

20 A. Almena, V. G. Pashakolaie, and M. Martin, "Unlocking BECCS viability through monetizing societal benefits by techno-socio-economic assessment," *Sustainable Production and Consumption*, vol. 59, pp. 255–273, 2025.

21 EFI Foundation, "Surveying the BECCS landscape," January 2022. Accessed: Jan. 21, 2024. [Online]. Available: https://efifoundation.org/wp-content/uploads/sites/3/2022/03/Survey-the-BECCS-Landscape_Report-v2.pdf

22 M. Abegg, Z. Clulow, L. Nava, and D. M. Reiner, "Expert insights into future trajectories: Assessing cost reductions and scalability of carbon dioxide removal technologies," *Frontiers in Climate*, vol. 6, p. 1331901, 2024.

23 L. Cretarola and F. Viganò, "Comparative performance analysis of bioenergy with carbon capture and storage (BECCS) technologies," *Energies*, vol. 18, no. 18, p. 4800, 2025.

24 Y. Gaucher, K. Tanaka, D. J. Johansson, O. Boucher, and P. Ciais, "Potential and costs required for methane removal to compete with BECCS as a mitigation option," *Environmental Research Letters*, vol. 20, no. 2, p. 024034, 2025.

25 M. Dufour and K. Möllersten, "Nordic net-zero: Counterbalancing residual emissions in the context of unevenly distributed BECCS potentials," *Climate Policy*, pp. 1–17, 2025. https://doi.org/10.1080/14693062.2025.2546116

26 L. G. Pereira et al., "Negative emission strategies to reduce the carbon intensity of Brazilian sugarcane ethanol under RenovaBio," *Cleaner Energy Systems*, vol. 12, p. 100208, 2025.

27 S. E. Tanzer, S. C. A. Cummings, O. M. Valenti, M. Junginger, and A. S. Duden, "So you want to build a BECCS plant: The patchwork policy context for bioelectricity with carbon capture and storage in Europe," *Environmental Research Communications*, vol. 7, no. 5, p. 051008, 2025.

28 Sylvera, "Direct air capture in 2025: The end of hype, the start of realism," 2025. Accessed: May. 21, 2025. [Online]. Available: https://www.sylvera.com/blog/direct-air-capture-dac-2025-progress-challenges-future

29 Allied Offsets, "Direct air capture (DAC) data summary," 2024. Accessed: Jan. 21, 2025. [Online]. Available: https://blog.alliedoffsets.com/the-current-state-of-direct-air-capture

30 J. Kotowicz, K. Niesporek, and O. Baszczeńska, "Advancements and challenges in direct air capture technologies: Energy intensity, novel methods, economics, and location strategies," *Energies*, vol. 18, no. 3, p. 496, 2025.

31 P. Agbo et al. (2024). Technological Innovation Opportunities for CO_2 Removal. U.S. Department of Energy's Carbon Negative Shot.

32 K. Lebling et al., "6 Things to know about direct air capture," World Resources Institute, September 2, 2025. Accessed: September. 21, 2025. [Online]. Available: https://www.wri.org/insights/direct-air-capture-resource-considerations-and-costs-carbon-removal

33 S. Ranevska, "Occidental's Stratos DAC hub to launch operations by the end of 2025," *Carbon Herald*, August 8, 2025. Accessed: September. 21, 2025. [Online]. Available: https://carbonherald.com/occidentals-stratos-dac-hub-to-launch-operations-by-the-end-of-2025/

34 K. Sievert, T. S. Schmidt, and B. Steffen, "Considering technology characteristics to project future costs of direct air capture," *Joule*, vol. 8, no. 4, pp. 979–999, 2024.

35 T. Daniel, A. Masini, C. Milne, N. Nourshagh, C. Iranpour, and J. Xuan, "Techno-economic analysis of direct air carbon capture with CO_2 utilisation," *Carbon Capture Science & Technology*, vol. 2, p. 100025, 2022.

36 T. Bush, "Direct air capture: The technology racing to scale carbon removal," *Decarbonfuse*, June 11, 2025. Accessed: September. 21, 2025. [Online]. Available: https://decarbonfuse.com/posts/direct-air-capture-the-technology-racing-to-scale-carbon-removal

37 S. Lilonfe, S. Rodgers, A. F. Abdul-Manan, I. Dimitriou, and J. McKechnie, "Technical, economic and lifecycle greenhouse gas emissions analyses of solid sorbent direct air capture technologies," *Carbon Capture Science & Technology*, vol. 15, p. 100380, 2025.

38 Y. Wang, L. Qu, H. Ding, P. Webley, and G. K. Li, "Distributed direct air capture of carbon dioxide by synergistic water harvesting," *Nature Communications*, vol. 15, no. 1, p. 9745, 2024.

39 C. Y. Chuah et al., "Applicability of adsorbents in direct air capture (DAC): Recent progress and future perspectives," *Industrial & Engineering Chemistry Research*, vol. 64, no. 8, pp. 4117–4147, 2025.

40 T. R. Alsenani and M. M. Amiri, "A survey on optimization techniques and modeling tools in sustainable energy planning," in *Sustainable Supply Chain of Renewable Energy Networks*, Springer Nature, 2025, pp. 175–204.

41 S. M. Kalfati and A. Abdulla, "An open-source dynamic model for direct air capture of carbon dioxide using solid sorbents," *Carbon Capture Science & Technology*, vol. 17, p. 100516, 2025.

42 N. Brazzola, A. Meskaldji, A. Patt, T. Tröndle, and C. Moretti, "The role of direct air capture in achieving climate-neutral aviation," *Nature Communications*, vol. 16, no. 1, p. 588, 2025.

43 V. Eke, T. Sahu, K. K. Ghuman, M. Freire-Gormaly, and P. G. O'Brien, "A comprehensive review of life cycle assessments of direct air capture and carbon dioxide storage," *Sustainable Production and Consumption*, vol. 55, pp. 217–241, 2025.

44 Y. M. Wei et al., "Unlocking the economic potential of direct air capture technology: Insights from a component-based learning curve," *Technological Forecasting and Social Change*, vol. 215, p. 124109, 2025.

45 R. Koch and R. Dittmeyer, "Comparative analysis of industrialization potentials of direct air capture technologies," *Frontiers in Climate*, vol. 7, p. 1558396, 2025.

46 S. Kar et al., "Direct air capture of CO_2 for solar fuel production in flow," *Nature Energy*, vol. 10, no. 4, pp. 448–459, 2025.

47 Q. Wang, C. Du, and X. Zhang, "Direct air capture capacity configuration and cost allocation based on sharing mechanism," *Applied Energy*, vol. 374, p. 124037, 2024.

48 L. Zhang, Y. Liang, A. Kioka, and T. Tsuji, “Economically viable geological CO_2 storage from direct air capture has critical threshold of 70% CO_2 concentration,” *Communications Engineering*, vol. 4, no. 1, p. 127, 2025.

49 J. Mota-Nieto, “Carbon dioxide removal (CDR): A key pillar of carbon management and sustainability,” Oxford Institute for Energy Studies, 2024.

50 S. Dell’Aversano, C. Villante, K. Gallucci, G. Vanga, and A. Di Giuliano, “E-fuels: A comprehensive review of the most promising technological alternatives towards an energy transition,” *Energies*, vol. 17, no. 16, p. 3995, 2024.

51 A. Soler et al., “E-fuels: A techno-economic assessment of European domestic production and imports towards 2050,” *Concawe, Brussels*, vol. 26, pp. 27–28, 2022.

52 S. M. Smith et al., *The State of Carbon Dioxide Removal—2nd Edition*, The State of Carbon Dioxide Removal, 2024.

53 University of Oxford, “New report states 7–9 billion tonnes of CO_2 must be sustainably removed per year to hit climate targets,” June 5, 2024. Accessed: September. 21, 2025. [Online]. Available: https://www.ox.ac.uk/news/2024-06-05-new-report-states-7-9-billion-tonnes-co2-must-be-sustainably-removed-year-hit

54 A. Pratt (2024). ARIES and ADMS Test Bed Overview and Updates. National Renewable Energy Laboratory (NREL), Golden, CO (United States), Report NREL/PR-5D00-92384.

55 F. Ding et al. (2022). Federated Architecture for Secure and Transactive Distributed Energy Resource Management Solutions (FAST-DERMS). National Renewable Energy Laboratory (NREL), Golden, CO (United States), No. NREL/TP-5D00-81566.

56 A. O. Ishola, O. G. Odunaiya, and O. T. Soyombo, “Framework for tailoring consumer-centric communication to boost solar energy adoption in US households,” *International Journal of Frontiers in Science and Technology Research*, vol. 7, no. 2, pp. 60–76, 2024.

57 NC Clean Energy Technology Center, “50 states of virtual power plants & supporting distributed energy resources,” 2024. Accessed: Jan. 21, 2025. [Online]. Available: https://nccleantech.ncsu.edu/wp-content/uploads/2025/02/2024-VPP-Report-Final.pdf

58 Cusln.org, “FERC ORDER 2222 & DER policy and implementation report,” July 2025. Accessed: September. 21, 2025. [Online]. Available: https://cuswebsite.blob.core.windows.net/2222tracker/Tracker-Report-July-2025.pdf

59 U.S. Department of Energy, “U.S. Department of Energy releases new reports highlighting benefits of consumer-centric solutions for households, businesses, and utilities,” 2024. Accessed: Jan. 21, 2025. [Online]. Available: https://www.energy.gov/technologytransitions/articles/us-department-energy-releases-new-reports-highlighting-benefits

60 NREL, "Impact of IEEE 1547 standard on smart inverters and the applications in power systems," 2024. Accessed: Jan. 21, 2025. [Online]. Available: https://www.nrel.gov/media/docs/libraries/grid/smart-inverters-applications-in-power-systems.pdf?sfvrsn=fae53014_0

61 QualityLogic, "IEEE 2030.5 takes off: The latest news on the IEEE 2030.5 standard," January 13, 2025. Accessed: September. 21, 2025. [Online]. Available: https://www.qualitylogic.com/knowledge-center/ieee-2030-5-takes-off

62 J. Wang, J. Comden, and A. Bernstein (2024). Real Time-Optimal Power Flow-based Distributed Energy Resource Management System (DERMS). National Renewable Energy Laboratory (NREL), Golden, CO (United States), No. NREL/TP-5D00-87767.

63 NREL, "Status of power system transformation leading topics of 2024," September 2024. Accessed: Jan. 21, 2025. [Online]. Available: https://docs.nrel.gov/docs/fy24osti/91357.pdf

64 Allied Offsets, "Q1 2025 carbon dioxide removal (CDR) market update & insights," 2025. Accessed: September. 21, 2025. [Online]. Available: https://blog.alliedoffsets.com/q1-2025-carbon-dioxide-removal-cdr-market-update

65 D. J. Beerling et al., "Transforming US agriculture for carbon removal with enhanced weathering," *Nature*, vol. 638, pp. 1–10, 2025.

66 IIASA, "The 2024 state of CDR report: Scaling up CO_2 removal to meet Paris targets," June 5, 2024. Accessed: Jan. 21, 2025. [Online]. Available: https://iiasa.ac.at/news/jun-2024/2024-state-of-cdr-report-scaling-up-co2-removal-to-meet-paris-targets

67 M. Bindl, M. R. Edwards, and R. Y. Cui, "Risks of relying on uncertain carbon dioxide removal in climate policy," *Nature Communications*, vol. 16, no. 1, p. 5958, 2025.

68 E. Cox, L. Waller, J. Palmer, and R. Bellamy, "Carbon removal support is tempered by concerns over whether biological methods are worth it," *Communications Earth & Environment*, vol. 6, no. 1, p. 711, 2025.

69 K. Motlaghzadeh, N. Craik, J. Moreno-Cruz, V. Schweizer, J. Fuhrman, and K. W. Hipel, "Applying equity principles leads to higher carbon removal obligations in Canada," *Communications Earth & Environment*, vol. 6, no. 1, p. 88, 2025.

70 T. Repke, F. Müller-Hansen, E. Cox, and J. C. Minx, "Attention and positive sentiments towards carbon dioxide removal have grown on social media over the past decade," *Communications Earth & Environment*, vol. 5, no. 1, p. 763, 2024.

71 World Economic Forum, "The new renewable revolution: Why carbon dioxide removal will transform the carbon market," April 10, 2025. Accessed: September. 21, 2025. [Online]. Available: https://www.weforum.org/stories/2025/04/carbon-dioxide-removal-carbon-credits/

72 A. Merfort et al., "Separating CO_2 emission from removal targets comes with limited cost impacts," *Nature Communications*, vol. 16, no. 1, p. 5298, 2025.

73 J. Alcalde, G. Johnson, and J. J. Roberts, "National climate strategies show inequalities in global development of carbon dioxide geological storage," *Communications Earth & Environment*, vol. 6, no. 1, p. 61, 2025.

74 S. Lück et al., "Scientific literature on carbon dioxide removal revealed as much larger through AI-enhanced systematic mapping," *Nature Communications*, vol. 16, no. 1, p. 6632, 2025.

75 A. Marcu and M. Varricchio, "Bioenergy with carbon capture and storage (BECCS) in the EU challenges and opportunities," European Roundtable on Climate Change and Sustainable Transition, 2025.

76 I. Martin (2025). Bioenergy with Carbon Capture and Storage Evidence on Potential Environmental Impacts. Chief Scientist's Group Report.

77 S. Cho, S. Baral, and D. Burlakoti, "Afforestation/reforestation and avoided conversion carbon projects in the United States," *Forests*, vol. 16, no. 1, p. 115, 2025.

78 S. Gorain, S. Dutta, S. Balo, A. Malakar, M. R. Choudhury, and S. Das, "Harnessing green wealth: A two-decade global assessment of forest carbon sequestration and credits and the economic implications of sustainable forest management practices," *Journal of Environmental Management*, vol. 393, p. 126987, 2025.

79 F. Ramos, "Support to the development of methodologies for the certification of permanent carbon removals," ICF S.A., 2025.

80 I. Butnar et al., "A review of life cycle assessment methods to inform the scale-up of carbon dioxide removal interventions," *Wiley Interdisciplinary Reviews: Energy and Environment*, vol. 13, no. 6, p. e540, 2024.

81 H. Lee et al., "Climate change 2023 synthesis report summary for policymakers," in *CLIMATE CHANGE 2023 Synthesis Report: Summary for Policymakers*, IPCC, 2024.

8

Evaluating Technological Innovations for Resilience of Decarbonized Communities with Energy Diversification

Justice Adjei Owusu[1], Gurcan Comert[1]*, Negash Begashaw[2], Ayobami Abiodun Taiwo[1], Marwan Bikdash[1], and Theresa Thompson[2]*

[1]*Department of Computational Data Science and Engineering, North Carolina Agricultural and Technical State University, Greensboro, NC, USA*
[2]*Department of Computer Science, Physics and Engineering, Benedict College, Columbia, SC, USA*

*Corresponding authors: jadjeiowusu1@aggies.ncat.edu, gcomert@ncat.edu

List of Acronyms

API	Application programming interface
CAIDI	Customer average interruption duration index
DOE	Department of Energy
EIA	US Energy Information Administration
ESS	Energy storage system
EV	Electric vehicle
GW	Gigawatt
HVAC	Heating, ventilation, and air conditioning
MW/MWh	Megawatt/megawatt-hour
PV	Photovoltaic
SAIDI	System average interruption duration index
SAIFI	System average interruption frequency index
SDG	Sustainable Development Goal
US	United States of America
UN	United Nations
V2G	Vehicle-to-grid
VRES	Variable renewable energy sources

8.1 Introduction

There is a growing interest in the resilience of power systems due to climate change, aging infrastructure, increasing demand, and growing geopolitical volatility [1]. Power system disruptions triggered by hurricanes, wildfires, extreme heatwaves, and cyber-physical threats are increasing among the US states and communities [2]. These disruptions have revealed systemic weaknesses in conventional grid structures and underscored the urgency of transitioning toward more robust, adaptable, and diversified energy portfolios.

Energy diversification is the strategic expansion of the energy mix to include multiple generation sources such as solar, wind, natural gas, nuclear, and hydrogen. Diversifying energy sources can greatly improve the resilience and robustness of the system for end users. Much emphasis is placed on energy diversification to avoid supply chain risks, climate vulnerability, and issues that could arise due to technical disruptions. Therefore, it is important to diversify power generation across different portfolios of renewable and low-carbon technologies.

In recent years, there has been a surge of technologies in the generation of power for transportation, food industries, and other sectors, from traditional energy sources such as fossil fuel to more eco-friendly sources to help reduce the emission of carbon dioxide (CO_2) into the atmosphere. This is in line with the United Nations (UN) Sustainable Development Goals (SDGs) 7, 11, and 13, which focus on promoting renewable energy to reduce dependence on fossil fuels, fostering sustainable cities and communities to reduce urban emissions, and taking urgent action to combat climate change through climate action initiatives.

Decarbonization is the deliberate effort to reduce net greenhouse emissions to mitigate impacts on climate change. Decarbonization of electricity is fundamental to deep decarbonization (decarbonization by 80–100%) because electricity can power many end users of energy and can be decarbonized at the power generation facility, rather than at every end-user device [3]. Decarbonization strategies should be integrated in a manner that preserves or enhances the overall resilience of communities.

The choice of the mix of generation for a particular state depends on factors like geographic location, climate, and socioeconomic conditions. States with an abundance of sun may favor solar energy, while coastal regions may rely heavily on wind energy. Balancing the mix of these sources is essential in improving the resilience and reliability of energy systems.

Resilience in electricity supply is defined as the prevention of disruption of power service caused by natural disasters and the restoration of generation as soon as possible after an inevitable outage [4]. A diversified low-carbon portfolio enhances community resilience by:

1. Mitigating the variability of renewable electricity generation sources: the variability of sources like wind and solar power makes it a priority to not depend on a single source. The impact of daily fluctuations in wind/solar power generation is a common theme in the field. Balancing this variation in peak and low seasons is very essential for the resilience of the system. Several methods, such as expanding transmission, improving the forecasting of variable renewable energy sources (VRES), increasing dispatchable backup power generation, and decoupling VRES generation from the grid, have been proposed to resolve variability [5].
2. Reducing single-point failures: one way to eliminate system vulnerability, thereby improving resilience, is to reduce single-point failures. The risk associated with equipment failure, such as hydroelectric turbines, weather anomalies for sources such as wind and solar costs, is reduced if we do not depend on a single source. Diversification and having multiple failure points improves community resilience.
3. Enhancing energy security: during global crises, there are many economic disruptions and instability due to high oil prices and fuel shortages, which then affect energy supplies, especially in economies that rely solely on fossil fuels for their electricity generation [6]. Depending on low-carbon sources like geothermal, wind, hydroelectric, and solar reduces the overreliance on fossils, which insulates states from geopolitical issues that may occur, and improves the resilience of the system.
4. Supporting economic stability: relying on natural resources such as the sun, wind and flowing water to spin turbines, and the heat of the earth will help channel funds that could have been used to purchase fossils into other sectors of the economy, thus improving the local economy. These diverse low-carbon sources will also buffer against market volatility and improve the system's resilience.

8.1.1 Chapter Objectives

1. To investigate the resilience of communities as a function of energy diversification levels.
2. To examine how energy diversification supports long-term decarbonization and strengthens climate and infrastructure resilience.
3. To explore innovative technologies that enhance the resilience of decarbonized communities through diverse and sustainable energy solutions.
4. To identify and analyze key decarbonization factors including emissions reduction, electrification, and low-carbon infrastructure within communities.

8.2 Problem Statement

The frequent increase in load growth, climate-induced disruptions due to hurricanes, wildfires, and heat waves poses a significant threat to the reliability and stability of energy systems across the United States. These events can cause power outages, economic losses, and public health risks, especially in more vulnerable communities. As these extreme events become more widespread, ensuring the resilience of the power infrastructure and energy systems has become a very critical challenge for policymakers, utilities, and urban planners [7]. A diversified energy portfolio for a state or city reduces the dependence on any single energy type, mitigates the impact of supply chain disruptions, and improves power restoration after disasters. Resiliency is defined as the ability of a system to quickly return to its intended or non-failure state [8]. The key element here is the ability of a system to bounce back to a desired state after disruptions. The actual relationship between energy diversification and system resilience remains inadequately explored, as existing studies often treat both topics as separate domains. Current reliability metrics, including SAIFI (system average interruption frequency index), CAIDI (customer average interruption duration index), SAIDI (system average interruption duration index), recovery time, etc., are typically utility-focused and may not fully represent resilience but can give a fair idea of the resiliency of an energy system. CAIDI is the measure of how long an interruption lasts or how quickly the utility responds to an interruption [9]. SAIFI indicates how often the average customer experiences a sustained interruption in a given period [10]. SAIDI, on the other hand, reflects the total duration of interruptions for the average customer over a defined period [10]. Socioeconomic vulnerabilities or adaptive capacities of communities also play a key role in the resilience of the power system.

These trends motivate a chapter linking the diversity of a state's power system, technology choices, decarbonization, and resilience (recorded CAIDI values). A lower CAIDI value translates to a more resilient energy sector. Incorporating utility outage data, energy mix profiles, and community vulnerability indices in different US states, how energy diversification influences the resilience of power systems and communities under evolving conditions was assessed.

8.3 Methodology

A data-driven approach was adopted to evaluate the relationship between energy diversification and power system resilience in the United States. This section integrates multiple data sets, preprocessing, and analytics to extract meaningful trends from the data.

8.3.1 Data Description

The power generation capability was sourced from the US Energy Information Administration (EIA) through its publicly available API (application programming interface) [11], which provides annual state-level statistics on installed capacity and generation type (Table 8.1). The dataset includes a wide range of energy sources, including:

- **Conventional sources:** coal, natural gas, nuclear, and petroleum.
- **Renewables:** wind, solar, hydroelectric, and geothermal.
- **Emerging technologies:** battery and hydrogen.

As summarized in Figure 8.1, complementary data on grid reliability metrics, specifically SAIDI, SAIFI, and CAIDI, were collected from the US Department of Energy (DOE) EIA (EIA Form 861) [13].

8.3.2 Energy Diversification Metric: Shannon Diversity Index

To quantify the diversity of the energy mix of each state over time, the *Shannon diversity index* (H') [14] was employed with a fixed logarithmic base of 25 (the number of categories of energy sources in our energy data), a widely used metric to capture the heterogeneity of a population or portfolio. Using base 25 normalizes the index to the unit interval, ensuring cross-year comparability even when some sources are absent in a given state-year. In this context, the index reflects how evenly electricity generation capacity is distributed among different energy types, accounting for both *variety* and *balance*,

$$H'_{25} = -\sum_{i=1}^{25} p_i \log_{25}(p_i), \log_{25}(x) = \frac{\ln x}{\ln 25} \tag{8.1}$$

Table 8.1 Description of power generation sources.

Energy source	Description
Solar (SOL)	Production of electricity through PV installations; converts sunlight directly to electricity
Wind (WND)	Generation of electricity via wind turbines; converts kinetic energy of moving air into electrical energy
Geothermal (GEO)	Production of electricity from thermal energy stored beneath the earth's surface
Hydroelectric conventional (HYC)	Generation of electricity using the flow of water to spin turbines

Source: Data from [12].

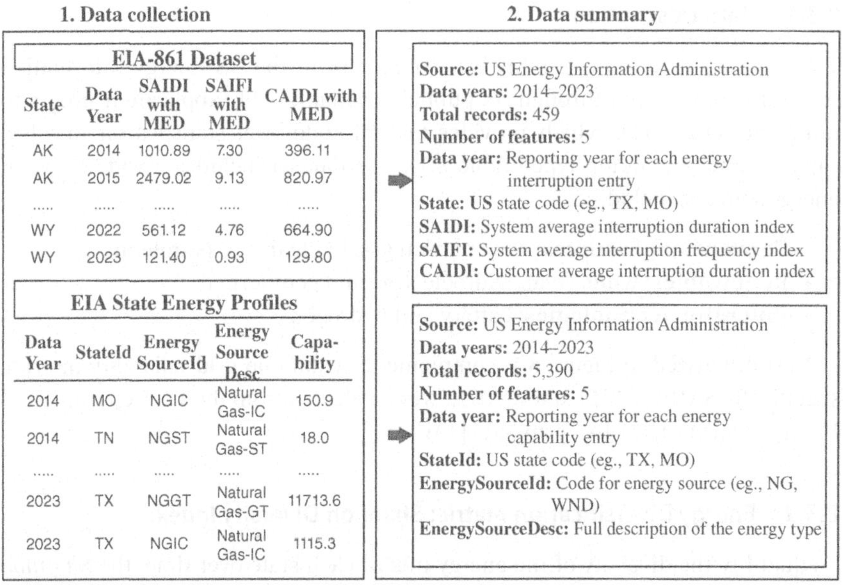

Figure 8.1 Data collection and variable summary from EIA-861 and state energy profiles.

where:

- p_i is the proportion of the total generation capacity of the source i in the given state-year (terms with $p_i = 0$ contributing 0 by continuity).
- The sum is taken over the entire set of 25 distinct energy sources (e.g., coal, natural gas, wind, and solar, among others).

Because $\log_{25}(25) = 1$, H'_{25} already lies in [0,1] with $H'_{25} = 1$ when the capacity is distributed uniformly across all 25 sources and $H'_{25} = 0$ when a single source dominates entirely.

8.3.3 Data Analysis

To analyze patterns in power generation, energy diversification, and system resilience across US states, an analytical framework was designed. Figure 8.2 illustrates the sequential steps undertaken in the analysis. Each step builds logically on the previous one, transitioning from descriptive trend analysis to diagnostic evaluations and economic implications.

First, the evolution of total generation capacity in different states of the United States was analyzed. Rapid increase in capacity is expected due to infrastructure modernization, population growth, or policy-driven investment. Compared with

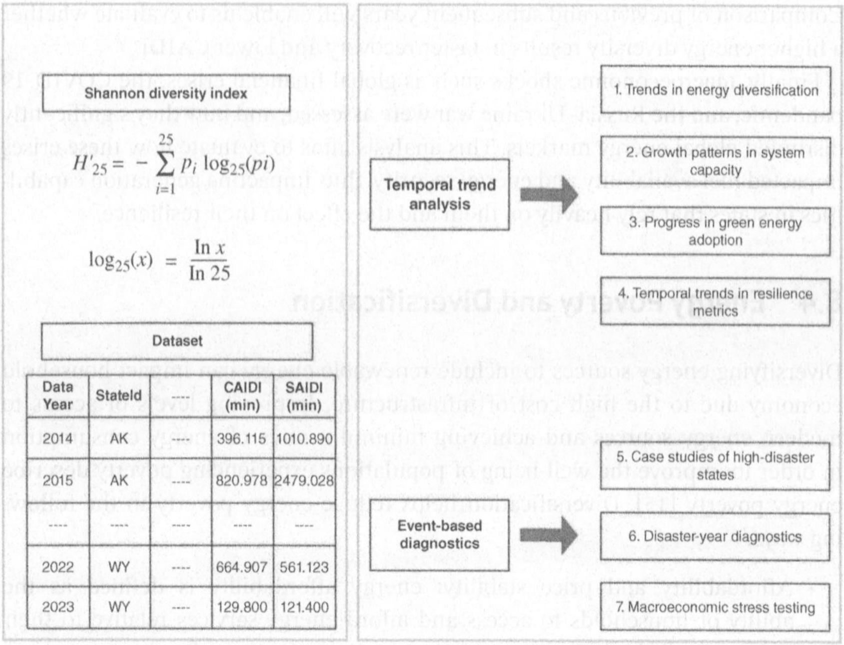

Figure 8.2 Assessing energy diversification and resilience.

CAIDI values over time, whether increased generation capacity corresponds to shorter outage durations was explored. Trends in the energy mix over time were also identified. The transition to low-carbon energy and the regional and temporal growth of clean energy sources such as wind, solar, and hydro were considered in states. This subsection quantifies the share of total generation from these renewables. A temporal trend in CAIDI values is analyzed across US states to identify whether improvements in resilience through the reduction of CAIDI coincide with increasing energy diversification. The underlying hypothesis is that a more distributed energy portfolio may enhance system redundancy and reduce average interruption delays.

High-impact disaster-prone states like Florida and California (CA) offer crucial test beds to evaluate the effectiveness of energy diversification. Florida, often hit by hurricanes, and CA, by wildfires and extreme heatwaves, offer the need to contrast the energy diversification index and resilience during these disasters. Isolated disaster-year diagnostics offer detailed information on how real-time system stress interacts with energy diversification. Assessing the disturbances during Hurricane Ian in 2022 and the Texas (TX) winter storm in 2021 will allow us to determine the resilience of the power infrastructure compared to other years.

Comparison of previous and subsequent years will enable us to evaluate whether a higher energy diversity results in faster recovery and lower CAIDI.

Finally, macroeconomic shocks such as global financial crises, the COVID-19 pandemic, and the Russia–Ukraine war were assessed, and how they significantly disrupted global energy markets. This analysis aims to evaluate how these crises impacted fuel availability and energy security, thus impacting generation capabilities in states that rely heavily on them and the effect on their resilience.

8.4 Energy Poverty and Diversification

Diversifying energy sources to include renewable energy can impact household economy due to the high cost of infrastructure. Improving levels of access to modern energy sources and achieving minimum rates of energy consumption in order to improve the well-being of populations experiencing poverty describe energy poverty [15]. Diversification helps reduce energy poverty in the following ways:

- Affordability and price stability: energy affordability is defined as the ability of households to access and afford energy services relative to their income levels [16]. The diversified portfolio will reduce the exposure to any single-point failure that can arise due to increases in fuel prices and a surge in demand. Stable grids will translate into lower bills. Price stability will help low-income customers plan well.
- Reliability and outage cost reduction: diversification reduces failure risks and enables resilient operation during grid disturbances. A reliable grid will reduce bill volatility and cost to vulnerable consumers when it comes to lost income, food spoilage, telehealth, etc.
- Access and inclusion: diversifying energy systems by including portable decarbonized sources like solar and battery will improve accessibility and extend service quality to underserved neighborhoods and remote areas.
- Development of the local economy: local jobs will be created, with the intention of boosting the local economy through income tax and other tax revenues, which will support bill assistance.

8.5 Innovative Technologies Enabling Resilience

Technological innovation is critical to ensure that decarbonized communities maintain resilience during environmental, economic, and technical stresses. Beyond reducing emissions and ensuring net-zero carbon, we have to ensure that these systems are stable and adaptable:

- Microgrids: a microgrid is a localized energy network and capable of islanding from the grid that operates as a controllable structure to the main grid and is constrained within well-defined electrical boundaries [17]. Microgrids can operate as backup power sources when there is an outage on the main grid and improve the power system's quality by sustaining operation of the local grid. A prime example is Brooklyn, New York's microgrid network that enables Brooklyn residents to generate, share, and trade their renewable electricity within their neighborhood. This enhances energy resilience and sustainability. Microgrids reduce carbon emissions since most of their prime mover technologies are from decarbonized sources like solar photovoltaic (PV) and windmills.
- Batteries: electrical energy storage systems (ESSs) such as batteries are required to store surplus renewable energy from solar PV and wind turbines during off-peak hours to prevent wastage. The stored energy is then supplied back to support the grid in peak time to assure a resilient power supply to consumers. An example is the Hornsdale Power Reserve, one of the world' s largest lithium-ion battery storage systems located in South Australia [18], which is very critical in their electricity grid. It stores excess energy generated from renewable sources like wind and solar and releases energy during peak demand or outages. It is regarded as a benchmark for large-scale battery integration in energy systems.
- Vehicle-to- grid (V2G) systems: to help balance the problems in demand and supply, one of the most efficient systems is V2G technology. The V2G technology involves a flow of energy between the electric vehicle (EV) owners, aggregators, and power grid to achieve a stable balance between demand and supply [19]. EVs release stored electricity back to the grid when needed. A practical example is what Nuvve Corporation is doing in Delaware, United States [20]. They have commercialized V2G technology by linking multiple EV batteries to form a virtual power plant that safely stores power when the grid has too much and supplies power to the electrical grid to balance demands at peak hours, all with zero emissions.
- AI for load balancing: using AI optimization methods to balance load to improve resilience and efficiency in microgrid operations when green energy isn't available and demand trend changes is a key technological innovation in recent times to improve the resilience of decarbonized systems. Machine learning and deep learning algorithms are used to study complicated patterns in data on parameters including weather change, energy use, etc. to develop models that help in decision-making by adjusting microgrid controllers to change the allocation of resources to meet the demands [21]. It is also able to identify faults and system malfeasances. This improves the resilience of microgrid systems that depend on green energy.

- Sustainable buildings: to enhance the decarbonization drive and boost resilience, modern buildings are being designed to generate as much energy as they consume, ESS-certified heat pumps are being utilized instead of traditional HVAC (heating, ventilation, and air conditioning) systems, and carbon-negative materials are being used in construction. In Whisper Valley, TX, United States, every home incorporates geothermal heating and cooling systems connected to a community-wide geogrid, which taps into the earth's natural underground temperature to minimize energy use and cost by more than 50% [22]. They have solar panels on rooftops to complement the geogrid while encouraging the use of energy-efficient appliances.

8.6 Decarbonization Factors in Communities

To further enhance the decarbonization strategies, there have to be strategies to reduce greenhouse gas emissions but not compromise resilience. Decarbonization will be effective if technology, government policies, and citizen buy-in align. Some notable approaches include:

1. Emission reduction strategies: there must be a deliberate effort to curb the heating, cooking, and industrial processes that depend on fossil fuels. They can be replaced with low-carbon and eco-friendly appliances and devices. The deployment of heat pumps and cooktops, as in the case of Whisper Valley, that utilize decarbonized energy sources will increase the penetration of renewable energy.
2. Low-carbon mobility and infrastructure: investing in pedestrian pathways, protected bike lanes, and efficient public transit will encourage locals to depend on scooters, electric bikes, and EVs to commute. The number of EV charging stations should be strategically increased to improve accessibility for EV users. This will drastically reduce the number of cars that depend on internal combustion engines. All public transport can be converted to battery electric.
3. Community engagement: an efficient way to encourage decarbonization and improve resilience is to educate local residents about their choice of appliances and their use. They can be encouraged to use electrical appliances instead of those that depend on fossils in their homes and turn off devices when not in use. This will help conserve energy and improve resilience during peak hours.
4. Policy and financing mechanisms: policies like time-of-use pricing and carbon pricing encourage people to use low-carbon appliances. Green bonds and targeted government subsidies will help low-income earners own EVs and benefit from decarbonization programs.

8.7 Results and Discussion

In this section, the analysis is organized into two broad thematic categories: long-term trends and event-based diagnostics. The first category explores the evolution of energy diversification and growth in system capacity, highlighting patterns in the adoption of green energy. The second category focuses on resilience, examining how the energy system responded to specific disruptions such as extreme weather events or macroeconomic shocks. This includes both temporal assessments of resilience metrics and targeted diagnostics for high-disaster states and critical years.

8.7.1 Temporal Trend Analysis

Assessment of how the US energy mix and infrastructure evolved over time across different states is presented in this subsection.

8.7.1.1 Trends in Energy Diversification

The evolution of energy diversification, measured by the normalized Shannon index, across selected US states from 2014 to 2023 was evaluated. Figure 8.3 illustrates how the diversity of the energy mix of each state evolved over time and how it may be related to the overall resilience of the system.

CA consistently demonstrated one of the highest diversification scores throughout the period, with values exceeding 0.70 after 2020. This reflects a consistently broad and balanced energy portfolio.

TX's trend persisted upward from 0.54 to 0.64 by 2023. Vermont (VT) also had a modest increase from 0.55 to 0.62 with a small drop in 2025 followed by gradual gains that indicated a mature and improving energy strategy.

North Carolina (NC) and South Carolina (SC) both exhibited moderate but steadily increasing diversification from 2014 to 2023. NC values rose from below 0.60 in 2014 to approximately 0.67 in 2023, indicating a gradual expansion to new energy sources. SC dipped in 2015 and rose to around 0.70 around 2020, easing slightly afterward, although it remained high.

Connecticut (CT) and Missouri (MO) showed slightly fluctuating trends. Although CT improved its diversification from approximately 0.63 to 0.68 in 2023, MO declined to 0.53 in 2019 followed by a sharp jump to 0.62 around 2020, then stabilized.

Nevada (NV) began with one of the lowest scores (0.53) but improved steadily, reaching 0.61 by 2023, indicating a significant diversification push in recent years.

Illinois (IL) had relatively lower and more variable Shannon indices, particularly in the mid-2010s, before rebounding closer to 0.62 in later years. This variability may reflect inconsistent shifts in the energy strategy.

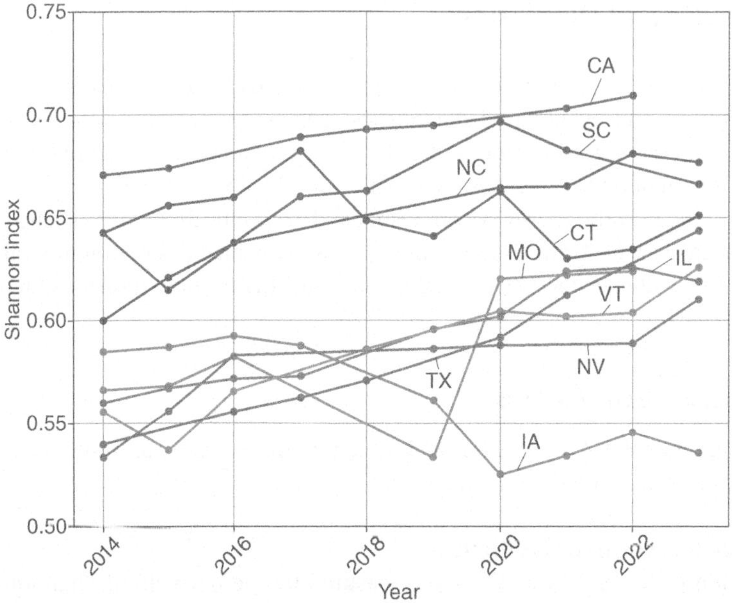

Figure 8.3 Trends in energy diversification.

Iowa (IA) consistently showed the lowest diversification among the group, with a Shannon index mainly around 0.52–0.59. This indicates a more concentrated energy mix, possibly dependent on a limited number of sources such as wind and coal, making it more vulnerable to supply disruptions.

In general, Figure 8.3 suggests that states with increasingly high Shannon index values, such as CA and Washington (WA), are likely to invest in various energy strategies, which can improve resilience. Meanwhile, states with flat or declining diversification, such as IA, may face elevated risk during disruptions due to energy concentration (Table 8.2).

8.7.1.2 Temporal Trends in Resilience

The national averages for CAIDI suggest a moderate improvement in the duration and frequency of the outage over time. However, as shown in Figure 8.4, there is still some volatility in the diversity index; states such as TX had a gradual increase in their diversity and little fluctuation compared to NC. There has also been a steady improvement in the CAIDI values in CA, demonstrating better resilience.

8.7.1.3 Growth Patterns in System Capacity

The evolution of total power generation capacity in selected US states and its potential correlation with resilience, as measured by CAIDI, was evaluated as shown in Figure 8.5.

Table 8.2 Characteristics of energy diversification by state (Shannon index).

State	Trend	Level by 2023	Characteristics
CA	Stable–increasing	High	Values exceed ≥ 0.70 after 2020; broad and balanced mix throughout the period
TX	Increasing	Moderate	0.54 → 0.64 (2014–2023); steady upward trajectory
VT	Increasing	Moderate	0.55 → 0.62 (2014–2023); small dip noted in 2022, then continued gains
NC	Increasing	Moderate–high	Below 0.60 (2014) to ≈ 0.67 (2023); gradual expansion into new resources
SC	Increasing (with early dip)	High	Dip in 2015; rose to ≈ 0.70 around 2020 and remained high
CT	Increasing	Moderate	≈ 0.63 → 0.68 (2014–2023); steady improvement
MO	Fluctuating, then stable	Moderate	Declined to 0.53 (2019), jumped to ≈ 0.62 around 2020, then stabilized
NV	Increasing	Low–moderate	0.53 → 0.61 (2014–2023); significant diversification push in recent years
IL	Variable, recent rebound	Moderate	Lower and more variable mid-2010s; rebounded to ≈ 0.62 in later years
IA	Low, slowly rising	Low	Lowest overall; mainly in the 0.52–0.59 range, indicating a concentrated mix

In Virginia (VA), generation capacity showed a gradual increase from 2014 through 2018, with a sharp spike in 2018 likely reflecting infrastructure expansion or major capacity additions. Meanwhile, CAIDI declined significantly after this peak, suggesting a potential benefit to resilience. There were fluctuations in CAIDI in later years despite relatively stable capacity.

Utah (UT) exhibited a similar mid-period capacity spike in 2018, followed by a stabilization trend. Despite the surge in capacity, CAIDI remained relatively volatile, with a noticeable peak in 2022.

In New Jersey (NJ), capacity was largely stable until a pronounced peak in 2018, then again in 2023. These spikes correspond to decreases in CAIDI, reinforcing the assumption that increased generation capacity can support the reduction of outages. However, the significant rise in CAIDI in 2020, despite modest capacity, suggests vulnerability to external stressors.

However, as shown in Figure 8.6, Pearson's correlation (r) of VA is −0.13 showing a weak correlation, UT has an r of − 0.05 showing a very weak correlation as well; however, r of NJ is 0.93 demonstrating that capacity growth alone does not universally improve CAIDI and that the relationship could be state specific.

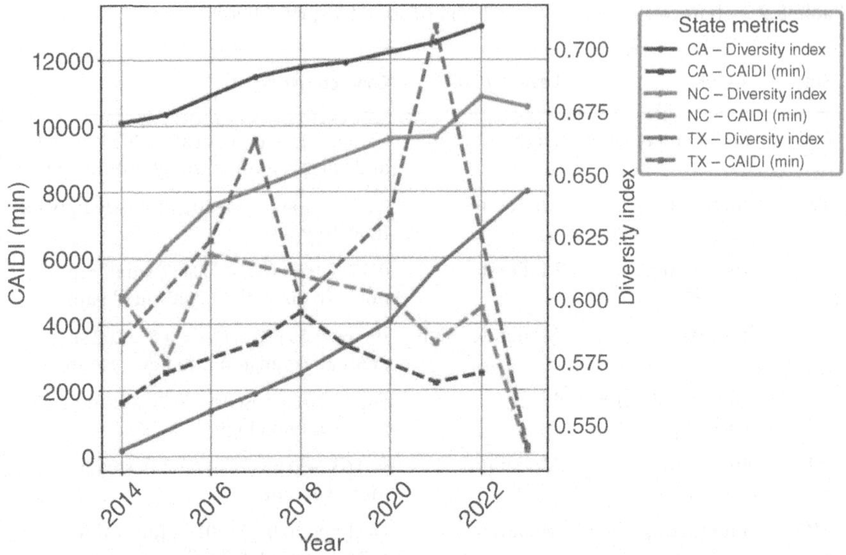

Figure 8.4 Temporal trends in resilience.

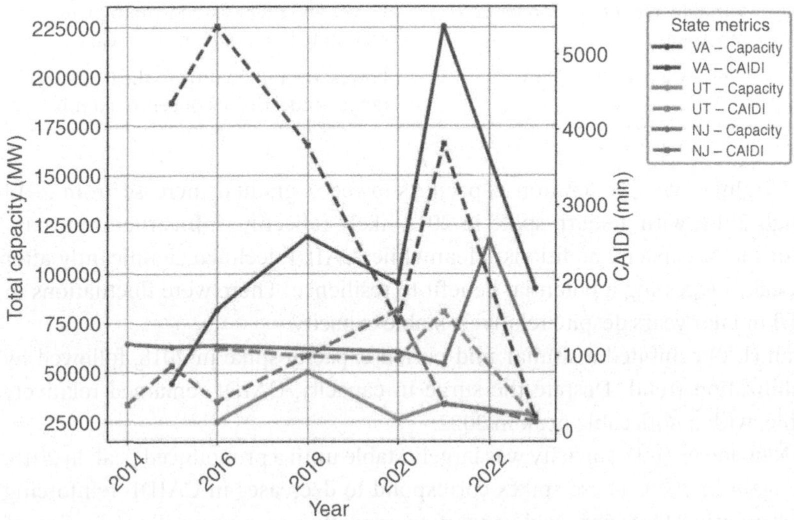

Figure 8.5 Growth patterns in system capacity and CAIDI.

8.7.1.4 Progress in Green Energy Adoption

In Figure 8.7, the share of renewable energy, such as solar, wind, and hydro, as a percentage of the total generation capacity is shown, and its association with resilience is evaluated, as reflected by CAIDI.

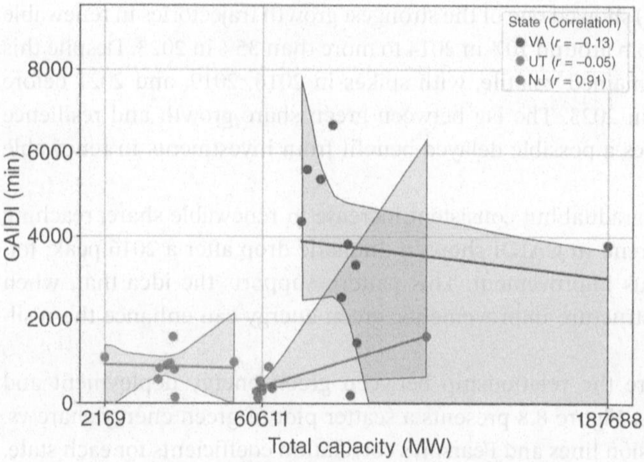

Figure 8.6 Growth capacity vs. CAIDI—correlation by state.

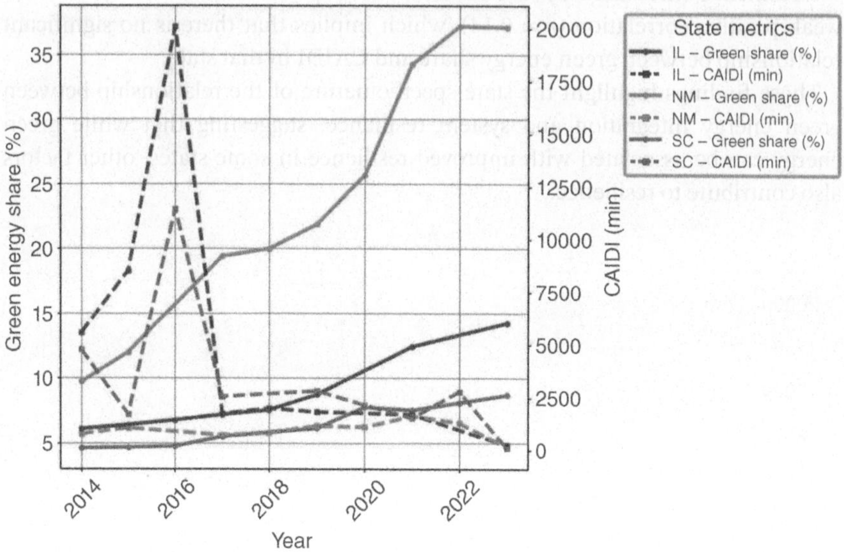

Figure 8.7 Green energy share and CAIDI.

In SC, the adoption of green energy exhibited steady growth from 2014 to 2023, increasing from less than 5% to nearly 9%. Notably, this upward trend was accompanied by a general decline in CAIDI values, suggesting that enhanced renewable penetration may contribute to improved resilience. However, CAIDI showed significant variability between 2015 and 2018, implying a likely influence of external disturbances.

New Mexico (NM) showed one of the strongest growth trajectories in renewable share, increasing from around 10% in 2014 to more than 35% in 2023. Despite this progress, CAIDI remained volatile, with spikes in 2016, 2019, and 2021 before a substantial drop in 2023. The lag between green share growth and resilience improvement implies a possible delayed benefit from investments in renewable integration.

IL experienced a gradual but consistent increase in renewable share, reaching 14% by 2023. The trend in CAIDI shows a dramatic drop after a 2016 peak, followed by continuous improvement. This pattern supports the idea that, when supported by infrastructure improvements, green energy can enhance the resilience of the system.

To further explore the relationship between green energy deployment and resilience outcomes, Figure 8.8 presents a scatter plot of green energy share vs. CAIDI, with regression lines and Pearson's correlation coefficients for each state. The results indicate a moderate to strong negative correlation for SC ($r = -0.64$) and IL ($r = -0.54$), suggesting that as the penetration of green energy increases, the resilience (measured by CAIDI) tends to improve. However, NM shows a weak positive correlation ($r = 0.14$), which implies that there is no significant relationship between green energy share and CAIDI in that state.

These findings highlight the state-specific nature of the relationship between green energy integration and system resilience, suggesting that while green energy can be associated with improved resilience in some states, other factors also contribute to resilience.

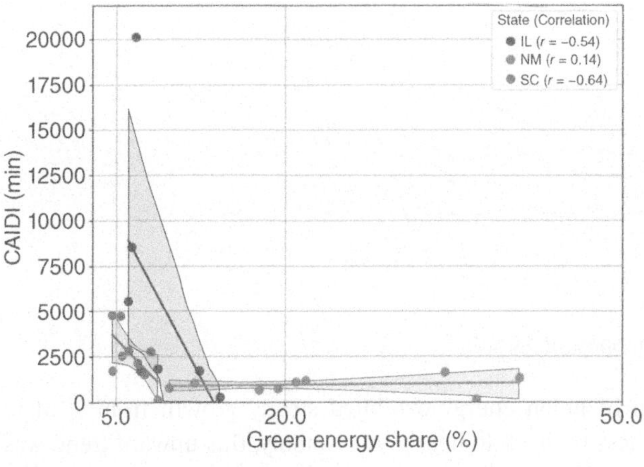

Figure 8.8 Green energy share vs. CAIDI—correlation by state.

8.7.2 Event-based Diagnostics

Specific years with extreme weather or macroeconomic events to assess real-time resilience performance were isolated.

8.7.2.1 Case Studies of High-disaster States

In Figure 8.9, Florida and CA were analyzed as representative high-disaster states to examine the interaction between energy diversity and system resilience. In Florida, the CAIDI metric fluctuated sharply between 2014 and 2023, with a dramatic spike in 2018 and again in 2022, the year of Hurricane Ian. Despite a relatively stable and slightly increasing diversity index after 2018, CAIDI remained highly variable, indicating that diversity alone did not shield the state from significant disruptions during disaster years.

In CA, a more gradual and consistent increase in the energy diversity index was observed over the same period. Although CAIDI peaked around 2018, it generally declined thereafter, showing signs of stabilization. This pattern may reflect the state's proactive investments in the modernization and diversification of the clean energy grid. However, CAIDI still showed sensitivity to external stress, suggesting that while diversity supports resilience, other factors also play a role.

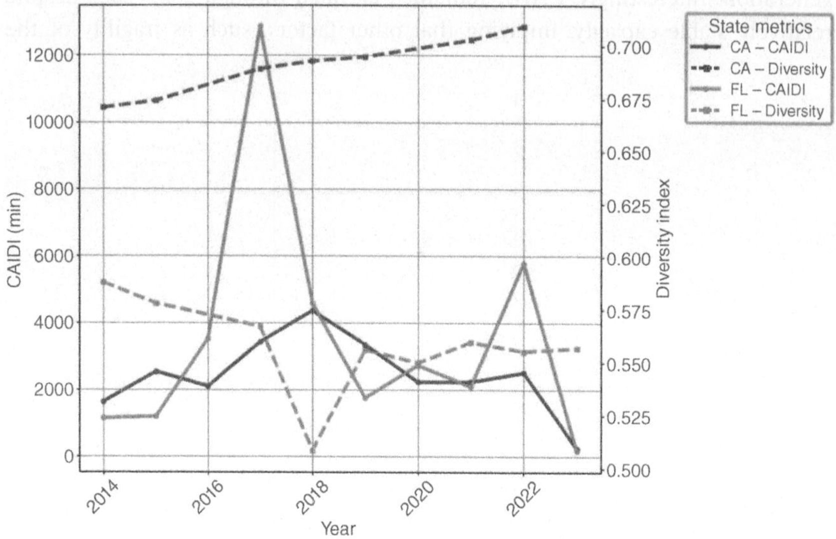

Figure 8.9 CAIDI vs. energy diversity in disaster-prone states (FL and CA).

8.7.2.2 Disaster-year Diagnostics

From Figure 8.10, a comparative analysis of TX (2021 winter storm) and Florida (2022 Hurricane Ian) reveals sharp spikes in CAIDI during their respective disaster years, indicating significant grid stress. In TX, CAIDI rose above 13,000 minutes despite the highest energy diversity index during the period, suggesting that the diversity of the fuel mix alone did not mitigate the vulnerability of the system to extreme cold.

In contrast, Florida's CAIDI rose during the hurricane year but declined rapidly thereafter, while the energy diversity improved slightly, pointing to a potential resilience benefit from ongoing diversification and regional preparedness.

8.7.2.3 Macroeconomic Stress Testing

The relationship between total generation capacity, foreign-fuel dependency, and CAIDI performance under macroeconomic and systemic stress was examined, with a focus on TX, WA, and Arkansas (AR), as shown in Figure 8.11.

In TX, the total generation capacity showed sharp fluctuations between 2014 and 2023, reaching its peak in 2020 and again in 2021. Generation based on foreign fuel (notably imported petroleum or gas sources) mirrored some of this volatility. The CAIDI values notably spiked during these same years, particularly in 2021, a period coinciding with the TX winter storm.

WA exhibited more modest fluctuations in both total capacity and foreign-fuel generation. Interestingly, CAIDI remained elevated through 2021–2022 despite relatively stable capacity, implying that other factors, such as fragility of the

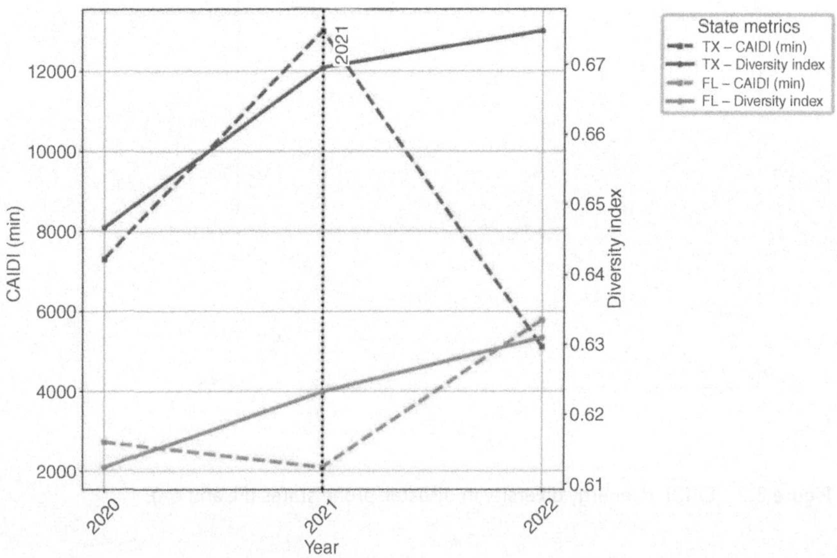

Figure 8.10 Disaster-year diagnostics.

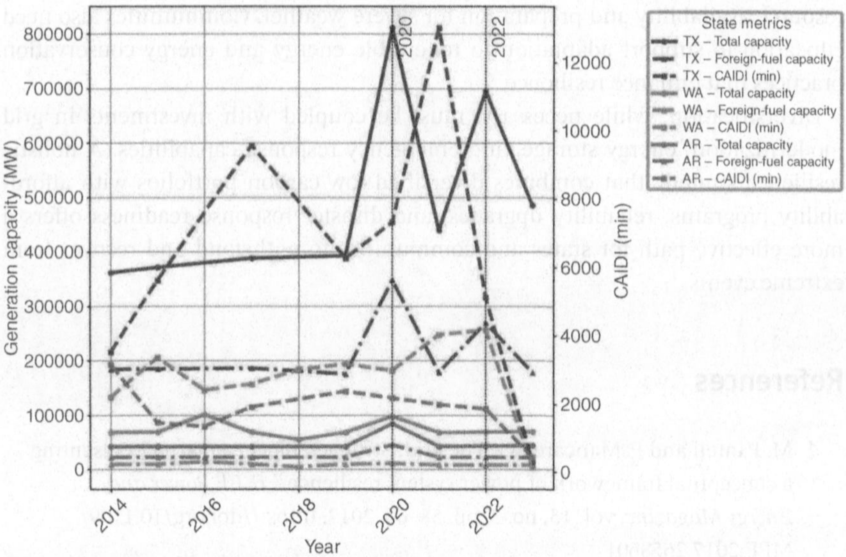

Figure 8.11 Macroeconomic stress test.

distribution or misalignment of policy, may have contributed more significantly to performance degradation under macroeconomic pressure.

In AR, the trend of generation capacity was more stable with periodic increases in foreign- fuel generation. CAIDI generally declined after 2020, despite observable volatility in external fuel dependence. This may indicate better contingency planning or adaptive system resilience in the face of external supply pressures.

8.8 Conclusion

Energy diversification plays a critical role in improving resilience, but it is not a panacea for all vulnerabilities of the energy system, particularly during extreme events. States follow different trajectories shaped by geography and infrastructure; some, such as NV, show consistent growth in diversifying their energy mix, while others exhibit volatility that contributes to fluctuations in resilience metrics. Increases in generation capacity often correspond to shorter outage durations, yet this relationship is not always linear or consistent across states. Infrastructure robustness, emergency preparedness, and exposure to disasters also play a critical role.

A major challenge in adopting green energy is the high initial cost, which limits access for low-income communities. Achieving resilient and decarbonized communities therefore requires deliberate policies and clear strategies to reduce emissions, bridge costs for low-income households, and align infrastructure with resource availability and preparation for severe weather. Communities also need education to support adaptation to renewable energy and energy conservation practices that enhance resilience.

Diversification, while necessary, must be coupled with investments in grid modernization, energy storage, and emergency response capabilities. A holistic resilience strategy that combines diversified low-carbon portfolios with affordability programs, reliability upgrades, and disaster response readiness offers a more effective path for states and communities to withstand and recover from extreme events.

References

1 M. Panteli and P. Mancarella, "The grid: Stronger, bigger, smarter? Presenting a conceptual framework of power system resilience," *IEEE Power and Energy Magazine*, vol. 15, no. 3, pp. 58–66, 2017. https://doi.org/10.1109/MPE.2017.2658601

2 C. Ji, Y. Wei, H. Liu, and H. Mei, "Resilience of the power grid: A modeling and evaluation approach," *IEEE Transactions on Smart Grid*, vol. 8, no. 2, pp. 808–817, 2017. https://doi.org/10.1109/TSG.2015.2468946

3 K. J. Holmes, E. Zeitler, M. Kerxhalli-Kleinfield, and R. DeBoer, "Scaling deep decarbonization technologies," *Earth's Future*, vol. 9, no. 11, p. e2021EF002399, 2021. https://doi.org/10.1029/2021EF002399

4 M. Bagheri, S. H. Delbari, M. Pizzamans, and C. A. Kennedy, "City-integrated renewable energy design for low-carbon and climate-resilient communities," *Applied Energy*, vol. 239 pp. 1212–1225, 2019. https://doi.org/10.1016/j.apenergy.2019.02.031

5 M. Beaudin, H. Zareipour, A. Schellenberglabe, and W. Rosehart, "Energy storage for mitigating the variability of renewable electricity sources: An updated review," *Energy for Sustainable Development*, vol. 14, no. 4, pp. 302–314, 2010. https://doi.org/10.1016/j.esd.2010.09.007

6 K. Khan, A. Khurshid, J. Cifuentes-Faura, and D. Xianjun, "Does renewable energy development enhance energy security?," *Utilities Policy*, vol. 87 p. 101725, 2024. https://doi.org/10.1016/j.jup.2024.101725

7 M. Panteli and P. Mancarella, "Influence of extreme weather and climate change on the resilience of power systems: Impacts and possible mitigation strategies," *Electric Power Systems Research*, vol. 127 pp. 259–270, 2015. https://doi.org/10.1016/j.epsr.2015.06.012

8 M. O. John and V. Adrian, "Resilience engineering: Concepts and precepts," in *Resilience Engineering in Practice: A Guidebook*, E. Hollnagel, J. Pariès, D. D. Woods, and J. Wreathall, Eds., Ashgate, 2013, pp. 3–17.

9 K. Sherish and S. Muhammad, "A resilience assessment model for power systems considering renewable energy sources," *Sustainable Energy Technologies and Assessments*, vol. 39 p. 100727, 2020. https://doi.org/10.1016/j.seta.2020.100727

10 M. Čepin, "Distribution and transmission system reliability measures," in *Assessment of Power System Reliability*, Springer, 2011, pp. 215–226. https://doi.org/10.1007/978-0-85729-688-7_14

11 U.S. Energy Information Administration, "State electricity profiles: Netsummer capacity byenergy source," 2025. Accessed: July 29, 2025. [Online]. Available: https://www.eia.gov/opendata/browser/electricity/state-electricity-profiles/capability?frequency=annual&data=capability&start=2014&end=2023&sortColumn=period&sortDirection=desc

12 R. Kleijn, E. van der Voet, G. J. Kramer, L. van Oers, and C. van der Giesen, "Metal requirements of low-carbon power generation," *Energy*, vol. 36, no. 9, pp. 5640–5648, 2011. https://doi.org/10.1016/j.energy.2011.07.003

13 U.S. Energy Information Administration, "Annualelectric power industry report (Form EIA-861) detailed data files," 2024. Accessed: July 29, 2025. [Online]. Available: https://www.eia.gov/electricity/data/eia861/

14 S. Pitkänen, "The use of diversity indices to assess the diversity of vegetation in managed boreal forests," *Forest Ecology and Management*, vol. 112, no. 1–2, pp. 121–137, 1998. https://doi.org/10.1016/S0378-1127(98)00319-3

15 S. Samarakoon, "A justice and wellbeing centered framework for analysing energy poverty in the Global South," *Ecological Economics*, vol. 165 p. 106385, 2019. https://doi.org/10.1016/j.ecolecon.2019.106385

16 P. Dato, S. Hekponhoue, L. Lawson, B. Mukhaya, and M. A. Okyere, "Implications of alleviating unreliable electricity supply for energy poverty: Evidence from households in Benin," *Energy Economics*, vol. 146 p. 108548, 2025. https://doi.org/10.1016/j.eneco.2025.108548

17 M. H. Saeed, W. Fangzong, B. A. Kalwar, and S. Iqbal, "A review on microgrids' challenges & perspectives," *IEEE Access*, vol. 9 pp. 166502–166517, 2021. https://doi.org/10.1109/ACCESS.2021.3135083

18 K. Field, "The world's biggest battery celebrates its first birthday," *Ecogeneration*, no. 110, pp. 42–47, 2019.

19 B. Bibak and H. Tekiner-Moğulkoç, "A comprehensive analysis of vehicle to grid (V2G) systems and scholarly literature on the application of such systems," *Renewable Energy Focus*, vol. 36 pp. 1–20, 2021. https://doi.org/10.1016/j.ref.2020.10.001

20 UDaily, "UD-developed vehicle-to-grid technology gains traction," *UDaily*, 2021. Accessed: Aug. 13, 2025. [Online]. Available: University of Delaware. https://www.udel.edu/udaily/2021/march/ud-developed-vehicle-to-grid-technology-gains-traction/

21 W. F. Faris, "Intelligent load balancing in microgrids with AI optimization," *Research Journal of Computer Systems and Engineering*, vol. 5, no. 1, pp. 83–94, 2024. https://doi.org/10.52710/rjcse.97

22 Whisper Valley, "The future of sustainable living: What's next for Whisper Valley," 2025. Accessed: Aug. 13, 2025. [Online]. Available: https://www.whispervalleyaustin.com/the-future-of-sustainable-living-whats-next-for-whisper-valley/

9

Artificial Intelligence Techniques for Carbon Removal of Modern Energy Networks

*Ali Darvishi, Mohammad Reza Mansouri, Reza Gharibi, and Rahman Dashti**

Clinical-Laboratory Center of Power System & Protection, Faculty of Intelligent Systems Engineering and Data Science, Persian Gulf University, Bushehr, Iran

*Corresponding author: r.dashti@pgu.ac.ir

List of Acronyms	
AI	Artificial intelligence
ANN	Artificial neural network
BECCS	Bioenergy with carbon capture and storage
CCUS	Carbon capture, utilization, and storage
CFD	Computational fluid dynamics
CNN	Convolutional neural network
COF	Covalent organic framework
DAC	Direct air capture
DACCS	Direct air carbon capture and storage
DFT	Density functional theory
DL	Deep learning
DT	Decision tree
EV	Electric vehicle
GA	Genetic algorithm
GAN	Generative adversarial network
GCMC	Grand canonical Monte Carlo
GNN	Graph neural network
GRU	Gated recurrent unit

List of Acronyms	
HVAC	Heating, ventilation, and air conditioning
KNN	*k*-Nearest neighbors
LSTM	Long short-term memory
MEN	Modern energy network
ML	Machine learning
MOF	Metal–organic framework
PCA	Principal component analysis
PINN	Physics-informed neural network
PMU	Phasor measurement unit
PSO	Particle swarm optimization
ReLU	Rectified linear unit
RF	Random forest
RL	Reinforcement learning
RNN	Recurrent neural network
SL	Supervised learning
SVDD	Support vector data description
SVM	Support vector machine
SVR	Support vector regression
USL	Unsupervised learning

9.1 Introduction

9.1.1 Motivation

With industrial progress and the emergence of technologies that rely significantly on electricity, energy consumption will inevitably rise [1]. Currently, servicing this need is primarily dependent on fossil fuels. While renewable energy supplies have grown significantly since prior times, they are still utilized mostly to supplement rather than replace fossil fuels. Current energy systems are unable to achieve carbon neutrality due to their reliance on fossil fuels [2]. Deep decarbonization needs a move toward decarbonizing technologies and their implementation in modern energy networks (MENs). Beyond achieving net-zero emissions, MENs must adopt carbon-negative strategies that incorporate carbon removal, carbon capture and utilization, and carbon storage technologies. Such strategies utilize surplus renewable energy to remove atmospheric carbon [3]. To achieve net-zero carbon goals, the use of negative emission technologies is inevitable. However, its execution confronts various challenges. Among the most important are technological constraints and a lack of large-scale development. Many negative carbon solutions, like bioenergy with carbon capture and storage (BECCS) and direct air carbon capture and storage (DACCS), need significant resources and energy. Furthermore, their life cycle assessments are highly ambiguous, making it impossible to generalize the findings. Addressing these difficulties would need extensive measures to guarantee that both public and commercial institutions actively embrace decarbonization initiatives. On the other hand, increasing investment and encouraging greater participation of researchers in developing negative carbon technologies and mitigating their side effects should be prioritized [4].

Today, artificial intelligence (AI)-based methods have made significant progress. The integration of AI with emerging technologies such as the Internet of Things, smart sensors, satellites, and high-performance computers has opened up a plethora of new prospects in a variety of disciplines. AI has the potential to significantly aid carbon removal operations due to its capacity to model processes, capture complicated and nonlinear correlations between variables, and produce data-driven forecasts. AI can contribute to direct carbon removal through the optimization of carbon capture, storage, and utilization processes, and through the design of high-efficiency adsorbents. Indirectly, it plays a key role in optimizing MEN operations and integrating renewable energy into the grid to substitute for fossil fuels. Furthermore, AI enables accurate carbon monitoring by analyzing data from satellites, drones, and sensors, and supports improved carbon planning and decision-making through enhanced reporting and validation [5]. Machine learning (ML) algorithms provide greater accuracy in predicting carbon capture effects than traditional methods. They are also extremely useful for assessing geological capabilities, selecting storage locations, and optimizing transportation routes. By merging simulation data with field observations, ML models may considerably speed up design and experimentation procedures. Moreover, AI tools can assist in managing technical uncertainties, economic considerations, and environmental constraints, thereby facilitating the transition from research to large-scale operational deployment [6].

The most important motivating factors associated with this issue can be summarized as follows:

- The sustained growth of global energy demand, driven by industrial expansion and the pervasive deployment of electricity-dependent technologies.
- The persistent dominance of fossil fuel-based energy systems, despite the accelerating penetration of renewable energy sources.
- The structural inadequacy of existing energy frameworks in achieving carbon neutrality and long-term climate mitigation targets.
- Substantial uncertainty in life cycle assessments and real-world performance metrics of carbon-negative technologies.

- The pressing requirement for advanced analytical and decision-support tools capable of addressing complexity, uncertainty, and nonlinear dynamics in coupled energy–carbon systems.
- The emerging and potentially transformative role of AI–based methodologies in enabling high-fidelity modeling, robust forecasting, and system-level optimization.

9.1.2 Literature Review

Existing researches highlight to major progress in MENs development and in leveraging AI for carbon removal. In [5], the authors examine the existing literature on the application of AI in carbon removal. The authors show that AI can help with many areas of decreasing and eliminating carbon emissions. AI may use data analysis to discover the most efficient carbon capture technologies, optimize industrial energy use, and assist in the design of carbon monitoring systems. Another notable strength of AI is its capacity to model complex processes. Study [7] invcstigates the role of AI in advancing carbon capture, utilization, and storage (CCUS) technologies. It demonstrates that applying ML and data analytics to adsorbent design can enhance efficiency while reducing both energy consumption and thermal requirements. The major focus of [8] is on using ML and predictive models to reduce the cost of decarbonization technology while also optimizing industrial processes and supply chains to reduce carbon emissions. It further differentiates AI applications based on pre-combustion, post-combustion, and oxy-combustion techniques. Paper [9] discusses on carbon capture using the post-combustion method, and the design of carbon capture columns is an important component of it. The research evaluates the efficacy of basic statistical methods, convolutional neural networks (CNNs), and graph neural networks (GNNs) in the construction and enhancement of carbon capture columns. In another research [10], the authors investigate AI's ability to reduce carbon emissions from oil and gas activities. They present possible AI-based ways to improve the overall efficiency of the CCUS process. References [11, 12] present the integration of AI algorithms with remote sensing technologies for monitoring environmental changes. Using AI and ML to look at satellite and drone images to find patterns in the spread of carbon in the atmosphere is a vital tool for making decisions. Using multispectral photos with modern algorithms makes detection more accurate, which leads to better outcomes and more reliable systems. The above cases highlight several studies that have examined various aspects of direct carbon removal.

Alongside direct removal approaches, indirect carbon removal solutions that focus on renewable energy development and carbon-free technologies are vital in preventing CO_2 from entering the atmosphere. In [13], AI-based energy management for the development of smart grids integrated with renewable resources is investigated. The study outlines AI's role in network optimization, generation and consumption forecasting, state estimation, anomaly detection, and maintenance. The results demonstrate that ML and deep learning (DL) approaches improve stability, resilience, and performance of the grid. In [14], a comprehensive analysis of renewable energy forecasting methodologies is offered, with a focus on solar and wind energy as the primary renewable sources. Forecasting approaches are classified according to variables, algorithms, and time periods. The application of ML and DL increases prediction accuracy and allows for modeling complex and nonlinear relationships. These improvements in the capacity of AI to make predictions make it easier to use renewable resources. In the field of consumer-side management and smart buildings, [15, 16] investigate the contributions of AI and ML to the improvement of building energy management systems. The research shows that powerful ML and reinforcement learning (RL) algorithms are very important for making the most of consumption and controlling equipment. Another significant indirect carbon removal strategy involves the development and implementation of electric vehicles (EVs) which reduce fossil fuel dependence. In [17, 18], the incorporation of EVs into smart grids and the function of AI are analyzed. These studies show that AI is quite good at improving the performance of EV charging, keeping the grid stable, and making the most of energy use. AI is also useful for figuring out when people will charge their devices, controlling loads, and synchronizing charging stations with renewable resources.

Based on a review of the existing literature and various studies on carbon, it is evident that AI and ML hold substantial potential to play an effective role in this field. Numerous studies have shown how AI may help remove carbon and cut down on emissions. AI is seen as an important tool for direct removal techniques such designing adsorbents, optimizing systems, capturing, storing, and using carbon, and keeping an eye on the environment. It is also a necessary technology for making the most use of renewable resources and connecting them to smart grids. AI and its continuing growth are also very important for the future of contemporary energy networks, EVs, smart home management, and more. However, alongside these significant opportunities, AI also presents challenges. This chapter addresses both the potentials and the challenges associated with AI in carbon removal of MENs. The issues addressed in this chapter are summarized below:

- In MENs, especially those of the future, the main signs of progress are the widespread use of renewable resources and the addition of EVs. The role of AI in these domains is explored from various angles.

- Understanding the technology used in carbon capture, storage, and use, as well as their strengths and limits, is critical for determining the role of AI in improving these areas. Accordingly, this chapter covers and explores carbon capture procedures, new technologies, and related AI concepts.
- Carbon capture and utilization technologies in industry play a significant role in carbon removal. This chapter reviews carbon-based technologies and their associated products as well as the role of AI in their development.
- Although AI offers significant opportunities for carbon removal and modern energy grid development, it also encounters major obstacles. Key challenges include computational limits, insufficient data, and occasional unreliability or opacity in results, which are explored in this chapter.

9.1.3 Chapter Organization

The remainder of the chapter is organized as follows: Section 9.2 introduces the fundamental concepts and key techniques of AI, outlines the characteristics of MEN, and explains its integration with AI in carbon removal. Section 9.3 discusses the applications of AI in direct carbon removal, including process optimization, material design, and monitoring. Section 9.4 focuses on indirect carbon removal through the integration of renewable energy, efficiency improvements, smart building, and EVs. The challenges and considerations of applying AI in carbon removal are examined in Section 9.5. Section 9.6 discusses and summarizes the criteria for selecting AI technologies and their role in decarbonization strategies. Finally, the chapter concludes by summarizing the potential of AI, reinforcing its connection to sustainability, and calling for further research and collaboration. The main objectives of this chapter are as follows:

- Explaining the fundamental concepts and key techniques of AI.
- Exploring applications of AI in direct carbon removal, including process optimization, material design, and monitoring.
- Exploring the role of AI in indirect carbon removal through the integration of renewable energy, improved efficiency, smart buildings, and EVs.
- Assessing the challenges and considerations of using AI in carbon removal.

9.2 Foundational Concepts: AI and MENs

Traditional energy networks are transitioning into MENs, primarily driven by technological advancements and the increasing need for sustainable and reliable energy supply. This transformation is motivated by the necessity to resolve the technical, economic, and environmental shortcomings of earlier networks.

MENs are designed to enable renewable energy sources work together better and to make clean energy generation as efficient as possible. These networks use cutting-edge technologies to fix the problems with old systems and make life better. They also stand out because they employ intelligent devices and integrated coordination systems [19].

The severe economic and environmental consequences of using fossil fuels have increased demand for sustainable energy. Even though renewable energy technologies have come a long way, their unpredictable nature is an important challenge. Therefore, MENs need to focus on two important traits: "reliability" and "resilience." Given the variability of renewable energy production, these networks require innovative strategies, careful monitoring, and resilient system designs that can quickly and consistently restore normal operations in the event of a disruption [20].

The use of AI-based systems as the central control and decision-making center is a novel way to improve the dependability, management, distribution, and regulation of energy production in MENs. These systems process and analyze data intelligently using technologies such as DL and ML. This strategy allows for the most effective integration of renewable energy resources, eliminates energy waste, and improves resource utilization. Furthermore, these technologies improve network stability by identifying and avoiding possible breakdowns, resulting in considerable economic gains from lower maintenance costs and energy losses [21]. Figure 9.1 presents a schematic representation of a MEN powered by AI technologies.

9.2.1 Brief Overview of Relevant AI Techniques

AI is a field of technology that tries to make systems that can see, think, and learn as people do. The main goal is to duplicate human intelligence and use it for tasks like planning surgeries, playing complex games, driving cars on their own, and making personalized shopping list suggestions. By leveraging AI, various software and processes can address real-world problems such as healthcare, transportation, marketing, and traffic management with greater accuracy and speed, effectively establishing an intelligent link between perception and action [22]. Over time, numerous algorithms have been developed for different purposes, including those initially designed for recreational applications such as IBM Deep Blue, IBM Watson, and DeepMind AlphaGo, as well as publicly accessible systems such as ChatGPT (GPT: Generative Pretrained Transformer). Today, three main and widely used branches of AI, natural language processing, ML, and DL, are frequently used in combination, and their interrelations within the overall structure of AI are illustrated in Figure 9.2.

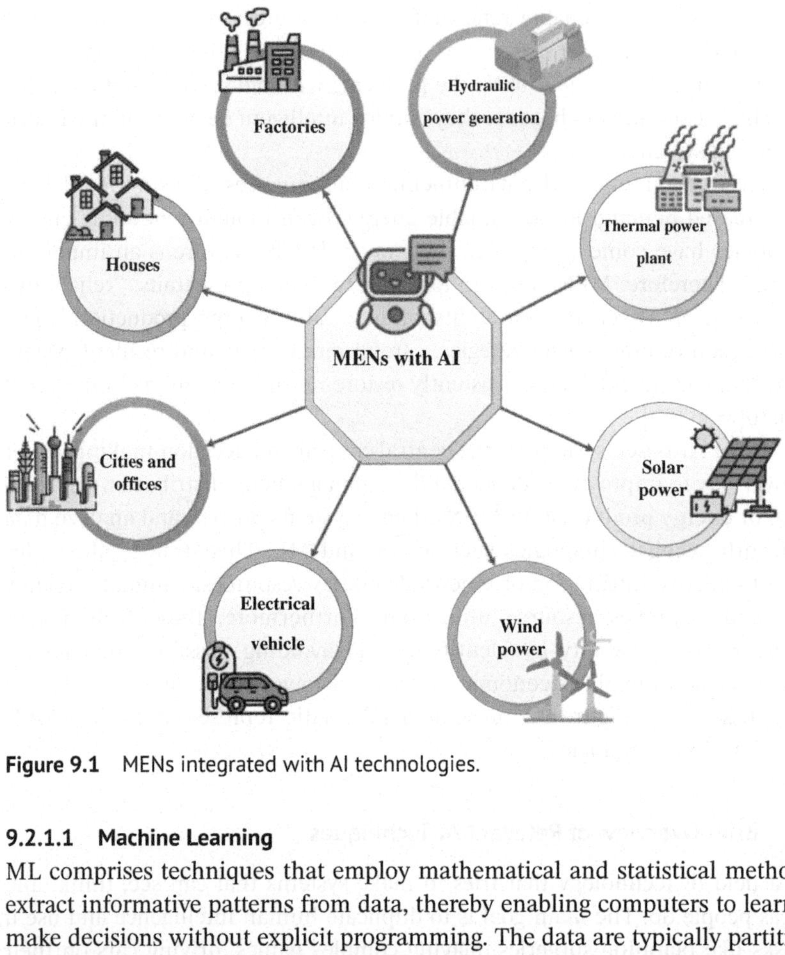

Figure 9.1 MENs integrated with AI technologies.

9.2.1.1 Machine Learning

ML comprises techniques that employ mathematical and statistical methods to extract informative patterns from data, thereby enabling computers to learn and make decisions without explicit programming. The data are typically partitioned into training and test sets: the model is trained on the former and evaluated on the latter, with accuracy determined by comparing predicted and actual outputs. When performance is inadequate, adjustments such as feature refinement or data scaling are applied. This methodology uncovers complex relationships and is applied across numerous domains. Broadly, ML methods are classified into supervised learning (SL), unsupervised learning (USL), RL, and semi-SL [23].

9.2.1.2 Supervised Learning

SL involves building a model from labeled training data that can make predictions about new or future data. The term "supervised" refers to a set of training examples in which the data inputs and desired outputs (labels) are specified in advance, and, in fact, SL is the process of modeling the relationship between the input data and the output labels; therefore, SL can be considered a type of "label-based learning."

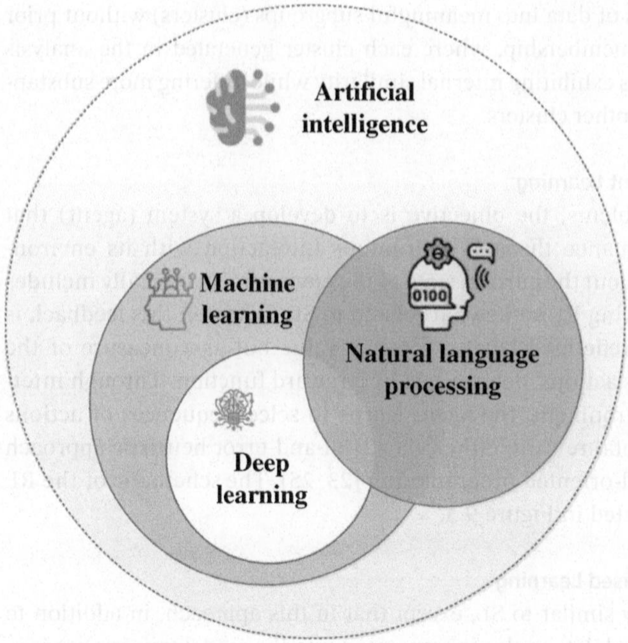

Figure 9.2 Relationship between natural language processing, ML, and DL in AI.

This approach includes two main categories of problems: classification and regression. In classification, the goal is to predict class labels for new data based on past observations, and these labels are discrete, unordered values that represent the membership of the examples in different groups; a common example of this type of problem is email spam detection, where a ML algorithm learns a set of rules to distinguish between two possible classes (spam and non-spam). However, the set of labels need not be binary, and a predictive model trained by an SL algorithm can assign any class label in the training data to new, unlabeled examples. In regression, we have a set of predictor variables (explanatory variables) and a continuous response variable (output variable), and the goal is to find a relationship between these variables that can predict continuous output values; for example, a linear regression model has been proposed for short-term system load forecasting, which takes into account factors such as holidays and temperature changes and uses them to improve forecast accuracy [24].

9.2.1.3 Unsupervised Learning

In this class of problems, the focus is on unlabeled data or data with unknown structure, and through USL techniques, it is possible to examine the data structure and derive meaningful information without the guidance of a known output variable. Clustering, a USL method for data analysis and pattern discovery, groups large volumes of data into meaningful subgroups (clusters) without prior knowledge of their membership, where each cluster generated in the analysis defines a set of objects exhibiting internal similarity while differing more substantially from objects in other clusters.

9.2.1.4 Reinforcement Learning

In this class of problems, the objective is to develop a system (agent) that improves its performance through continuous interaction with its environment. Information about the current state of the environment typically includes a reward signal, making RL somewhat related to SL; however, this feedback is not provided as a predefined label or correct value but as a measure of the quality of the agent's actions, determined by a reward function. Through interaction with the environment, the agent learns to select sequences of actions that maximize the total reward, either via a trial-and-error heuristic approach or by employing goal-oriented programming [23, 25]. The schematic of the RL framework is illustrated in Figure 9.3.

9.2.1.5 Semi-supervised Learning

These issues are very similar to SL, except that in this approach, in addition to labeled data, unlabeled data is also incorporated into the model training process. In traditional SL, labeled data plays a central role in model construction; however, preparing and labeling training data for real-world applications is often a challenging, expensive, and time-consuming process, as it requires the participation of human labelers who possess specialized knowledge and experience within specific domains. Obtaining these labels from experts not only imposes significant time constraints but also results in substantial financial costs. These challenges become particularly critical in applications involving a large number of class labels and considerable similarities between them. Semi-SL addresses this

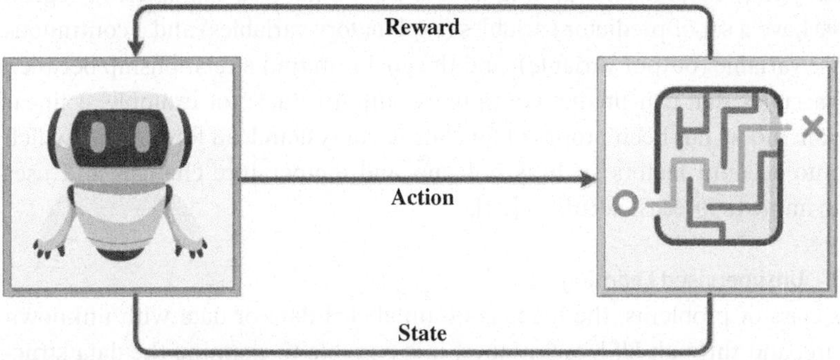

Figure 9.3 Schematic of an RL.

inherent bottleneck by enabling some or all of the available unlabeled data to be used in the learning process and aims to improve model performance by leveraging this data while reducing the extensive human effort required for labeling [26].

9.2.1.6 Deep Learning

In 1986, the term DL was first used in the context of ML, and it was later expanded to include artificial neural networks (ANNs). DL, a specialized field of ML, employs numerous layers of nonlinear processing to extract data features at various levels of abstraction and to learn complex concepts derived from simpler ones. This approach is used in both supervised and USL modes to solve problems, including feature representation, classification, and pattern recognition. ANNs are the most popular DL algorithms; however, CNNs, recurrent neural networks (RNNs), and other newer approaches have also contributed significantly to the field's evolution. These advancements have resulted in DL gaining more attention than ML as a whole [27]. The following subsections provide an in-depth exploration of three widely used DL architectures: ANN, CNN, and RNN, highlighting their structures, working principles, and key applications.

9.2.1.7 Artificial Neural Network

ANNs are inspired by the structure of the human brain, which employs numerous neurons to process vast amounts of information. Input data are represented as nodes in the input layer, while output data are represented as nodes in the output layer, with one or more hidden layers containing a large number of nodes situated between them. During data transmission between nodes, weight and bias coefficients are applied to determine the influence of each input on solving the problem. The simplest model of this network is the single-layer perceptron (SLP), which is limited to classifying linearly separable problems. A more advanced model, the multilayer perceptron (MLP), is capable of solving nonlinear problems and predicting experimental data with high accuracy by employing the backpropagation algorithm for training. Figure 9.4 shows an ANN with one hidden layer [28].

Each neuron in an ANN has an activation function responsible for determining its output. By applying the activation function to the weighted sum of input values, the linear combination of inputs is transformed into a nonlinear output. Among the most often employed activation functions in fully connected layers, the rectified linear unit (ReLU) is well known for its ability to improve network performance by boosting convergence speed and decreasing computing complexity. On the other hand, the Softmax function is often used in output layers to turn scores from the network into a probability distribution over several classes, which helps make sure that classification decisions are accurate. Table 9.1 shows some of the most common activation functions used in neural networks [29].

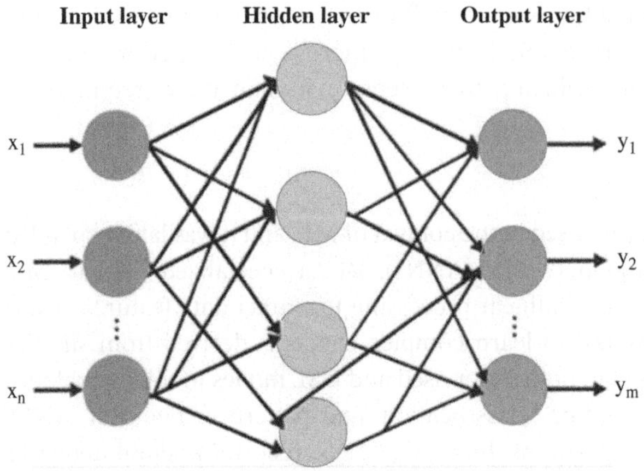

Figure 9.4 Structure of an ANN.

Table 9.1 Common activation functions in ANN.

Activation function	Formula	Main usage
ReLU	$f(w) = \max(0, w)$	Most common for hidden layers; improves training speed and reduces vanishing gradient issues
Sigmoid	$f(w) = \dfrac{1}{1 + e^{-w}}$	Used for binary classification and probabilistic outputs
Hyperbolic tangent	$f(w) = \dfrac{e^{w} - e^{-w}}{e^{w} + e^{-w}}$	Used where zero-centered outputs are preferred, e.g., hidden layers
Softmax	$f(z_i) = \dfrac{e^{z_i}}{\sum_{j=1}^{k} e^{z_j}}$	Used in the output layer for multi-class classification

9.2.1.8 Convolutional Neural Network

CNN have become highly popular in image-related applications due to their high accuracy and optimal performance and are also employed in fields such as speech recognition, image analysis, and autonomous vehicles. These networks have an input layer, convolutional layers, fully connected layers, activation functions, and phases for feed-forward and backpropagation, much like regular ANNs. The convolutional layer is the most important part of the CNN. It is where most of the calculations happen using a collection of filters that create feature maps. Adding more convolutional layers to a network makes it deeper and more accurate at predicting things. At first, random weights are given to inputs, activation functions turn them into nonlinear outputs, and error values are figured out during training. After this forward pass, error backpropagation changes the weights to make the model more accurate and lower the error. The cycle is repeated multiple times until optimal performance is achieved, which constitutes the fundamental working principle of CNNs [30].

9.2.1.9 Recurrent Neural Network

The RNN is one of the most common neural networks used in DL. It is used for tasks where the order of the data is important, such translating languages and predicting electric load. In these networks, the way neurons are connected in a sequence and the way they change over time make it possible to simulate data that changes over time, which means that past outputs might affect current decisions. RNNs are made to do the error backpropagation process over time, however, they have an issue with forgetting information from the past when the sequences are too long. The long short-term memory (LSTM) model was created to solve this problem. It can remember things for a long time and make more accurate predictions. Memory blocks are the basic building blocks of the recurrent hidden layer in LSTM. They help store information for a long time. There are three multiplicative gates in each memory block: an input gate that controls the entrance of new information, a forget gate that decides whether to keep or throw away old information, and an output gate that controls the final output of the cell. This architecture makes LSTM a better sort of RNN that can simulate long and complicated sequences well, as illustrated in Figure 9.5 for a clearer understanding of the difference [30].

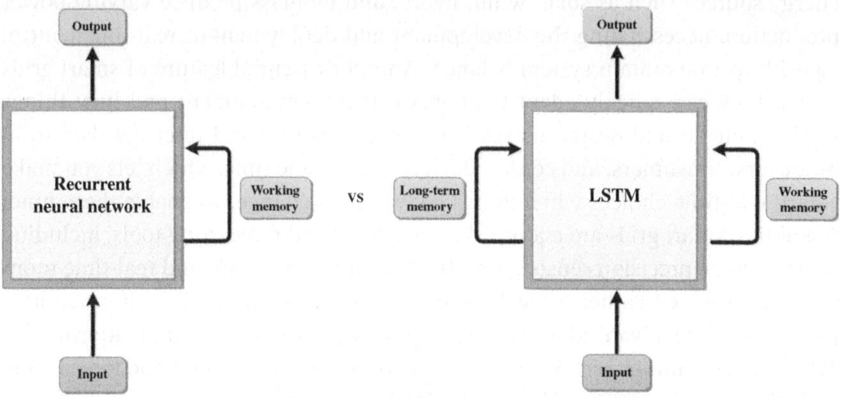

Figure 9.5 Comparison between an RNN and an LSTM model.

9.2.1.10 Natural Language Processing

NLP is a field that combines computer science, linguistics, and AI to help computers understand and interact with human language. The technology helps with things like semantic interpretation, machine translation, and finding patterns in language data. In contemporary NLP systems, words and sentences are frequently converted into numerical vector representations (embeddings) using models like Word2Vec, followed by statistical analysis, ML, neural networks, and advanced text analytics. A major example is power demand forecasting, where using text-derived variables such as word frequency, emotion indicators, and subject distributions enhances forecast accuracy dramatically. These textual resources, which capture social dynamics such as pandemics, political events, international conflicts, and transportation factors, not only reveal their impact on electricity consumption but also broaden traditional analytical paradigms and emphasize the importance of unstructured text data in predictive modeling [31].

9.2.2 AI-based Characteristics of MENs

Smart grids, also known as MENs, are a new type of energy infrastructure that can fulfill the complex, varied, and changing needs of energy use thanks to new communication, processing, and computing technology. Smart grids are different from traditional grids because they are decentralized, multilayered, and adaptable. This means they can respond to changes in output and demand in real time. One important part of smart networks is that they employ distributed generation. This means that energy is made by a lot of different sources, not only big power plants. For example, residential solar panels, tiny wind turbines, and electric cars may all make electricity. This advanced system requires intelligence and coordinated management across both local and national levels. The integration of renewable energy sources such as solar, wind, hydro, and biomass produce varying power production, necessitating the development and deployment of real-time control algorithms to maintain system balance. Another essential feature of smart grids is that they can quickly adapt to changes in the environment and how things work. Their layered design makes it possible to gather and interpret data from producers, consumers, and control devices at the same time, which lets you make smart, real-time choices when things go wrong or change. To enable these functionalities, smart grids are equipped with control and forecasting tools, including smart meters, precision sensors, weather forecasting systems, and real-time monitors, all of which gather extensive data across the network. This information is processed using advanced analytical frameworks such as big data platforms, AI, ML, and data mining, which facilitate consumption analysis, demand forecasting, and disruption detection. Ultimately, MENs represent a transformative framework for intelligent, sustainable, and reliable energy production, distribution, and use in advanced societies [32].

AI and its related disciplines have significantly influenced the development and operation of energy systems, giving rise to AI-driven MENs. AI is a big aspect of making smart grids perform better, be more dependable, and be more adaptable. It can process large volumes of data, learn about consumption patterns, forecast demand, and make real-time choices. The key parts of AI-based MENs are shown in Figure 9.6. They are divided into three main areas: monitoring, optimization and control, and energy management.

9.2.3 Synergy Between AI and MENs for Decarbonization

Green energy generation has become a primary focus in shaping energy policies aimed at decarbonization, as continued reliance on fossil fuels not only increases carbon and other greenhouse gas emissions but also poses serious and lasting threats to both environmental and economic sustainability. Renewable energy technology has advanced quickly in response to these problems, allowing MENs to deliver a larger amount of clean, carbon-neutral electricity. Even if they help a lot to minimize carbon footprints, these sources still have big problems with availability and supply consistency. Wind, solar, and hydropower, although technologically mature, remain subject to variations in weather and seasonal cycles, limiting their ability to deliver continuous and adjustable output. Achieving the decarbonization objectives of MENs, therefore, depends on the development and implementation of innovative measures that strengthen grid flexibility, expand energy storage capacity, and support intelligent, optimized consumption management [20].

AI methods can significantly contribute to overcoming these challenges by enhancing the predictability, stability, and efficiency of integrating renewable energy sources into modern energy grids. In particular, AI has shown strong potential in renewable energy generation forecasting, a process that supports essential activities such as power system planning, operation, and management.

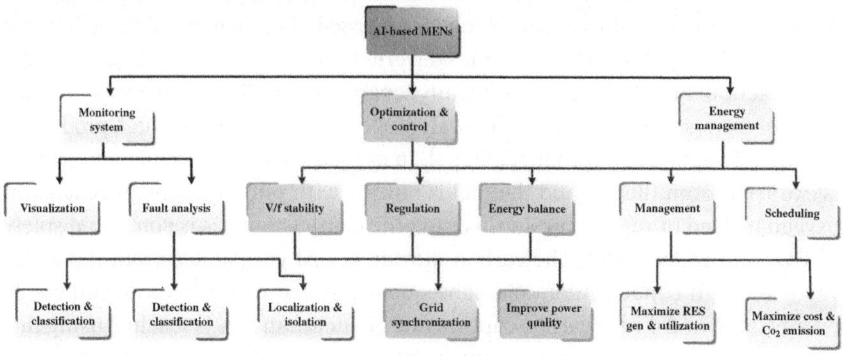

Figure 9.6 Functional architecture of AI-based MENs.

Yet, because resources like wind and solar are naturally variable and affected by random weather patterns, producing accurate forecasts continues to be a difficult task [33].

9.3 AI Applications for Direct Carbon Removal Techniques

Increasing concentrations of carbon dioxide are associated with significant and harmful effects on the environment and public health. Between 15 and 40% of the carbon dioxide that is emitted into the atmosphere is kept in the atmosphere for many years. Of this, around 10–25% is retained for tens of thousands of years. Air quality, soil, water resources, forest cover, and agricultural yields are directly influenced by the escalation of carbon dioxide emissions and the accumulation of carbon in the atmosphere. Hence, the necessity of both carbon emission reduction and carbon removal has been emphasized as two fundamental strategies in combating climate change. Multiple approaches and techniques have been proposed for the removal of atmospheric carbon dioxide, including direct air capture (DAC), BECCS, sequestration, continuous monitoring, and the application of satellite imagery. In all of these methods, AI has been used as an effective tool to achieve the goals of carbon neutrality, carbon-negative, and ultimately, net-zero. ML, neural networks, and RL are methods that can help manage, optimize, and improve carbon capture, storage, and removal systems [5]. Applications of AI in carbon dioxide management are illustrated in Figure 9.7.

9.3.1 Optimization of Carbon Capture Technologies

In industries, carbon capture approaches have been classified and examined under three levels: pre-combustion, oxygen-enriched combustion, and post-combustion. In the pre-combustion, fossil fuels such as coal, natural gas, oil, and biomass are gasified prior to being burned and consumed. During the gasification process, fossil fuels are converted into syngas, which is composed of carbon monoxide (CO) and hydrogen (H_2). After that, the CO is converted into CO_2 and H_2 via a water–gas shift reaction. Finally, pure CO_2 is readily separated from the gas stream and is utilized for storage or other applications. Hydrogen is combusted as a fuel, producing only water vapor as a byproduct [33, 34]. In oxygen-enriched combustion, oxygen is extracted from the air, and the fuel is burned with pure or high-concentration oxygen instead of regular air. In this case, the combustion gas is composed solely of CO_2 and water vapor. The carbon dioxide is easily separated from the water vapor through condensation and subsequently stored or utilized. Regardless of energy input and cost, a carbon capture rate greater than 97% is reached using this method. The post-combustion capture method is presently recognized as the most

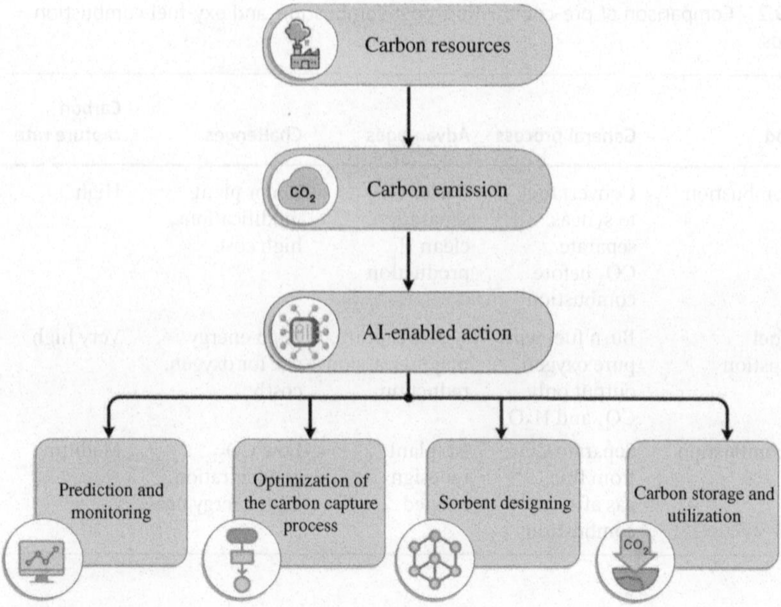

Figure 9.7 AI application in carbon management.

common approach. Challenges such as low CO_2 concentrations and significant energy demands are encountered. However, no structural alterations to existing power plants or industrial facilities are required. In post-combustion, the fuel is burned through the conventional combustion process. Following combustion, the flue gas separation process is conducted. In addition to CO_2 the flue gas is composed of impurities such as water, nitrogen, and sulfur. After cooling, the flue gas is absorbed by chemical solvents, filters, and adsorbents [35]. Table 9.2 summarizes the main features of pre-combustion, post-combustion, and oxy-combustion procedures.

Furthermore, DAC technology has become a popular approach to get carbon dioxide out of the air. Unlike previous technologies, DAC may be used on more than only fixed sources like power plants, flues, or combustion chambers. This technology can be deployed in various locations and is capable of extracting CO_2 from the atmosphere even at very low concentrations. CO_2 is absorbed from the air using solid or liquid adsorbents through physical or chemical adsorption mechanisms, and the captured CO_2 is then released by applying heat or vacuum for storage or further utilization. DAC has been highlighted as a potentially critical instrument for becoming carbon-neutral or carbon-negative. Still, things like energy usage and cost must be taken into account.

Table 9.2 Comparison of pre-combustion, post-combustion, and oxy-fuel combustion methods.

Method	General process	Advantages	Challenges	Carbon capture rate
Pre-combustion	Convert fuel to syngas, separate CO_2 before combustion	Easier CO_2 separation, clean H_2 production	Major plant modifications, high cost	High
Oxy-fuel combustion	Burn fuel with pure oxygen, output only CO_2 and H_2O	> 97% capture, major emission reduction	High energy use for oxygen, costly	Very high
Post-combustion	Separate CO_2 from flue gas after combustion	No plant redesign needed	Low CO_2 concentration, high energy use	Medium

There are two basic types of carbon capture mechanisms: physical and chemical adsorption. The numerous ways of doing this depend on how the CO_2 molecule interacts with the adsorbent material. In physical adsorption, CO_2 is retained on the surfaces of materials such as zeolites, activated carbon, or metal–organic frameworks (MOFs) through weak van der Waals interactions or covalent bonds. In this method, the process is more high reversibility and low energy demand. In contrast, chemical adsorption involves a reaction between the CO_2 molecule and the surface of the adsorbent, forming stable compounds such as carbamates or bicarbonates [36].

AI can play a significant role across various carbon capture technologies, including the design of chemical and physical adsorbents as well as DAC, post-combustion, pre-combustion, and oxy-combustion methods. The combination of ML and statistical approaches is an effective tool for material screening. ML, along with grand canonical Monte Carlo (GCMC) simulations, has been used to screen covalent organic framework (COF) structures, allowing for the quick and precise identification of better materials from among the available carbon capture options. Generative models are one of the important AI tools used in MOF design. DiffLinker, variational autoencoders, and generative adversarial networks (GANs) have been employed to generate thousands to millions of new MOF combinations and identify the optimal candidate for carbon capture. Additionally, transformer models have been used to assess and rank MOFs and COFs in order to construct membrane units capable of separating carbon dioxide from other gases. Another use of AI in adsorbent material design is the use of ML to forecast material heat capacity. Based on density functional theory (DFT) data, researchers constructed an ML model to forecast the thermophysical properties of adsorbent solids. The primary objective was to predict the heat capacity of zeolites, MOFs, and COFs to facilitate the effective design of physical adsorbents. Furthermore, ANNs have been used to model nonlinear reaction behaviors. Metaheuristic optimization techniques, including genetic algorithm (GA) and particle swarm optimization (PSO), have been adopted to optimize parameters and balance adsorption and release properties. Likewise, random forest (RF) and support vector machine (SVM) algorithms have been employed for material classification and separation tasks [7].

AI-based methods play a significant role in the manufacturing processes and equipment associated with adsorbent production. CNNs have been used to monitor important equipment by analyzing thermal and visual data, allowing for faster identification of failures, leaks, and overheating. This optimizes equipment and system maintenance by minimizing wear, corrosion, and efficiency loss. RL methodologies have also been used to control processes in real time. Based on sensor feedback, these systems regulate the solvent flow rate, pressure, and temperature to enhance adsorption efficiency while minimizing energy usage. LSTM, RNNs, and gated recurrent units (GRUs) are data-driven algorithms that use time-series data from carbon capture systems, such as temperature, pressure, and concentration, to anticipate carbon capture rates and energy consumption. These approaches further enable error prediction and performance degradation detection, thereby supporting equipment maintenance. Transformer models have the ability to predict processes and control systems by determining optimal parameters. When combined with other methods, they provide a powerful approach for maintaining and optimizing carbon capture equipment [8].

In post-combustion processes, CNNs and GNNs have been employed to design and assess the geometry of packings and absorption columns using computational fluid dynamics (CFD) data. These models enable faster prediction of performance indicators, such as interfacial area and the wetted area, and allow for generalization of results across different scales [9]. Researchers have also used DL and SVM models for the classification and quality assessment of biomass pellets and waste management. These models can be applied to fuel management in oxy-combustion power plants [37]. AI techniques are broadly useful in material design, process control, equipment maintenance, rapid screening of adsorbents, predictive modeling, real-time optimization, and improving system reliability. AI-based approaches enhance productivity, reduce energy consumption, and support the transition to carbon-neutral and carbon-negative solutions. Table 9.3 summarizes the applications of AI in carbon capture technologies.

Table 9.3 Applications of AI in carbon capture technologies.

Application area	AI methods	Studied case
Adsorbent material design (MOF/COF/Zeolite)	ML, GAN, transformer, GCMC simulation	Rapid material screening, new structure design, heat capacity prediction
Carbon capture process optimization	ANN, RL, LSTM, RNN, GRU	Control of flow rate, temperature, and pressure; energy reduction; efficiency improvement
Equipment monitoring and maintenance	CNN, transformer	Fault and leak detection, failure prediction, performance degradation prevention
Post-combustion and oxy-fuel management	CNN, GNN, SVM	Packing and column design, biomass and waste fuel management

9.3.2 Enhancing Carbon Utilization

Carbon capture is the first step in meeting carbon-neutral and carbon-negative targets. Following this, carbon use has arisen as a critical topic of study that has received substantial attention. Carbon dioxide is utilized extensively in the oil industry to improve oil recovery. ANNs are capable of predicting pressure, saturation, and production rates in oil and gas reservoirs using data from CO_2 injection and production operations. In general, ML models can be applied to determine optimal oil well locations, establish injection rates, and plan operational schedules, all to maximize oil recovery [10]. Another important area for AI methods is the design of catalysts and materials for converting carbon dioxide into valuable products. Feature-based ML models can predict fundamental catalyst properties using structural databases, reducing the need for costly and time-consuming laboratory experiments. Furthermore, ML aids in the best selection of adsorbents and catalysts by modeling their thermodynamic and chemical characteristics [38].

Carbon dioxide is also utilized in chemical industries, including urea and methanol production. Furthermore, carbon dioxide can react with minerals to form stable bicarbonates, benefiting the building materials and cement industries. These applications contribute to economic value creation and lay the foundation for CO_2-based markets. To support the development of carbon markets, AI methods such as RL are applied to optimize carbon transport networks and design integrated CCUS systems. In addition, ANN, CNN, and LSTM models are applied to monitor carbon storage sites, detect leaks, and evaluate

Table 9.4 AI applications in carbon utilization.

Application	Objective	What AI does
Injection in oil reservoirs	Enhanced oil recovery	ANN-based prediction of reservoir parameters and optimization of well placement
Urea and methanol production	Chemical product manufacturing	ML and feature-based models for catalyst selection and thermodynamic simulation
Building materials and cement	Carbonation and long-term carbon storage	Material design and CO_2 mineralization using ML and RL
Catalyst and adsorbent production	Conversion of CO_2 into value-added products	Catalyst property prediction and optimal material selection using ML and GAN

reservoir stability [39]. The applications of AI in carbon utilization are categorized and presented in Table 9.4. Captured carbon can be reused in different industrial applications, providing environmental and economic benefits beyond greenhouse gas reduction. In the oil and gas industry, injecting captured CO_2 increases oil recovery and enables partial storage of carbon in depleted reservoirs. In the chemical industry, captured CO_2 is used as a feedstock for producing value-added products such as methanol, urea, and synthetic hydrocarbons, thereby reducing reliance on fossil-based raw materials. In the construction sector, the use of CO_2 in mineralization and carbonization processes supports the production of durable building materials with long-term carbon storage benefits. Overall, these applications help reduce emissions, improve resource efficiency, and establish CO_2-based value chains. AI enhances these impacts by optimizing process conditions, supporting catalyst and material selection, and aligning carbon utilization with modern energy systems, thus enabling scalable and economically viable decarbonization solutions.

9.3.3 Monitoring and Verification of Carbon Sequestration

Carbon monitoring, together with emission source monitoring, is an essential aspect of carbon management. With adequate knowledge of temporal–geographic trends and carbon behavior, planning can be both effective and efficient. Using satellite imagery and remote sensing data integrated with AI and ML methods provides additional analytical capabilities that are highly effective for carbon sequestration processes. Forests are an important natural source of carbon sequestration.

Tree biomass includes a substantial quantity of carbon, which helps to improve soil fertility while also reducing erosion and deterioration. As a result, soil carbon monitoring is directly related to atmospheric CO_2 management and climate trend tracking. ML techniques such as ANNs, SVMs, and decision trees (DTs) are effective for modeling relationships between soil properties and soil organic carbon content with high accuracy based on remote sensing data [40]. CNNs are used to extract temporal and spatial characteristics of forest cover, count trees, and monitor biomass volume through the analysis of satellite images. RF and DT models assist in forest cover classification and the analysis of time-series satellite imagery [11].

Industrial activities, urban traffic, and vehicles are significant contributors to CO_2 emissions. For good planning and decision-making, it's important to keep an eye on various sources of emissions and measure carbon levels at certain places. It is increasingly common to employ satellites and drones, together with multispectral images, to keep an eye on carbon emissions in cities and factories. RNN and LSTM models analyze time-series data and images to evaluate daily pollutant emissions and identify behavior patterns. RF and DT algorithms are also used to identify pollution hotspots and detect areas with elevated carbon accumulation [12]. The relationship between the components of carbon monitoring is shown in Figure 9.8.

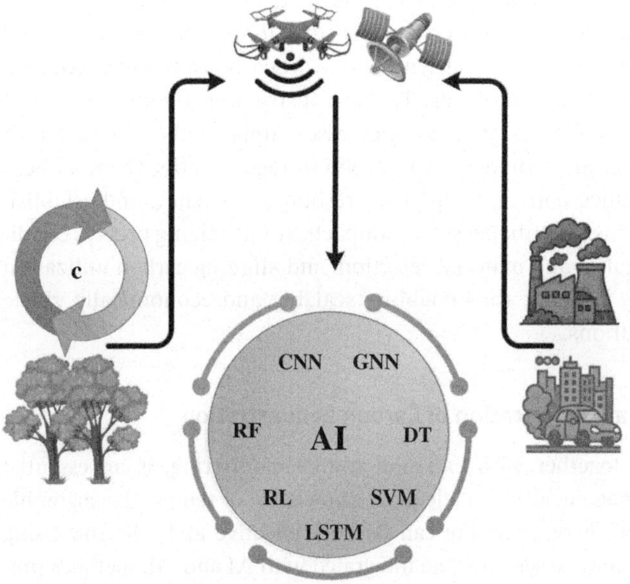

Figure 9.8 The relationship between the components of carbon monitoring.

9.4 AI Applications for Indirect Carbon Removal/Reduction in MENs

Fossil fuels, which are the main source of energy, cause a lot of pollution by releasing a lot of carbon dioxide and other greenhouse gases. Cutting down on our reliance on unclean sources of energy and fossil fuels is a key step that helps remove carbon in an indirect way. This reduction in dependency can take several forms, one of which is switching to renewable energy sources. In this context, AI is used to predict, forecast, and manage renewable resources, as well as integrate them into current energy systems. Part of cutting emissions is controlling how much consumers use and making things more efficient. AI algorithms help with this by making energy systems work better, supporting demand response programs, and predicting patterns in production and consumption [41].

Another important aspect of reducing dependence on traditional energy resources is the adoption of technologies focused on emission reduction. EVs are currently among the most prominent technologies in this regard. The application of AI in EVs and their integration with renewable energy sources represent a significant step toward achieving carbon reduction goals. AI and ML methods are highly effective for optimizing battery performance, managing charging and discharging cycles, and improving charging infrastructure [17]. Figure 9.9 shows an overview of what is covered in this subsection.

9.4.1 Optimizing Renewable Energy Integration

Expanding the usage of renewable energy and integrating it into existing power infrastructures is critical for achieving carbon neutrality. The uncertain and nonlinear nature of renewable energy sources such as wind and solar poses challenges for their planning, scheduling, and management [42]. However, this process presents other challenges, which AI and ML methods can help address and overcome. One of the important applications of AI is its ability to manage uncertainty and capture nonlinear relationships between variables in renewable energy systems. ANNs are extensively used in solar and wind energy applications because of their ability to model complex, nonlinear patterns and their self-organizing nature. They are used to predict how much solar radiation, wind speed, and total power will be produced by renewable energy systems. Long-term predictions and capturing patterns that vary over time can be done with LSTM networks, which can model time-series data. Also, combining them with other models makes them work much better, making it possible to reliably predict things like temperature and radiation in the environment. Compared to LSTM, GRUs offer a lower computational burden, making them a less

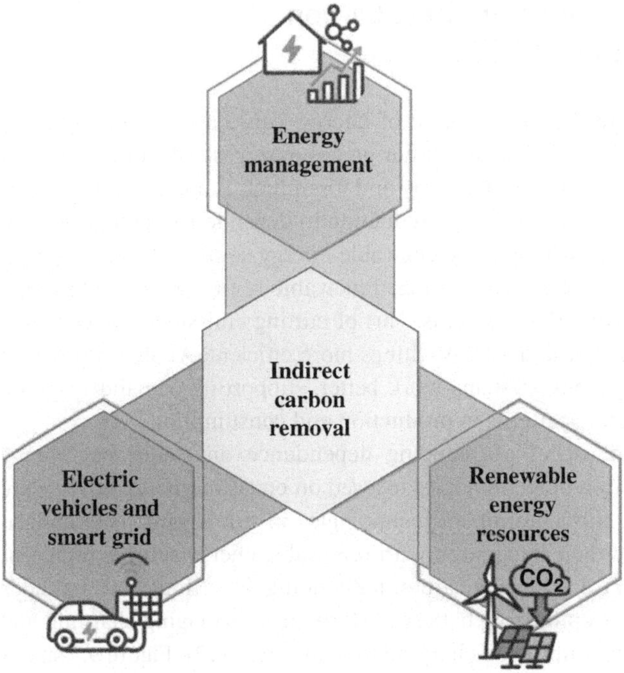

Figure 9.9 Outlook of indirect carbon removal.

expensive choice. When integrated with appropriate preprocessing and complementary techniques, GRU can be effective for forecasting renewable energy output and modeling fluctuating data. SVM, integrated with other AI methods, optimization techniques, and mathematical models, is applied to predict energy storage efficiency, photovoltaic production rates, and wind fluctuations. CNNs are among the most significant methods, in addition to the aforementioned techniques. CNNs are great at finding patterns in space and time, whether they work alone or in hybrid models. For instance, CNN can be combined with LSTM to predict solar radiation across different time intervals. Alternatively, it can be used on its own to generate short-term solar radiation forecasts using infrared sky imagery. RNNs are also important architecture that are effectively used for time-series modeling, as well as short-term and long-term forecasting [14]. AI models for optimizing renewable energy systems are shown in Figure 9.10.

Maintaining performance in modern networks relies heavily on striking a balance between supply and demand. Supply and demand may be better managed with the evolution of smart hardware such as meters, sensors, and control instruments, as well as advancements in AI technologies. AI-based methods can make demand response strategies better by using both past data and current network

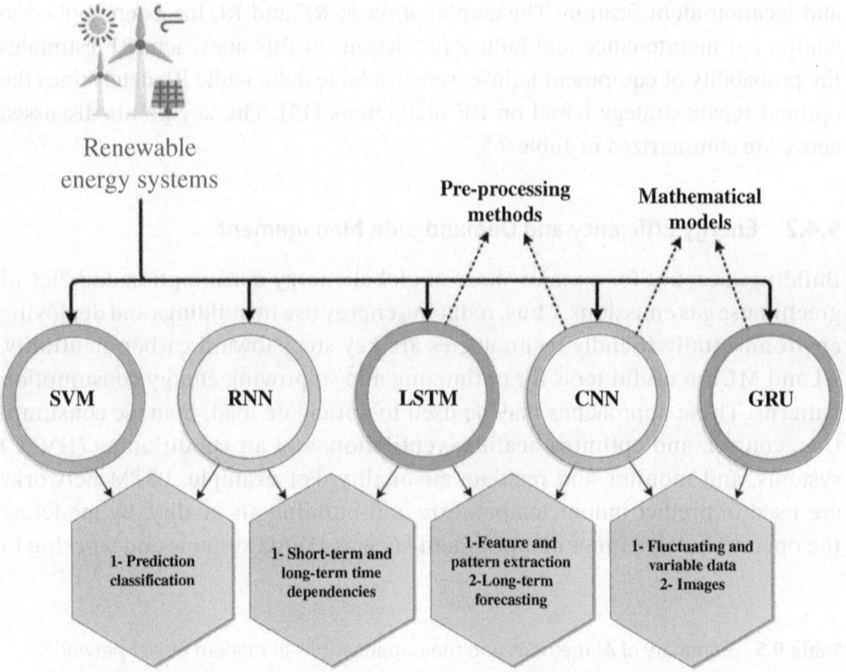

Figure 9.10 AI models for optimizing renewable energy systems.

status data. The integration of AI with smart meters enables the automatic application of strategies such peak load shifting and load shaving. SVM models assess and categorize consumer load data to anticipate consumption patterns, which the management system subsequently uses to change the operating schedules of flexible loads. In addition, RL is capable of learning optimal strategies and reacting in real time to network signals, including voltage, current, and price. It can adapt strategies and update policies dynamically in response to changing network conditions. GNNs are highly effective for monitoring and determining the operational state of power networks. Using phasor measurement unit (PMU) data, they can model network topology as a graph and determine node connectivity. LSTM networks, trained on SCADA and PMU data, are valuable for estimating node voltage and phase. Furthermore, LSTMs are used in anomaly detection and prediction because their capacity to learn patterns allows for the discovery of slight deviations from normal circumstances, which helps to improve network dependability. Autoencoders are also very good at identifying anomalous network behaviors, including voltage dips, surges, and harmonic disturbances. The combination of CNN with wavelet transform improves the model's capacity to capture transient and long-term fluctuations, resulting in better fault type and location identification. The combination of RF and RL has been applied to equipment maintenance and failure prediction. In this approach, RF estimates the probability of equipment failure from available data, while RL determines the optimal repair strategy based on RF predictions [13]. The key points discussed above are summarized in Table 9.5.

9.4.2 Energy Efficiency and Demand-side Management

Buildings account for a major share of global energy consumption and 26% of greenhouse gas emissions. Thus, reducing energy use in buildings and deploying environmentally friendly technologies are key steps toward carbon neutrality. AI and ML are useful tools for optimizing and improving energy consumption patterns. These approaches may be used to anticipate load, manage consumption, control, and optimize heating, ventilation, and air conditioning (HVAC) systems, and monitor and regulate air quality. For example, LSTM networks are used to predict indoor temperature and building air quality. By modeling the operational dynamics of equipment such as HVAC systems and working in

Table 9.5 Summary of AI methods and their applications in modern power networks.

AI method	Application	Key features
SVM	Forecasting consumption; classifying consumption behavior	Analyzes load data for accurate prediction; supports demand-side management
RL	Real-time strategy learning for demand response and maintenance decisions	Dynamically adapts to network signals (e.g., voltage, price); learns optimal policies
GNN	State estimation of power network states using PMU data	Models grid topology as a graph; identifies node connectivity and system status
LSTM	Voltage and phase estimation; anomaly detection and prediction	Captures time-series dependencies; detects deviations from normal behavior
Autoencoder	Abnormal behavior detection	Learns normal operational patterns; flags unseen or unexpected disturbances
CNN + wavelet transform	Fault type detection and fault location	Enhances both transient and steady-state feature extraction
RF + RL	Predictive maintenance and failure management	RF estimates failure probability; RL defines best repair actions

conjunction with RL-based methods, LSTM enables optimal management of these systems. RL algorithms, when combined with techniques such as LSTM, CNN, and attention mechanisms and trained on building data, can effectively balance air quality with energy consumption. Empirical results show that this method achieves a minimum of 15% energy savings and a 16% increase in thermal comfort. Combining RL with GRUs is effective for achieving optimal and cost-efficient HVAC control. GRUs, with a lighter architecture than LSTMs, perform time-series predictions of temperature or energy consumption, which RL then uses to determine optimal control actions. ANNs and MLPs are also effective in controlling air quality and reducing costs, as they can learn nonlinear and complex relationships between environmental variables such as temperature and humidity. GNNs can also describe how different sensors in a building interact with each other since they have a graphical framework that can capture spatial and temporal connections. This capability enhances the accuracy of predictions and decision-making. SVM, support vector regression (SVR), DT, and RF can also perform well when used in combination with each other or with other methods for building management-related predictions [15]. Table 9.6 presents major AI and ML methods for building energy optimization, HVAC control, and air quality management, with their main applications and features.

AI offers multiple approaches for maintaining building equipment, predicting performance, and detecting faults. This can be classified as predictive maintenance.

Table 9.6 AI and ML methods for energy optimization in buildings.

Method	Main application	Effect
LSTM	Temperature and air quality prediction	Models HVAC dynamics for optimal control
RL + LSTM/CNN/attention	Energy–comfort balance	~15% energy savings, 16% comfort gain
RL + GRU	Cost-efficient HVAC control	GRU for time-series prediction, RL for control
GRU	Energy/temperature forecasting	Lighter architecture than LSTM
ANN/MLP	Air quality and cost control	Learns nonlinear relations (temperature, humidity)
GNN	Sensor relationship modeling	Improves prediction and decision accuracy
SVM, SVR, DT, and RF	Building condition prediction	Works alone or in hybrid approaches

HVAC systems represent one of the most important operational components in buildings. Any disruption in this system negatively affects energy consumption, operational costs, and the thermal comfort of the user. RL can be applied for fault diagnosis and troubleshooting in HVAC operations. By optimally and dynamically adjusting system parameters, RL not only ensures efficient device control but also helps maintain system functionality under fault conditions [16]. The LSTM model is suitable for tracking consumption patterns and monitoring performance trends in building equipment. When used with principal component analysis (PCA) and support vector data description (SVDD), it can find deviations from normal operation, which lowers the chance of equipment failures and service disruptions. Also, using LSTM networks with RNNs can help plan repairs and maintenance by predicting when equipment will break down. Sensor data like vibration, temperature, current, and voltage can be used to find or forecast mechanical and electrical problems. To forecast defects and provide repair times, ML methods including DT, RF, and XGBoost are used. Supervised algorithms such as k-nearest neighbors (KNN) and SVM have demonstrated notable performance in HVAC fault detection and classification [43]. Figure 9.11 categorizes the use of AI methods for cost and energy savings, key variable prediction, maintenance, and enhancing user comfort.

9.4.3 Electric Vehicle Integration and Charging Optimization

In recent years, there has been a surge of interest in electric automobiles. Growing environmental concerns, as well as an international effort to limit carbon emissions from fossil fuels, have fueled the rise in popularity of clean-energy technology such as EVs. Integrating EVs into the grid and increasing their use might cut carbon emissions by up to 28%. However, an expanding EV fleet may overwhelm the grid and raise expenses, jeopardizing grid sustainability. These

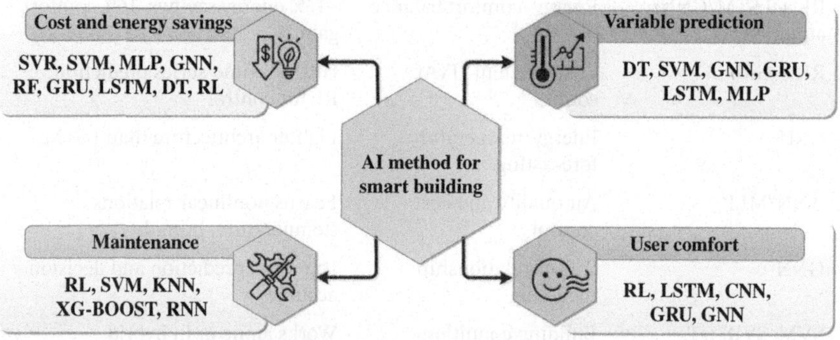

Figure 9.11 AI method for smart building.

difficulties need the implementation of a robust and integrated grid management strategy. Effective smart management must include economic, environmental, and user comfort concerns, for which AI is an unavoidable instrument. AI can achieve high accuracy in predicting charging behaviors by learning from charging and discharging patterns. In addition, optimal planning of these processes using AI can reduce costs and ease grid stress. AI systems that can make decisions in real time can assist keep the network stable and safe. AI can also plan when to charge and discharge batteries so that renewable energy generation is more efficient and less reliant on fossil fuels. This cooperation makes charging more efficient from clean sources and cuts down on carbon emissions. Enhancing user comfort is a key application of AI. By accounting for user preferences and priorities, AI can coordinate actions across different users to preserve network stability while maintaining comfort [18]. The role of AI in EV integration within the network is shown in Figure 9.12.

EVs are classed as all-electric (EV), plug-in hybrid (PHEV), or fuel cell electric cars (FCEV). Their properties are shown in Table 9.7. In the smart grid, EVs can operate in three modes: V2G (vehicle-to-grid), V2V (vehicle-to-vehicle), and V2H (vehicle-to-home). Based on these groupings and the unique opportunities and challenges presented by each network, AI methods should be designed to maintain network stability and support optimal planning [17, 44]. Neural networks

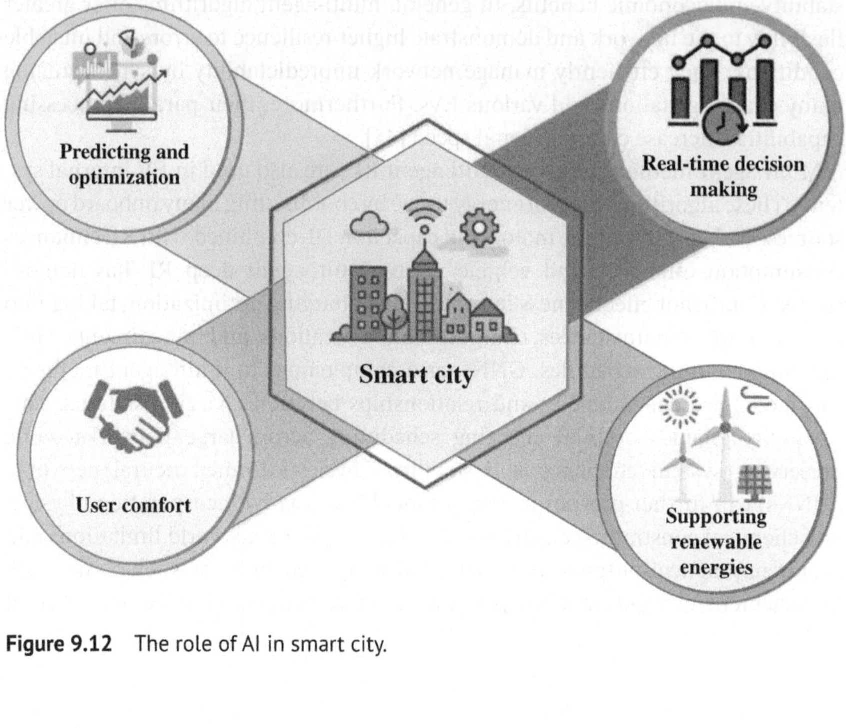

Figure 9.12 The role of AI in smart city.

Table 9.7 Basic characteristics of EV, FCEV, and PHEV.

Type	Energy source	Emissions
EV	Battery (electricity)	Zero tailpipe
FCEV	Hydrogen fuel cell	Zero tailpipe
PHEV	Battery + fuel	Low (depends on fuel use)

are employed to predict EV charging demand and manage network operations under uncertainty. The behavior of EV drivers and the variability of renewable energy production are the primary sources of uncertainty that AI must address. In V2G and V2H operations, ANNs can manage power sharing to keep vehicle battery levels within safe limits and maintain grid stability. This power sharing also has an economic dimension, allowing AI to decide whether to supply power to the grid or draw power based on energy prices. An RL agent can optimally plan power exchanges (charging and discharging) by monitoring variables such as battery charge level, current electricity price, renewable energy forecasts, and energy demand. This strategy reduces cost by charging while prices are low and renewables are available, then discharging when prices rise. Deep RL algorithms, with their increased capacity to manage many EVs, can provide both network stability and economic benefits. In general, multi-agent algorithms offer greater flexibility to the network and demonstrate higher resilience to errors and unstable conditions. They efficiently manage network unpredictability by synchronizing many charging stations and various EVs. Furthermore, their parallel processing capabilities increase computational speed [45].

Multi-agent methods, such as multi-agent RL, are also used in EV internal systems. These algorithms control energy usage by coordinating many onboard power sources, such as the battery, motor, and capacitor. DL combined with RL enhances consumption efficiency and vehicle safety. Multi-agent deep RL has demonstrated significant effectiveness in routing and charging optimization, taking into account traffic circumstances, charging station locations, and impediments while planning charging schedules. GNNs, as a complement to multi-agent methods, can model the dependencies and relationships between charging stations. This capability enables optimal charging scheduling across large networks while preserving system efficiency and stability. Physics-informed neural networks (PINNs) can further prevent incorrect model learning by incorporating physical and chemical constraints, ensuring the model respects real-world limitations and avoids impractical outputs. In general, AI is applied in areas such as network management, charging station operation, and forecasting [46]. In vehicle internal

Table 9.8 AI methods for EV and smart grid optimization.

Approach	Application area	Key functions/benefits
Neural networks (ANNs)	V2G and V2H operations	Predict demand, manage power sharing, maintain safety and stability
Reinforcement learning	Charging and discharging planning	Charge low/discharge high price, reduce costs
Deep RL and multi-agent methods (multi-agent RL, multi-agent deep RL)	EV network, charging station, and internal system management	Coordinate EVs/stations, optimize routing and charging, improve stability and efficiency
GNNs	Charging network scheduling	Model station links, optimize scheduling
PINNs	Model training accuracy	Enforce physical constraints, prevent unrealistic outputs

systems, it is effective for estimating battery life, optimizing energy management, planning routes, and enhancing user comfort. Table 9.8 presents a summary of the key points discussed.

9.5 Challenges and Considerations for AI in Carbon Removal of MENs

AI is a key part of lowering greenhouse gas emissions because it can look at large amounts of data and make accurate predictions. It does this directly by helping to design and run carbon capture and storage systems, and indirectly by reducing energy waste and dependence on fossil fuels. A good example is how smartly combining renewable energy sources like wind and solar into the electrical system. Advanced ML and DL models can predict resource production with high accuracy, allowing grid operators to better coordinate generation, storage, and consumption. This informed coordination makes the grid more reliable, makes operating schedules more efficient, and makes the grid more resistant to variations in generation and the impacts of climate change, speeding up the switch to low-carbon, sustainable infrastructure [47]. However, despite these benefits, implementing AI techniques in current energy systems to obtain sustainable energy has obstacles, which will be discussed below.

9.5.1 Data Requirements

AI in MENs achieves its potential only when the information feeding its algorithms is both trustworthy and well structured. Accurate, current, full, and interference-free records are necessary for CO_2 mitigation systems to function. This contribution can be accomplished by direct action, such as expanding carbon capture and utilization technology, or indirectly by increasing operating efficiency in energy networks and lowering dependency on emission-intensive resources. When the information is flawed, outdated, incomplete, or inconsistent, the resulting models risk producing misguided forecasts, lowering overall system performance, and missing critical opportunities for emission reduction.

Applying AI to carbon management in the petrochemical industry, particularly in direct removal initiatives, faces obstacles on several fronts. Many promising concepts, such as the formulation of novel capture materials, the engineering of specialized catalysts, integrated process simulations, and complete life-cycle evaluations, are still confined to research settings with no large-scale deployment. Advanced computational methods are expensive and hard to understand, and there are not enough complete, high-fidelity operating and experimental samples to make progress toward real use. Models are not very reliable because they are hard to adapt to different business situations, and the results from simulations and the real world don't always match up. Financial restrictions and a lack of clear assessment criteria for carbon capture facilities exacerbate the difficulty [48].

In renewable energy platforms, where AI contributes indirectly to lowering emissions, data limitations remain a decisive barrier. Information from these kinds of systems is often inconsistent, noisy, or skewed toward normal working conditions. For example, problems aren't shown as much as they should be. This mismatch could make training harder and make predictions less accurate. Sensor types, environmental factors, and recording methods that are not all the same cause more problems that need to be fixed before the data can be used successfully. The amount of data that can be analyzed is also limited by restrictions on access, which may be due to privacy concerns, private control, or the high cost of transfer and storage. Getting this data clean and putting it into the right categories takes time and money, which slows down the use of AI in green energy operations [49].

A further challenge with using AI in MENs is that it relies on private datasets a lot, especially accurate readings from smart meters. These records, which are necessary for predicting load consumption and making the network work better, can give you a lot of information about how people live and what they do. If they are widely used without strict privacy protections, they could violate user rights and be used in the wrong way. Still, these same data are essential for making networks more efficient, cutting down on wasted energy, and indirectly helping to lower carbon emissions. Therefore, AI systems must be conceived and executed in a manner that addresses the need for high-quality data while preserving consumer privacy in a complete and enduring way [50].

9.5.2 Complexity and Reliability of AI

The "black-box" challenge in AI, especially in DL and high-parameter neural networks, stems from internal complexity that obscures how inputs are transformed into outputs for human observers. Models with millions of automatically tuned, nonlinear parameters can be highly accurate, yet explaining why a particular input leads to a specific output is often infeasible. As shown in Figure 9.13, this opacity means the decision-making pathway is not transparent to end users. In sensitive settings, including modern energy systems and carbon removal, such opacity can translate into serious economic, environmental, or safety consequences when decisions are incorrect or ambiguous. Transparency deficits weaken stakeholder trust, slow adoption, and complicate validation, raising doubts about whether the model has captured true physical and logical structure or merely overfit to spurious correlations. Explainable AI has therefore advanced methods including LIME and SHAP for feature importance, saliency and sensitivity maps for visualization, intrinsic interpretability for designing transparent models, and mechanistic interpretability to reveal deep-network logic. These approaches aim to make decisions comprehensible and to bolster trust in critical systems, yet the black-box problem remains a central, unresolved barrier to deploying complex models in critical infrastructure.

On the other hand, AI systems in MENs cannot always produce predictions with complete certainty [50]. This kind of uncertainty can happen when the need for electricity changes quickly, when green energy sources like wind and solar power change, or when the quality of the data coming in isn't good enough [51]. When faced with rare events or limited information, even models that are very good at what they do may make statements that can't be trusted. Without giving decision makers clear information about this unpredictability, there is a greater chance that they will make choices that hurt the economy, the environment, or

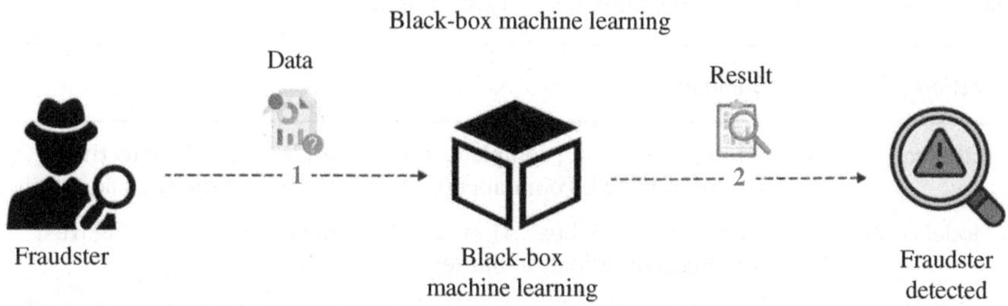

Figure 9.13 Schematic of the AI "black-box" challenge.

grid security. To keep people's trust in these systems, it's important to correctly measure and clearly describe uncertainty [50].

9.5.3 Integrating AI into Existing Network and Computational Challenges

Computer power is one of the primary challenges with using powerful AI models. A lot of the time, training these models means using a lot of GPUs at once and working with very big datasets. As computing jobs get bigger, they use a lot more energy, which puts a lot of stress on the gear and makes it harder to keep and run. Models with millions of parameters require much longer training times and use far more resources, which in turn increases both costs and energy use. These intensive requirements can become a major obstacle to the development and deployment of such systems, with the availability of sufficient infrastructure often deciding whether they can be implemented at all [52].

On the other hand, when energy systems shift to modern designs, one of the most critical obstacles in implementing AI is the need to match it with legacy infrastructures developed many years ago. These infrastructures, which are frequently built on traditional technologies and operating methodologies, are not entirely compatible with the hardware and software requirements of AI systems. The ensuing incompatibilities make merging harder and, in many cases, require major changes to the tools and methods that are already in place. Furthermore, the absence of consistent standards and data formats throughout the network hinders effective data exchange. In other circumstances, AI deployment necessitates infrastructure modifications or complete replacements, which are both resource- and time-expensive. If such coordination is not established, the likelihood of system malfunctions, decreased dependability, and operational interruptions rises. The difficulty intensifies when advanced AI models require high computational power that legacy systems cannot deliver. Therefore, both technical modernization and careful balancing of computational demands with infrastructure constraints are crucial to successfully implementing AI in MENs [53]. Table 9.9 summarizes key AI challenges in carbon removal.

Table 9.9 Summary of key AI challenges in carbon removal.

Section	Summary
Data requirements	AI needs accurate, complete, and accessible data to be effective in carbon removal. Poor data quality limits performance and adoption
Model complexity	Complex "black-box" AI models lack transparency, reducing trust and making validation harder
Existing infrastructure	Integration with old energy infrastructure and high computational demands create major adoption challenges

9.6 AI in Decarbonization Strategies

AI plays a decisive role on both fronts of decarbonization: in the direct pathway, it optimizes CCUS, advances materials design, and strengthens monitoring and verification; in the indirect pathway, it supports renewable integration, improves building efficiency, and manages load and EV charging. Practical success, however, depends on selecting an AI strategy that fits the specific problem. Each algorithmic family has distinctive strengths and success conditions, and the right choice emerges when those traits align with the nature of the data, the safety and explainability requirements, and the deployment horizon. On this basis, practitioners can make a guided selection across a spectrum of methods, from transparent and lightweight models to advanced sequence, graph, and RL approaches, and match method to mission with confidence [5].

Begin by characterizing the data so you can align it with the right model family. For operational time series such as SCADA and PMU readings, or sensor measurements of temperature, pressure, and concentration in absorption processes, sequence-based models are the natural choice: LSTMs and GRUs are efficient and reliable for capturing short- to mid-term patterns, whereas temporal transformers achieve higher accuracy when the forecasting horizon is longer, the number of variables is larger, or the patterns are nonstationary [14]. Whenever network topology plays a crucial role, for example, in power system state estimation or in modeling interactions among feeders and substations, GNNs provide a clear advantage through explicit representation of nodes and edges, and they can also be coupled with sequence models to capture temporal dynamics [13]. For image- and spatial-based data, ranging from satellite and drone imagery to CFD fields and material microstructures, CNNs represent the most effective starting point; if temporal dynamics are also relevant, combining CNNs with LSTM or GRU facilitates the discovery of spatiotemporal patterns. In low-quality or high-noise imaging tasks, incorporating wavelet preprocessing or classical filters can improve feature extraction. For tabular, feature-based data such as material descriptors, billing and electricity market records, or laboratory outputs, tree-based methods including RF, XGBoost, and SVMs typically provide strong, fast, and relatively interpretable baselines, making them reliable choices for industrial applications. When complex nonlinear dependencies reduce their effectiveness, a shallow MLP with proper regularization and early stopping may serve as the next modeling option. In practice, if computational resources are limited or edge deployment is required, GRUs are generally preferable to LSTMs [5].

Beyond the alignment between data and models, the ultimate success of AI in industrial and energy applications also depends on how well the chosen approach conforms to project-level constraints. In many safety-critical grids and process control environments, interpretability and trust become as important as accuracy, making transparent learners or hybrid data- and physics-informed designs a sensible starting point. When computational resources and energy consumption are constrained, especially under edge deployment, compact models and compression techniques such as knowledge distillation, quantization, and pruning can substantially reduce costs without compromising performance. Moreover, in cases where data are fragmented across sites and direct transfer is impractical or risky, federated learning provides a way to train effective global models while safeguarding confidentiality and preserving local autonomy [50].

In AI-driven decarbonization strategies, selecting the right algorithm is only one part of the process; evaluation is just as important. Metrics should not be limited to accuracy but also reflect reliability and readiness for real deployment. Accuracy points to the technical strength of results, reliability builds confidence in their consistency, cost-effectiveness shows economic viability, and user satisfaction indicates how well solutions address real needs. Together, these measures create a balanced basis for both technical and economic decision-making, making evaluation metrics a central pillar of AI strategies for decarbonization rather than a secondary detail. Building on this view, while earlier tables focused on specific domains such as carbon capture, utilization, grid operation, building systems, and electric vehicles, Table 9.10 brings these insights together in a broader perspective. It offers an integrative summary that connects scenarios, data types, algorithm

Table 9.10 Practical guide to selecting AI strategies for decarbonization.

Scenario	Dominant data	Primary objective	Recommended family	Evaluation metrics
Absorber column control (post/DAC)	Process time series (T, P, concentrations)	↑ Capture rate, ↓ reboiler energy	RL/deep RL + LSTM/GRU; Hybrid/PINN	Capture rate/purity, regeneration energy, safety-constraint violations [8]
Materials screening/ design (MOF/COF/ membrane)	Structural/ thermophysical features + simulations	Identify high-capacity/ selective candidates	VAE/GAN/ Diffusion + rankers; with GCMC/DFT	Capacity/ selectivity, synthesizability, cost [7]
Monitoring and verification for storage and leakage	Satellite/UAV imagery + sensors	Leak detection/ estimation with spatial accuracy	CNN/ autoencoder + RNN/LSTM	AUROC/ rare-event recall, detection latency [11]

Scenario	Dominant data	Primary objective	Recommended family	Evaluation metrics
Wind/solar forecasting	SCADA data, weather records	↓ Generation uncertainty	LSTM/GRU or temporal transformer	RMSE/MAE, uncertainty calibration [14]
HVAC	Temp/humidity/CO_2/occupancy	Energy–comfort balance	RL + LSTM/CNN/attention; GNN for sensor graphs	kWh savings, comfort index, fault-tolerance [15]
Multi-site EV/V2G	Load/price/state of charge/traffic	↓ Cost and feeder peak/congestion	Multi-agent deep RL + GNN	Total cost, feeder peak, user satisfaction [47]

families, and evaluation metrics, serving as a practical guide to align methods with missions across diverse decarbonization tasks [15].

9.7 Conclusion

Achieving deep decarbonization of MENs is a critical step in confronting the global climate crisis. Reaching net-zero emissions demands the adoption of effective carbon removal strategies. This chapter emphasizes the capacity of AI to accelerate both decarbonization and the creation of new energy networks. It investigates the many uses of AI in direct and indirect carbon removal, including ML and optimization. Specifically, AI can enhance the operation of direct removal technologies and accelerate the discovery of adsorbent materials. In carbon utilization and sequestration, AI enhances monitoring and verification processes. For indirect carbon reduction, AI supports renewable energy integration through precise forecasting and smart grid management, while also increasing energy efficiency and enabling demand-side management in buildings, industrial processes, and EV charging systems. The chapter further addresses the challenges of applying AI in this context, including data requirements, model complexity, and existing infrastructure limitation. To fully leverage AI's potential for a low-carbon and sustainable future, coordinated efforts in interdisciplinary research, sustainable investment, and regional cooperation are essential. Without such collaboration, a vital opportunity to address the climate crisis may be lost. Beyond current approaches, the next generation of AI technologies should be integrated into decarbonization strategies. Large language models and foundation models, with their capacity to process vast and heterogeneous data, can deliver advanced analytics and decision-making in complex environments. Digital Twins is another technology that improves decision making and planning by generating accurate, continuous representations of energy systems. Federated learning, by enabling distributed data processing without the need for centralized storage, safeguards privacy and lowers security risks, expanding AI applicability on a wider scale. Focusing resources on these technologies can substantially improve the efficiency, resilience, and scalability of decarbonization initiatives and MENs. The main achievements of this work are listed as follows:

- Demonstration of AI's role in direct and indirect carbon removal, including the enhancement of removal technologies, material discovery, monitoring, and verification.
- Identification of AI-driven support for indirect carbon reduction, such as renewable energy integration, energy efficiency improvement, and demand-side management across multiple sectors.
- Discussion of challenges and requirements for AI adoption in decarbonization, highlighting data availability, model complexity, infrastructure limitations, and optimal strategies.
- Introduction of next-generation AI technologies for future decarbonization strategies, including large language models, digital twins, and federated learning to enhance scalability, resilience, and decision-making.

References

1 N. Talebpour, H. Behrooz, and M. Ilbeigi, "Optimizing hybrid power networks to facilitate equitable access to electricity," *Sustainable Cities and Society*, vol. 126, p. 106413, 2025.

2 R. Gharibi, B. Vahidi, and R. Dashti, "Importance of energy justice and equity in grid modernization," in *Communities for Clean Energy Justice and Equity in Grid Modernization*, John Wiley & Sons, 2025, pp. 11–50.

3 S. Önnered and I. Bravić, "Systemic innovation in the European energy sector: A collective outlook," *Journal of Innovation Management*, vol. 12, no. 2, pp. 50–73, 2024.

4 S. Oh, J. Greene, M. Honegger, and A. Michaelowa, "Review of economics and policies of carbon dioxide removal," *Current Sustainable/Renewable Energy Reports*, vol. 12, no. 1, p. 6, 2025.

5 F. A. S. Islam, "The convergence of AI and nature: Advancing carbon dioxide capture, removal, and storage technologies through integrated ecosystem-based strategies," *International Journal of Applied and Natural Sciences*, vol. 3, no. 1, pp. 90–130, 2025.

6 X. Du, M. N. Khan, and G. C. Thakur, "Machine learning in carbon capture, utilization, storage, and transportation: A review of applications in greenhouse gas emissions reduction," *Processes*, vol. 13, no. 4, p. 1160, 2025.

7 M. Tawalbeh, M. Sabri, H. Kazim, A. Al-Othman, and F. Almomani, "Artificial intelligence and material design in carbon capture and utilization: A review of emerging synergies," *Carbon Capture Science & Technology*, vol. 16, p. 100470, 2025.

8 S. Manikandan et al., "Artificial intelligence-driven sustainability: Enhancing carbon capture for sustainable development goals–A review," *Sustainable Development*, vol. 33, no. 2, pp. 2004–2029, 2025.

9 Y. Hu et al., "Comparative study of machine learning techniques for post-combustion carbon capture systems," *Frontiers in Artificial Intelligence*, vol. 7, p. 1441934, 2024.

10 J. B. Asere, A. M. Ijeoma, H. Agboro, T. Kareem, S. T. Adebisi, and I. K. Egbuna, "The integration of Artificial Intelligence in carbon capture, utilization and storage for environmental sustainability in the oil and gas industry," *World Journal of Advanced Research and Reviews*, vol. 26, no. 2, pp. 359–369, 2025.

11 T. Bossy et al., "State of the art and for remote sensing monitoring of carbon dynamics in African tropical forests," *Frontiers in Remote Sensing*, vol. 6, p. 1532280, 2025.

12 J. O. Ojadi, C. S. Odionu, E. Cynthia, and O. A. O. Onukwulu, "AI-powered computer vision for remote sensing and carbon emission detection in industrial and urban environments," *Iconic Research and Engineering Journals*, vol. 7, no. 10, pp. 490–505, 2024.

13 I. K. Egbuna et al., "Advances in AI-powered energy management systems for renewable-integrated smart grids," *World Journal of Advanced Engineering Technology and Sciences*, vol. 15, no. 2, pp. 2300–2325, 2025.

14 R. Teixeira, A. Cerveira, E. J. S. Pires, and J. Baptista, "Advancing renewable energy forecasting: A comprehensive review of renewable energy forecasting methods," *Energies*, vol. 17, no. 14, p. 3480, 2024.

15 D. M. T. E. Ali, V. Motuzienė, and R. Džiugaitė-Tumėnienė, "AI-driven innovations in building energy management systems: A review of potential applications and energy savings," *Energies*, vol. 17, no. 17, p. 4277, 2024.

16 M. Mshragi and I. Petri, "Fast machine learning for building management systems," *Artificial Intelligence Review*, vol. 58, no. 7, p. 211, 2025.

17 M. J. M. Louis and M. Jackson, "AI integration for enhanced smart grids and electric vehicle efficiency," ResearchGate Preprint, pp. 1–12, 2024.

18 T. Senthilkumar, S. S. Sivaraju, T. Anuradha, and C. Vimalarani, "An intelligent electric vehicle charging system in a smart grid using artificial intelligence," *Optimal Control Applications and Methods*, vol. 46, no. 3, pp. 1180–1192, 2025.

19 I. Worighi, A. Maach, A. Hafid, O. Hegazy, and J. Van Mierlo, "Integrating renewable energy in smart grid system: Architecture, virtualization and analysis," *Sustainable Energy, Grids and Networks*, vol. 18, p. 100226, 2019.

20 A. Golshanfard and Y. Noorollahi, "Coordinated operation and planning of the modern heat and electricity incorporated networks," in *Coordinated Operation and Planning of Modern Heat and Electricity Incorporated Networks*, John Wiley & Sons, 2022, pp. 177–209.

21 Z. Shi et al., "Artificial intelligence techniques for stability analysis and control in smart grids: Methodologies, applications, challenges and future directions," *Applied Energy*, vol. 278, p. 115733, 2020.

22 P. K. Garg, "Overview of artificial intelligence," in *Artificial Intelligence*, Chapman and Hall/CRC, 2021, pp. 3–18.

23 T. O. Ayodele, "Types of machine learning algorithms," in *New Advances in Machine Learning*, IntechOpen, 2010, pp. 19–48.

24 A. D. Papalexopoulos and T. C. Hesterberg, "A regression-based approach to short-term system load forecasting," *IEEE Transactions on Power Systems*, vol. 5, no. 4, pp. 1535–1547, 1990.

25 Z. Yadollahi, R. Gharibi, R. Dashti, and A. T. Jahromi, "Optimal energy management of energy hub: A reinforcement learning approach," *Sustainable Cities and Society*, vol. 102, p. 105179, 2024.

26 Y. Reddy, P. Viswanath, and B. E. Reddy, "Semi-supervised learning: A brief review," *International Journal of Engineering and Technology*, vol. 7, no. 1.8, p. 81, 2018.

27 A. Voulodimos, N. Doulamis, A. Doulamis, and E. Protopapadakis, "Deep learning for computer vision: A brief review," *Computational Intelligence and Neuroscience*, vol. 2018, no. 1, p. 7068349, 2018.

28 M. Ramezanizadeh, M. H. Ahmadi, M. A. Nazari, M. Sadeghzadeh, and L. Chen, "A review on the utilized machine learning approaches for modeling the dynamic viscosity of nanofluids," *Renewable and Sustainable Energy Reviews*, vol. 114, p. 109345, 2019.

29 A. Apicella, F. Donnarumma, F. Isgrò, and R. Prevete, "A survey on modern trainable activation functions," *Neural Networks*, vol. 138, pp. 14–32, 2021.

30 M. M. Forootan, I. Larki, R. Zahedi, and A. Ahmadi, "Machine learning and deep learning in energy systems: A review," *Sustainability*, vol. 14, no. 8, p. 4832, 2022.

31 Y. Bai, S. Camal, and A. Michiorri, "News and load: A quantitative exploration of natural language processing applications for forecasting day-ahead electricity system demand," *IEEE Transactions on Power Systems*, vol. 39, no. 5, pp. 6222–6234, 2024.

32 J. J. Moreno Escobar, O. Morales Matamoros, R. Tejeida Padilla, I. Lina Reyes, and H. Quintana Espinosa, "A comprehensive review on smart grids: Challenges and opportunities," *Sensors*, vol. 21, no. 21, p. 6978, 2021.

33 H. Wang, Z. Lei, X. Zhang, B. Zhou, and J. Peng, "A review of deep learning for renewable energy forecasting," *Energy Conversion and Management*, vol. 198, p. 111799, 2019.

34 R. Gharibi, B. Vahidi, R. Dashti, and R. Khalili, "From electrons to solutions: Power-to-X strategies for energy systems integration," in *Power-to-X in Regional Energy Systems*, CRC Press, 2025, pp. 201–227.

35 P. Vaziri, M. R. Rasaei, S. Seyfoori, S. Zamani, M. Mahmoodi, and B. Sedaee, "Advancing carbon capture technologies in CCS: A comprehensive review of pre-combustion processes," *Gas Science and Engineering*, vol. 131, p. 205481, 2024.

36 H. Xu, L. Yu, C. Chong, and F. Wang, "A comprehensive review on direct air carbon capture (DAC) technology by adsorption: From fundamentals to applications," *Energy Conversion and Management*, vol. 322, p. 119119, 2024.

37 A. Vasileiadou, "Advancements in waste-to-energy (WtE) combustion technologies: A review of current trends and future developments," *Discover Applied Sciences*, vol. 7, no. 5, p. 457, 2025.

38 E. G. Al-Sakkari, A. Ragab, H. Dagdougui, D. C. Boffito, and M. Amazouz, "Carbon capture, utilization and sequestration systems design and operation optimization: Assessment and perspectives of artificial intelligence opportunities," *Science of The Total Environment*, vol. 917, p. 170085, 2024.

39 A. Saxena et al., "Innovative pathways in carbon capture: Advancements and strategic approaches for effective carbon capture, utilization, and storage," *Sustainability*, vol. 16, no. 22, p. 10132, 2024.

40 D. Triantakonstantis and A. Karakostas, "Soil organic carbon monitoring and modelling via machine learning methods using soil and remote sensing data," *Agriculture*, vol. 15, no. 9, p. 910, 2025.

41 B. I. Oladapo, M. A. Olawumi, T. O. Olugbade, and S. O. Ismail, "Data analytics driving net zero tracker for renewable energy," *Renewable and Sustainable Energy Reviews*, vol. 208, p. 115061, 2025.

42 R. Gharibi, R. Khalili, B. Vahidi, A. F. Nematollahi, R. Dashti, and M. Marzband, "Enhancing energy hub performance: A comprehensive model for efficient integration of hydrogen energy and renewable sources with advanced uncertainty management strategies," *Journal of Energy Storage*, vol. 107, p. 114948, 2025.

43 A. Adhikari, T. Karunaratne, and N. Sumanarathna, "Machine learning techniques for predictive maintenance of building services: A comprehensive review and research outlook," *Facilities*, vol. 43, no. 11–12, pp. 803–817, 2025.

44 R. Gharibi, B. Vahidi, and R. Dashti, “Green transportation systems,” in *Interconnected Modern Multi-Energy Networks and Intelligent Transportation Systems: Towards a Green Economy and Sustainable Development*, John Wiley & Sons, 2024, pp. 8–38.

45 H. S. Daliri, M. Dejagah, H. R. Baghaee, H. A. Abyaneh, and A. Bakhshai, “HMM-based feature extraction and machine learning methods for event detection and classification in microgrids,” *IEEE Transactions on Smart Grid*, vol. 16, no. 6, pp. 4366–4379, 2025.

46 A. Kermansaravi, S. S. Refaat, M. Trabelsi, and H. Vahedi, “AI-based energy management strategies for electric vehicles: Challenges and future directions,” *Energy Reports*, vol. 13, pp. 5535–5550, 2025.

47 M. Kiasari, M. Ghaffari, and H. H. Aly, “A comprehensive review of the current status of smart grid technologies for renewable energies integration and future trends: The role of machine learning and energy storage systems,” *Energies*, vol. 17, no. 16, p. 4128, 2024.

48 A. K. Priya, B. Devarajan, A. Alagumalai, and H. Song, “Artificial intelligence enabled carbon capture: A review,” *Science of The Total Environment*, vol. 886, p. 163913, 2023.

49 Y. S. Afridi, K. Ahmad, and L. Hassan, “Artificial intelligence based prognostic maintenance of renewable energy systems: A review of techniques, challenges, and future research directions,” *International Journal of Energy Research*, vol. 46, no. 15, pp. 21619–21642, 2022.

50 F. Henao, R. Edgell, A. Sharma, and J. Olney, “AI in power systems: A systematic review of key matters of concern,” *Energy Informatics*, vol. 8, no. 1, p. 76, 2025.

51 A. Darvishi, B. Ranjbar, R. Gharibi, R. Khalili, and R. Dashti, “Multi-objective optimization of a socio-economic energy hub with demand response program and considering customer satisfaction,” *Journal of Energy Storage*, vol. 100, p. 113624, 2024.

52 P. Huber et al., “Energy consumption in parallel neural network training,” arXiv Preprint, arXiv:2508.07706, 2025.

53 M. S. S. Danish, “AI in energy: Overcoming unforeseen obstacles,” *AI*, vol. 4, no. 2, pp. 406–425, 2023.

10

Intelligent Cyber-physical Systems in Deep Decarbonization of Modern Energy Networks

Reza Sepehrzad[1], Hossein Hosseinalibeiki[2], Zahra Yousefi[3], Ahmed Al Durra[4], and Mahdieh S. Sadabadi[5]*

[1] *Electrical and Computer Engineering Department, University of Canterbury, Christchurch, New Zealand*
[2] *DISA-MIS Department, University of Salerno, Fisciano, Campania, Italy*
[3] *Department of Chemical and Polymer Engineering, Yazd University, Yazd, Iran*
[4] *Advanced Power & Energy Center, EECS Department, Khalifa University, Abu Dhabi, UAE*
[5] *Department of Electrical and Electronic Engineering, The University of Manchester, Manchester, UK*

*Corresponding author: sepehrzad.reza@gmail.com

List of Acronyms

AI	Artificial intelligence
CESN	Climate–energy–sustainability nexus
CNNs	Convolutional neural networks
DDPG	Deep deterministic policy gradient
DERs	Distributed energy resources
DoS	Denial-of-Service
FL	Federated learning
GAEs	Graph autoencoders
GCN	Graph convolutional network
GNN	Graph neural network
HVAC	Heating, ventilation, and air conditioning
ICPS	Intelligent cyber-physical systems
IDS	Intrusion detection systems
IoT	Internet of Things
IT	Information technologies
LLMs	Large language models

List of Acronyms	
LSTM	Long short-term memory
MDP	Markov decision process
MENs	Modern energy networks
ML	Machine learning
OT	Operational technologies
P2P	Peer-to-peer
PMU	Phasor measurement unit
RL	Reinforcement learning
SCADA	Supervisory Control and Data Acquisition

10.1 Introduction: Toward an Intelligent Nexus Infrastructure

The transformation of modern energy networks (MENs) into resilient, adaptive, and decarbonized systems is an urgent operational mandate. Climate change, resource limitations, and sustainability targets demand a fundamental redesign of energy networks. This challenge is encapsulated in the climate–energy–sustainability nexus (CESN), a framework that forces us to confront the tight coupling between environmental, energy, and socioeconomic systems [1]. Achieving the deep decarbonization required by CESN necessitates a re-engineering system of sensing, control, and, most importantly, securing the entire energy value chain [2]. This is where the role of intelligent cyber-physical systems (ICPS) becomes critical. Unlike traditional energy system automation and control, which mainly rely on centralized controls, ICPS fundamentally change the energy systems' architectures [3]. They embed computational intelligence directly into physical infrastructure, enabling a distributed network capable of self-awareness, autonomous response to disturbances, and real-time management of decentralized resources. This shift is enabled by leveraging distributed sensing and processing, the use of edge-cloud, and deploying artificial intelligence (AI)-enabled decision logic that operates across the vast geographic dimensions of the energy ecosystem [4].

The promise of ICPS is a system that can simultaneously optimize for technical efficiency, climate resilience, and decarbonization goals. However, this very integration of data-driven intelligence and autonomous operation introduces new vulnerabilities. In this chapter, we will discuss the conceptual architecture of ICPS for MENs, arguing that their strategic deployment is the principal enabler for deep decarbonization within CESN [5]. We will also critically examine the cybersecurity risks inherent in this data-centric autonomy and propose resilience constructs that are robust against both cyber threats and climate-induced disruptions.

10.1.1 CESN: A Strategic Imperative for MENs

CESN emerges from the critical interdependence of climate change, energy security, and sustainability. This framework recognizes that environmental health, energy transitions, and social equity issues are fundamentally interconnected and must be tackled together. Traditional policy approaches treated energy security and environmental protection as separate challenges. However, CESN requires integrated planning that brings together infrastructure development, regulations, and technical expertise across all these domains. The central goal is achieving deep decarbonization, significantly reducing greenhouse gas emissions at every stage of our energy systems, from power generation and grid transmission all the way to how we use energy in our daily lives [6]. However, deep decarbonization cannot be achieved through technological substitution alone. It requires a new category of intelligent, adaptive, and cyber-aware infrastructures that can sense, learn, adapt, and work across distributed environments. This transformation points to the role of ICPS as an integrated framework where physical assets and digital intelligence co-evolve to provide safe, responsive, and sustainable operations. Augmenting ICPS in MENs provides the operational substratum to achieve CESN principles. By means of ICPS, this vision becomes practical. This is achieved by enabling real-time adjustments, coordinating distributed resources, and automating and controlling processes while considering resilience and reliability constraints. These technological features transform CESN from being an abstract policy goal into something that can be built into our energy infrastructure [7].

In CESN, there should be operational consistency across a diverse set of energy carriers, renewables, battery energy storage systems, flexible loads, and legacy grid assets. In this multi-vector setting, ICPS aim to provide the interface of subsystems and components in CESN through open communication protocols, sensor fusion, and AI-enabled data analyses. By embedding intelligence in the structure of MENs, ICPS can create a base in which environmental boundaries and system objectives can be continuously and autonomously aligned [8]. CESN goes beyond just being a smart technology, as it fundamentally changes how we understand energy systems. Instead of seeing them as purely technical infrastructure, this approach recognizes that energy systems are inherently social, requiring us to address climate adaptation, equity, and economic sustainability all at once. Managing these multiple dimensions needs more than a smart technology. It requires new institutions that can learn, adapt, and make ethical decisions over time. ICPS support this vision by ensuring humans remain central to decision-making and creating meaningful ways for communities to participate in how these systems work [9]. This becomes increasingly crucial as public trust, transparency, and legitimacy determine whether energy transitions succeed or fail.

10.1.2 Rethinking MENs: An Intelligence and Security Perspective

The energy systems have changed from inflexible and centralized infrastructures to flexible, decentralized, and sustainability-centered systems. Conventional energy systems did not consider flexibility for energy flows, predictable demand profiles, or hierarchical control structures. However, the integration of intermittent renewable energies, such as wind and solar, bidirectional energy flows, demand-side participation, and multi-agent markets, has introduced significant complexity and uncertainty in MENs [10]. There are limits to what conventional control structures of energy networks and human decision-making can achieve. MENs must be built on intelligent systems that perceive, interpret, and act within rapidly changing and uncertain environments with little or no external inputs [11]. This intelligent characteristic must not be separated from the parallel consideration of security. As MENs become more digitalized, they are more exposed to cyber-physical threats that can adversely impact system stability and performance, data integrity and availability, and the trustworthiness of operations.

The convergence of operational technologies (OT) and information technologies (IT) means that the attack surface is large, covering grid edge devices, communication protocols, and AI-enabled control functions. Without a security-by-design plan, intelligence will be a liability, from which adversarial agents can be used or corrupted. As any reconfiguration of MENs must include a plan for embedding ICPS for adaptive, self-organizing function, it also must include a plan for becoming resilient against cyber intrusion, miscoordination, and cascading failure [11]. ICPS supply a coordinated response to the twofold requirement of integrating sensing, decision-making, and actuation at both secure computing layers simultaneously throughout the system. By use of distributed edge intelligence and cloud-assisted coordination, ICPS can localize disruption, limit propagation, and restore operation in near real-time. The design of the ICPS can integrate support for optimal energy dispatch and carbon-aware scheduling, intrusion detection, anomaly classification, and even cooperative responses between spatially distributed assets. In this way, ICPS can transform MENs from simple delivery pipelines into intelligent, responsive systems that can advance sustainability goals while enhancing operational security [12].

10.1.3 Role of ICPS in Creating Decarbonized, Secure, and Adaptive Infrastructures

Moving toward deep decarbonization means fundamentally changing how energy infrastructure operates. This requires shifting from reactive control systems to intelligent networks that can actively manage operations. ICPS are not just a technological upgrade. They are essential building blocks for making the climate–energy–sustainability vision a reality. These systems can process real-time data, learn from changing conditions, and autonomously adjust operations as needed. ICPS are becoming the backbone of next-generation energy networks by helping coordinate decisions about energy use, emissions reduction, and enhancing system resilience [13].

One of the main advantages of ICPS is their ability to embed the intelligence feature across energy systems with both spatial and functional layers. On the device layer, sensors and embedded processors provide real-time, high-resolution monitoring of generation units, distributed energy resources (DERs), and energy consumption (load) patterns. Meanwhile, sophisticated algorithms and AI models running on edge computing systems and in the cloud help optimize power dispatch, forecast renewable intermittency, and identify even marginal productivity losses. The increasing intelligence allows us to more efficiently control physical assets. This implies that optimizing scheduling and alignment with carbon reductions can happen in real time. In this way, ICPS represent the computational base layer for emissions-aware energy transitions [3, 4]. However, in the integration of ICPS, cybersecurity must be considered as an equally important design factor. As ICPS bring digitalization and communication technologies in MENs, they simultaneously create new cyberattack vulnerabilities in communication, coordination, and data integrity. Therefore, ICPS must be integrated with cybersecurity protection surfaces, such as anomaly detection frameworks, adversarial learning, encryption protocols, and decentralized trust models. They can detect, categorize, and mitigate cyber-physical threats in real time. This enables the achievement of intelligence, automation, and cybersecurity [14].

Due to increasing demand with climate variability, technological shifts, and changing policies, energy networks need to change and adapt. ICPS help energy networks adapt to changes. They provide this adaptation by self-healing control loops that operate cooperatively using feedback optimization. For example, in the presence of faults, ICPS can restore their stability, redistribute the load during emergencies, and evolve the network topology to adapt to external conditions, enhancing the energy systems' reliability. It is worth mentioning that these adaptive features ensure that the intended CESN goals are pursued in future conditions in the event of cascaded or external disturbances. As a result, ICPS fuse the static infrastructure with intelligent system behavior that seeks to align with goals [15].

10.1.4 Chapter Goals and Structure

The rapid electrification of end-use sectors, large-scale integration of inverter-based resources, and increased reliance on digital coordination have turned modern energy infrastructures into tightly coupled cyber-physical systems. In this setting, operational performance and decarbonization targets are no longer determined solely by physical assets, but also by sensing, communication, data-driven control, market coordination, and the cyber trustworthiness of these layers. This chapter is motivated by a practical gap: many discussions treat power engineering, digitalization, and cybersecurity as separate topics, while real-world failures and resilience outcomes emerge from their interaction.

The scope of this chapter is to provide a structured, system-level view of cyber-physical energy systems in complex energy system networks, with an emphasis on how digital intelligence and security mechanisms co-determine reliability, flexibility, and decarbonization performance. The chapter focuses on integrated cyber-physical system architectures, digital twins, and data-driven operational intelligence, and the evolving threat surface created by distributed assets, market participation, and multi-stakeholder connectivity. At the same time, it remains deliberately non-prescriptive about any single vendor stack and does not attempt to replace detailed standards documents or device-level implementation manuals; instead, it synthesizes principles, design patterns, and evaluation criteria that remain valid across platforms.

The chapter's contributions are threefold. First, it consolidates key architectural building blocks of integrated cyber-physical energy systems and clarifies the functional role of sensing, communication, computation, and control in complex energy system networks. Second, it connects digital intelligence methods, including digital twins, machine learning (ML), and emerging large language model (LLM) capabilities, to concrete operational tasks such as situational awareness, forecasting, anomaly detection, maintenance planning, and decision support under uncertainty. Third, it frames cybersecurity and resilience not as add-on requirements, but as measurable system properties, introducing CESN-aware considerations that link threat models, mitigation strategies, and resilience metrics to operational goals and governance needs. This positioning is intended to help readers understand not only what technologies exist, but also when they are appropriate, what they enable, and where their limits and risks appear in practice.

Chapter at a glance: this chapter explains why modern energy infrastructures should be treated as cyber-physical systems, where physical reliability and decarbonization performance depend on digital coordination and cyber trust. It outlines the architecture of integrated cyber-physical energy systems, then shows how digital twins, ML, and LLM capabilities can support operational decision-making, monitoring, and planning. It also clarifies how cyber threats propagate through data, communication, and market layers, and how resilience-oriented defenses can be evaluated using system-relevant metrics. The intended takeaway is a unified map that links architecture, intelligence, and security to measurable outcomes, providing readers with a clear entry point into specialized topics without requiring deep prior expertise.

In summary, the main contributions of this chapter are:

- To evaluate ICPS concepts and configure their critical role in CESN structure based on the requirements of smart cities, and improve the reliability, security, sustainability, and resilience indicators of modern energy distribution systems.
- To evaluate and configure concepts based on intelligent models such as ML, digital twins, LLM, and their role in developing cybersecurity approaches and improving energy distribution security based on supporting deep decarbonization policies, resilience, and adaptive policies.
- To provide a role for the path to integrate a safe and sustainable ICPS structure into modern energy systems, considering the challenges ahead, research gaps, as well as policy implications based on the requirements of future energy systems.

The rest of the chapter is organized as follows: Section 10.2 provides the foundation of ICPS in MEN contexts, covering a layered framework, sensing-actuation properties, and frameworks for semantic data integration. Section 10.3 focuses on enabling digital twins and associated ML algorithms and LLMs for responsive, autonomous decision environments. Section 10.4 focuses on the cybersecurity threat landscape and deficiencies of legacy defense paradigms in ICPS-embedded infrastructures. Section 10.5 defines proactive methods for intelligent anomaly detection, enhancing the resilient operations of ICPS, and autonomous response coordination. Section 10.6 considers the challenge of aligning operational intelligence with sustainability metrics. It examines how ICPS can think carbon-aware, balance performance with emissions, and enable automation while preserving trust. Section 10.7 discusses remaining challenges for ICPS within MENs and remaining gaps for policy, especially in the areas of standardization, human-in-the-loop control, and ethical design. The chapter concludes by exploring how intelligence, security, and sustainability work together in decarbonized energy networks, highlighting why ICPS are essential for driving long-term systemic change in MENs. The chapter hopes to provide a state-of-the-art review and a conceptual road map for researchers, engineers, and policymakers working on the design of modern energy systems that are as efficient, cyber-secure, and sustainable.

10.2 Foundations of ICPS in Energy Contexts

The increasing augmentation of ICPS into MENs initiates a paradigm shift in the design, operation, and governance of energy systems consistent with CESN requirements. While automation in energy systems has centered on a conventional approach based on central logic and deterministic control, ICPS leverage decentralized intelligence, adaptability in real time, and seamless coupling between cyber and physical components. This section presents the architectural basis of ICPS. In particular, it covers sensing and actuation architecture, embedded intelligence through Internet of Things (IoT) architecture, and the capacity that semantic, scalable, and interoperable data architecture offers in enabling action. We examine each aspect of the architecture to establish a foundational understanding of how ICPS creates an environment for intelligent decisions and, ultimately, to leverage complex and multi-agent energy systems to secure sustainability transitions.

10.2.1 Architectural Overview: Sensing, Actuation, and Control Layers

The ICPS architecture deployed within MENs consists of multiple layers of configuration. Each layer embraces physical elements and computational intelligence layers to execute real-time decision-making, control in a secure manner, and optimize across the system. The architecture can be reduced to three core layers: (i) the sensing and perception layer, (ii) the control and actuation layer, and (iii) the cyber coordination and intelligence layer, each are performing their own roles but coordinates their functions to achieve the objectives for CESN, particularly navigating through high uncertainty, volatile and distributed environments [14, 15]. The sensing layer is the interface between the physical world and digital intelligence. It consists of a dense network of smart meters, phasor measurement units (PMUs), voltage sensors, thermal cameras, and IoT-enabled devices dispersed across generation units, substations, distribution feeders, and, critically, at end users' premises. These sensors collect high-resolution spatiotemporal data on voltage, current, frequency, power quality, temperature, weather, and emission data in real time. The extensive data flow is the raw material that the next layer can use to analyze, predict, and act. The integrity of the sensing layer's devices and the data obtained as well as their resilience to latency and interoperability remain critical properties for facilitating fast and accurate action by the system [3, 4].

The actuation and control layer transforms analytical insight into operational action. It includes Supervisory Control and Data Acquisition (SCADA) systems, digital protection relays, control systems of DERs, and demand response actuators. Traditional systems had to configure these control elements manually or could only use fixed logic. ICPS architectures utilize adaptive control approaches (with reinforcement learning [RL], model predictive control, or rule-based decision trees), allowing the system to reconfigure setpoints automatically, warrant DERs dispatch, and identify faults based on real-time changing conditions. Hence, this layer represents the critical part of operationalizing emissions-reduction schedules, voltage and frequency fluctuations stabilization, and operational security [16].

The cyber coordination layer is responsible for the total behavior of the system through distributed intelligence, communications, and security logic. The coordination layer includes edge-cloud architecture, blockchain or consensus protocols, federated learning (FL) agents, and semantic data engines. The coordination layer determines decision flows across assets, balances local and global optimization, and mitigates potential cyber threats associated with intrusion detection and autonomous responses [3]. For example, the coordination layer also enables system-wide coordination among market participants, utilities, and grid assets to promote dynamic pricing and peer-to-peer (P2P) energy trading, as well as actively enforce energy policy. The intelligence of the layer is not just computational but systemic, synchronizing technical performance with higher value ideals within CESN, including equity, transparency, and sustainability [17]. Table 10.1 shows the architectural layers of ICPS embedded within MENs, and the main functions and contributions to CESN themes.

10.2.2 IoT-based Perception and Embedded Intelligence

The first part of any ICPS is the situation and environmental awareness of real physical systems. Over the last decade, the IoT has more or less created a distributed architecture of connected energy systems that draw on an abundance of low-cost, sensor-based devices that are connected to distributed infrastructure and collect and wirelessly transfer high-resolution data, and in some cases, use pre-modeling

Table 10.1 Functional architecture of ICPS in MENs [3, 4, 16, 17].

Layer	Components	Functions	Objectives for CESN
Sensing and perception	Smart meters, PMUs, emission monitors, weather stations, and IoT sensors	Data acquisition in real time, system state assessment, and anomaly reporting	Supports emissions tracking, climate adaptation, and distributed observability
Control and actuation	Protection relays, DER controllers, smart inverters, and SCADA systems	Grid stability management, real-time dispatch, fault isolation, and managed or planned demand response	Enables low-cost energy optimization and responsive sustainability initiatives
Cyber coordination	Edge-cloud systems, semantic data engines, blockchain, and FL agents	Distributed decision-making, system optimization, and cyber defense coordination	Facilitates secure, adaptive, and trust-meaningful behavior of the system in transition to CESN

or preprocessing, across a spatially distributed asset to inform user or stakeholder action. Within MENs, IoT systems are connected to generation assets, substations, distribution lines, microgrids, prosumer interfaces, and smart buildings, continuously monitoring current, voltage, frequency, environmental conditions, asset health, and carbon pollution. These sensing functions serve dual purposes in their support of operational situational awareness and input of resolution data that is essential for intelligent decision-making that adheres to a CESN value [18]. Furthermore, IoT-based intelligence supports a more granular and participatory form of energy governance. Prosumer households equipped with smart meters, appliances, and controllers can monitor consumption patterns, optimize usage based on real-time price signals, and trade surplus energy within P2P markets. This decentralization of intelligence contributes to the CESN framework by creating several features, including fostering energy democratization, demand-side flexibility, and local sustainability optimization. Moreover, it enhances social resilience, as a localized control architecture reduces dependency on central coordination and enables community-driven response in the face of disruptions [19].

The intelligent capability of IoT systems makes modern ICPS different from conventional data collection architectures. IoT devices use lightweight microcontrollers, on-board analytics, and edge-optimized ML models to provide first-layer signal processing, event detection, and local decision support where no cloud solutions are available. For example, a smart inverter in a distributed solar system may detect instability in grid frequency and initiate an autonomous islanding procedure. Similarly, an emission sensor may report noncompliance with carbon thresholds and notify localized controllers. Distributed capabilities reduce latency and improve fault tolerance (reliability), compared to centralized ones. Hence, they yield quicker responses that also tend to consume less energy, which are critical factors for systems tasked with reconciling deep decarbonization with real-time reliability [19, 20]. However, the integration of IoT faces challenges in terms of scalability, interoperability, and cybersecurity issues. Many IoT devices are limited in computational, memory, and power resources. This compromises their ability to perform secure protocols through strong encryption and/or dynamic negotiation of communication protocols. Therefore, in the process of ICPS design, securing communications and ensuring interoperability among heterogeneous devices while maximizing embedded intelligence must be dealt with. To achieve these aims, there will be shifts to standardize IoT architectures, strong protocols for edge-cloud coordination, and middleware layers designed to mitigate trust. IoT-based perception could have a significant social and environmental impact as a catalyst for CESN-aligned intelligent and resilient MENs, but only if it is assessed and designed holistically [18–20]. Table 10.2 provides an overview of the fundamental dimensions of IoT-based perception and embedded intelligence in ICPS.

Table 10.2 Multidimensional role of IoT-based perception and embedded intelligence in ICPS [18–20].

IoT function layer	Technologies	Embedded intelligence	Objectives for CESN	Main implementation constraints
Sensing and data acquisition	PMUs, thermal sensors, smart meters, CO_2 monitors, and weather stations	High-frequency sampling, edge preprocessing	Real-time emissions profiling, load tracking, and climate operating	Non-homogeneity of data, signal noise, and calibrating the sensor
Local analytics and response	Local gateways, smart inverters, and microgrid controllers	Threshold alerts, event detection, and localized fault clearance	Decentralized control, localized robustness, and low-latency carbon scheduling	Resource constraints, firmware limitations, and limited ML capabilities
Distributed decision support	Fog computing nodes, edge AI devices	Predictive maintenance, anomaly streamlining, and prosumer enhancements	Efficiency opportunities, prosumer participation experience, and decentralized energy governance solutions	Model compression, explanation of algorithms, and edge-cloud consistency
Communication and interoperability	MQTT, 5G NR, ZigBee, LoRaWAN, and IPv6-based mesh protocols	Secure routing, protocol switching, and associated metadata	Inter-device coordination, energy democratization, and social inclusion	Fragmentation of protocol, latency-sensitive, and vulnerabilities in security
Security and trust layer	Blockchain modules, hardware root of trust, and distributed identity	Device authentication, encrypted telemetry, and tamper detection	Trust-preserving automation, privacy-by-design feature, and CESN models accountability	Crypto-limited resources, interoperability with legacy systems

10.2.3 Edge-cloud Systems, AI Integration, and Semantic Data Flows as Key Enablers in ICPS

The successful integration of ICPS for transforming MENs and delivering CESN-compliant outcomes relies on a set of integrated enablers. These key enablers mediate how data is processed, interpreted, and acted upon at different spatial and temporal scales. Among these enablers, three of the most important ones include:

1. Edge-cloud architectures that enable latency-sensitive computations to run locally with scalable analytics in the cloud.
2. AI pipelines that convert raw sensor data to actionable intelligence.
3. Semantic data flows providing interoperability across diverse systems.

These building blocks form a digital nervous system for the ICPS, enabling resilient, automated, intelligent, and coordinated behavior across DERs [21]. Edge-cloud systems represent a distinct shift from a centralized approach to data processing, toward a more hierarchical, layered, and flexible architecture. In ICPS, edge nodes, e.g., local controllers, gateways, or embedded processors, perform low-latency actions such as filtering, fault detection, and implementing local control actions. Thus, cloud platforms conduct heavy workloads that require offline computations such as training a deep learning model, making long-term forecasts, or conducting large-scale optimizations. This structure also permits fast real-time responses at the local level while allowing the edge-cloud system to be aware of real-time global trends to orchestrate more strategic actions. Moreover, the edge-cloud organization alleviates bandwidth bottlenecks and limits cybersecurity risks from transmitting all data to a hub [21].

The AI pipeline is the analytical engine of ICPS, and it is responsible for uncovering patterns, determining anomalies, and forecasting states of systems. AI pipelines typically consist of steps including data cleansing, feature extraction, model training, and inferencing, while these pipelines can run across edge or cloud platforms depending on the application context. Within MENs, AI models are used to support applications like load forecasting, predicting renewable generation, managing congestion, and optimizing carbon emissions. Critically, these models must be retrained continually to account for nonstationary conditions such as changes in federal or state policy, variations in weather, or shifts in user behavior. Therefore, automation of pipelines and integration of feedback are essential to help with the ongoing performance of the CESN [22].

In addition to the physical and computational layers, there is a semantic data layer. This layer determines how information is structured, labeled, and understood between devices and platforms. Semantic models allow ICPS to understand the meaning behind the data streams, rather than simply algorithmically

Table 10.3 Core enablers of ICPS in MENs: architecture, intelligence, and semantics [21–23].

Enabler	Main technical components	ICPS functionality	Objectives for CESN	Main implementation constraints
Edge-cloud architecture	Cloud servers, fog nodes, and edge devices (controllers, gateways)	Distributed processing, latency-aware control, and hybrid coordination	Local emission response, grid stability, and scalable decarbonization	Synchronization, bandwidth constraints, and edge-cloud consistency
AI pipelines	ML/DL models, feature engineering tools, and data ingestion modules	Load forecast, anomaly detection, and predictive control	Process efficiency, emission reduction, and adaptive operation	Model drift, retraining time frame, and transparency of ML decisions
Semantic data flows	Data catalogs, knowledge graphs, ontologies, and metadata tagging engines	Interoperability, reasoning with contextualized data, and traceability across the system	Alignment with sustainability metrics, human–machine interpretability, and transparency across domains	No standard ontologies, overhead of semantic integration, and computation costs
Hybrid integration layer	FL agents, middleware, APIs, and message brokers (MQTT, Kafka)	Bridging the intelligent device layer, cloud coordination, and semantic integration	Holistic decision-making, trust-preserving automation, and sustainability policy integration	Latency–security trade-offs, interoperability across vendors, and privacy safeguarding

processing digitized numeric values, e.g., temperatures. As a result, they can be semantically connected to equipment failure probabilities, energy efficiency indices, and indicators of grid stress. This is important for two reasons. First, cross-layer reasoning and interoperability are requirements of complex systems like multi-vendor distributed MENs with many disparate protocols. Second, semantic tagging and ontologies allow for a seamless integration into regulatory dashboards, sustainability reporting, and citizens' facing platforms where transparency, trust, and alignment with CESN metrics are necessary [23]. Table 10.3 shows that each of the technological enablers adds to the functionalities associated with ICPS and to the operational realization of CESN design principles, which guide the operation of complex, dynamic, and distributed MENs.

10.2.4 Challenges in ICPS Integration Within Dynamic Energy Environments

Although ICPS have the potential to fundamentally transform MENs, there are considerable integration issues with the implementation of ICPS across energy systems due to the complexity of energy and the new requirements under the CESN agenda. These issues cover governance, interoperability, scalability, and resilience issues. The integration of ICPS in evolving heterogeneous energy networks showcases a series of structural mismatches between established energy infrastructure and the requirements of intelligent, cyber-physical coordination [14, 15].

A main challenge is the interoperability gap, which exists between existing energy systems and next-generation intelligent components. Some MENs continue to rely on hardware and protocols that were not designed for real-time (bidirectional) data exchange, decentralized control, or to interface with AI pipelines. These incompatibilities prevent us from seamless integration of ICPS and moving on to the second phase of deployment. The ability to deploy ICPS is limited by expensive re-engineering, retrofitting, and the creation of protocol translation layers and middleware to allow exquisite latency and efficiency to be lost. Moreover, integrating these systems becomes particularly challenging when standards vary widely between vendors and jurisdictions, especially as ICPS span regional grids, microgrids, and cross-sector connections, such as transportation and heating networks [16, 17].

Another important challenge is the computational and data infrastructure required for ICPS. Although edge-cloud coordination and AI pipelines have the capacity to scale, their deployment requires a high level of ICT maturity, consistent communication networks, and an uninterrupted power supply. Several MENs, especially in developing regions and rural areas, lack reliability in these prerequisites. Moreover, large-scale data ingestion and ML workloads come with energy overhead and associated carbon footprints. Therefore, they create a paradox where initiating intelligence for decarbonization can potentially increase emissions if the intelligence is not deployed and powered sustainably. These considerations highlight the need for energy-aware AI architectures or green digital infrastructure [18, 19].

The other obstacle in ICPS adoption within MENs is cybersecurity and operational trust. With increasing digitalization and interconnection in energy infrastructures, they are becoming more vulnerable to cyberattack vectors, such as data spoofing, ransomware, adversarial ML input attacks, or coordinated attacks on distributed devices. The complexity of the ICPS environment is further magnified, as decision-making may be fully autonomous, can involve agents that are geographically dispersed, and we cannot rely upon well-established human processes to observe ambiguities and complex decision processes. Building trust in these systems is much more complex and requires a multilayered security approach with encrypted communications, anomaly detection, zero-trust architectures, and explainable AI. Failing to address these security concerns properly can lead to operational disruptions, public backlash, and regulatory implications [20, 21].

Additionally, institutional inertia and regulatory uncertainty hinder the integration of ICPS into MENs. Many regulatory frameworks were derived from centralized company control, a static framework of rates, and predictable load patterns. However, ICPS typically leverage flexibility, dynamic real-time rates, nonregulated prosumption, and adaptive coordination. Until regulations such as regulator-sanctioned sandbox arrangements, digital energy codes, and mechanisms for cohort data governance exist, it is possible to undercut, halt, or misalign ICPS initiatives and project intended decarbonization nationally. Effectively addressing this issue requires deeper collaborations among technical experts, regulators, and civil society to establish institutional constructs for CESN-aligned ICPS that follow the functional logic of its potentials [22, 23]. A real CESN entails layered transformations of institutions, a portfolio of infrastructures, and a kind of change of epistemic belief. Therefore, ICPS implementation is not a matter of just technical deployments, as it is a design and governance problem. Table 10.4 summarizes the challenges of ICPS integration in MENs.

Table 10.4 Integration challenges of ICPS in MENs [14–23].

Domain of challenges	Barriers	Impact on ICPS deployment	Objectives for CESN	Mitigation requirement
Legacy infrastructure and interoperability	Vendor lock-in, lack of real-time capability, and incompatible protocols	Limits the seamless integration of sensors, controllers, and AI models	Strangles coordination for decarbonization, real-time adaptation, and grid efficiency	Standardization, protocol translation, and optimized middleware
Data and computation infrastructure	Energy-intensive AI workloads, insufficient edge/cloud resources, and unreliable connectivity	Poor intelligence, higher latency, and higher operational energy costs	Contradicts sustainability objectives if data processing was carbon intensive	Green data centers with energy-aware ML and cloud infrastructure
Cybersecurity and trust	Adversarial ML threats, vulnerability to attacks (e.g., spoofing, ransomware)	Weakens autonomy, reliability, and trust by users in ICPS systems	Threatens secured coordination, privacy, and explainable automation	Zero-trust architecture, secure firmware, and continuous real-time intrusion detection systems

(Continued)

Table 10.4 (Continued)

Domain of challenges	Barriers	Impact on ICPS deployment	Objectives for CESN	Mitigation requirement
Institutional and regulatory constraints	Lack of digital energy legislation, static regulatory frameworks, and fragmented jurisdiction	Misalignment of ICPS capabilities with operational logic allowed	Prevents scaled deployment of adaptive price, demand-side flexibility, and carbon-based signals	Regulatory innovation, sandbox projects, and participatory governance
Socio-technical complexity	System heterogeneity, emergent behaviors, and integration of multi-agent dynamics	Exacerbates uncertainty in control strategies and the risk of instability and unintended feedback	Challenges adaptive management needs to achieve performance aligned with the CESN	Agent-based modeling, resilient control frameworks, and system-of-systems engineering

10.3 Digital Twins, ML, and LLMs

The confluence of digital modeling, algorithmic learning, and language-based intelligence is reconstituting the architecture and operating logic of ICPS within MENs. This section describes how digital twins, ML models, and LLMs are computational enablers that provide capabilities for energy transitions aligned with CESN. Digital twins serve as high-fidelity real-time digital representations of physical assets and systems that provide operators and engineers with predictive diagnostics, virtual experimentation, and sustainability scenario planning. ML techniques, such as supervised classification and RL, can support event prediction, energy optimization, anomaly detection, and adaptive control in dynamic energy domains. Emerging LLMs provide entirely new capabilities in naturalistic multimodal interaction, contextual interpretation, and decision-support systems that have transcended human supervision and autonomy toward machine using both model-based and knowledge-based methods of supervision. Together, these technologies are providing increased intelligence, transparency, and enhanced coordination capabilities for ICPS in the pursuit of decarbonization, operational resilience, and human–machine collaboration in complex, decentralized, and distributed energy architectures.

10.3.1 Digital Twins for Real-time Modeling, Simulation, and Sustainability Forecasting

Digital twins play a critical role in the digital intelligence layer of ICPS. They provide a continuous connection between physical energy systems and virtual, data-driven constructs. Digital twins develop dynamic and real-time simulation environments (replicas) using measurements from sensors that are distributed throughout the MENs infrastructure and those located in places such as substations and DERs. Digital twins have the capabilities to provide real-time state estimation, fault detection, and forecasting without actually disrupting the physical system. In CESN contexts, digital twins are simulation environments used to test decarbonization scenarios, resilience actions, and policy responses using adaptive control techniques prior to implementation in the physical system. In operational scenarios, digital twins help decision makers identify risk components, maximize the use of the assets, and refine energy dispatch during uncertainty [24]. For example, the digital twin of a distribution feeder can calculate overload scenarios under extreme EV charging demand or model the effects of renewable intermittency on voltage stability.

Digital twins can harness carbon intensity signals and policy constraints over multi-day simulation horizons, helping grid operators find paths to go from risk to reduced costs and emissions. The ability to conduct these types of exercises with technical optimization to provide improved environmental outcomes is a vital component of the CESN paradigm. In addition to technical performance, digital twins can provide transparency and trust among stakeholders, considering all technology and policy options by delivering visual and interactive representations of system behavior. For instance, cities, utilities, and citizens can engage with digital dashboards that illuminate energy flows, emissions profiles, and what-if scenarios under different regulatory or investment decisions. Digital twins can provide machine-level operational control and become a governance tool for participatory planning and CESN-aligned accountability [25, 26]. Table 10.5 provides an overview of how digital twins operate across the various ICPS layers and how they are used as operational intelligence, real-time monitoring, and long-term sustainability planning.

10.3.2 ML for Energy Event Prediction, Optimization, and Adaptation

ML is central to enabling ICPS to act, learn, and adapt in MENs. As these systems grow more complex with renewable integration, many energy prosumers, and increasingly time-varying sustainability goals, data-driven learning models are a focal point. ML techniques can learn from historical data (training) and

Table 10.5 Strategic roles of digital twins in CESN-aligned ICPS applications [24–26].

ICPS layer	Digital twin function	CESN alignment	Type of data sources	Application
Physical asset layer	Fault identification, real-time tracking	Increase equipment life and reliability regarding climate-related stresses	Sensor data (temperature, voltage, and vibration) equipment downtime logs	Asset health and modeling at transformers and inverters
System operation layer	Dynamic dispatching, load forecasting, and fault comparison	Support emissions-focused grid reliability and operational flexibility	Demand profiles, DERs generation data, and weather predictions	Prioritizations for low-carbon microgrid operations
Sustainability planning layer	Scenario testing, compliance alignment, and emissions modeling formats	Design options and policies with designed scenarios for CESN	CO_2 intensity curves, investment scenarios, and regulatory goals	City-wide decarbonization strategy to support digital twin creation

real-time data (learning) to address distributional shift and optimize control performance across stochastic environments. For CESN-aligned infrastructures, using ML will be a mainstay for forecasting and control, but it can also provide crucial meta-learning capabilities to continually realign energy behavior with carbon-load reduction and resilience objectives. One of the most identified applications of ML in MENs is providing short-term energy consumption forecasts: to provide demand fluctuations to plan resource allocation, and avoid demand overload scenarios [27, 28]. Supervised learning models can generate forecasts [29]. The generalized regression-based model to predict energy demand is presented as follows:

$$\widehat{E}(t+\Delta t) = f(E(t), X(t), \theta) \tag{10.1}$$

where $\widehat{E}(t+\Delta t)$ is the predicted energy consumption at time $(t+\Delta t)$, $E(t)$ is the historical energy consumption, $X(t)$ is a vector of exogenous variables (e.g., weather, socioeconomic factors, and occupancy), and θ represents model parameters learned during training. Comprehensive models enhance forecasting quality by incorporating seasonality, user activity, DERs output, and even constraints from policies. Random forests, gradient boosting, and neural networks can be used to forecast with rich data, particularly long short-term memory (LSTM)-based models that have performed well.

In addition to forecasting, ML also allows for event classification and fault detection. In the supervised classification scenario, supervised classifiers are trained to identify category labels such as cyberattacks and equipment degradation from patterns derived from historical sensor data. In unsupervised scenarios, clustering algorithms or autoencoders are used to detect anomalous behavior without being given explicit labels, identifying when a distributed asset's behavior significantly deviates from normal. This aspect of unsupervised detection is critical for early mitigation of failures and loss of sustainability. Finally, RL is a robust approach for adaptive control problems. Unlike forecasting, RL agents are generating interactions in an environment to learn or calculate policies that will maximize their long-term cumulative reward [30]. A base equation calculation of RL is the Bellman optimality condition for the action-value function:

$$Q^*(s,a) = E\left[R(s,a) + \gamma \,.\, \max a' Q^*(s',a')\right] \tag{10.2}$$

where $Q^*(s,a)$ is the optimal value of taking action a in state s, $R(s,a)$ is the immediate reward (e.g., emission reduction, energy cost savings), $\gamma \in [0, 1]$ is the discount factor, s' is the next state, and a' are possible future actions. In MENs, RL is applied to smart heating, ventilation, and air conditioning (HVAC) control, energy storage management, and demand response systems dealing with constraints on local energy usage vs. carbon intensity (or cost) on the electricity grid. Model-free approaches such as Q-learning and deep deterministic policy gradient (DDPG) methods are well fitted for decentralized and nonlinear energy situations [31].

There are barriers to the incorporation of ML into ICPS for CESN objectives. First, model drift caused by changing climate conditions, new regulations, or changes to user behavior can degrade performance quickly, necessitating, in some cases, constant retraining or domain adaptation. Second, establishing the basis for trust is related to explainability in the models, which is important in legally or ethically sensitive applications. Shapley additive explanations and local interpretable model-agnostic explanations are approaches that more frequently form part of ICPS pipelines. Third, additional layers of cybersecurity are needed in conjunction with ML algorithms to mitigate data poisoning, adversarial attacks, and issues related to untrusted FL agents [32, 33]. Table 10.6 provides a general appreciation of how the different paradigms of ML: supervised, unsupervised, and RL, are leveraged in ICPS in MENs.

Table 10.6 ML applications in ICPS for CESN-oriented MENs [27–33].

ML framework	Use in MENs	Contribution for CESN	Main challenges
Supervised learning	Energy forecasting, DERs output forecasting, and fault classification	Enhanced energy efficiency, proactive fault response, and low-carbon scheduling	Lack of labeled data, generalization across systems
Unsupervised learning	Anomaly detection, sensor drift detection, and detection of cyber-physical intrusions	Early threat mitigation, self-healing infrastructure enabled	False positives, interpretability, and data imbalance
RL	Battery scheduling, HVAC control, and real-time demand response	Adaptive and carbon-aware decision-making in the face of uncertainty	Reward design, converging reliably, and safety assurances

10.3.3 Role of LLMs in Interface Naturalization, Coordination, and Decision Automation

LLMs, which are based on the transformer neural architecture, are emerging as tools for facilitating intelligent coordination, naturalistic interaction, and policy-aligned reasoning in ICPS in the energy domain. In the case of MENs, these models are not simply linguistic processors; they are cognitive intermediaries that take the raw system data, meaningfully integrate stakeholder input, and provide useful, actionable insight. One of the cornerstone elements of LLMs is the self-attention mechanism, which facilitates token-to-token contextualization within a sequence [34]. Self-attention is defined as:

$$\text{Attention}\,(Q,K,V) = \text{softmax}(QK^T / \sqrt{d}) \cdot V \tag{10.3}$$

where Q, K, and V are matrix projections representing the queries, keys, and values, respectively, derived from the model's input tokens, and d is the dimensionality of the key vectors, acting as a normalization factor to prevent gradient explosion. The softmax function converts the attention score set into a probability distribution over the input sequence. This structure allows the LLM to emphasize semantic relatedness to long inputs to internally reason about causal relations, states of operation, and their actions in a MEN. For example, when querying a fault event in a microgrid, the LLM can extract data relationships, including across time, topology, and control layers, producing human-interpretable narrative explanations. In addition to interpretability, LLMs provide the potential for automating logic generation from language to code [35]. This potential is expressed by the autoregressive decoding equation:

$$\hat{y} = \arg\max yP(y|x,\Theta) \tag{10.4}$$

where x is a natural language input, such as a user command, regulatory condition, or performance criterion, y is a structured target output (e.g., decision rule, executable code, and structured configuration), Θ represents the trained parameters of the language model, and $\hat{y}$ is the most probable structured output generated based on the learned conditional distribution.

This transformation enables operators, planners, or agents to articulate high-level goals in plain language while ensuring executable logic that meets operational limits and that CESN goals are delivered to ICPS control components. For example, the prompt "minimize charging during peak grid emissions hours" can be defined as a control procedure encoded in a local DER agent. In addition, LLMs can act as inter-agent mediators in a multilayered ICPS. Even when the components are from different vendors or ontologies, LLMs can accept semantics as terms and standardize them based on energy vocabularies, legal taxonomies, and control schemas. This is especially beneficial for achieving energy system integration across national boundaries or industry sectors, which is an increasingly important qualification for CESN interoperability and compliance. LLMs can support automated documentation and compliance workflows, data sources, parse emissions logs, map regulatory requirements to telemetry outputs, and deliver sustainability reports following ISO 50001, ISO 14064, or national climate template formats. Their narrative-generating and regulatory-translating functions contribute to the auditability and institutional transparency of ICPS operation, both of which are crucial to CESN legitimacy [36]. Table 10.7 summarizes the operational integration of LLMs in the use of ICPS, highlighting the importance of their role in enabling CESN-compatible operational functions.

10.3.4 Case Examples: Intelligent Scheduling, Energy Sharing, and Predictive Maintenance

The tangible integration of ICPS into MENs can best be shown by looking at operational domains where intelligence, adaptation, and sustainability are not theoretical aspirations but ongoing constraints on performance. In this subsection, three domains relevant to CESN are examined, including intelligent scheduling, P2P energy sharing, and predictive maintenance. The case examples demonstrate how digital twins, ML, and LLMs merge into an integrated, cognitive ecosystem for both real-time decision support and long-term optimization.

Table 10.7 Applications of LLM in CESN-aligned ICPS [34–36].

Functional role	LLM capability	Contribution to CESN in MENs	Implementation considerations
Interface naturalization	Contextual translation, synthesis for human consumption, and query-to-insight	Improves interpretability and alone trust across tech boundaries	Apprenticeship in domain-specific prompts, lag times, and hallucinations
Semantic coordination	Protocol alignment, ontology reasoning, and multi-agent natural language mediation	Encourages interoperability and cross-system participation	Conformance to standards (e.g., CIM, IEC 61850)
Decision automation	Constraints interpreting and fallback generating, sequence to policy	Affords low-carbon adaptive responses through natural language-driven inference	Safety verification, deterministic control conformance
Compliance and reporting	Emissions narrative generation, policy rule verification, and audit log generation.	Assists in transparency and regulatory conformity with the CESN suites	Citation sourcing, temporal fidelity, and reporting traceability

10.3.4.1 Intelligent Scheduling and Decentralized Coordination

In traditional energy systems, scheduling is limited to the use of remote, static optimization approaches to define goals. ICPS-enabled scheduling and time control tied to monitoring of emissions and behavioral actions of multi-agents sharing regional distributed assets, require continual feedback and coordination between previously mentioned components [37, 38]. RL agents operating in DER digital twins adjust scheduling policy by considering emissions variability, load, and resource restrictions. Instead of performing a fixed control path, RL agents will adjust scheduling policy relative to environmental variability, system contingencies, and regulatory reporting in the form of a human-defined or directed cognitive translation via LLM. Natural language policy directives such as emission window cutoffs or time-of-day priority request indications can be translated into structured input that ICPS agents can monitor, execute, and report on performance and behavioral changes in, while aggregations of telemetry with cognitive agent learning translated via LLM could be rendered into constructively explainable summaries providing a system operator or planner an insight to assess, review, confirm or change operating status without disrupting RL decision-making of the underlying framework [35, 36].

10.3.4.2 P2P Energy Sharing and Semantic Market Automation

The development of decentralized engagement within MENs requires transaction architectures that are structurally scalable and semantically robust. In ICPS-mediated P2P networks, prosumers articulate preferences, constraints, and pricing rules through LLM-assisted interfaces that then translate those preferences into machine-readable contract structures that keep consistency with grid codes, interoperability standards, and CESN-equity principles [11, 39]. Digital twin models represent local system conditions and establish the feasibility of exchange scenarios, and resolve conflicts or opportunities related to neighborhood-scale transactions. At the same time, ML algorithms examine transaction history, forecast acceptance likelihood, and handle offers while ensuring fairness and system-wide security. The result is a highly responsive transaction environment where human intentions get translated into action through smart interpretation, real-time analysis, and automated grid decisions. Equity, transparency, and resilience are features of the market behaviors as opposed to external regulatory constraints [13, 14].

10.3.4.3 Predictive Maintenance and Condition-based Foresight

Failure forecasting and proactive measures are fundamental to maintaining the availability of reliability-critical components of MENs. Currently, integrated diagnostic logic based on ML models for pattern recognition and anomaly detection pipelines, and high-fidelity digital twins are available in relation to ICPS systems. These aspects come together to sense behavior drifting from baseline operating parameters, to detect subtly before further developments lead to asset failure, or system instabilities [33]. The cognitive interface layer, referencing LLMs, represents these multivariate behaviors as contextual narratives and action protocols. Operators receive an understandable firm of advisories, structured with causal understanding, backtracking symptom-response mapping, and associated system dependencies, and time-sensitive, administered escalation pathways [34]. Predictive maintenance is therefore more than a system for condition-based alerts; it is also an intelligent, communicative agent that serves the wider objectives of CESNs that include resilience, resource efficiency, and life cycle management [35]. Table 10.8 presents ICPS-enabled intelligence across three important CESN-oriented operational domains.

Table 10.8 ICPS-enabled intelligence across CESN-oriented operational domains [11, 33–39].

Application	Technologies	Functionality	Contribution for CESN	LLM interface role
Intelligent scheduling	RL agents, DER twins, and carbon-aware forecasting	Flexible load sharing, emissions-sensitive scheduling	Peak carbon avoidance, flexible optimization, and policy compliance	Translating policy objectives and reports agent behaviors
P2P energy sharing	Semantic contracts, ML-based transaction filters, and local grid simulation	Distributed negotiation, conditional matching, and fairness maintenance	Participatory governance, local sustainability, and distributed equity	Translating intent into actionable offers and monitoring for semantic integrity
Predictive maintenance	Asset twins, anomaly classifiers, and pattern-recognition pipelines	Early failure detection, hazard classification, and planned maintenance	Carbon-saving system dependability, infrastructure longevity, and failure avoidance	Escalation protocols, creating diagnostic, and causal interpretation

10.4 Intelligent Cybersecurity Challenges in MENs

Cybersecurity is a strategic imperative in ICPS-enabled MENs. It highlights that intelligent infrastructures not only recognize cyber threats but can also reshape our entire approach to cyber-physical risk management. Integration of automation, AI-driven decision-making, and ubiquitous sensing in ICPS creates additional, more difficult to detect, model, and contain disruption pathways in energy infrastructures beyond our standard cybersecurity frameworks. This section presents a systemic, behavioral, and resilient framework for cybersecurity from a hardware- or perimeter-based model of defense. This framework directly supports CESN's aim to look at cybersecurity as an enabler, rather than a constraint, for mutually beneficial, sustainable energy transitions that preserve trust.

10.4.1 Expanding Cyberattack Surfaces in ICPS-integrated MENs

The adoption of ICPS into MENs has fundamentally changed the perimeter, complexity, and behavior of the cyber threats to MENs. Compared with the legacy infrastructures operating locally in an isolated, centralized, and physically separated manner, ICPS-integrated MENs consist of thousands of entities, which are interrelated, from edge sensors, distributed controllers, aggregations, to federated clouds, ML pipelines, and back. As a result, modern energy systems have considerably expanded their cyberattack surface, and the conventional cybersecurity approaches have shifted irreversibly [40].

Three main architectural elements drive this expansion. First, many edge devices, smart meters, PMUs, DER controllers, and IoT-based sensors produce a startling number of low-power, often ratio-constrained endpoints. Edge devices are generally installed with minimal firmware protection and limited capabilities for cryptography, typically beyond patch-level vulnerability. This makes them targets for lateral movement or distributed cyberattacks. Second, the federation of intelligence architecture is layered and utilizes cloud assistance with other frameworks, leading to different trust zones with distinct control, data integrity, and cyberattack exposure. Third, the heavy reliance on information and real-time communications for DERs, substations, and instances of market interfaces necessitates a continuous bidirectional flow of data. This automatically increases cyberattack surfaces, such as man-in-the-middle attacks, protocol/spoofing, and time series injection vulnerabilities [41].

In addition, ICPS change the way we conceptualize logic from deterministic logic to learning, and experience, as opposed to software code, opens new cyberattack vectors based on exploiting model inference behavior. Adversaries could generate inputs that deceive the ML model in a specific way by mixing or distorting the input. Therefore, when the model makes a prediction, it alters the optimization of the prediction objectives or creates a cascade of failures without crossing legacy thresholds capable of detecting this anomalous behavior [40]. In RL control architectures, this could involve corrupting the reward function or policy gradient used by agents to construct energy usage behaviors, all while compromising the grid stability objectives for the CESN. The idea of a "security perimeter" will no longer apply to developing MENs using ICPS systems. Rather than thinking of an ICPS as a system, the concept should instead be envisioned as a dynamic multi-agent surface with limitless exposure, where adversaries can utilize technical vulnerabilities, but just as likely, not to mention even more insidious, semantic ambiguity, data uncertainty, and misalignments and incompatibilities of protocol [41]. The key point of the current subsection is that the most significant cybersecurity vulnerabilities in ICPS are not actually from malware, intrusions, and the like, but rather from the structural design choices where connectivity, flexibility, and scalability adequately drive consideration of any embedded mechanisms of trust. Addressing these vulnerabilities requires nothing less than a wholesale redesign of the architecture of cybersecurity, not as purely a compliance function, but as a governance layer of an ICPS, and in a distributed adaptive fashion [42]. Table 10.9 presents a summary of the expanding layers of cyber-physical attack surfaces

Table 10.9 Multilayered expansion of cyber-physical attack surfaces in ICPS-enabled MENs [40–42].

ICPS layer	Technologies	Vulnerabilities	Vectors of attacks	Impact on CESN objectives
Perception and sensing	IoT sensors, environmental monitors, smart meters, and PMUs	Possible firmware extraction, unencrypted communication, and ability to spoof input measurements	Sensor spoofing, signal injection, and impersonation attacks	Unreliable emissions data, falsified load profiles, and loss of observability
Control and actuation	DER controllers, SCADA systems, microgrid automation devices, and protection relays	Privilege escalation for commands, ability to intercept commands, and ability to corrupt control logic	Command spoofing, man-in-the-middle (MitM), and Denial-of-Control (DoC)	Disrupted grid stability, unplanned dispatch of DERs, and uncontrolled oscillation throughout the system
Communication infrastructure	Mesh routing protocols, MQTT, ZigBee, 5G, and LoRaWAN	Protocol mismatch, mesh traffic sniffing, perform routing changes, and hijacking traffic	Protocol downgrade, packet replay, and link-layer injection	Delays of real-time controls, communication blackout, and loss of coordination
Edge-cloud integration	Model repositories, fog nodes, containerized ML services, and cloud APIs	Authentication bypass, insecure API exposed, and change or corrupt ML model	API hijacking, adversarial inputs, and credential theft	Misoptimized performance, emissions-inefficient outcomes, and compromised privileged data

AI and learning models	RL agents, federated training pipelines, and supervised/ unsupervised models	Poison training data to build in bad triggers, attack the reward function, and could cause model inversion	Backdoor injection, gradient manipulation, and false feedback loops	Misaligned energy decisions, carbon-blind scheduling, and lost knowledge equity
Human–machine interface	Natural-language command interpreters, LLM-driven dashboards, semantic agents	Misinterpretation, prompt injection, and prompt injection for a potential example of accessing to correlated data from UI manipulation	Interface spoofing, prompt injection, and semantic fuzzing	Regulatory noncompliance, operational misalignment, and human trust erosion

embedded in a MEN but integrated with an ICPS framework, indicating technology assets, emergent vulnerabilities, and CESN-based impacts across all layers of architecture.

10.4.2 Vulnerabilities in Communication, Data Integrity, and Control Coordination

The increasing attack surfaces of ICPS-enabled MENs increase potential vectors for intrusion, but the biggest weaknesses often exist in the internal logic of the system, associated with the communication protocols, data integrity mechanisms, and control coordination. These are not edge effects—they are happening in the internal operations of the system where decisions regarding energy flow, DERs scheduling, emissions accounting, and market interactions are governed by cyber-physical feedback loops. Therefore, while it is convenient to look at threats to these layers of the MEN from a technical risk management perspective, they represent risks to the system and to the reliability, accountability, and sustainability goals of the CESN [43].

On the communication layer, ICPS depend on a dense collection of machine-to-machine (M2M) communications across assets that are geographically and topologically distributed. While data from DERs are often exchanged via lightweight protocols (e.g., MQTT, CoAP, Modbus, or LoRaWAN), which are developed to efficiently transmit between two machines in constrained environments, they often lack the mechanisms for mutual authentication, encryption, or payload validation. Therefore, adversaries rely on protocol-level vulnerabilities to inject malicious commands, disrupt timing synchronization, and misroute control signals. Even minimal latencies and discrepancies in DER control loops can unintentionally produce incorrect dispatch decisions, load imbalances, or emissions noncompliance on an adaptive, ML-coordinated system [44]. Simultaneously, the integrity of the ICPS data pipeline is also at risk of corruption and drift. Many real-time decisions in ICPS architectures are predicated upon high-frequency telemetry, state estimations, or semantic metainformation with the belief that these inputs are trustworthy and contextual. Telemetry streams can be compromised, sensor readings may be delayed or scrambled, and metadata tags may have been replaced. Attackers don't need to gain direct control of a system to disrupt functionality; rather, it is often enough to manipulate the perception of system state to trigger cascading suboptimal decisions. For instance, the system may dispatch assets on the basis of the system perceiving that DERs availability or carbon intensity is lower than it actually is, thereby increasing emissions or destabilizing a feeder segment [45].

More precisely and equally important are shortcomings in control coordination among distributed agents. ICPS (and RL agents or local twin-driven schedulers) rely broadly on decentralized decision-making, particularly in environments where agents need to act semi-independently but still respect overall system coherency. These agents often have locally defined assumptions, update asynchronously, and receive varying bits of global state. If synchronization is absent, or if any agent is misinformed for any reason due to data feeds being compromised, or through adversarial policy updates, the coordination process fails. The outcome of this failure is perhaps not a system failure, but instead a slow loss of service level. As service or interpretive degradation can be difficult to track down and expand increasingly through semantic layers and time binding, this type of failure is difficult to identify and mitigate [46].

In CESN-aligned energy networks, security vulnerabilities create a dual problem. First, they undermine the transparency and accountability of energy operations, leading to regulatory violations and erosion of public trust. Second, they cause sustainability impacts, including an increase in emissions, a weakening of system resilience, and a deterioration of fairness. These impacts are not necessarily from malicious attacks; they are mainly because the system can no longer guarantee integrity in its cyber-physical interactions. To address these vulnerabilities, it is required to go beyond just securing communication channels. It is critical to protect the meaning and interpretation of data, ensuring systems can verify not only that information has not been tampered with, but more critically, that it is being understood and acted upon correctly across a distributed network working toward shared goals [45, 46].

10.4.3 Dynamic Risk Modeling and Adaptive Security Auditing Frameworks

The increasing cyber-physical complexity of ICPS in MENs brings an urgent need to evolve from reactive defense mechanisms to proactive, predictive, and continuous cybersecurity assessment. In CESN-enabled infrastructures, cybersecurity issues involve more than just privacy and availability. They are tied to energy sustainability, emissions compliance, and systems equity. To accommodate these multiple and layered choices, ICPS require dynamic risk modeling and adaptive auditing frameworks that operate across time scales, decision layers, and asset types. This subsection introduces the main equations and principles that support this type of architecture and allows formal evaluations of exposure and impact with respect to trustworthiness in CESN-enabled systems.

10.4.3.1 Quantifying Risk Exposure

The basic unit of cyber-physical risk in ICPS is often represented as the expected risk value $\mathscr{R}_i$ associated with a specific attack scenario A_i. A foundational formulation is:

$$\mathscr{R}_i = P(A_i) \cdot C(A_i) \tag{10.5}$$

where $\mathscr{R}_i$ is the expected risk value for attack A_i, $P(A_i)$ is the estimated probability of attack A_i occurring within the system, and $C(A_i)$ is the quantified consequence or cost of attack A_i, which could be in the form of system downtime, energy inefficiency, emissions deviation, or regulatory breach. This expression assumes probabilistic independence and is simply a means for prioritizing response to threats under resource constraints. For ICPS contexts, $C(A_i)$ should be multidimensional, encompassing not only economic loss but also degradation in CESN performance indicators, e.g., increased CO_2 intensity, decreased grid resilience [41–43].

10.4.3.2 Aggregated System-wide Risk

At the network level, a full ICPS deployment will host dozens or hundreds of concurrently active risk nodes. To assess total system exposure, an aggregate metric R_{total} is defined:

$$\mathscr{R}_{\text{total}} = \sum (w_i \, . \, \mathscr{R}_i) \tag{10.6}$$

where $\mathscr{R}_{\text{total}}$ is the total cyber-physical risk value across n individual risks. $w_i \in [0, 1]$ is a weight coefficient reflecting the criticality or sensitivity of the system component associated with A_i. The weight w_i may be computed using metrics such as asset redundancy, operational priority, DERs scheduling influence, or environmental impact coefficients [26, 28]. This formulation supports security budgeting and system-wide defense prioritization under CESN-aware conditions.

10.4.3.3 Anomaly Scoring in Telemetry Streams

ICPS audit pipelines frequently use anomaly detection models to assess data trustworthiness. A simplified statistical anomaly score $S(x)$ can be formulated as:

$$S(x) = \frac{|x - \mu|}{\sigma} \tag{10.7}$$

where $S(x)$ is the anomaly score of input x, μ is the historical or expected mean of the monitored telemetry stream, e.g., voltage, DERs output, and carbon intensity, and σ is the standard deviation of that stream over a stable operating window. A threshold (typically > 3) is set to signal anomalies when $S(x)$ is exceeded, signaling statistically significant divergence. In the CESN context, this function is used to identify not just cyber intrusions, but also to indicate systemic misalignment between intention (sustainability/environmental) and outcomes (actual operations) [3, 4].

10.4.3.4 Trust Auditing Across ICPS Agents

For distributed ICPS, it is essential to quantify inter-agent trust, especially when agents operate semi-autonomously. A time-weighted trust function $T_{ij}(t)$ for agent j as evaluated by agent i at time t can be represented as:

$$T_{ij}(t) = \alpha \cdot T_{ij}(t-1) + (1-\alpha) \cdot \tau_{ij}(t) \tag{10.8}$$

where $T_{ij}(t) \in [0, 1]$ is the current trust level that agent i assigns to agent j, $\tau_{ij}(t)$ is the immediate observed behavior rating of j by i, based on compliance with coordination rules, $\alpha \in [0, 1]$ is a memory decay factor indicating how much weight is given to historical trust, and $T_{ij}(t-1)$ is the previous trust value at the last audit time step. This function allows agents to adapt their interaction frequently, adapting based on reliability, coordination behavior, and audit feedback. Whenever trust levels decline below a lower threshold (set by systems), fallback routines, redundancy activation, and quota capabilities are activated to promote alignment toward system credible designs, stability, and balance through CESN in ambiguous situations, even when external manipulation creates restrictions on direct or individual agents to comply [23, 24].

10.4.4 Limitations of Legacy Cybersecurity Solutions in Modern Energy Infrastructures

As MENs gradually shift toward hybrid ICPS-enabled ecosystems, archaic cybersecurity paradigms, designed for siloed, deterministic, human-supervised infrastructures, cannot provide adequate protection. Traditional cybersecurity paradigms are mostly perimeter-focused, using static firewalls, hierarchical access control, signature-based intrusion detection, and episodic auditing. They have worked in traditional SCADA environments, but they do not have the flexible, semantic granularity, and temporal flexibility that hybrid ICPS systems require. Perhaps more importantly, they fit poorly with the principles of sustainability and resilience and the cross-agent collaboration inherent in CESN-oriented infrastructures. One of the primary shortcomings of legacy solutions is their reactive posture. Classic intrusion detection systems (IDS) operate on predefined rules, or attack signatures [3, 4]. They are entirely blind to all new, adaptive, or stealthy attacks such as adversarial ML inputs, false data injection in learning loops, or semantic prompt perturbations in LLM-driven interfaces. In ICPS, as a set of components that utilizes online learning agents and autonomous feedback controllers, attacks may change as the system changes. In these systems, rule-based logic is not always adequate: the threat landscape is no longer static but co-evolving [34, 35].

Furthermore, legacy architectures use centralized trust and static zoning models, categorizing assets into fixed security domains. However, trust for ICPS must be computed dynamically across agents, data sources, and devices that can operate in fluid configurations. DERs, mobile assets, fog nodes, and LLM interfaces operate across, not within, operational zones, creating static policy enforcement that is brittle and incomplete [36]. A DER, for example, might switch from participating in a local energy market as an active load during peak load one hour to ancillary service coordination in another region. Legacy cybersecurity systems do not audit, authenticate, or trace these types of transitions and changes in trust in real time. Legacy systems are also limited with respect to semantic awareness. They can see aberrations, or irregularities, in things like malformed packets or onto unauthorized logs, but cannot "see" higher-order aberrations such as "emissions-target violations under deceptive telemetry circumstances." Or "intent drift in control routines responding to adversarial prompts." Because these behaviors have semantic reasoning and cross-modal anomaly fusion and context-dependent thresholds, legacy systems cannot interpret static defenses [26].

As LLMs, ML models, and digital twins define the "new normal" within workflows, cybersecurity must become context-dependent so that cybersecurity can detect not only a technical situation but also a conceptual noise in a goal misalignment. The shortcomings of the legacy solution from CESN's view are strategic. Failing to update outdated defense postures can lead to inaccurate emissions statements for the utility, waste of community resources, and reduced trust in publicly funded infrastructure. A relevant example might be a breach that changes the way a dispatch is processed and does not lead to an immediate outage. The result would be, however, an unclear carbon output increase or absolute loss of equal access to community-regulated DERs. In this regard, cybersecurity failure is another avenue to degrading sustainability. To combat this, we must have intelligent, self-auditing, and policy-aware defense architectures that provide real-time reasoning, self-explanation for anomalies, and trust for multi-agent operation [31–37].

10.5 Advanced Detection, Resilience, and Adaptive Response

Modern advanced cyber-physical threats are becoming more sophisticated and distributed than ever. For ICPS-integrated MENs, a transition must occur from human-operated static defense architectures to self-defending adaptive systems that can detect, adapt, and recover without human intervention. This section discusses new approaches to deep learning-based intrusion detection mechanisms, federated anomalous classification, coordinated multi-agent responses, and resilience to stressed adversarial conditions (e.g., sensor spoofing, Denial-of-Service [DoS]). By focusing on system outcomes from the perspective of CESN, these approaches position cybersecurity as a dynamic capacity of the system, internalized, embodied in the control plane, distributed among trust-aware agents, and continuously evaluated and adapted based on real-time feedback and learning. The section illustrates resilience and detection as integrated capabilities in intelligent and adaptive infrastructure design, not as distinct or separate subsystems.

10.5.1 Intrusion Detection with Deep Learning and Graph-based Methods

The new threat landscape of ICPS-integrated MENs has changed the field of pattern and rule-based IDS systems. Today's attacks are often multistage, multidomain, stealthy, and contingent on context. Adversaries today often attack the communication, but they also will want to attack control planes, learning agents, and digital twins. Detection systems need to overcome current pattern-matching technology and function more like context-aware classifiers that are capable of learning based on changing behaviors and states of a distributed system. Deep learning and models in graph theory can conduct detection in this way, making access to both known and zero-day threats possible in real time, and they can both dynamically recognize patterns and be aware of topology or structural position.

10.5.1.1 Deep Learning for Contextual Pattern Recognition

Convolutional neural networks (CNNs) and LSTM architectures, and transformer-based encoders have been shown to perform well in cybersecurity use cases for ICPS. CNNs, LSTMs, and transformer models take as input multivariate telemetry streams made up of voltage severity, timing of traffic, logs of devices, and operating control commands, and learn temporal or spatial patterns associated with normal behavior and anomalous behavior [47]. In particular, the standard loss function for binary classification tasks such as these is cross-entropy loss presented as follows:

$$L_{CE} = -\left[\delta \log\left(\hat{\delta}\right) + (1 + \delta)\log\left(1 - \hat{\delta}\right)\right] \tag{10.9}$$

where L_{CE} is cross-entropy loss used for binary classification, δ is the ground truth binary label indicating attack (1) or normal (0), and $\hat{\delta}$ is the model's predicted probability of a data point being an attack.

10.5.1.2 Graph-based Models for Structural Anomaly Detection

Because of the interrelatedness of MENs, graph neural network (GNN), graph convolutional networks (GCNs), and graph autoencoders (GAEs) have become popular for identifying structural anomalies because these models work with graph-structured data in which nodes represent system components (e.g., DERs and sensors) and edges identify communication or control dependencies [48]. A usual anomaly score in a GAE setting will be:

$$S_i = \left\|G_i - \widehat{G}_i\right\|_2 \tag{10.10}$$

where S_i is anomaly score assigned to node i in the graph, G_i is original attribute or connection vector for node i, $\widehat{G}_i$ is reconstructed vector for node i predicted by the GAE, and $\|\cdot\|_2$ is Euclidean norm, indicating reconstruction error magnitude.

10.5.1.3 Relevance to CESN

Detection systems in CESN-based ICPS must weigh availability, sustainability, and equity. Detection must not delay emission-aware optimization, nor create false alarms that disrupt demand-response scheduling. Interpretability is paramount. Meaning deep and graph-based techniques are used to create a spatially conscious, time-sensitive, and objective-aware detection logic. In this way, cybersecurity shifts from a static, surveillance perspective to an intelligent, resilient embedding of technology in the infrastructure [47, 48].

10.5.2 Federated Anomaly Detection Across Distributed Assets

In MENs that are ICPS-enabled, anomaly detection is necessary under dispersed geography, data sovereignty, and communication constraint conditions. Increasingly, centralized monitoring is no longer viable in MENs for reasons of data volume and latency, and for reasons of privacy, regulation, and resilience. FL may be one viable alternative because it facilitates the collective training of distributed anomaly detection models via edge devices without transferring raw data. The federation retains local context, privacy, and responds quickly, which is especially useful in CESN-compliant energy infrastructures with a global commitment to transparency and a local commitment to decentralization [49, 50].

The federated update rule for anomaly detection is presented as follows:

$$w_{t+1} = \sum_{k=1}^{K} \frac{n_k}{n} \cdot w_t^{(k)} \tag{10.11}$$

where w_{t+1} is updated global model parameters after round t+1, $w_t^{(k)}$ is a locally trained model from node k at round t, n_k is number of samples at node k, n is the total number of samples across all nodes, and K is the total number of participating nodes. The local anomaly score for edge detection is characterized as follows:

$$S(x) = \|x - \hat{x}\|_2 \tag{10.12}$$

where $S(x)$ is anomaly score for input $x, \hat{x}$ is the reconstructed version of x produced by the model, and $\|\cdot\|_2$ is the Euclidean norm measuring reconstruction error.

10.5.2.1 CESN-relevant Implications

Federated anomaly detection advances CESN priorities across multiple dimensions. First, it maintains data locality and transparency of algorithms so that operational or emissions data is not exposed during training. Second, it allows for context-dependent detection given that each node learns anomalies that are meaningful to its own operational context, especially in areas with varied DER profiles, control policies, or emission constraints. Third, FL enhances resilience in the distributed model of collaboration, diffuses risk if a point of failure or directed attack leads to compromising a centralized model, or creates a cascading failure. Moreover, FL architectures can adopt additional security measures like secure aggregation, differential privacy, or consensus scoring to augment trust and fairness across multiple multi-agent operations. In the context of MENs geared toward CESN, this approach shifts anomaly detection from passive monitoring to a collaborative, participatory intelligence layer that is entirely congruous with both operational heterogeneities, as well as the democratic ideals at the heart of decentralized energy systems [49, 50].

10.5.3 Autonomous Response Systems and Multi-agent Coordination

As energy infrastructures transition to intelligent, distributed ecosystems, autonomous response plays a critical role in the timely mitigation of cyber-physical threats. Unlike traditional systems that rely on human incident response capabilities or rule-triggered detection systems to safeguard incidents, MENs involve a whole new challenge. The challenge is that ICPS have been integrated into the different components of MENs and do not permit autonomous agents to interpret anomalies, evaluate systemic risk, and develop mitigating countermeasures in milliseconds. They reside as embedded agents in controllers, DERs, and substations, operating independently or as a team, responding to not only local disruption (inferred risks) but generating an informed response that may include compromised emissions control, energy equity considerations, and grid resilience [47, 48].

Coordination among agents is frequently regulated by decentralized decision policies based on Markov decision process (MDP) or consensus-based logics. Agents undertake negotiating control actions with limited information about a global state and trust metrics that can change in real time. For example, if an edge agent in a microgrid determines that a neighboring generation unit is spoofing the frequency, it may island, meanwhile notifying its neighboring agents and modifying the inverter dispatch to maintain CESN targets. The reliability of such a structure depends on the use of adaptively propagating trust, minimal latency among agents, and the semantic interpretation of reported responses, acknowledging agents in the request for action to not only contain but to rekindle and preserving sustainability options [49, 50].

10.5.4 Resilience Against Cascading Failures, Sensor Spoofing, and DoS Attacks

Cascading failures in ICPS-integrated MENs often originate from localized cyber-physical disturbances. The disturbances might be a result of a spoofed sensor reading or a transient DoS attack that propagates through hard-linked control and communication layers. Cascading cyber-physical disturbances/disruptions occur in critical systems with high levels of DERs and dynamic load balancing. In such systems, a small perturbation might change optimization decisions, induce incorrect grid states, and/or violate emissions constraints. In contrast to fault propagation as described in previous subsections, these cascades exploit both the technical and semantic dependencies. These technical and semantic dependencies do not lend themselves to legacy models of failure modeling, as it may be difficult to isolate and predict failure uses utilizing legacy means of being able to tell when failures occur and proceed to determine root causes [3].

In this context, the resilience feature requires being more than fault-tolerant, as it necessitates adaptive structural flexibility. Infrastructures of ICPS must allow for real-time reconfiguration of communication topologies and selective trust discounting for compromised inputs. Moreover, they should have fail-operational control schemes that take CESN objectives into account and permit communication to be prioritized for the objectives, even during degraded operation. This includes, but is not limited to, dynamic rescheduling of DERs themselves, reinforcement-based fallback policies, and resilient telemetry fusions from redundant sources. What is critical in that resilience framework is considered a continuous property of the system, and not merely as recovery from an incident. It is assumed that resilience is incorporated into the learning, sensing, and actuation layers of the system from design [40].

10.6 CESN-aware System Intelligence and Sustainability Metrics

The evolution of MENs toward intelligent and adaptive infrastructures does not simply place importance on efficiency and resilience. From an institutional and societal perspective, there are still critical carbon reduction and sustainability goals to be achieved. This section will explore how CESN-aligned metrics, covering emissions, energy equity, and other sustainability boundaries, can be directly integrated into each of the decision layers for ICPSs. Moreover, this chapter moves beyond traditional performance indicators to describe how intelligent systems need to navigate trade-offs in real time, remain explainable to humans and institutions, and align with broader sustainability frameworks, not limited to, for example, planetary boundaries [37, 38]. Finally, it demonstrates how the intelligence and design of such systems are normative, as they reflect values regarding transparency, fairness, and long-term ecological health.

10.6.1 Embedding Carbon and Sustainability Goals into System Objectives

In infrastructures with CESN as a high priority, sustainability needs to be considered as a fundamental objective of intelligent control systems rather than a constraint stacked onto technical optimization; its impact on traditional frameworks cannot stop at being considered an external constraint. Instead, ICPS should model carbon intensity, externalities, and social equity as first-class decision variables. To advance from efficiency-oriented control policies, or cost-minimization policies, or loss-minimization policies, it is required to adopt multi-objective frameworks that incorporate life cycle emissions, renewable deployment priority, and equity of distribution. The better ICPS controllers, learning agents, and scheduling engines can treat sustainability metrics as first-class citizens in their reward system, the more likely their daily operational decisions can align with their long-term accountabilities to the climate [26].

For this embedding to be operationalized, ICPS must embed real-time, high-resolution emissions data, e.g., nodal carbon intensity, generation portfolio makeup, or upstream supply chain footprints, within the ICPS planning and optimization layers. For example, RL controllers and predictive scheduling agents could embed real-time emissions data within their reward function that incentivizes decisions with successive minimal emissions while maintaining stability and reliability. In distributed systems, the reward will be able to guarantee the local social context, e.g., abide by the varying carbon pricing schemes, representative energy mix, and use local emission and sustainability targets. In this way, energy agents can respond autonomously to modifying ecological limits, enforcing sustainability is the nature of the operation itself rather than retrospective accountability through audits around standards and policy [27–29].

10.6.2 Dynamic Trade-offs Among Performance, Efficiency, and Carbon Emissions

Recent deployments of ICPS in MENs work in a constrained optimization space, where system-level objectives such as performance stability, energy efficiency, and carbon reduction can be competing. Traditional infrastructure arranged these dimensions in a hierarchy such that stability always comes first, economic optimization comes second in an unconditionally optimizable domain, and sustainability is measured as a tertiary consideration, or exogenously. In contrast, CESN architectures have suitable designs that can handle multiple objectives and seek to optimize, making decisions in real time via dynamic and adaptive scenarios in response to disturbances. Intelligent agents must assess situational trade-offs in real time, rather than establishing fixed-temporal hierarchies, and therefore continuously reframe operational priorities [12].

A significant difficulty that arises in these trade-offs is the contextual nonlinearity of sustainability impacts. For example, running a storage unit that provides frequency support may provide a benefit for the system, but if the electricity used to recharge that unit is highly carbon intensive, it can result in an overall increase in carbon emissions. Similarly, significant load-shifting that reduces the cost of electricity by using off-peak loads may greatly reduce costs as well, but if those loads are in a part of the day dominated by fossil fuels, then that leads to another trade-off between short-term cost and long-term carbon-optimal dispatch. This is the reality for ICPS, which must balance immediate performance needs (like voltage stability and response times) with long-term sustainability goals (such as cumulative CO_2 emissions and environmental impact). This requires these systems to operate sustainably by using multi-objective optimization approaches or by adjusting their priorities based on current system conditions [13].

To make these decisions, agents and control systems need to include real-time feedback loops and predictive modeling. These models can help estimate the marginal emissions outcome of possible actions, factor in latency costs or resilience risks, and assess trade-offs not independently, but as system-level Pareto frontiers. The ability to add predictive sustainability indicators, such that controlling possible signals have forward-looking carbon curves or emissions elasticity, allows agents to reason over time horizons and to optimize holistically [14, 15]. Table 10.10 builds out the key performance–emissions–efficiency tensions that exist in different subsystems of ICPS and addresses the CESN-aware resolution strategies that are needed to best deal with them.

Table 10.10 Trade-off dimensions in CESN-oriented ICPS operations [12–15].

Subsystem and function	Performance criteria	Objective	Emissions impact	CESN-aware resolution strategy
Frequency regulation via BESS	Deviation minimization, response latency	Optimal SoC profile, minimal cycling loss	Relies on the carbon intensity of the recharge	Emissions-based scheduling; collectively referenced carbon-adjusted reward signals
DERs dispatch scheduling	Phase balance, power quality	Optimal asset utilization	The emissions profile of the electricity generation varies depending on when it was dispatched	Continued time-of-use carbon pricing; marginal intensity controlled
Demand response	User compliance, load stability	Peak shaving efficiency	May shift load to higher emission time windows	Fusing integrated forecasted carbon signals; recalibrating incentive-based
Thermal control (HVAC, water)	Temperature deviation, comfort level	Energy consumption minimization	Emissions are acute to grid policy and so the instantaneous grid mix	Adaptive comfort margin that considers emission-weighted thermal inertia; control of gross thermal inertia
Grid islanding/ microgrid ops	Black-start latency, transition smoothness	Local resource optimization	Potentially lose access to clean grid energy	Carbon compensation through resilience-weighted representations; reentry validation logic to verify no adverse effects
EV smart charging	Charge time, availability compliance	Energy cost optimization	High at night when using fossil-heavy grids	Carbon charge window assignment with carbon awareness for users; feedback for users suitable for the carbon emission categories.

10.6.3 Toward Explainable, Trust-preserving Intelligent Operations

As ICPS more extensively utilize ML algorithms, distributed coordinating agents, and adaptive optimization practices, the explainability and trustworthiness of their functioning become fundamental issues, especially to MENs that are CESN-focused, whose system's operation interacts with social accountability, regulatory compliance, and sustainability expectations. The former deterministic control systems could provide some rule-based traceability but intelligent agents, especially those derived from deep RL or federated configurations, severely limit interpretability; without being explainable, outputs from a systems point of view could not be audited, defended, or corrected, particularly with respect to emissions reporting, DERs prioritization, or load balancing utilizing social equity [49, 50].

To ensure trust and legitimacy, ICPS must employ human-understandable reasoning layers that can reason not only in technical terms but in ethical and policy-relevant terms as well. Understanding which inputs most influenced an AI agent's decisions, which sustainability constraints limited its options, and how confident it was in its predictions are all crucial for building trust. This can be accomplished through techniques like attention mapping, feature analysis, scenario testing, and tracing decision pathways. However, trustworthy integrated systems must go beyond just explaining decisions after the fact. They need to build transparency directly into how they make decisions, incorporating interpretability as part of their core optimization goals. In CESN contexts, this transparency is essential. It enables effective coordination among multiple stakeholders, societal acceptance for systems, and ensures adaptive scope management in environmental policies [46–48].

10.6.4 Integrative Effects of Intelligent Technologies on CESN

The smart technologies discussed throughout this chapter, digital twins, ML pipelines, and LLM interfaces, are key to bridging the interlinkages of climate change, energy, and sustainability. Digital twins capture dynamic forecasting environments that recreate decarbonization pathways, offering the opportunity for real-time experimentation of emissions pathways and adaptive strategies to bolster resilience. ML analytics extend these possibilities, enabling predictions on renewable intermittency, demand response optimization, and adaptive decision-making that folds sustainability measures directly into the equation. LLMs further contribute to transparency and inclusiveness by translating technical results into narratives that stakeholders can understand, while also capturing compliance reports, therefore allowing operational intelligence to remain operationally tied to ecological limits and policy constraints. Together, these technologies promote a heuristic framework in which climate impacts, energy savings, and sustainability targets are measured not independently, but rather as co-evolving within a living system.

Nonetheless, the effectiveness of these technologies depends on many key factors, such as the availability and granularity of real-time emissions data, strength of cyber protection, and the interoperability of multi-agent systems that operate on distributed infrastructures. Another key factor is policy alignment and regulatory flexibility, as sustainability-focused control signals and carbon-aware optimization algorithms will not be effective without supporting governance structures. Lastly, social acceptance and participation of stakeholders supported by explainable AI and participatory digital twin dashboards are important to ensure technological systems translate into sustainable behavioral and institutional change. When taken together, these enabling factors illustrate not only the technical feasibility of the ICPS framework but also indicate its potential to act as a unifying mechanism to adjust systemic misalignments between climate objectives, energy operations, and long-term sustainability agendas.

10.7 Research Challenges and Policy Implications

The integration of ICPSs into MENs aligned with CESN provides both unique opportunities and challenges. Intelligent automation can enhance efficiency, resilience, and sustainability. However, the success of intelligent automation relies on the successful navigation of significant structural, organizational, and ethical limitations. In this section, three critical issues are identified: (i) technical standards and disintegration; (ii) ethical and governance concerns regarding opaque and automated decision-making; (iii) a demand for forward-looking policy frameworks for secure, equitable, and carbon-aligned digital energy transitions. The interconnected nature of these challenges means all three concerns must be addressed if ICPS are going to succeed as genuine solutions rather than remain technological possibilities that cannot withstand serious climate pressures.

10.7.1 Standardization Gaps and Integration Challenges

While substantial strides have been made in smart grid integration and energy digitalization, the integration of ICPS is still limited by competing standards, overly defined protocols, and a lack of interoperability across vendors, platforms, and jurisdictions. This fragmentation impedes seamless coordination between DERs, grid operators, and intelligent agents due to inefficiencies and duplicative efforts, which may also limit scalability. For example, the absence of standardized data models, consistent semantic labeling, and secure communication interfaces prevents effective system-wide optimization and coordinated operational control.

These technical incompatibilities become even more problematic in multi-vendor settings or cross-border CESN initiatives, where energy, climate, and digital systems must develop together [38, 40].

A change needs to happen towards open, modular, ontology-aligned standards. Standardization bodies will need to move away from only hardware-level compliance (e.g., IEEE 2030.5, IEC 61850) to focus on addressing interoperability for AI agents, explainable logic for interface actions, and context-sensitive semantics for control. They must ensure that the integration framework allows for adaptive layering, where local standards can be maintained alongside GAMER-based global alignment mechanisms without inhibiting system consistency. Open testbeds, cross-silo governance compliance sandboxes, and federated certification regimes will be critical to enabling fast and compatible scaling of intelligent infrastructures that comply with CESN [38, 40].

10.7.2 Ethics, Transparency, and Human-in-the-loop Control in ICPS

Critical control functions can be legally delegated to autonomous ICPS agents, leading to fundamental ethical questions about responsibility, particularly where decisions involve access to resources, responsibility for emissions, and public trust. Black-box systems may reinforce inequities by unintentionally excluding low-consumption users and distorting sustainability metrics, especially if their training data are biased or incomplete. Furthermore, the lack of transparency in these decision-making processes compounds accountability challenges. When autonomous agents take control and override their human operators, or act on uncertain predictions, the ability to identify who should be held responsible is compromised, and reduced legitimacy and governance capacity ensue [23, 24].

Incorporating human-in-the-loop controls, transparent audits, and multilayered explanations is critical. ICPS require auditing beyond simple code inspection. They need to provide contextual explanations for their decisions, especially those impacting environmental, economic, and social outcomes. Ethical considerations must extend to upstream data governance and policy alignment. Establishing cross-disciplinary ethics panels, AI safety protocols, and sustainability-focused design principles will create necessary boundaries to keep system intelligence aligned with human values and ensure continued human participation [17].

10.7.3 Policy Enablers for Secure and Sustainable Digital Energy Transitions

The successful integration of ICPS for CESN-compliant energy systems depends on proactive policy frameworks that can react to emerging technical and social developments. Regulatory regimes designed for fixed infrastructures with episodic market oversight have insufficient agility to govern systems that operate based on learning and adaptation. In the absence of an agile response by regulators, progress on innovation is delayed, and systems may operate with regulatory gaps in relation to cybersecurity, emissions responsibility, and data sovereignty. In addition, the inconsistencies inherent in fragmented jurisdictional mandates inhibit coherence in frameworks, especially in federated or transnational energy systems. Policymakers, in addressing these risks, should promote nimble, anticipatory models of governance that integrate dynamic risk assessment, stakeholder engagement, and mission-driven digital infrastructure plans. Their institutional support for regulatory sandboxes, AI impact assessments, and climate-linked (e.g., emissions-linked) procurement incentives can promote innovation while protecting the public interest. Policy cannot be a step behind technology; it must co-evolve to ensure ICPS deployment speeds decarbonization and ultimately systemically supports democratic control, transparency, and systemic resilience [3–9].

10.8 Conclusion

This chapter has provided a framework for the conceptual, technical, and operational architecture necessary for ICPS to be implemented in support of the CESN embedded in MENs. ICPS can advance beyond conventional control systems by integrating ML, edge intelligence, secure federated architectures, and normative sustainability logic. This enables real-time, collaborative adaptation, multi-objective optimization, and carbon-aware intelligence, allowing MENs to retain the capacities of decarbonization, resilience, and equity. The evolution from rule-based infrastructures to learning-based cyber-physical systems embedded with trust establishes a foundational shift in how energy will be managed, secured, and governed.

This chapter showed how ICPS must integrate CESN priorities as both restrictions and fundamental operational requirements, which are integrated into system goals, incentive structures, and design principles. The examples provided throughout the chapter illustrated this conceptualization of intelligence from collaborative anomaly detection to multi-agent coordinating responses to emissions-centric scheduling. This intelligence feature is to be achieved through the convergence of sustainability, performance, and security. We have redefined intelligence in these situations based on several factors, including the consideration of the quantity and speed of answers to a problem and the interpretability, fairness, and alignment with objectives. This places ICPS as a future socio-technical institutional sustainability innovation. However, its success relies on the sophistication of the algorithms, the ability to engage with existing legal, ethical, and regulatory conditions to govern adaptive and distributed systems. The most important achievements of this chapter are:

- It discusses the important role of ICPS architectures within current approaches to CESN. The chapter addresses the significant cybersecurity threats facing existing energy systems as they transform from traditional distribution models to modern networks focused on accounting, delivery, system engineering, deep decarbonization, resilience, adaptive policy, and other key priorities.
- It provides a critical review of enabling technologies, which include but are not limited to IoT-based perception, digital twins, ML, LLMs, edge-cloud architectures, and the cybersecurity aspects of anomaly detection, FL, and adaptive responses.
- The chapter highlights effective resilience strategies, such as self-healing control loops, smart scheduling, P2P energy sharing, and predictive maintenance. Moreover, it frames them alongside the CESN-related priorities of carbon reduction, equity and social justice, and openness in the energy systems.
- It identifies critical research gaps and policy implications that remain unexplored. The particular focus is on stronger standardization, ethical governance frameworks, and forward-looking regulatory reforms needed for sustainable and socially responsible ICPS deployment.

CESN-aligned ICPS provide a strategic approach to reimagining energy governance for modern challenges. This transformation requires parallel advancement in technology and policy development, innovative metrics for assessing system intelligence, and inclusive mechanisms that embed democratic values in autonomous infrastructure. With climate transition becoming increasingly urgent, this ICPS framework offers both a practical guide for digitally transforming energy infrastructure and the groundwork for more resilient, ethical, and sustainable energy networks.

References

1 R. Sepehrzad, M. E. Hassanzadeh, A. R. Seifi, and M. Mazinani, "An efficient multilevel interconnect control algorithm in AC/DC micro-grids using hybrid energy storage system," *Electric Power Systems Research*, vol. 191 p. 106869, 2021.

2 S. H. A. Aleem, M. E. Balci, and M. J. H. Rawa (Eds.), *Energy Efficiency of Modern Power and Energy Systems*. Elsevier, 2024.

3 R. Sepehrzad, M. J. Faraji, A. Al-Durra, and M. S. Sadabadi, "Enhancing cyber-resilience in electric vehicle charging stations: A multi-agent deep reinforcement learning approach," *IEEE Transactions on Intelligent Transportation Systems*, vol. 25, no. 11, pp. 18049–18062, 2024.

4 R. Sepehrzad, A. S. G. Langeroudi, A. Khodadadi, S. Adinehpour, A. Al-Durra, and A. Anvari-Moghaddam, "An applied deep reinforcement learning approach to control active networked microgrids in smart cities with multi-level participation of battery energy storage system and electric vehicles," *Sustainable Cities and Society*, vol. 107 p. 105352, 2024.

5 X. Yu, N. Liu, and Y. Xue, "The role of cyber-physical-social systems in smart energy future," *IEEE Transactions on Industrial Cyber-Physical Systems*, vol. 2 pp. 35–42, 2024.

6 Y. Zhong and Y. Li, "Statistical evaluation of sustainable urban planning: Integrating renewable energy sources, energy-efficient buildings, and climate resilience measures," *Sustainable Cities and Society*, vol. 101 p. 105160, 2024.

7 H. Kanso, A. Noureddine, and E. Exposito, "A review of energy aware cyber-physical systems," *Cyber-Physical Systems*, vol. 10, no. 1, pp. 1–42, 2024.

8 A. Parizad, H. R. Baghaee, and S. Rahman, "Overview of smart cyber-physical power systems: Fundamentals, challenges, and solutions," in *Smart Cyber-Physical Power Systems: Fundamental Concepts, Challenges, and Solutions*, vol. 1 John Wiley & Sons, 2025, pp. 1–69.

9 U. Bhadani, "Smart grids: A cyber–physical systems perspective," *International Research Journal of Engineering and Technology (IRJET)*, vol. 11, no. 06, p. 801, 2024.

10 R. Sepehrzad, A. S. G. Langeroudi, A. Al-Durra, A. Anvari-Moghaddam, and M. S. Sadabadi, "Demand response-based multi-layer peer-to-peer energy trading strategy for renewable-powered microgrids with electric vehicles," *Energy*, vol. 320 p. 135206, 2025.

11 G. Hu and F. You, "AI-enabled cyber-physical-biological systems for smart energy management and sustainable food production in a plant factory," *Applied Energy*, vol. 356 p. 122334, 2024.

12 S. Norbu et al., "Cyber-physical energy systems: Enabling tailored decarbonization and net-zero pathways for communities," in *IET Powering Net Zero (PNZ 2024)*, IET, 2024, pp. 1–7.

13 R. Sepehrzad, H. Hosseinalibeiki, N. Taghinezhad, N. Khosravi, A. Al Durra, and M. S. Sadabadi, "Enhanced cyber-resilience in flexible energy markets for microgrids: A trust-aware federated deep reinforcement learning framework," *Applied Energy*, vol. 406 p. 127282, 2026.

14 T. Jelinek et al., "International collaboration: Mainstreaming artificial intelligence and cyberphysical systems for carbon neutrality," *IEEE Transactions on Industrial Cyber-Physical Systems*, vol. 2 pp. 26–34, 2024.

15 M. R. Almassalkhi and S. Kundu, "Intelligent electrification as an enabler of clean energy and decarbonization," *Current Sustainable/Renewable Energy Reports*, vol. 10, no. 4, pp. 183–196, 2023.

16 R. Sepehrzad, M. Moafi, A. Al Durra, and M. S. Sadabadi, "Supervisory-learning microgrid control strategy in the presence of energy storage systems based on uncertainties and energy market interactions to improve stability," *Energy*, vol. 342 p. 139738, 2026.

17 C. Mu et al., "A light blockchain for behind-the-meter peer-to-peer energy transactions in cyber-physical power systems," *IEEE Transactions on Smart Grid*, vol. 15, no. 1, pp. 1063–1074, 2023.

18 U. Tariq, I. Ahmed, A. K. Bashir, and K. Shaukat, "A critical cybersecurity analysis and future research directions for the internet of things: A comprehensive review," *Sensors*, vol. 23, no. 8, p. 4117, 2023.

19 A. Salam, "Internet of Things for sustainability: Perspectives in privacy, cybersecurity, and future trends," in *Internet of Things for Sustainable Community Development: Wireless Communications, Sensing, and Systems*, Springer International Publishing, 2024, pp. 299–326.

20 T. Zhukabayeva, L. Zholshiyeva, N. Karabayev, S. Khan, and N. Alnazzawi, "Cybersecurity solutions for Industrial Internet of Things–edge computing integration: Challenges, threats, and future directions," *Sensors*, vol. 25, no. 1, p. 213, 2025.

21 D. Sabella, K. Maloor, N. Smith, M. Vanderveen, and A. Kourtis, "Edge computing cybersecurity standards: Protecting infrastructure and applications," *IEEE Access*, vol. 12 pp. 185328–185335, 2024.

22 R. Sepehrzad et al., "A critical overview of local energy communities: State-of-the-art, real-life applications & challenges and tackling the academia-industry gap," *Renewable and Sustainable Energy Reviews*, vol. 226 p. 116165, 2026.

23 A. K. Jameil and H. Al-Raweshidy, "Enhancing offloading with cybersecurity in edge computing for digital twin-driven patient monitoring," *IET Wireless Sensor Systems*, vol. 14, no. 6, pp. 363–380, 2024.

24 M. R. Kabir, D. Halder, and S. Ray, "Digital twins for IoT-driven energy systems: A survey," *IEEE Access*, vol. 12 pp. 177123–177143, 2024.

25 W. Xu and S. Liu, "Novel economic models for advancing urban energy management and transition: Simulation of urban energy system in digital twin," *Sustainable Cities and Society*, vol. 101 p. 105154, 2024.

26 S. Reza, H. Atefeh, G. Javid, and A. D. Ahmed, "Multilevel energy management and optimal control system in smart cities based on deep machine learning," in

Applications of Deep Machine Learning in Future Energy Systems, Elsevier, 2024, pp. 265–314.

27 K. Aygul, M. Mohammadpourfard, M. Kesici, F. Kucuktezcan, and I. Genc, "Benchmark of machine learning algorithms on transient stability prediction in renewable rich power grids under cyber-attacks," *Internet of Things*, vol. 25 p. 101012, 2024.

28 Y. Li, W. Ma, Y. Li, S. Li, Z. Chen, and M. Shahidehpour, "Enhancing cyber-resilience in integrated energy system scheduling with demand response using deep reinforcement learning," *Applied Energy*, vol. 379 p. 124831, 2025.

29 S. Kapp, J.-K. Choi, and T. Hong, "Predicting industrial building energy consumption with statistical and machine-learning models informed by physical system parameters," *Renewable and Sustainable Energy Reviews*, vol. 172 p. 113045, 2023.

30 A. Khodadadi, S. Adinehpour, R. Sepehrzad, A. Al-Durra, and A. Anvari-Moghaddam, "Data-driven hierarchical energy management in multi-integrated energy systems considering integrated demand response programs and energy storage system participation based on MADRL approach," *Sustainable Cities and Society*, vol. 103 p. 105264, 2024.

31 E. H. Sumiea et al., "Deep deterministic policy gradient algorithm: A systematic review," *Heliyon*, vol. 10, no. 9, p. e30697, 2024.

32 Z. Song, S. Cao, and H. Yang, "An interpretable framework for modeling global solar radiation using tree-based ensemble machine learning and Shapley additive explanations methods," *Applied Energy*, vol. 364 p. 123238, 2024.

33 M. Yang, C. Xu, Y. Bai, M. Ma, and X. Su, "Investigating black-box model for wind power forecasting using local interpretable model-agnostic explanations algorithm," *CSEE Journal of Power and Energy Systems*, vol. 11, no. 1, pp. 227–242, 2023.

34 M. A. K. Raiaan et al., "A review on large language models: Architectures, applications, taxonomies, open issues and challenges," *IEEE Access*, vol. 12 pp. 26839–26874, 2024.

35 R. Gong et al., "A survey of low-bit large language models: Basics, systems, and algorithms," *Neural Networks*, vol. 192 p. 107856, 2025.

36 Y. Chang et al., "A survey on evaluation of large language models," *ACM Transactions on Intelligent Systems and Technology*, vol. 15, no. 3, pp. 1–45, 2024.

37 R. Sepehrzad, A. R. Moridi, M. E. Hassanzadeh, and A. R. Seifi, "Intelligent energy management and multi-objective power distribution control in hybrid micro-grids based on the advanced fuzzy-PSO method," *ISA Transactions*, vol. 112 pp. 199–213, 2021.

38 R. Sepehrzad, A. Mahmoodi, S. Y. Ghalebi, A. R. Moridi, and A. R. Seifi, "Intelligent hierarchical energy and power management to control the voltage and frequency of micro-grids based on power uncertainties and communication latency," *Electric Power Systems Research*, vol. 202 p. 107567, 2022.

39 R. Sepehrzad, A. Al-Durra, A. Anvari-Moghaddam, and M. S. Sadabadi, "Short-term and probability scenario-oriented energy management of integrated energy distribution systems with considering energy market interactions and end-user participation," *Energy*, vol. 322 p. 135691, 2025.

40 R. Sepehrzad, A. Khodadadi, S. Adinehpour, and M. Karimi, "A multi-agent deep reinforcement learning paradigm to improve the robustness and resilience of grid connected electric vehicle charging stations against the destructive effects of cyber-attacks," *Energy*, vol. 307 p. 132669, 2024.

41 N. Kharlamova, C. Træhold, and S. Hashemi, "Cyberattack detection methods for battery energy storage systems," *Journal of Energy Storage*, vol. 69 p. 107795, 2023.

42 H. R. Abdolmohammadi, R. Sepehrzad, E. A. Farsani, A. Oubelaid, and N. Khosravi, "A new gradient projection active set approach for reliable and economic operation of combined energy systems," *Energy Conversion and Management: X*, vol. 29 p. 101518, 2026.

43 R. Sepehrzad, H. Hosseinalibeiki, M. S. Sadabadi, and A. Al Durra, "Microgrid secure pricing and real-time flexible energy market integrated with trust-based demand response: A multi-layer FDRL architecture," *International Journal of Electrical Power & Energy Systems*, vol. 174 p. 11153, 2026.

44 B. S. Murthy, P. B. V. Raja Rao, M. Prasad, P. K. Sree, P. J. R. S. Raju, and K. S. Kumar, "Hybrid security framework and machine learning based anomaly detection for machine-to-machine communications," in *2025 International Conference on Computational Robotics, Testing and Engineering Evaluation (ICCRTEE)*, IEEE, 2025.

45 R. Sepehrzad, A. Hedayatnia, M. Amohadi, J. Ghafourian, A. Al-Durra, and A. Anvari-Moghaddam, "Two-stage experimental intelligent dynamic energy management of microgrid in smart cities based on demand response programs and energy storage system participation," *International Journal of Electrical Power & Energy Systems*, vol. 155 p. 109613, 2024.

46 A. Hedayatnia, J. Ghafourian, R. Sepehrzad, A. Al-Durrad, and A. Anvari-Moghaddam, "Two-stage data-driven optimal energy management and dynamic real-time operation in networked microgrid based on a deep reinforcement learning approach," *International Journal of Electrical Power & Energy Systems*, vol. 160 p. 110142, 2024.

47 W. Lee, K. Kim, J. Park, J. Kim, and Y. Kim, "Forecasting solar power using long-short term memory and convolutional neural networks," *IEEE Access*, vol. 6 pp. 73068–73080, 2018.

48 A. D. Pazho, G. A. Noghre, A. A. Purkayastha, J. Vempati, O. Martin, and H. Tabkhi, "A survey of graph-based deep learning for anomaly detection in distributed systems," *IEEE Transactions on Knowledge and Data Engineering*, vol. 36, no. 1, pp. 1–20, 2023.

49 J. Jithish, B. Alangot, N. Mahalingam, and K. S. Yeo, "Distributed anomaly detection in smart grids: A federated learning-based approach," *IEEE Access*, vol. 11 pp. 7157–7179, 2023.

50 B. Dong, D. Chen, Y. Wu, S. Tang, and Y. Zhuang, "FADngs: Federated learning for anomaly detection," *IEEE Transactions on Neural Networks and Learning Systems*, vol. 36, no. 2, pp. 2578–2592, 2024.

11

Cybersecure Digital Twin Architectures and Quantum-augmented Analytics for Intelligent Cyber-physical Energy Systems in Deep Decarbonization

Ehsan Naderi

Department of Electrical Engineering, College of Engineering and Computer Science, Arkansas State University, Jonesboro, AR, USA

Corresponding author: enaderi@astate.edu

List of Acronyms	
CESN	Climate–energy–sustainability nexus
CPES	Cyber-physical energy system
CPS	Cyber-physical system
DER	Distributed energy resource
DT	Digital twin
DT-QC	Digital twin-quantum computing
EVO	Energy valley optimizer
FDIA	False data injection attack
FPR	False positive rate
HQEPSO	Hybrid quantum-enhanced particle swarm optimization
HQEPSO-EVO	Hybrid quantum-enhanced particle swarm optimization-energy valley optimizer
IoT	Internet of Things
ML	Machine learning
PMU	Phasor measurement unit
PSO	Particle swarm optimization
PV	Photovoltaic
QADT	Quantum-augmented digital twin

List of Acronyms	
QC	Quantum computing
QKD	Quantum key distribution
RES	Renewable energy source
RTU	Remote terminal unit
SAIDI	System average interruption duration index
SAIFI	System average interruption frequency index
SCADA	Supervisory Control and Data Acquisition
TPR	True positive rate

11.1 Introduction

11.1.1 Chapter Context and Motivation

As power systems transition toward deeply decarbonized, decentralized, and digitally managed architectures, cybersecurity and operational resilience have emerged as critical challenges alongside sustainability objectives. High penetrations of renewable energy sources (RESs), increased reliance on distributed energy assets, and the pervasive use of data-driven control platforms expose modern grids to new classes of cyber-physical vulnerabilities. Addressing these challenges requires not only advanced monitoring and control strategies but also computational paradigms capable of handling high-dimensional system complexity, real-time decision-making, and adversarial conditions. This chapter is motivated by the need to bridge these emerging challenges through an integrated, forward-looking cyber-physical framework aligned with the climate–energy–sustainability nexus (CESN).

11.1.2 Scope and Conceptual Approach

Within this context, the chapter introduces a quantum-augmented digital twin (QADT) framework that synergistically combines digital twin (DT) technology with quantum-enhanced analytics to strengthen the intelligence, security, and adaptability of modern energy systems. DTs provide a continuously synchronized virtual representation of the physical grid, enabling predictive analysis, anomaly detection, and real-time situational awareness. Quantum computing (QC), in turn, augments these capabilities by addressing computational bottlenecks in secure optimization, large-scale analytics, and threat mitigation. By integrating these technologies into a unified cyber-physical architecture, the proposed framework aims to enhance resilience against cyberattacks, improve operational efficiency under renewable uncertainty, and support secure decision-making in next-generation low-carbon power networks.

11.1.3 Background, Definitions, and Foundations

The global push toward deep decarbonization, aligned with CESN, is about to reshape the landscape of modern energy systems [1]. This transformation is characterized by the increasing penetration of RESs, proliferation of both renewable- and nonrenewable-based distributed energy resources (DERs), and the digitalization of grid operations [2]. While these advancements support climate goals as well as energy justice and equity, they also introduce significant challenges related to system complexity, cyber vulnerability, and operational uncertainty [3]. In this context, intelligent cyber-physical systems (CPSs) have emerged as a cornerstone of next-generation energy infrastructure, empowering real-time monitoring, autonomous control, and adaptive resilience. However, the convergence of physical assets with digital technologies necessitates a rethinking of cybersecurity-related policies and system integrity strategies [4]. DTs, defined as dynamic virtual representations of physical energy systems, offer a promising pathway for continuous situational awareness and proactive threat responses [5]. Simultaneously, QC presents transformative potential in addressing computational bottlenecks in high-dimensional analytics and secure communications [6]. Motivated by the urgent need to build secure, low-carbon (e.g., ideally zero), and intelligent energy networks, this chapter investigates the synergistic integration of cybersecure DT (CDT) architectures with quantum-augmented analytics, aiming to establish a resilient framework for future-ready energy systems operating under CESN principles and codes of conduct.

11.1.4 Literature Review

Graham et al. investigated cybersecurity challenges in decentralized smart grids, emphasizing real-world vulnerabilities highlighted by the Ukraine blackout and recommending a layered NIST-based (National Institute of Standards and Technology) defense strategy [7]. Alcaraz and Lopez provided a comprehensive survey of DT security threats, identifying attack vectors across data integrity, model validation, and control loop vulnerabilities [8]. Josifovska et al. proposed a general reference framework for DTs within CPS, detailing the mapping between physical and virtual domains and implications for security monitoring [9]. Lin et al. introduced DT4I4-Secure, a DT architecture tailored for Industry 4.0, featuring real-time intrusion detection and secure data management [10]. Şen et al. developed a test bed oriented toward cyber-physical-twin concepts for smart grid cybersecurity experiments, replicating microgrid intrusion scenarios and evaluating detection efficacy [11]. Mohsin et al. outlined the concept of security for DTs, framing "Shift-Left/Shift-Right" paradigms to embed cybersecurity from design through runtime [12]. Ajagekar and You explored the use of QC for energy system optimization, benchmarking quantum algorithms against classical methods [13]. Fernández-Campoamor et al. applied quantum annealing to community detection in IEEE test systems, demonstrating improvements in partition optimality and speed [14]. Kandasamy et al. constructed a DT test platform for cybersecurity training in power grids, illustrating attack and defense workflows in scalable virtual environments [15]. Homaei et al. surveyed AI-enhanced DT cybersecurity, categorizing techniques like learning automata and rule-based detection across Internet of Things (IoT)/CPS deployments [16].

IEEE SA recently released DER cybersecurity standards (i.e., IEEE 1547.3-2023, P2658), underscoring protocols for securing distributed renewable assets [17]. Research in quantum key distribution has begun targeting grid-scale infrastructure, leveraging BB84 and QRNGs to strengthen CPS communications [18]. Studies in post-quantum cryptography (e.g., Shor's threats, ring-LWE, and NIST PQC standards) have outlined the timeline and methods for upgrading CPS encryption [19]. Toward that end, Wang et al. surveyed human-cyber-physical systems in energy internet contexts, highlighting digital-human twins and their relevance to security-aware CPS design [20]. Senthilnathan and Annapoorani reviewed the cyber twin paradigm for smart grids, mapping data flows, loop architectures, and edge security layers [21]. Saad et al. implemented an IoT-based DT for microgrids with Amazon Web Services (AWS) edge integration, showing potential for secure monitoring in distributed systems [22]. A real-time DT of an Alaskan microgrid was demonstrated using Supervisory Control and Data Acquisition (SCADA) and phasor measurement units (PMUs), offering a test bed for resilient control and cybersecurity improvements [23]. Studies on resilient control systems (e.g., fuzzy-neural data fusion engines) explored diagnostic architectures for sensor integrity within CPS [24]. Quantum circuit automation research addressed hardware-aware mapping and machine learning (ML)-driven gate optimization, essential for scalable quantum analytics in CPS [25]. Moreover, quantum cybersecurity products (e.g., Quantinuum's Quantum Origin) have begun combining quantum-generated randomness with conventional grid-scale encryption methods [26].

Together, these selected studies precisely show a research trajectory converging around CDTs for smart power grids, augmented by quantum-oriented analytics for optimization and defense, setting a strong foundation for the proposed integration within decarbonized network architectures.

11.1.5 Research Gaps and Contribution of This Work

Despite significant advancements in smart grid modernization, state-of-the-art DT frameworks, and early-stage QC applications, critical research gaps persist at the *intersection of cybersecurity, real-time system intelligence, and quantum-resilient architectures in deeply decarbonized energy systems*. Existing DT models, while effective for simulation and operational oversight, often lack built-in, adaptive cybersecurity mechanisms capable of defending against sophisticated threats such as coordinated false data injection attacks (FDIAs), zero-day exploits, and adversarial ML attacks. Moreover, most DT implementations fail to scale efficiently in highly decentralized environments with high penetrations of renewables and bidirectional power flows, where latency, data authenticity, and integrity are under the spotlight of researchers. Simultaneously, the integration of QC into modern energy systems has largely remained exploratory, with limited convergence between quantum algorithms and operational control of cyber-physical energy infrastructure. The lack of a unified framework that synergizes quantum-augmented analytics with CDT architectures under the CESN paradigm severely hampers the evolution toward resilient, intelligent, and net-zero (eco-friendly) power systems. This work directly addresses these limitations by proposing a novel hybrid framework, combining QC capabilities (e.g., quantum algorithms for secure system optimization) with intelligent DT architectures, tailored to modern and distributed energy networks. Introducing a modified version of the IEEE 39-bus New England test system as a case study, this chapter scrutinizes the proposed QADT architecture under realistic attack/non-attack scenarios and renewable penetrations, demonstrating significant improvements in *detection latency*, *system resilience*, and *operational efficiency*.

Accordingly, the main contributions and objectives of this chapter can be itemized hereunder:

- Propose a unified QADT framework for enhancing the intelligence, security, and resilience of modern cyber-physical energy systems (CPESs).
- Integrate QC methods into DT architectures to enable secure system optimization and real-time threat mitigation.
- Demonstrate scalable DT deployment in decentralized, renewable-rich environments with high data and operational complexity.

- Evaluate the proposed framework through a case study based on the modified IEEE 39-bus system, under varying cyber-physical threat and renewable penetration scenarios.

11.1.6 Chapter-at-a-glance Summary Box

Motivation	Increasing cyber vulnerabilities in renewable-rich, decentralized power systems demand new frameworks that combine real-time intelligence, adaptive control, and advanced computational methods
Core technologies	• DTs for real-time system replication, monitoring, and anomaly detection • QC for secure optimization, accelerated analytics, and cyber defense
Key contributions	• A unified QADT architecture for CPESs • Integration of quantum optimization and quantum ML into DT-based control loops • A comprehensive cyber-threat evaluation under FDIA, communication latency, and adversarial learning scenarios • Validation using a modified IEEE 39-bus New England system with varying renewable penetration levels
Key outcomes	• Significant reduction in cyberattack detection latency • Improved dispatch efficiency under communication constraints • Enhanced resilience and stability in renewable-dominated grid environments
Relevance	The proposed framework supports CESN objectives by enabling secure, intelligent, and scalable operation of future low-carbon power systems

11.1.7 Organization of the Chapter

The remainder of this chapter is structured as follows. Section 11.2 introduces the proposed QADT framework, outlining its main layers and major components. Section 11.3 formulates the relevant optimization problems, including objective functions and relevant constraints. Section 11.4 introduces the case study employed in this chapter, including all modifications made to the standard version of the IEEE 39-bus New England test system. Section 11.5 discusses the initialization procedures, presents the obtained results, and provides detailed analyses. Finally, Section 11.6 concludes the chapter by highlighting the limitations of the study and suggesting possible policy updates and directions for future work.

11.2 Proposed Framework

11.2.1 Overview

The proposed QADT framework of this chapter introduces a modular, layered architecture designed to enhance the cyber resilience, predictive intelligence, and operational efficiency of decarbonized energy systems. Furthermore, the proposed framework advances DT technology with QC capabilities in order to enable real-time system replication, threat detection, and secure decision-making under the CESN concept.

11.2.2 Framework Components

The following provides detailed descriptions of the core layers and key components of the proposed QADT framework:

Physical layer (cyber-physical energy infrastructure): this layer comprises physical assets (or field devices) such as RESs (solar panels and wind turbines), DERs, storage units, smart meters, PMUs, remote terminal units (RTUs), and IoT sensors/actuators. In this regard, the standard version of the IEEE 39-bus New England test system is highly modified and remodeled as the physical backbone of this study, providing a benchmark for simulating frequency synchronization, load dynamics, generation fluctuations, and fault events, all under FDIA scenarios.

Digital twin layer: this layer is in fact the virtual replica of the physical system, continuously synchronized via SCADA, PMUs, and IoT sensors, that supports real-time monitoring, forecasting, and simulation of system behavior, integrating load flow, asset health diagnostics, and operational parameters (e.g., voltage stability, power quality, renewable curtailment).

Quantum analytics layer: this layer integrates quantum algorithms through a quantum-classical interface.

Control and decision-making layer: this layer ingests analytics from the unified DT-QC (digital twin-quantum computing) engine to support autonomous and semi-autonomous decisions in the form of load shedding and/or DER control to modify DT parameters based on QC inferences and operator inputs to expeditiously respond to abnormalities.

Rules, safety, and compatibility layer: this part makes sure everything follows the right rules and ethical guidelines. It helps protect private data, makes sure systems from different regions can work together, and can also connect with blockchain technology in the future to keep secure records.

To obtain a deeper perspective about the details and interconnections of the aforementioned layers, Figure 11.1 illustrates the proposed framework in this chapter, where a comprehensive, multilayered approach is proposed to enhance CPESs under the CESN paradigm. From this figure, one can infer that the entire process begins with macro-level CESN drivers supporting the evolution of smart grids characterized by high renewable integration, decentralization, and digital complexity. Besides this, a cyber-threat landscape analysis identifies key risks arising from cyber vulnerabilities and stochastic system behaviors. In response, a hybrid architecture is introduced, integrating CDT systems with QC capabilities. The DT layer provides real-time simulation, anomaly detection, and system control, while QC, via a quantum optimization algorithm, augments predictive analytics and cyber defense to recover the targeted power grid and bring it back to normal operation. These components converge in a hybrid intelligence core that enables adaptive optimization, threat mitigation, and secure operations. To indicate the clear relationships among different layers, introduced in Figure 11.1, the proposed QADT framework is structured as an integrated, multilayered system in which each layer plays a distinct yet interdependent role in enhancing the intelligence, security, and resilience of modern energy networks. The physical layer (see Layer 1) serves as the foundational source of real-time data, feeding the DT layer (i.e., internal Layer 2), which continuously mirrors system states and supports predictive simulations. Outputs from the DT are analyzed by the quantum analytics layer (see the internal Layer 3), which applies quantum-augmented algorithms to extract insights for risk assessment and optimization. These insights flow into the control and decision-making layer (i.e., Layer 4), enabling adaptive operational responses and autonomous control actions. Overarching the entire architecture, the rules, safety, and compatibility layer (see Layer 5) ensures compliance, data integrity, and interoperability across all components. Collectively, these interconnected layers form a closed-loop system where real-time data, quantum-enhanced analytics, and intelligent decision-making work in harmony to support CESN-aligned, cyber-resilient energy infrastructures.

As mentioned earlier, the proposed framework is validated using a modified version of the IEEE 39-bus New England test system, simulating and validating the proposed QADT's performance under various cyberattack and renewable penetration scenarios. Detailed information about the necessary modifications applied on IEEE 39-bus system is provided in Section 11.4.

According to Figure 11.1, although economic and sustainability aspects are not represented as standalone layers in the proposed QADT framework, they are functionally integrated across multiple components. Economic optimization is addressed within the quantum analytics layer and control and decision-making

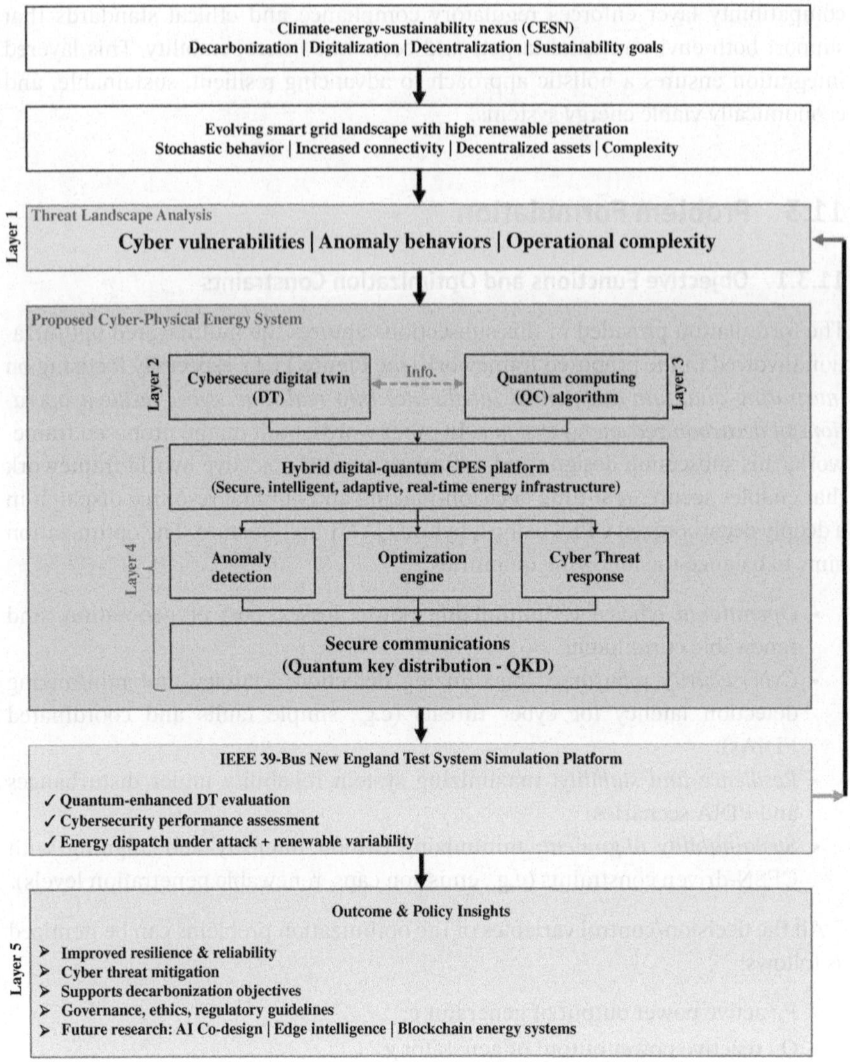

Figure 11.1 Proposed framework of this chapter, integrating QC and DT technologies for intelligent CPESs oriented toward deep decarbonization in energy infrastructure.

layer, where quantum algorithms support cost-efficient resource dispatch and operational planning. Similarly, sustainability objectives are inherently considered within the physical and DT layers, which model, monitor, and optimize the use of RESs in alignment with CESN goals. Additionally, the rules, safety, and compatibility layer enforces regulatory compliance and ethical standards that support both environmental targets and economic interoperability. This layered integration ensures a holistic approach to advancing resilient, sustainable, and economically viable energy systems.

11.3 Problem Formulation

11.3.1 Objective Functions and Optimization Constraints

The formulation provided in this subsection captures the multilayered optimization involved in the proposed framework (see Figure 11.1), especially focusing on *integrating quantum-augmented intelligence into real-time, cyber-resilient operations of decarbonized energy systems*. In other words, built on the proposed framework, this subsection designs and optimizes a multi-objective hybrid framework that enables secure, real-time decision-making and optimal resource dispatch in a deeply decarbonized CPES using a hybrid QADT architecture. The optimization aims to balance the following quantities:

- *Operational efficiency*: minimizing power losses, cost of generation, and renewable curtailment.
- *Cybersecurity robustness*: maximizing detection accuracy and minimizing detection latency for cyber threats (e.g., simple faults and coordinated FDIAs).
- *Resilience and stability*: maximizing system reliability under disturbances and FDIA scenarios.
- *Sustainability alignment*: minimizing carbon intensity and aligning with CESN-driven constraints (e.g., emission caps, renewable penetration levels).

All the decision/control variables of the optimization problems can be itemized as follows:

- P_g: active power output of generator g;
- Q_g: reactive power output of generator g;
- V_b: voltage magnitude at bus b;
- θ_b: voltage phase angle at bus b;
- $\delta_{a,t}$: binary variable indicating detection of anomaly a at time t;
- $k_{b\acute{b}}$: quantum key status (valid/invalid) between DT nodes b and $\acute{b}$;
- q_r: qubit allocation or state parameters for quantum subroutine r; and
- λ_m: weight assigned to quantum hybrid inference for ML model m.

The (to be minimized) operational efficiency objective function can be written in (11.1), where $C_g(P_g)$ denotes the cost function of gth generator; Loss_g shows the total network loss from bus g; and Curtailment_g is the renewable curtailment associated with gth generator [27],

$$\min\left\{\sum_{g\in G}\left(C_g(P_g) + \alpha \times \text{Loss}_g + \beta \times \text{Curtailment}_g\right)\right\} \tag{11.1}$$

The robustness of the cybersecurity layer can be mathematically formulated as an objective function to be maximized, which is provided in (11.2) [28],

$$\max\left\{\sum_{a\in\mathcal{A},t}\left(\text{TPR}_{a,t} - \gamma \times \text{FPR}_{a,t} - \eta \times \text{Latency}_{a,t}\right)\right\} \tag{11.2}$$

where TPR and FPR are, respectively, the true positive rate and false positive rate for the anomaly detection; and Latency denotes the latency of the detection stage.

The objective function associated with resilience and stability can be written in (11.3), where $\omega \in \Omega$ shows the evaluation under the stochastic sets of FDIA scenarios [29],

$$\max\left\{\mathbb{E}_{\omega\in\Omega}(\text{SAIFI}_{\omega} + \text{SAIDI}_{\omega} - \text{RT}_{\omega})\right\} \tag{11.3}$$

where SAIFI stands for system average interruption frequency index; and SAIDI stands for system average interruption duration index.

The quantum-secured communication objective function can be mathematically written as (11.4), where it incorporates key refresh mechanisms and communication delays. This objective function focuses on ensuring post-quantum secure communication between distributed components of the CPES, particularly between the DT, control centers, and field devices [30],

$$\max\{\text{ValidQK}_{b\acute{b}} \times \text{ChannelSecurity}_{b\acute{b}} - \xi \times \text{KeyRefreshLatency}_{b\acute{b}}\} \tag{11.4}$$

where $\text{ValidQK}_{b\acute{b}}$ is a binary variable indicating whether the quantum distributer between buses b and $\acute{b}$ has successfully established a valid quantum key; $\text{ChannelSecurity}_{b\acute{b}} \in [0,1]$ is a continuous score reflecting the security quality of the quantum communication channel between buses b and $\acute{b}$; $\text{KeyRefreshLatency}_{b\acute{b}}$ is the time delay (in terms of milliseconds or seconds at worst) required to refresh or renegotiate a valid quantum key between buses b and $\acute{b}$; and ξ is a penalty coefficient that controls the trade-off between maximizing communication security and minimizing latency.

To ensure the deeply decarbonization involved in the proposed framework, (11.5) indicates the sustainability metric in this chapter, which is central to aligning the proposed QADT architecture with the CESN paradigm. It is noted that the first term of (11.5) highlights the minimization of total carbon-weighted power dispatch, the second term minimizes the proportion of total generated power that comes from non-RESs, the third term reduces the renewable energy curtailment, and the last term enables the flexibility of the demand in reflecting the priority of renewable energy to supply the load [31],

$$\min\left\{\sum_{g\in G}\left(\epsilon_g \times P_g + \delta_1 \times \frac{\sum_{g\in \acute{G}} P_g}{\sum_{g\in G} P_g} + \delta_2 \times \text{Curtailment Renewables} - \delta_3 \times \text{FLS}\right)\right\} \tag{11.5}$$

where ϵ_g is the emission rate associated with the gth generator that promotes clean dispatch decisions under CESN targets; $g \in \acute{G}$ shows the subset of the nonrenewable DERs in the system (i.e., diesel generators); δ_1 is a tunable parameter that adjusts how strongly the optimization penalizes reliance on nonrenewables, for example, higher amounts of δ_1 refers to stricter sustainability preference to ensure the deep decarbonization of the proposed framework; CurtailmentRenewables is the unused renewable energy due to network constraints or demand mismatch; and FLS is the flexible load that is shifted in time to align with renewable generation.

Each of the objective functions (11.1)–(11.5) needs to be optimized subject to fundamental and immediate constraints and limitations. These constraints are presented in (11.6)–(11.11), where (11.6) and (11.7) are the power balance constraints to be considered for minimizing objective function (11.1), (11.8) indicates the cybersecurity detection threshold, which needs to be taken into account for maximizing objective function (11.2), (11.9) denotes the quantum resource limit for maximizing objective function (11.3), (11.10) shows the renewable penetration constraint, which has to be considered in maximizing objective function (11.4), and (11.11) ensures the data authenticity via quantum distribution processes for minimizing objective function (11.5) [32],

$$P_b - D_{pb} = \sum_{\acute{b}} V_b \times V_{\acute{b}} \times (G_{b\acute{b}} \cos\theta_{b\acute{b}} + B_{b\acute{b}} \sin\theta_{b\acute{b}}) \tag{11.6}$$

$$Q_b - D_{qb} = \sum_{\acute{b}} V_b \times V_{\acute{b}} \times (G_{b\acute{b}} \sin\theta_{b\acute{b}} - B_{b\acute{b}} \cos\theta_{b\acute{b}}) \tag{11.7}$$

$$\text{TPR}_{a,t} \geq \tau_a, \ \forall a, t \tag{11.8}$$

$$\sum_r q_r \leq Q_{\max}, \ \lambda_m \in [0,1] \tag{11.9}$$

$$\frac{\sum_{g\in\tilde{\tilde{G}}} P_g}{\sum_{g\in G} P_g} \geq R_{\min} \tag{11.10}$$

$$\text{Integrity}_{b\acute{b}} \geq \sigma_{\min}, \text{ for all communication links between buses } b \text{ and } \acute{b} \tag{11.11}$$

where D_{pb} and D_{qb} are, respectively, the active and reactive powers demand at bth bus; V_b and $V_{\acute{b}}$ are the voltage magnitudes at buses b and $\acute{b}$, respectively; $G_{b\acute{b}}$ and $B_{b\acute{b}}$ are, respectively, the conductance and susceptance of power line connecting buses b and $\acute{b}$; τ_a is the threshold of the false data detection mechanism; $Q_{\max}$ represents the total available QC resources that can be used across all quantum analytics tasks in the system; $\tilde{\tilde{G}}$ is a subset of generators that are RESs (e.g., solar, wind, and hydro); $R_{\min}$ is the minimum required renewable energy share, a policy-driven parameter; and $\sigma_{\min}$ denotes the minimum acceptable integrity threshold, which is a system-wide policy parameter that dictates how strict the integrity requirement is, for example, 95% can be used for mission-critical grid protection data, and 70% can be used for lower priority data forecasting.

The rest of the constraints to ensure safe, reliable, and secure operation of the power grid either in the normal operation or while being targeted by FDIAs are written in (11.12)–(11.20), where $P^{\text{Th}}_{g(\acute{g})}$ is the output power of gth thermal (nondiesel) generator and $\acute{g}$th diesel generator; $P^{\text{Th}}_{g(\acute{g}),\min}$ and $P^{\text{Th}}_{g(\acute{g}),\max}$ are, respectively, the minimum and maximum boundaries of output active power associated with gth thermal (nondiesel) generator and $\acute{g}$th diesel generator; likewise $Q^{\text{Th}}_{g(\acute{g})}$, $Q^{\text{Th}}_{g(\acute{g}),\min}$, and $Q^{\text{Th}}_{g(\acute{g}),\max}$ indicate the lower and upper limits of reactive power for gth thermal (nondiesel) generator and $\acute{g}$th diesel generator; $P^{\text{Solar}}_{\tilde{\tilde{g}}}$ is the contribution of $\tilde{\tilde{g}}$th photovoltaic (PV) plant, and Ψ is a subset of solar-based RESs; $P^{\text{Wind}}_{\tilde{\tilde{g}}}$ denotes the contribution of $\tilde{\tilde{g}}$th wind turbine, and Φ is a subset of wind-based RESs; ϑ_{PV} and ϑ_{WT} are, respectively, the control parameters limiting the contribution of solar plants and wind farms; V_b, $V_b^{\min}$, and $V_b^{\max}$ are the voltage magnitude of bth bus and its minimum and maximum boundaries; $I_{\text{trans},\varrho}$, $I^{\min}_{\text{trans}}$, and $I^{\max}_{\text{trans}}$ are the magnitude of current associated with ϱth transformer and its lower and upper boundaries; I^{line}_l is the current magnitude for lth transmission line; $I^{\text{line}}_{l,\max}$ is the maximum of current that can be tolerated by the lth line without triggering the overcurrent relays; likewise, S^{line}_l and $S^{\text{line}}_{l,\max}$ indicate the apparent power and its maximum limit associated with lth line; z is the set of measurements coming from the smart meters; $\tilde{x}$ is the estimated state variables of the power system; h is a nonlinear function that links the measurements sets to the state variables; and finally, π is the threshold of power system state estimator [33],

$$P^{\mathrm{Th}}_{g(\acute{g}),\min} \leq P^{\mathrm{Th}}_{g(\acute{g})} \leq P^{\mathrm{Th}}_{g(\acute{g}),\max}, \ \forall g \in G, \acute{G} \tag{11.12}$$

$$Q^{\mathrm{Th}}_{g(\acute{g}),\min} \leq Q^{\mathrm{Th}}_{g(\acute{g})} \leq Q^{\mathrm{Th}}_{g(\acute{g}),\max}, \ \forall g \in G, \acute{G} \tag{11.13}$$

$$\sum_{\tilde{\tilde{g}} \in \Psi} P^{\mathrm{Solar}}_{\tilde{\tilde{g}}} \leq \vartheta_{\mathrm{PV}} \times P_{\mathrm{Demand}}, \ \forall \tilde{\tilde{g}} \in \Psi \tag{11.14}$$

$$\sum_{\tilde{\tilde{g}} \in \Phi} P^{\mathrm{Wind}}_{\tilde{\tilde{g}}} \leq \vartheta_{\mathrm{WT}} \times P_{\mathrm{Demand}}, \ \forall \tilde{\tilde{g}} \in \Phi \tag{11.15}$$

$$V_b^{\min} \leq V_b \leq V_b^{\max}, \ \forall b \tag{11.16}$$

$$I^{\min}_{\mathrm{trans}} \leq I_{\mathrm{trans},\varrho} \leq I^{\max}_{\mathrm{trans}}, \ \forall \varrho \tag{11.17}$$

$$\left|I_l^{\mathrm{line}}\right| \leq I^{\mathrm{line}}_{l,\max} \tag{11.18}$$

$$\left|S_l^{\mathrm{line}}\right| \leq S^{\mathrm{line}}_{l,\max} \tag{11.19}$$

$$\|z - h(\tilde{x})\| \leq \pi \tag{11.20}$$

It is noted that (11.12) and (11.13) denote the boundaries of output power for thermal power plants and nonrenewable DERs (i.e., diesel generators) in the system, (11.14) limits the contribution of all PV plants to the ϑ_{PV} percent of the total demand of the system, (11.15) limits the power generated by all wind turbines to a fraction of the total system demand, which is denoted by ϑ_{WT}, (11.16)–(11.19) ensure the rest of security constraints of the system, and finally, (11.20) limits the data manipulations into the sensors' measurements are to a threshold of π, which is the threshold of bad data detection algorithms embedded in the state estimation.

11.3.2 Proposed Quantum-based Algorithm to Optimize the Objective Functions

This subsection presents a quantum-enhanced variant of the standard particle swarm optimization (PSO) to solve the introduced optimization problems in Section 11.3. Hence, the proposed optimizer in this chapter is called hybrid quantum-enhanced particle swarm optimization-energy valley optimizer (HQEPSO-EVO), which is mathematically written in (11.21)–(11.28), in which

(11.21) and (11.22) presents the main equations of the standard PSO algorithm [34], (11.23) enhances the standard PSO algorithm via uniformly sampled stochastic term simulating quantum perturbations having small random shifts [35], and (11.24)–(11.28) provides the fundamentals of the energy valley optimizer (EVO) [36]. It is noted that in the qubit representation, as presented in (11.23), particles do not own directly positions/velocities but rather probability amplitudes Θ_i. In other words, a particle's state is encoded as a qubit, updated via quantum gate rotation, enabling exploration over a probabilistic state space rather than fixed coordinates,

$$\chi_{i(\tau+1)} = \chi_{i(\tau)} + w \times v_{i(\tau)} + \mathrm{lc}_1 \times \mathrm{rd}(\cdot) \times \left(\chi_{i(\tau)}^{\mathrm{best}} - \chi_{i(\tau)}\right) + \mathrm{lc}_2 \times \mathrm{rd}(\cdot) \times \left(\chi_{(\tau)}^{\mathrm{glob}} - \chi_{i(\tau)}\right) \tag{11.21}$$

$$w = \overline{w} - \tau \times \left(\frac{\overline{w} - \underline{w}}{\overline{\tau}}\right) \tag{11.22}$$

$$\Theta_{i(\tau+1)} = \Theta_{i(\tau)} + \sigma \times \left(\sin(\sqrt{\mathrm{GBP}})^{-1} - \Theta_{i(\tau)}\right) \tag{11.23}$$

$$\chi = \begin{bmatrix} \chi_1 \\ \chi_2 \\ \vdots \\ \chi_i \\ \vdots \\ \chi_{\mathrm{pop}} \end{bmatrix} = \begin{bmatrix} \chi_{1,1} & \cdots & \chi_{1,d} \\ \vdots & \ddots & \vdots \\ \chi_{\mathrm{pop},1} & \cdots & \chi_{\mathrm{pop},d} \end{bmatrix} \tag{11.24}$$

$$\chi_{i,j(\tau)} = \underline{\chi}_{i,j(\tau)} + \mathrm{rd}(\cdot) \times \left(\overline{\chi}_{i,j(\tau)} - \underline{\chi}_{i,j(\tau)}\right) \tag{11.25}$$

$$\rho = \frac{\sum_{i=1}^{\mathrm{pop}} \mu_i}{\mathrm{pop}} \tag{11.26}$$

$$v_i = \frac{\mu_i - v_i^{\mathrm{best}}}{v_i^{\mathrm{worst}} - v_i^{\mathrm{best}}} \tag{11.27}$$

$$\chi_{i,j(\tau+1)} = \chi_i\left(\chi_{v_i^{\mathrm{best}}}\left(\chi_{i,j(\tau)}\right)\right) \tag{11.28}$$

where $\chi_{i(\tau)}$ and $\chi_{i(\tau+1)}$ are, respectively, the positions of ith particle at τth and $(\tau+1)$th iterations; w, $\underline{w}$, and $\overline{w}$, are, respectively, the inertia weight of particles and its minimum and maximum boundaries; lc_1 and lc_2 are, respectively, the individual and social learning factors of particles; $\chi_{i(\tau)}^{\text{best}}$ and $\chi_{(\tau)}^{\text{glob}}$ are, respectively, the best position for ith particle and the best position obtained by the entire population at τth iteration; rd(·) is a random number in the range of (0,1); $\overline{\tau}$ is the maximum number of iterations preset by the operator; $\Theta_{i(\tau)}$ and $\Theta_{i(\tau+1)}$ are, respectively, quantum angles encoding the state for ith particle at τth and $(\tau + 1)$th iterations; GBP is the global best probability amplitude for ith particle; σ is the learning rate or quantum rotation factor; pop is the size of the population; d denotes the dimension of the optimization problem; $\underline{\chi}_{i,j(\tau)}$ and $\overline{\chi}_{i,j(\tau)}$ are, respectively, the minimum and maximum boundaries associated with ith particle in the jth dimension at τth iteration; ρ is the enrichment limit for the particles; μ_i is the neutron enrichment level associated with ith particle; υ_i, $\upsilon_i^{\text{worst}}$, and υ_i^{best} are, respectively, the stability level of ith particle and its worst and best levels so far; and $\chi_{i,j(\tau+1)}$ is the final position of the ith particle in the jth dimension at $(\tau + 1)$th iteration.

To obtain a better perspective, Figure 11.2 presents the flowchart of the proposed HQEPSO-EVO in handling the optimization problems in this chapter.

11.3.3 Handing Multi-objective Optimization Problems

In this study, to enable simultaneous minimization of conflicting objective functions (11.1)–(11.5) within the unified framework, an aggregation-based scalarization approach is utilized. This allows the multi-objective problems to be transformed into a single-objective optimization model, while still capturing the trade-offs among all objectives. Generally, the original set of objective functions can be written as (11.29), where X represents the vector of decision variables; and OF_n denotes the nth objective function,

$$\text{opt}\{\text{OF}(X)\} = [\text{OF}_1(X), \text{OF}_2(X), \ldots, \text{OF}_n(X)] \tag{11.29}$$

To simultaneously optimize more than one objective function in this study, an aggregated objective function needs to be formed, as presented in (11.30), where $\widetilde{\text{OF}}_n(X)$ indicates the normalized form of the objective function $\text{OF}_n(X)$ to ensure scale comparability; μ_n is the penalty factor assigned to nth objective function in such a way that $\sum_n \mu_n = 1$,

$$\text{OF}(X) = \sum_{n=1}^{N_{\text{OF}}} \mu_n \times \widetilde{\text{OF}}_n(X) \tag{11.30}$$

Accordingly, to avoid different units and magnitudes, each objective function $\text{OF}_n(X)$ is first normalized using a min-max normalization strategy, as provided in (11.31), where $\text{OF}_n^{\min}$ and $\text{OF}_n^{\max}$ are, respectively, the minimum and maximum values of the nth objective function. It is noted that $\text{OF}_n^{\min}$ is obtained via

- Zero-day exploit simulations are launched against DER control at buses 21 and 32.
- Adversarial ML attacks are tested on the classical anomaly detection pipeline of the DT layer.

11.4.5 DT Layer Configuration

The DT platform is initialized to run continuous simulations of load flow and contingency analysis, asset health monitoring, renewable output forecasting, and voltage and frequency stability. In addition, cybersecurity modules, embedded into the proposed framework, include quantum algorithms for FDIA detection and for integrity verification.

11.4.6 Quantum Computing Interface

A quantum simulator interfaces with the DT via a classical-quantum bridge. Toward that end, the quantum analytics tasks initialized include anomaly detection using variational quantum optimization for optimal power dispatch during attacks or contingencies for buses with control coordination links (e.g., between buses 1↔3, 10↔15, 25↔29, and 36↔39).

11.4.7 Control and Decision-making Setup

Autonomous decisions simulated include grid reconfiguration in response to contingencies or malicious cyberattacks, DER activation and curtailment, and dynamic system adjustments to flatten the load curve. In addition, the proposed QADT loop adjusts DT parameters in real time based on the outputs of the quantum optimization layer and also system operator feedback.

11.4.8 Governance, Compliance, and Interoperability Simulation

To simulate governance, compliance, and interoperability aspects within the proposed QADT framework, the modified IEEE 39-bus test system incorporates several CESN-aligned regulatory and ethical constraints. In this regard, the emission caps and renewable penetration targets are enforced through policy-driven parameters, such as maintaining a minimum of 60% renewable share in total generation (see constraints [11.14] and [11.15]). Data privacy is modeled via selective access control across different system layers, reflecting jurisdictional boundaries and operator roles. Interoperability between legacy grid infrastructure and modern digital components is emulated by introducing heterogeneous communication protocols and control interfaces. Additionally, for extensibility toward future research directions, a blockchain-based data integrity verification mechanism can simply be included in the proposed QADT framework to simulate transparent and decentralized trust in energy transactions. Together, these elements provide a realistic testing ground for evaluating how the proposed hybrid framework complies with sustainability mandates while ensuring secure and interoperable grid operations.

To obtain a better perspective about the introduced case study, Figure 11.3 illustrates the single-line diagram of the modified IEEE 39-bus New England test system. In addition, Table 11.2 presents detailed information about each bus in the system and its assets.

11.4.9 Necessary Data for Simulating the Optimization Problems

The solar systems integrated into the case study, as illustrated in Figure 11.3, are modeled based on the specifications of Kyocera solar modules [38], each with a rated power output of 135 W and a cell efficiency of 12.9%. These modules offer a balanced trade-off between performance and cost, making them suitable for distributed energy integration. Each solar plant installation is allocated a land area

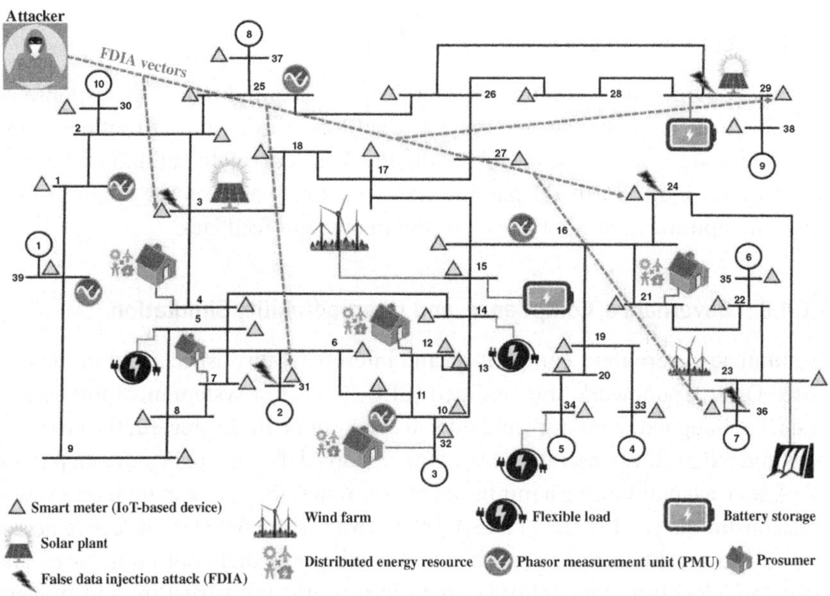

Figure 11.3 The modified IEEE 39-bus New England test system, including renewable energy, energy storage systems, flexible loads, PMUs, etc.

Table 11.2 Outline of the modified IEEE 39-bus New England's assets.

Bus no.	Modification/role
1	PMU installed; QKD link to bus #3
3	100 MW PV plant installed; SCADA + IoT; target of FDIA; QKD with bus #1
4	DER integration (rooftop PV modules); prosumer behavior modeled
5	Flexible load enabled for demand response simulation
7	DER integration (rooftop PV modules); prosumer bus
10	PMU installed; SCADA target of FDIA; QKD with bus #15
12	DER integration: dynamic load management
14	Flexible load modeled
15	120 MW wind farm installed; battery energy storage; QKD with bus #10
16	PMU installed; critical for observability
21	Prosumer simulation with smart meter; zero-day attack target
23	80 MW hydro plant installed; seasonal dispatch flexibility
25	PMU installed; QKD with bus #29
29	75 MW PV plant installed + battery energy storage; QKD with bus #25
32	DER bus with variable demand/generation mix
34	Flexible load for renewable alignment
36	100 MW wind farm installed; QKD with bus #39
39	PMU installed; QKD with bus #36

of 5,000 m^2, which reflects a realistic spatial requirement that accounts for optimal panel spacing to reduce shading effects and to allow for maintenance access.

To convert the DC output from the solar arrays into usable AC power, Xantrex inverters [39] are employed. These inverters are selected based on their high operational efficiency of 92% and their rated capacity of 1.5 kW, ensuring minimal energy loss during conversion and compatibility with the distributed scale of the solar plants. The inverter specifications are chosen to maximize energy yield while maintaining operational reliability in decentralized power systems.

For wind energy integration, wind turbines manufactured by Bonus [40] are considered, with a rated power of 300 kW and a rotor diameter of 33.4 m. The chosen turbine model provides an effective compromise between swept area and energy capture, particularly in medium-wind-speed regions, thereby ensuring consistent generation performance across the case study. The turbine parameters are selected to align with practical deployment scenarios within prosumer-based distribution systems.

To accurately model the renewable energy output, site-specific meteorological data—including solar irradiance and wind speed—are sourced from [41, 42]. This empirical data ensures a realistic simulation of renewable energy variability and supports the accurate forecasting of generation profiles over time.

In this study, five prosumer communities are integrated into the modified IEEE 39-bus system at buses #4, #7, #12, #21, and #32, as shown in Figure 11.3. Each community comprises 10 residential or mixed-use prosumers equipped with rooftop solar panels, a small-scale wind turbine, and battery energy storage. The installed solar/wind capacity per prosumer ranges between 30 and 90 kW, resulting in an aggregated solar/wind capacity of 0.6–1 MW per bus. Battery storage capacities are assumed to be 50 kWh per prosumer with a maximum discharge rate of 8.5 kW. The daily load profile for each prosumer follows a realistic consumption pattern: a base load of 0.5 kW during night hours (00:00–06:00 a.m.), morning peak of 3 kW (07:00–09:00 a.m.), midday dip to 1.5 kW (10:00 a.m.–02:00 p.m.), evening peak of 4 kW (06:00–09:00 p.m.), and gradual decline thereafter. Solar module generation is modeled using a bell-shaped profile with a peak output between 11:00 a.m. and 02:00 p.m., corresponding to clear-sky solar irradiance conditions, with zero generation during nighttime. Geographically, these buses represent diverse areas within the modified IEEE 39-bus network (i.e., the case study), with buses #4 and #7 assumed to be in urban zones with high load density, while buses #12, #21, and #32 are modeled as semiurban or rural areas with greater rooftop solar/wind adoption potential. These prosumer parameters are synthetically generated to support the simulation of DERs integration, and localized demand-side flexibility within transmission-level power system studies.

In addition to the prosumer communities, two large-scale battery storage systems are integrated at buses #15 and #29 to support grid flexibility and peak load management. Each unit has a rated energy capacity of 11 MWh and a maximum charge/discharge power of 4.5 MW, with a round-trip efficiency of 92%. The battery unit at bus #15 is modeled to operate in a load-heavy urban zone, providing peak shaving and frequency support, while the one installed at bus #29, located in a generation-rich area, primarily facilitates renewable energy firming and congestion management. Both systems are assumed to operate under a daily cycle, charging during off-peak hours (typically 00:00–06:00 a.m.) and discharging during peak demand periods (05:00–09:00 p.m.).

Finally, the foundational electrical parameters of the IEEE 39-bus test system—such as line impedances, bus connectivity, and transformer ratings—are adopted from [37], ensuring that the test system maintains consistency with standard benchmarking practices in power system studies.

11.5 Initialization, Simulation Results, and Analysis

11.5.1 Initialization Parameters and Simulation Setup

Table 11.3 presents the initialization of the case study to evaluate the effectiveness of the proposed QADT framework in this chapter.

It is noted that the performance of each simulation scenario, introduced in the following subsection, is evaluated against the multi-objective functions defined in Section 11.3, with particular attention to cybersecurity robustness (see objective function [11.2]), resilience (see objective function [11.3]), and sustainability (refer to objective function [11.5]).

11.5.2 Simulation Scenarios, Obtained Results, and Analysis

To demonstrate the effectiveness of the proposed QADT framework, it is important to design simulation scenarios that reflect realistic and strategically challenging conditions. Table 11.4 tabulates three scenarios tailored to the problem formulation, presented in Section 11.3.

The objective of Scenario I is to evaluate the proposed QADT's capability to detect and mitigate advanced cyberattacks in a highly renewable integrated system. To this end, a coordinated FDIA targets state estimation and control signals in the modified IEEE 39-bus test system (see Figure 11.3), considering variable renewable energy penetration (e.g., 60–80% from wind and solar). Hence,

Table 11.3 Parameter setting for all simulated scenarios.

Parameter	Value/configuration
Renewable penetration target ($R_{\min}$)	60% of total generation
QKD refresh latency threshold (ξ)	50 ms
FDIA detection threshold (τ_a)	≥95% TPR
Qubit resource limit ($Q_{\max}$)	128 logical qubits (simulated)
Carbon intensity weights (ϵ_g)	Scaled by generator type (coal > gas > hydro/solar)
Simulation time horizon	24 h (time resolution: 15 min)
Attack injection schedule	Randomized over a 12-h period

Table 11.4 Outline of the performance evaluation scenarios in this chapter.

Scenario #	Focus area	Cyberattack or challenge	Evaluation metrics/remarks
I	Cyberattack	Coordinated FDIA	Threat detection and system resilience
II	Operational optimization	Communication delay	Dispatch efficiency and control robustness
III	ML security	Zero-day ML attack	Model robustness and quantum ML accuracy

the attacker attempts to manipulate load or generation data by injecting small amounts of false data into the smart meters without triggering traditional anomaly detection, provided in (11.20).

In Scenario II, the proposed QADT's real-time optimization and control robustness in the presence of communication latency and decentralized prosumer actions are evaluated. In this regard, time-varying communication delays and partial data loss between distributed DT agents within the modified IEEE 39-bus test system are scrutinized.

Finally, Scenario III investigates the proposed QADT's resilience against unknown, sophisticated threats leveraging adversarial ML techniques via introducing misleading data to cause misclassification or delayed threat detection.

11.5.2.1 Scenario I: Coordinated FDIA Under High Renewable Penetration

In this scenario, a coordinated FDIA is simulated under high renewable penetration (i.e., 70%) across several buses in the IEEE 39-bus system. The attack targets state estimation processes, aiming to mislead the control center without triggering conventional alarms. The objective is to compare the performance of a traditional DT model with the proposed QADT in terms of detection accuracy, latency, and false positives. Table 11.5 presents the simulation results. According to this table, one can perceive that the proposed QADT significantly outperforms traditional DT systems for the following justifications: detection accuracy increased by approximately 14%, while detection latency decreased by more than 60%, showcasing the capability of quantum-enhanced anomaly detection models to rapidly and accurately identify cyber intrusions. Additionally, the FPR was reduced to 2.1%, indicating that the proposed QADT offers not only speed and accuracy, which are key common features in cybersecure energy systems, but also reliability in filtering out benign fluctuations caused by renewable intermittency to enhance the power quality for the entire system.

Table 11.5 FDIA detection performance comparison.

Method	Detection accuracy (%)	Detection latency (ms)	False positive rate (%)
Traditional DT + ML	84.7	195	8.3
Proposed QADT	96.5	73	2.1

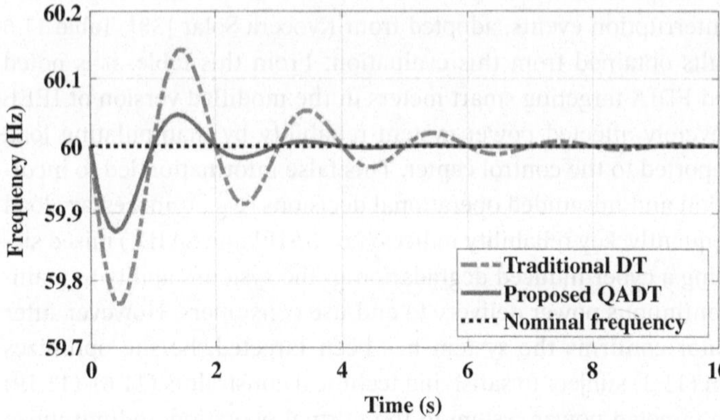

Figure 11.4 Frequency response during coordinated FDIA in Scenario I.

To obtain a better perspective, Figure 11.4 depicts system frequency dynamics during a coordinated FDIA on a highly penetration of renewable energy, as outlined earlier. Although frequency itself is not a direct target of data manipulations, the physical frequency of the modified IEEE 39-bus test system dynamically responds to the consequences of the FDIA targeting measurement data associated with smart meters (see Figure 11.3). According to Figure 11.4, it can be gathered that with traditional DT systems, the frequency shows slower damping and higher deviation from nominal values, indicating delayed detection and inadequate response coordination. Contrary to this, the proposed QADT-enabled system exhibits a significantly faster return to nominal frequency, reflecting improved detection latency and real-time corrective control via optimizing objective functions (11.2) and (11.4). From a power system operation standpoint, this enhanced frequency recovery underlines the critical importance of rapid cyber-physical response in maintaining system stability and avoiding under-frequency load shedding during cyber-induced disturbances.

In order to investigate how the coordinated FDIA, targeting smart meters, affects power system reliability, measured by SAIFI and SAIDI, in the IEEE 39-bus test system, a typical daily load profile in the modified IEEE 39-bus system (i.e., Figure 11.3) is simulated. Toward this end, the coordinated FDIA is launched manipulating smart meters data (i.e., load date) associated with buses #3, #15, #17, # 18, and # 24 and causing incorrect load reporting to the system operator. This misleads the control center into re-dispatching generation and making faulty protection decisions, resulting in extended customer interruptions. Using historical assumptions for number of customers per load bus, duration of each outage event, and number of interruption events, adopted from Kyocera Solar [38], Table 11.6 presents the results obtained from this evaluation. From this table, it is noted that a coordinated FDIA targeting smart meters in the modified version of IEEE 39-bus system severely affected power system reliability by manipulating load measurements reported to the control center. This false information led to incorrect state estimation and misguided operational decisions (e.g., unnecessary load shedding). Consequently, key reliability indices (i.e., SAIFI and SAIDI) raised significantly, reflecting a cyber-induced degradation in the system's ability to maintain stable and continuous power delivery to end-use consumers. However, after the system operator confirms the system has been targeted, he/she optimizes objective function (11.3) subject to satisfying technical constraints (11.6)–(11.19) to bring back the targeted power system to the normal operation and minimize the inconvenience of end-use customers by at least 80%.

Furthermore, Figure 11.5 depicts the voltage profile of the modified IEEE 39-bus test system before the coordinated FDIA targeting smart meters by injecting false load data, after the cyberattack poses operational issues, and after the

Table 11.6 SAIFI and SAIDI indices on the modified IEEE 39-bus test system before and after the coordinated FDIA, as well as after optimizing the system reliability by the operator.

Metric	Before FDIA (normal operation)	After FDIA	After handling system optimization on the targeted system
SAIFI (interruptions/customer/year)	0.3	2.8	0.5
SAIDI (min/year)	34	219	41
Buses affected	—	Buses #3, #15, #17, # 18, and # 24	—
Load shedding status (MW)	—	120 (unnecessary)	—

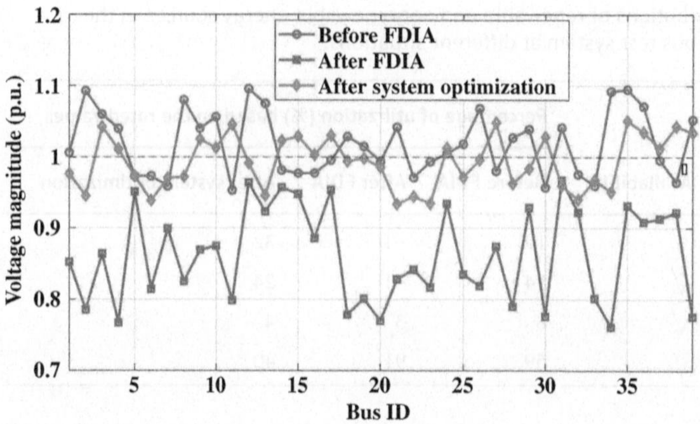

Figure 11.5 Voltage profile of the modified IEEE 39-bus test system from different standpoints.

system operator optimizes the targeted system via solving objective functions (11.1), (11.3), and (11.5). From Figure 11.5, it can be perceived that before the FDIA, all bus voltages are within the nominal range of [0.95–1.1] pu, indicating a healthy and well-regulated power grid. However, following the FDIA, voltage levels across most buses drop to the range of [0.76–0.97] pu, reflecting a system under operational issue due to inaccurate control measures caused by falsified measurements. However, after handling the objective functions, the voltage profile improves substantially, with all buses restored to a range of [0.93–1.06] pu.

Once again, Figure 11.5 demonstrates the potential of advanced optimization techniques to not only recover system performance but also enhance voltage stability beyond pre-attack periods.

To show the effectiveness of the proposed QADT framework toward deep decarbonization via optimizing objective functions (11.1) and (11.5), Table 11.7 presents the utilization of the available renewable energy in the in the modified IEEE 39-bus test system after the coordinated FDIA and after handling the cyberattack by the system operator. According to this table, it can be inferred that before the coordinated FDIA, renewable energy played a vital role in serving the system demand, reducing dependence on fossil fuels and lowering emissions. However, the attack disrupted this balance, leading to a temporary decline in clean energy integration due to the involvement of false information in the process of state estimation and forcing greater reliance on thermal power plants to meet the system demand. Nevertheless, following the system optimization process via minimizing objective functions (11.1) and (11.5), the availability of RESs improved, allowing for reduced dependence on thermal generation and a return to cleaner, more sustainable energy production.

Table 11.7 Contributions of renewable and nonrenewable energy sources in the modified IEEE 39-bus test system at different situations.

		Percentage of utilization (%) based on the rated values		
Energy sources	**Availability**	**Before FDIA**	**After FDIA**	**After system optimization**
Solar plant	√	19	~1	32
Wind farm	√	14	~1	24
Diesel engine	√	8	3	4
Thermal plant	√	59	91	40

Table 11.8 Dispatch optimization under communication delay.

Optimization metric	Classical optimization approach	Proposed HQEPSO-EVO
Total generation cost ($/h)	14,875	13,620
Voltage stability index	0.87	0.94
Avg. load shedding (MW)	12.3	5.7
Computation time (ms)	147	42

11.5.2.2 Scenario II: Real-time Optimal Dispatch with Communication Delay

This scenario addresses the proposed QADT's ability to maintain efficient and stable dispatch in a decentralized environment subject to communication delays. DERs and prosumers were integrated across the grid (see Figure 11.3), and latency ranging from 150–250 ms is introduced in control signals. The QADT uses the proposed HQEPSO-EVO, as provided in (11.21)–(11.28), for dispatching the system, while the baseline used conventional linear programming. The obtained simulation results are tabulated in Table 11.8, illustrating that the proposed QADT achieved an 8.4% reduction in generation cost compared to traditional optimization. The voltage stability index improved from 0.87 to 0.94, indicating better voltage regulation under delayed communications. Moreover, the average load shedding dropped by over 50%, showing the ability of the proposed QADT to better utilize distributed assets despite delayed data. The proposed quantum optimizer (HQEPSO-EVO) demonstrated a significantly shorter computation time—over 70% faster than the classical optimization method—thereby improving the real-time responsiveness of the framework for power system operators in

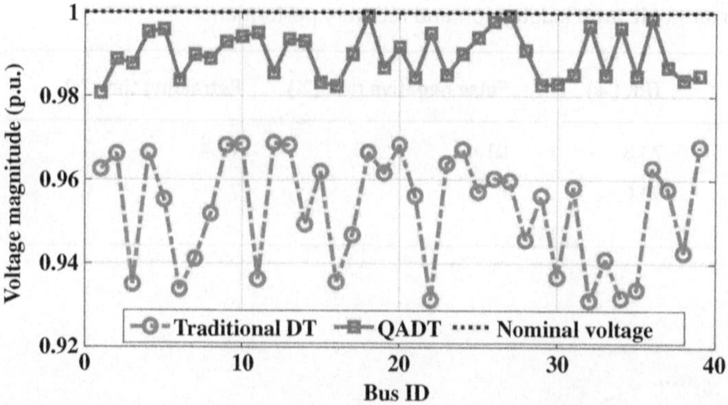

Figure 11.6 Voltage profile comparison in the modified IEEE 39-bus test system under communication delay.

targeted power systems, where seconds matter in recovering the system to normal operation.

In addition, Figure 11.6 illustrates the voltage profile of the system buses, highlighting the impact of communication latency on optimal dispatch and voltage regulation of the modified IEEE 39-bus test system. According to this figure, it can be perceived that the traditional DT approach showed slight undervoltages across multiple buses, with values drifting below 0.96 pu, likely due to delayed data processing and reactive power imbalance. In contrast, the proposed QADT framework maintains voltages closer to the nominal 1 pu, showcasing superior dispatch decisions and real-time optimization despite communication delays. Operationally, this improved voltage regulation is essential to prevent voltage collapse, maintain power quality, and reduce unnecessary reactive compensation in increasingly decentralized grids with active prosumers and DERs.

11.5.2.3 Scenario III: Zero-day ML Attack on Forecasting Models

In the third scenario, a zero-day adversarial ML attack is simulated, targeting the forecasting models embedded in the DT environment. Small, imperceptible perturbations are injected into the input data to mislead the model into underestimating future load demand. The scenario tested the detection and retraining capabilities of the proposed QADT framework. As provided in Table 11.9, the proposed QADT showed a substantial improvement in adversarial threat detection. To obtain a better perspective, the TPR increased by nearly 19%, and the false negative rate was reduced by more than 70%, highlighting the robustness of quantum-enhanced classification against adversarial ML attacks. Furthermore, the retraining time was significantly shorter, enabling near real-time recovery of forecasting accuracy, which is critical for systems that rely heavily on predictive

Table 11.9 Adversarial ML attack detection and recovery performance.

Detection method	TPR (%)	False negative rate (%)	Retraining time (s)
Classical ML	73.8	21.4	19.4
QADT with quantum ML	93.6	6.2	7.2

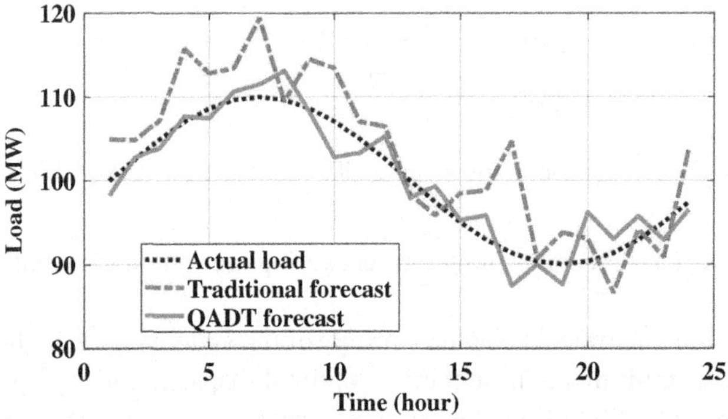

Figure 11.7 Forecast accuracy on a sample load profile before and after QADT retraining associated with Scenario III.

analytics for operational planning, such as the future smart microgrids integrated with renewable energy.

To obtain a better view about Scenario III, Figure 11.7 depicts 24-hour load forecasts before and after the QADT system retrained its prediction model in response to an adversarial ML attack. From this figure, one can infer that the traditional forecast deviates significantly from actual load, with persistent overestimation due to the injected adversarial bias. This misprediction can lead to overcommitment of generation resources and inefficient unit dispatch. Post-retraining, the proposed QADT-enhanced forecast closely tracks actual load, with reduced variance and bias. From a system operations perspective, accurate load forecasting under the threats of cyber incidents ensures economically optimal dispatch, reduced reserve requirements, and increased reliability, particularly vital in grids relying on variable renewables and intelligent forecasting engines.

11.5.3 Discussions

Across the three simulated scenarios in Tables 11.5–11.9, the proposed QADT framework consistently demonstrated substantial performance improvements over traditional DT approaches in key areas related to cybersecurity, operational optimization, and system resilience. In the first scenario, where a coordinated FDIA was launched under conditions of high renewable energy penetration, the proposed QADT reduced detection latency by approximately 63%, enabling faster and more accurate anomaly identification and response. In the second scenario, which introduced communication delays in a decentralized grid environment, the proposed QADT achieved an 8.4% reduction in total generation cost through quantum-enhanced dispatch optimization and lowered average load shedding by 53.6%, indicating improved utilization of DERs and enhanced voltage stability. The quantum optimizer also required significantly less computation time, which is critical for real-time control in latency-sensitive systems. In the third scenario, the proposed QADT's use of quantum ML enabled a dramatic reduction of over 70% in the false negative rate during an adversarial ML attack, while retraining time for compromised forecasting models was reduced by more than 63%, ensuring rapid system adaptation and recovery. These integrated results highlight the versatility and robustness of the proposed QADT across different operational stress conditions, confirming its value as a forward-looking solution for securing and optimizing CPESs under the complex dynamics of decarbonized, digitalized, and decentralized power networks.

In addition, Figures 11.4–11.7 collectively illustrate the multifaceted benefits of the proposed QADT framework in enhancing the cyber-physical performance of modern energy systems aligned with CESN objectives. To be specific, the superior frequency recovery under coordinated FDIAs clearly demonstrates the proposed QADT's enhanced anomaly detection and real-time corrective action capabilities. As a case in point, the accelerated stabilization of frequency not only confirms the framework's resilience to cyber threats but also underscores its ability to preserve dynamic stability in high-renewable environments where inertia is inherently lower. This aligns directly with the cybersecurity and operational integrity goals of the proposed QADT framework. In addition, the voltage profile comparison across the IEEE 39-bus test system under communication delays highlights the proposed QADT's strength in maintaining optimal dispatch and voltage regulation in decentralized, latency-prone networks such as the case study of this chapter. The improved voltage consistency underlines the framework's capability to support DER integration and real-time control in the face of data latency, which is a critical requirement for future smart grids under increasing digitalization. From a system operation standpoint, this translates to lower power losses, reduced voltage violations, and improved load serving capability across the network. Moreover, the proposed QADT framework not only reduces forecast error but also adapts swiftly to data manipulation, showcasing a novel defense mechanism that is absent in traditional DT architectures. This is crucial in energy systems where accurate forecasting drives renewable scheduling and demand response.

Collectively, these obtained results offer evidence of how the proposed QADT framework strengthens predictive accuracy, dispatch efficiency, and cyber resilience, three key segments of intelligent and adaptive energy infrastructure in deep decarbonization of modern power systems. More importantly, they reinforce the central claim of this chapter that the convergence of QC with secured DTs offers a robust, scalable, and promising solution for securing and optimizing CPESs under the imperatives of the CESN.

11.5.4 Positive Effects of the Proposed QADT Framework on Deep Decarbonization

The transition toward deeply decarbonized energy systems, in alignment with the CESN, demands solutions that can simultaneously address cybersecurity vulnerabilities, operational complexity, and renewable integration challenges. The proposed QADT framework directly supports these decarbonization imperatives by enhancing the intelligence, resilience, and adaptability of modern CPESs. This subsection consolidates the key impacts of the proposed QADT architecture on deep decarbonization across multiple dimensions, referencing the simulation results and analyses presented in Section 11.5.

Strengthening cybersecurity in-rich environments: decarbonized power systems are inherently more vulnerable to cyber threats due to increased connectivity, data exchange, and distributed control mechanisms. The proposed QADT framework introduces quantum-enhanced anomaly detection along with quantum optimization algorithms, which significantly improve the detection and response to FDIAs in real time. The key features are itemized hereunder:

- Reduction in detection latency by 63% during coordinated FDIAs in high renewable penetration scenarios (see Table 11.5).
- Over 70% reduction in FNR under adversarial ML attacks, indicating enhanced detection accuracy (see Table 11.9).
- Figures 11.4 and 11.5 verify improved frequency recovery post-FDIA, validating QADT's contribution to dynamic cyber-physical resilience. Therefore, the decarbonization impact statement is as follows:
  - Enhancing cybersecurity preserves system reliability, which is essential for integrating volatile RESs and avoiding fossil-fuel-based fallback operations during system disruptions.

Improving operational efficiency and resource optimization: high levels of renewable integration introduce uncertainty in generation patterns, requiring real-time optimization for dispatch and control. The quantum-augmented optimization engine within the proposed QADT yields measurable gains such as:

- The reduction of 8.4% in total generation cost under communication delay scenarios (see Table 11.7).

- The reduction of 53.6% in load shedding, indicating improved DER coordination and load-serving capability (see Table 11.7).
- Faster quantum-assisted computation times support latency-sensitive operational environments. Accordingly, the following statement outlines the impact of decarbonization:
  - Lowering generation costs and reducing load shedding allows for more effective utilization of RESs and minimizes reliance on peaking fossil units, thereby reducing overall carbon emissions.

Enhancing system resilience against dynamic instabilities: renewable-dominated grids typically suffer from low inertia, making them more vulnerable to frequency and voltage violations. The proposed QADT framework demonstrates enhanced system stability. As a case in point:

- Figures 11.4–11.6 show accelerated frequency and voltage recovery during and after cyber-physical stress events, reflecting improved resilience.
- Enhanced control under delayed communication scenarios supports robust real-time voltage regulation, which is critical for secure DER operation (see Table 11.6). Therefore, decarbonization impact is outlined below:
  - Improved dynamic stability ensures that high levels of renewables can be integrated without compromising system integrity, reducing the need for backup fossil-fuel support.

Enabling scalable deployment in decentralized architectures: deep decarbonization is accompanied by increased decentralization—both in energy generation (e.g., rooftop PV, and wind farms) and control mechanisms. The proposed QADT framework is designed for scalable deployment. For instance:

- Demonstrated effectiveness in the modified IEEE 39-bus test system, reflecting realistic decentralized, renewable-rich grid conditions.
- Quantum-enhanced DT components support real-time decision-making even under high data throughput and complex system topologies. Therefore, the resulting decarbonization impact is summarized as follows:
  - Scalable DT deployment in decentralized grids supports widespread adoption of RESs, peer-to-peer trading, and prosumer participation—key elements of a net-zero energy ecosystem.

Enhancing predictive capabilities and forecasting accuracy: accurate forecasting is fundamental in scheduling renewables and managing demand-response programs. The proposed QADT framework leverages quantum algorithms to improve prediction reliability and adaptability. For example:

- The reduction of 63% reduction in retraining time for compromised models under adversarial conditions (see Table 11.9).

Table 11.10 Summary of deep decarbonization benefits.

Perspective	QADT impact summary	Decarbonization benefit
Cybersecurity	Faster detection, fewer false negatives, adaptive defenses	Reliable integration of RESs and digital assets
Operational efficiency	Lower costs, reduced load shedding, optimized DER usage	Minimization of fossil-fuel dispatch
System resilience	Faster recovery under cyber-physical stress, better voltage/frequency regulation	Stability in low-inertia, high-RES environments
Scalability and decentralization	Real-time control in distributed architectures	Support for prosumers and localized generation
Forecasting and adaptability	Robustness against data attacks, rapid model retraining	Accurate renewable scheduling and demand response

- Enhanced robustness to data manipulation improves trust in forecasting tools, as illustrated in Figure 11.7. Therefore, here is the statement detailing the decarbonization impact:
  - Reliable forecasting ensures optimal scheduling of renewable assets, reducing curtailment and enabling accurate grid balancing without carbon-intensive reserves.

Table 11.10 tabulates the provided information to facilitate readers' access to the core decarbonization effects of this chapter.

11.6 Conclusion, Policy Insights, Future Directions

This chapter presented a QADT framework as a promising solution for increasing cybersecurity and efficiency of CPESs. Through three designed simulation scenarios, comprising coordinated FDIAs, communication delays, and zero-day ML threats, the proposed framework consistently outperformed traditional systems. In this regard, the main outcomes include a 63% reduction in cyberattack detection latency, an 8.4% cut in dispatch costs under latency-prone environments, and a 70% drop in false negatives during adversarial ML attacks. This is where the importance of the proposed framework in meeting the critical demands of modern, renewable-integrated power systems under the concept of CESN comes under the spotlight. Accordingly, the main achievements of this chapter can be itemized hereunder:

- Demonstrated the feasibility and effectiveness of integrating QC techniques into DT architectures for CPESs, forming the novel QADT framework.
- Validated the QADT's adaptability across diverse operational stress scenarios, including cyberattacks, communication delays, and adversarial ML threats.
- Showcased a hybrid approach that leverages quantum optimization and quantum ML to enhance system security, control performance, and operational efficiency simultaneously.
- Highlighted the framework's alignment with future energy system demands, specifically targeting challenges introduced by decentralization, decarbonization, and digitalization.
- Provided empirical support for real-time control improvements using quantum methods, contributing toward latency-sensitive, DER-rich grid environments.
- Introduced a novel cyber-physical resilience strategy capable of maintaining grid stability, minimizing load shedding, and ensuring robust system adaptation in the presence of both physical and cyber disruptions.
- Supported the QADT's role in achieving CESN's objectives, particularly in enhancing anomaly detection, reducing forecasting errors, and ensuring secure integration of renewable resources.
- Established a scalable, future-ready architecture for intelligent grid management that can adapt to increasing complexity in data, control, and threat vectors.

From a policy and regulatory standpoint, the findings of this study support the urgent need to revise cybersecurity/operational standards for smart power grids to comprise hybrid architectures such as the proposed QADT in this chapter. Traditional reliability metrics (e.g., SAIFI, SAIDI) experienced severe degradation under coordinated FDIAs, emphasizing the growing importance of cyber-resilience as a dimension of system reliability. Regulatory agencies should prioritize frameworks that integrate three major aspects of anomaly detection, adaptive optimization, and real-time control under high penetrations of renewable energies. Moreover, the confirmed ability of the proposed QADT to reduce load shedding, enhance frequency recovery, and accelerate retraining times offers a compelling case for investing in next-generation digital infrastructure.

Although the proposed framework in this chapter was the first step, looking ahead, future steps of this research will explore integrating AI co-design, edge intelligence, and blockchain-based energy systems to further enhance the scalability and autonomy of the proposed QADT framework. Toward that end, the AI co-design will enable the collaborative development of DTs alongside AI and quantum models, ensuring that control, detection, and forecasting algorithms are holistically optimized across both physical and digital domains. In addition, edge intelligence, via deploying privacy-preserving inference models, will minimize system latency and maintain critical functionalities during central system failures or network congestion, caused by severe FDIAs bypassing the detection stage. Furthermore, the concept of blockchain energy systems will be leveraged to ensure secure, transparent, and tamper-free energy transactions and data exchanges among all the available energy resources (e.g., both renewable and nonrenewable), prosumers, aggregators, and control centers in deregulated environments. The integration of these technologies with the proposed QADT in this chapter will be crucial to building an adaptive, decentralized, and cyber-resilient framework that align with the long-term objectives of grid decarbonization.

References

1 M. Agabalaye-Rahvar, A. Mirzapour-Kamanaj, K. Zare, and A. Anvari-Moghaddam, "Hybrid interval-stochastic optimal operation framework of a multi-carrier microgrid in the presence of hybrid electric and hydrogen-based vehicles intelligent parking lot," in *Energy Systems Transition*, Power Systems, Springer, Cham, 2023, pp. 209–236. https://doi.org/10.1007/978-3-031-22186-6_8.

2 M. Daneshvar, B. Mohammadi-Ivatloo, and K. Zare, "An innovative sustainable multi-energy management framework for full renewable grid-tied microgrids under the transactive energy paradigm," *Sustainable Cities and Society*, vol. 120, p. 106167, 2025.

3 K. Taghizad-Tavana, H. S. Kheljani, S. H. Hosseini, M. Tarafdar-Hagh, and M. Daneshvar, "Multi-dimensional management of smart distribution networks: comparative analysis of box and polyhedral methods for modeling uncertainties," *Sustainable Cities and Society*, vol. 108, p. 105488, 2024.

4 E. Naderi and A. Asrari, "A remediation framework against false data injection cyberattacks targeting ULTC transformers to avoid voltage collapse in smart distribution networks," *IEEE Transactions on Industry Applications*, vol. 61, no. 3, pp. 5148–5160, 2025.

5 E. Naderi, A. Asrari, and M. Saffari, "Digital twin-enabled resilient network reconfiguration for cybersecurity of decarbonized power distribution systems," in *7th Global Power, Energy and Communication Conference (GPECOM)*, IEEE, 2025, pp. 768–772.

6 T. Morstyn, "Annealing-based quantum computing for combinatorial optimal power flow," *IEEE Transactions on Smart Grid*, vol. 14, no. 2, pp. 1093–1102, 2023.
7 J. Graham, M. Turner, and A. Elmaghraby, "Cybersecurity for the smart grid," *IEEE Smart Grid Bulletin*, 2016. [Online]. Available: https://smartgrid.ieee.org/bulletins/july-2016/cybersecurity-for-the-smart-grid
8 C. Alcaraz and J. Lopez, "Digital twin: a comprehensive survey of security threats," *IEEE Communications Surveys and Tutorials*, vol. 24, no. 3, pp. 1475–1503, 2022.
9 K. Josifovska, E. Yigitbas, and G. Engels, "Reference framework for digital twins within cyber-physical systems," in *IEEE/ACM 5th International Workshop on Software Engineering for Smart Cyber-Physical Systems (SEsCPS)*, IEEE, 2019, pp. 25–31.
10 Y.-Z. Lin, S. Shao, M. H. Rahman, M. Shafae, and P. Satam, "DT4I4-secure: digital twin framework for Industry 4.0 systems security," in *IEEE 14th Annual Ubiquitous Computing, Electronics & Mobile Communication Conference (UEMCON)*, IEEE, 2023, pp. 200–209.
11 Ö. Sen, F. Schmidtke, F. Carere, F. Santori, A. Ulbig, and A. Monti, "Investigating the cybersecurity of smart grids based on cyber-physical twin approach," in *IEEE International Conference on Communications, Control, and Computing Technologies for Smart Grids (SmartGridComm)*, IEEE, 2022, pp. 439–445.
12 A. Mohsin, H. Janicke, S. Nepal, and D. Holmes, "Digital twins and the future of their use enabling shift left and shift right cybersecurity operations," arXiv Preprint, arXiv:2309.13612, 2023.
13 A. Ajagekar and F. You, "Quantum computing for energy systems optimization: challenges and opportunities," *Energy*, vol. 179, pp. 76–89, 2019.
14 M. Fernández-Campoamor, C. O'Meara, G. Cortiana, V. Peric, and J. Bernabé-Moreno, "Community detection in electrical grids using quantum annealing," arXiv Preprint, arXiv:2112.08300v2, 2021.
15 N. K. Kandasamy, S. Venugopalan, T. K. Wong, and N. J. Leu, "An electric power digital twin for cyber security testing, research and education," *Computers and Electrical Engineering*, vol. 101, p. 108061, 2022.
16 S. Homaei, S. Band, A. Pescape, and A. Mosavi, "The role of artificial intelligence in digital twin's cybersecurity," *IEEE Access*, vol. 9, pp. 63131–63148, 2021.
17 IEEE SA, "Protecting our power: cybersecurity standards for distributed energy resources," *IEEE SA Blog*, 2024.
18 S. Pirandola et al., "Advances in quantum cryptography," *Advances in Optics and Photonics*, vol. 12, no. 4, pp. 1012–1236, 2024.

19 P. W. Shor, "Polynomial-time algorithms for prime factorization and discrete logarithms on a quantum computer," *SIAM Journal on Computing*, vol. 26, no. 5, pp. 1484–1509, 1997.

20 B. Wang, H. Zhou, G. Yang, X. Li, and H. Yang, "Human digital twin-driven human-cyber-physical systems: key technologies and applications," *Chinese Journal of Mechanical Engineering*, vol. 35, p. 11, 2022.

21 K. Senthilnathan and I. A. K. Annapoorani, "Cyber twin approach for smart grids," *IEEE Smart Grid Bulletin*, 2019. [Online]. Available: https://smartgrid.ieee.org/bulletins/november-2019/cyber-twin-approach-for-smart-grids

22 A. Saad, S. Faddel, and O. Mohammed, "IoT-based digital twin for energy cyber-physical systems: design and implementation," *Energies*, vol. 13, no. 18, p. 4762, 2020.

23 M. Panwar, R. Hovsapian, M. Mohanpurkar, and C. Koplin, "Real-time digital twin for an Alaskan microgrid using SCADA, distribution PMUs, and smart meter data," *IEEE Smart Grid Bulletin*, 2022. [Online]. Available: https://smartgrid.ieee.org/bulletins/october-2022/real-time-digital-twin-for-an-alaskan-microgrid-using-scada-distribution-pmus-and-smart-meter-data

24 D. Wijayasekara, O. Linda, M. Manic, and C. Rieger, "FN-DFE: fuzzy-neural data fusion engine for enhanced resilient state-awareness of hybrid energy systems," *IEEE Transactions on Cybernetics*, vol. 44, no. 11, pp. 2065–2075, 2014.

25 T. Fösel, M. Y. Neu, F. Marquardt, and L. Li, "Quantum circuit optimization with deep reinforcement learning," arXiv Preprint, arXiv:2013.07585, 2021.

26 Quantinuum, "Quantum Origin," Product Release, 2024.

27 M. R. Maghami, E. Yaghoubi, M. Mohamed, E. Yaghoubi, M. Z. Jahromi, and T. K. Fei, "Multi-objective optimization of unbalanced power distribution systems: A comprehensive approach to address uncertainties and enhance performance," *Energy Conversion and Management: X*, vol. 27, p. 101087, 2025.

28 S. Sayyed, M. Zhang, S. Rifat, A. Swami, M. D. Lucia, and F. Restuccia, "Resilience and security of deep neural networks against intentional and unintentional perturbations: survey and research challenges," arXiv Preprint, arXiv:2408.00193. https://doi.org/10.48550/arXiv.2408.00193

29 R. Z. Fanucchi et al., "Stochastic indexes for power distribution systems resilience analysis," *IET Generation, Transmission, and Distribution*, vol. 13, no. 12, pp. 2507–2516, 2019.

30 R. Mandill, S. DiAdamo, B. Qi, and A. Shabani, "Quantum key distribution in a packet-switched network," *Quantum Information*, vol. 9, p. 85, 2023.

31 H. Han et al., "A low-carbon dispatch strategy for power systems considering flexible demand response and energy storage," *Frontiers in Energy Research*, vol. 10, p. 883602, 2022.

32 E. Naderi, L. Mirzaei, M. Pourakbari-Kasmaei, F. V. Cerna, and M. Lehtonen, "Optimization of active power dispatch considering unified power flow controller: application of evolutionary algorithms in a fuzzy framework," *Evolutionary Intelligence*, vol. 17, pp. 1357–1387, 2024.

33 E. Naderi, L. Mirzaei, J. P. Trimble, and D. A. Cantrell, "Multi-objective optimal power flow incorporating flexible alternating current transmission systems: application of a wavelet-oriented evolutionary algorithm," *Electric Power Components and Systems*, vol. 52, no. 5, pp. 766–795, 2024.

34 H. Narimani, A. Azizivahed, E. Naderi, M. Fathi, and M. R. Narimani, "A practical approach for reliability-oriented multi-objective unit commitment problem," *Applied Soft Computing*, vol. 85, p. 105786, 2019.

35 M. Chen, H. Song, D. Zhu, W. Chen, and C. Cai, "Data collection in underwater acoustic sensor networks using multiple AUVs with sequential greedy-QPSO and fast marching algorithms," *IEEE Sensors Journal*, vol. 25, no. 11, pp. 20433–20444, 2025.

36 M. Azizi, U. Aickelin, H. A. Khorshidi, and M. B. Shishehgarkhaneh, "Energy valley optimizer: A novel metaheuristic algorithm for global and engineering optimization," *Scientific Reports*, vol. 13, p. 226, 2023.

37 T. Athay, R. Podmore, and S. Virmani, "A practical method for the direct analysis of transient stability," *IEEE Transactions on Power Apparatus and Systems*, vol. PAS-98, no. 2, pp. 573–584, 1979.

38 Kyocera Solar Website. Accessed: Jul. 10, 2025. [Online]. Available: https://www.kyocerasolar.eu/

39 Xantrex Power Inverter Website. Accessed: Jul. 10, 2025. [Online]. Available: https://www.xantrex.com/

40 Wind Turbine Models Website. Accessed: Jul. 10, 2025. [Online]. Available: https://en.wind-turbine-models.com/

41 EIA Website. Accessed: Jul. 10, 2025. [Online]. Available: https://www.eia.gov/

42 Weather Website. Accessed: Jul. 10, 2025. [Online]. Available: https://www.meteoblue.com/

12

Innovative Carbon Capture Cycles in Zero-emission Resilient Power Generation for Sustainable Deep Decarbonization of Modern Energy Networks

*Elahe Soleymani and Hadi Ghaebi**

University of Mohaghegh Ardabili, Ardabil, Iran

*Corresponding author: hghaebi@uma.ac.ir

List of Acronyms	
4SC	Four-stage compressor
CCS	Carbon capture and storage
CCUS	Carbon capture, utilization, and storage
CO_2	Carbon dioxide
NO_x	Nitrogen oxides
Symbols	
$\dot{E}x$	Exergy rate (kW)
h	Specific enthalpy (kJ/kg)
LHV	Lower heating value (kJ/kg)
$\dot{m}$	Mass flow rate (kg/s)
$\dot{n}$	Molar rate (mol/s)
P	Pressure (bar)
PR	Pressure ratio
$\dot{Q}$	Heat transfer rate (kW)
$\overline{R}$	Universal gas constant (J/mol·K)
r_{4SC}	Four-stage compressor pressure ratio
s	Specific entropy (kJ/kg·K)
$\dot{W}$	Net power (kW)

Subscripts and Superscripts	
CC	Combustion chamber
D	Destruction
en	Energetic
F	Fuel
FC	Fuel compressor
HE	Heat exchanger
HPGT	High-pressure gas turbine
mix	Mixer
net	Net value
out	Outlet
P	Product
ph	Physical
Pum	Pump
tot	Total value
1, 2, ...	Cycle locations
0	Dead state
Greek Symbol	
η	Efficiency

12.1 Introduction

The global energy sector is undergoing a critical transformation, shaped by the interconnected goals of energy security, environmental sustainability, and long-term economic development. Global energy demand is increasing due to industrial growth, population expansion, and urbanization. This rise has intensified dependence on fossil fuels, which remain the primary energy source. At the same time, this reliance has intensified environmental degradation, air pollution, and the accumulation of greenhouse gases, especially carbon dioxide (CO_2), which remains the leading factor in human-induced climate change [1]. Escalating energy costs, the gradual exhaustion of fossil fuels, and mounting evidence of climate change emphasize the urgent need for innovative energy conversion systems that combine high efficiency with low emissions. Currently, coal, oil, and natural gas supply about 85% of the world's primary energy. This heavy reliance on conventional fuels contributes significantly to greenhouse gas emissions and environmental degradation. Developing technologies capable of meeting growing energy demand while minimizing ecological consequences is therefore critical to ensuring sustainable and secure energy for the future [2–4]. Although fossil fuels continue to be economically viable and technologically established, they are also the primary source of CO_2 emissions, directly intensifying the greenhouse effect and contributing to climate instability. The Intergovernmental Panel on Climate Change warns that, without strong mitigation measures, rising CO_2 levels could cause irreversible climate changes in global temperatures, ocean currents, and ecological systems. As a result, reducing the carbon intensity of energy systems has emerged as a key priority in both research and policy development. Recent investigations indicate that combining oxy-fuel combustion with hydrogen-enriched fuels can achieve near-zero emissions while sustaining high thermodynamic efficiency, presenting a promising strategy for clean and sustainable power generation [5]. In response to these challenges, renewable energy technologies have expanded significantly over the last 20 years. Energy sources such as solar, wind, hydropower, geothermal, and biomass have steadily increased their contribution to total final energy consumption, propelled by ongoing technological advancements, falling production costs, and robust policy incentives [6, 7]. Moreover, integrating oxy-fuel cycles with advanced multi-objective optimization techniques has been shown to improve both economic and environmental outcomes, simultaneously lowering operational costs and reducing CO_2 emissions [8]. The integration of renewable energy sources into power generation, combined with a gradual reduction in reliance on traditional biomass, has steadily increased the contribution of renewables to global electricity production. However, these sources continue to face limitations due to intermittency, restricted storage capabilities, and geographical variability. As a result, transitioning to a fully decarbonized energy system in the near term requires technologies that can reduce emissions from conventional fuels while supporting the incremental growth of renewable infrastructure.

Among the strategies for carbon mitigation, carbon capture and storage (CCS) and carbon capture, utilization, and storage (CCUS) have become crucial for bridging the gap between existing fossil-based systems and future carbon-neutral energy frameworks. In this context, oxy-fuel combustion power cycles have attracted considerable research interest due to their inherent capability to enable near-complete CO_2 capture [9]. The core concept of oxy-fuel combustion involves substituting atmospheric air with high-purity oxygen during the combustion process. This replacement removes nitrogen from the oxidizer, producing a flue gas primarily composed of CO_2 and H_2O. After combustion, the water vapor can be condensed, allowing for the extraction of a nearly pure CO_2 stream suitable for compression, transportation, and geological storage. This built-in separation capability enables oxy-fuel systems to operate with near-zero emissions without requiring complex post-combustion capture equipment. Recent studies have further demonstrated that coupling oxy-fuel combustion cycles with CCUS technologies and renewable energy sources can effectively facilitate the deep decarbonization of modern energy networks (MENs), while simultaneously enhancing grid stability, operational flexibility, and overall sustainability [10, 11]. From a thermodynamic perspective, oxy-fuel power cycles present clear advantages over conventional combustion systems. By eliminating nitrogen from the oxidizer, these cycles not only simplify CO_2 separation but also allow for better control of combustion temperatures and a reduction in nitrogen oxide (NO_x) formation, thereby enhancing both efficiency and environmental performance. Nevertheless, the deployment of oxy-fuel systems faces several technical and economic obstacles, including the substantial energy required for oxygen production, material degradation under high-temperature, CO_2-rich conditions, and the design and optimization of heat recovery networks. Overcoming these challenges has become a central focus of current research aimed at developing more advanced and practical oxy-fuel power cycles.

Over the past 20 years, a variety of oxy-fuel cycle designs have been developed to improve thermodynamic performance, CO_2 capture efficiency, and overall system integration. Notable examples include the S-Graz cycle [12–14] and the MATIANT cycle [15, 16]. The S-Graz cycle, an advanced evolution of the original Graz cycle, utilizes supercritical CO_2 as the working fluid, achieving higher thermal efficiency through improved heat transfer and reduced compression work. Its design incorporates sophisticated heat recuperation and high-pressure turbine operation, allowing effective energy recovery while maintaining a CO_2-rich exhaust stream suitable for capture and storage. The MATIANT cycle introduces novel turbine arrangements, combustion chamber configurations, and heat exchanger layouts to optimize energy recovery and minimize exergy losses. Comparative studies show that these oxy-fuel cycles can attain competitive efficiencies relative to conventional combined-cycle gas turbines, with the added benefit of simplified CO_2 separation and lower environmental impact. Ongoing research focuses on integrating oxy-fuel systems with renewable and nuclear heat sources, enhancing material durability under high-temperature CO_2 conditions, and advancing cost-effective oxygen production methods such as ion transport membranes and chemical looping combustion. The coordinated implementation of these technologies promises to improve both efficiency and economic feasibility, supporting the shift toward sustainable, low-carbon power generation. In summary, transitioning to high-efficiency, low-emission energy systems remains a critical technological and environmental challenge. Oxy-fuel power cycles, with their combination of thermodynamic efficiency, operational flexibility, and inherent CO_2 capture capability, offer a scalable and promising pathway for deep decarbonization. Continued development of cycles such as S-Graz and MATIANT enhances understanding of high-temperature CO_2-based thermodynamics and highlights the essential role of oxy-fuel technologies in shaping a sustainable, resilient global energy future.

Saeidi et al. [17] performed an optimization study on an integrated power system combining the MATIANT cycle, a biomass-fueled gas turbine, and an organic Rankine cycle, explicitly taking economic factors into account. Their research aimed to determine the optimal system configuration that balances overall efficiency, power output, and cost-effectiveness, highlighting the potential benefits of multi-cycle integration for improving performance within low-carbon energy frameworks. Damen et al. [18] investigated hybrid coal–natural gas power plants equipped with CO_2 capture, achieving energy and exergy efficiencies of 52.06 and 47.02%, respectively, under optimized turbine operating conditions. Their findings emphasized the significant impact of turbine parameters on overall system performance and underscored the importance of precise thermodynamic optimization in multi-fuel oxy-combustion systems. Zhao et al. [19] carried out a detailed optimization study of a modified MATIANT cycle, focusing on key thermodynamic performance parameters, and reported net efficiencies approaching 45.6%, demonstrating the effectiveness of design and operational optimization strategies. Beyrami et al. [3] combined an oxy-fuel power plant with the MATIANT cycle through a multi-objective optimization framework that accounted for both energy and economic objectives. This configuration was capable of producing around 96 MW of net power with a payback period of four years. Sensitivity analysis highlighted combustion temperature and turbine outlet pressure as the most influential factors for both energy output and economic performance, providing essential guidance for system design. Barba et al. [20] conducted a comprehensive evaluation of zero-emission multi-cycle systems, considering technical, economic, environmental, and social aspects. Their results indicated that the E-MATIANT cycle exhibited superior economic efficiency compared with alternative configurations, confirming its potential as a practical low-emission power generation solution.

Rogalev et al. [21] performed a thermodynamic comparison among the semi-closed oxy-fuel combustion combined cycle (SCOC-CC), the E-MATIANT cycle, and the Allam cycle. They found that reducing the coolant liquid temperature could enhance net efficiency by 3.2% for SCOC-CC and 0.8% for E-MATIANT. Furthermore, the Allam cycle achieved net efficiencies 5.6% higher than SCOC-CC and 11.5% higher than E-MATIANT, demonstrating its thermodynamic advantage under optimized conditions.

12.1.1 The MATIANT Cycle as a Pathway for Greenhouse Gas Mitigation and Reliable Energy System Operation

The MATIANT cycle marks a notable progress in low-carbon power production, providing substantial opportunities to curb greenhouse gas emissions and improve energy reliability worldwide. As a sophisticated thermodynamic configuration, it combines staged expansion, high-pressure CO_2 compression, and efficient heat recovery processes, enabling superior energy utilization relative to conventional power cycles. Its closed-loop design, with carbon dioxide as the primary working fluid, allows near-complete-emission containment and facilitates effective CCUS, directly lowering atmospheric CO_2 levels.

From an environmental standpoint, the MATIANT cycle achieves high energy and exergy efficiencies, approximately 47 and 42%, respectively, reducing fuel consumption per unit of electricity produced and consequently lowering direct CO_2 emissions. Utilizing supercritical and transcritical CO_2 conditions further minimizes waste heat losses and enhances thermal recovery, decreasing the carbon intensity of generated power. Coupled with renewable or low-carbon fuels such as bio-syngas or hydrogen-enriched mixtures, the cycle can operate near-zero emissions. Regarding energy security, the MATIANT cycle strengthens system stability and resilience through flexible operation across diverse thermal sources and fuel types. Its modular configuration supports deployment in distributed generation networks, industrial cogeneration, and hybrid renewable fossil energy systems. High-pressure CO_2 as the working medium enables compact design, rapid startup, and improved load-following capabilities, which are essential for integrating intermittent renewable sources such as solar and wind into existing grids. These features reduce dependence on imported fossil fuels and enhance national energy autonomy.

Furthermore, the cycle's adaptability to integrate with waste heat recovery and carbon capture infrastructures supports the transition toward a sustainable, low-carbon energy economy. By delivering high efficiency, low emissions, and operational flexibility, the MATIANT cycle serves as a key enabling technology for achieving carbon neutrality while ensuring a reliable energy supply. Overall, it contributes not only to greenhouse gas mitigation but also to long-term energy security through technological innovation, resource efficiency, and environmental stewardship.

12.1.2 Advantages and Applications of MATIANT Cycle Technology

With high operational efficiency and reduced emissions, the MATIANT cycle stands out as a versatile technology for sustainable power production and industrial use. A primary advantage of this cycle is its high energy and exergy efficiencies. Its staged expansion design, combined with high-pressure CO_2 compressors, allows multi-level energy recovery, enhancing overall system efficiency while reducing fuel consumption. This not only lowers operational costs but also contributes directly to the reduction of greenhouse gas emissions. Another important strength of the MATIANT cycle is its fuel and operational flexibility. The system can operate efficiently with conventional fuels, bio-derived fuels, or hydrogen-enriched mixtures and can integrate with renewable thermal sources or industrial waste heat streams. This adaptability enables the cycle to respond to variable power demands, improving both the resilience and security of energy systems. Regarding applications, the MATIANT cycle is particularly well suited for high-efficiency power generation, CHP systems, and industrial waste heat recovery in heavy industries, refineries, and chemical plants. Its modular and compact design allows deployment in distributed generation networks, small- to medium-scale power plants, and hybrid energy systems. Additionally, the cycle can be combined with CCUS technologies, making it a viable solution for low-carbon and sustainable energy systems.

12.1.3 Key Points/Objectives Box

- Present and evaluate the MATIANT cycle as a zero-emission power generation technology with inherent CO_2 capture capability.
- Conduct a thermodynamic analysis from energy and exergy perspectives to identify key sources of irreversibility and improve system performance.
- To identify and quantify the primary sources of exergy destruction in major components such as compressors, turbines, and heat exchangers.
- Assess the impact of key operational parameters, including fuel molar flow rate, pump outlet pressure, four-stage compressor (4SC) pressure ratio, and expander inlet temperature, on system efficiency and performance.
- Propose strategies to enhance combustion chamber and compressor design, as well as optimize heat recovery processes, to reduce exergy destruction and increase efficiency.
- Highlight the role of the MATIANT cycle in greenhouse gas mitigation, strengthening global energy security, sustainable deep decarbonization, and enhancing the stability and resilience of MENs.

12.2 Decarbonized System Description

In the MATIANT cycle, the process initiates with liquid carbon dioxide, which is first preheated using the exhaust from the low-pressure turbine under specific temperature and pressure conditions. The preheated CO_2 then expands through the high-pressure turbine, where it is mixed with oxygen and methane fuel before entering the first combustion chamber. The combustion products subsequently expand in the medium-pressure turbine, undergo a second mixing with oxygen and methane, and exit the second combustion chamber at a controlled temperature. These gases continue their expansion through the low-pressure turbine,

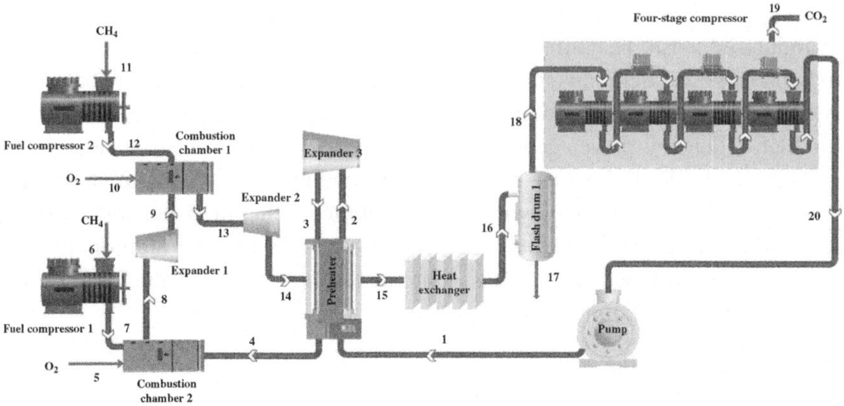

Figure 12.1 Visual representation of the MATIANT cycle.

contributing to the net power output. To maximize energy recovery from the high-temperature working fluid exiting the turbines, the flow is directed into a heat exchanger where heat from both turbine inlet and outlet streams is transferred to the carbon dioxide. The fluid is partially cooled in a subsequent heat exchanger before entering the steam separation unit. Due to the significant difference in specific heat capacities between CO_2 and vaporized water, the carbon dioxide within the combustion products, acting as the working fluid for certain portions of the cycle, is effectively separated and stored during water condensation. The collected water in the flash separator is then directed to the water electrolysis section, while the isolated CO_2 is sent to a 4SC. Within the compression system, carbon dioxide is pressurized and circulated throughout the closed-loop cycle. Inter-stage cooling is applied between compressor stages to regulate CO_2 temperature and maintain optimal compression efficiency. At each compression stage, the incoming stream is raised to a specific pressure target, cooled, and passed to the next stage. Finally, a high-pressure pump further pressurizes the CO_2, completing its circulation within the cycle. Figure 12.1 illustrates the zero-emission MATIANT cycle.

12.3 Mathematical Modeling and Key Assumptions for the MATIANT Cycle Analysis

During the pre-simulation phase, the assumptions for evaluating the proposed system are defined, as outlined in [22]. These assumptions include:

- The maximum ambient temperature is considered to be 30°C.

- The temperature difference across heat exchangers is assumed to be 10 °C, resulting in a minimum process temperature of 40 °C, given the availability of cooling water at 30 °C.
- Expansion in the expander is treated as an isentropic process with an efficiency of 87%.
- Compression and pumping operations in both compressors and pumps are also assumed to be isentropic, with a designated efficiency of 75%.
- Natural gas is modeled as pure methane.

Additionally, the key parameters required for the foundational modeling of the system are summarized in Table 12.1. Based on data reported in [17, 23, 24] as well as the values listed in Table 12.1, these assumptions are considered appropriate and sufficient to support the objectives of this study.

12.3.1 Mass and Energy Balance

Under the assumption of steady-state operation for each control volume depicted in Figure 12.1, the mass and energy balances for individual components of the system are formulated as follows [25–28]:

$$\sum \dot{m}_{\text{in}} = \sum \dot{m}_{\text{out}} \tag{12.1}$$

$$\sum \dot{E}_{\text{in}} = \sum \dot{E}_{\text{out}} \tag{12.2}$$

which $\dot{E}_{\text{in}}$ and $\dot{E}_{\text{out}}$ represent the amount of input energy and the amount of output energy to the control volume, respectively.

The balance equations for each component are presented in Table 12.2.

Table 12.1 Reference conditions and fixed parameters employed for thermodynamic simulation of the system.

Parameter	Value	Source
Environmental temperature, T_0 (K)	298.15	[17]
Atmospheric pressure, P_0 (bar)	1.013	[17]
Maximum ambient temperature, T_{ma} (°C)	30	[23]
Heat exchangers approach temperature, $T_{app,HT}$ (°C)	10	[23]
Minimum available temperature in processes, T_{min} (°C)	40	[17]

(Continued)

Table 12.1 (Continued)

Parameter	Value	Source
Cooling water available temperature, T_{CW} (°C)	30	[17]
Outlet temperature of the combustor, T_{CC} (°C)	1,300	[24]
HP turbine inlet temperature, $T_{in,HPT}$ (°C)	600	[24]
High pressure of the cycle, P_{High} (bar)	300	[24]
Low pressure of the cycle, P_{Low} (bar)	1	[24]
Expanders isentropic efficiency, $\eta_{is,EX}$ (%)	87	[23]
Compressors isentropic efficiency, $\eta_{is,Com}$ (%)	75	[23]
Pump's isentropic efficiency, $\eta_{is,Pum}$ (%)	75	[23]
Lower heating value of CH_4, LHV_{CH4} (MJ/kmol)	740	[23]

Table 12.2 The governing energy and exergy balance equations for the major components of the MATIANT cycle.

Component	Energy balance equations	Exergy balance equations
Fuel compressor 1	$\dot{W}_{FC1} = \dot{m}_6(h_7 - h_6)$ $\eta_{is,FC1} = (h_{7s} - h_6)/(h_7 - h_6)$	$\dot{E}_{D,FC1} = \dot{W}_{FC1} - (\dot{E}_7 - \dot{E}_6)$
Combustion chamber 1	$\dot{m}_4 + \dot{m}_5 + \dot{m}_7 = \dot{m}_8$ $\dot{m}_4 h_4 + \dot{m}_5 h_5 + \dot{m}_7 h_7 = \dot{m}_8 h_8$	$\dot{E}_{D,CC1} = (\dot{E}_4 + \dot{E}_5 + \dot{E}_7) - \dot{E}_8$
Expander 1	$\dot{W}_{Ex1} = \dot{m}_8(h_8 - h_9)$ $\eta_{is,Ex1} = (h_8 - h_9)/(h_8 - h_{9s})$	$\dot{E}_{D,Ex1} = (\dot{E}_8 - \dot{E}_9) - \dot{W}_{Ex1}$
Fuel compressor 2	$\dot{W}_{FC2} = \dot{m}_{11}(h_{12} - h_{11})$ $\eta_{is,\,FC2} = (h_{12s} - h_{11})/(h_{12} - h_{11})$	$\dot{E}_{D,FC2} = \dot{W}_{FC2} - (\dot{E}_{12} - \dot{E}_{11})$
Combustion chamber 2	$\dot{m}_9 + \dot{m}_{10} + \dot{m}_{12} = \dot{m}_{13}$ $\dot{m}_9 h_9 + \dot{m}_{10} h_{10} + \dot{m}_{12} h_{12} =$ $\dot{m}_{13} h_{13}$	$\dot{E}_{D,CC2} = (\dot{E}_9 + \dot{E}_{10} + \dot{E}_{12}) - \dot{E}_{13}$
Expander 2	$\dot{W}_{Ex2} = \dot{m}_{13}(h_{13} - h_{14})$ $\eta_{is,Ex2} = (h_{13} - h_{14})/(h_{13} - h_{14s})$	$\dot{E}_{D,Ex2} = (\dot{E}_{13} - \dot{E}_{14}) - \dot{W}_{Ex2}$
Preheater	$\dot{m}_1 + \dot{m}_3 + \dot{m}_{14} = \dot{m}_2 + \dot{m}_4 + \dot{m}_{15}$ $\dot{m}_1 h_1 + \dot{m}_3 h_3 + \dot{m}_{14} h_{14} = \dot{m}_2 h_2 +$ $\dot{m}_4 h_4 + \dot{m}_{15} h_{15}$	$\dot{E}_{D,PH} = (\dot{E}_1 + \dot{E}_3 + \dot{E}_{14})$ $- \dot{E}_2 - \dot{E}_4 - \dot{E}_{15}$

Component	Energy balance equations	Exergy balance equations
Expander 3	$\dot{W}_{Ex3} = \dot{m}_2(h_2 - h_3)$ $\eta_{is,Ex3} = (h_2 - h_3)/(h_2 - h_{3s})$	$\dot{E}_{D,Ex3} = (\dot{E}_2 - \dot{E}_3) - \dot{W}_{Ex3}$
Heat exchanger	$\dot{m}_{30} + \dot{m}_{50} = \dot{m}_{31} + \dot{m}_{51}$ $\dot{m}_{15} h_{15} + \dot{m}_{in,CW} h_{in,CW} =$ $\dot{m}_{16} h_{16} + \dot{m}_{out,CW} h_{out,CW}$	$\dot{E}_{D,HE} = (\dot{E}_{15} + \dot{E}_{in,CW})$ $- \dot{E}_{16} - \dot{E}_{out,CW}$
Flash drum	$\dot{m}_{16} h_{16} = \dot{m}_{17} h_{17} + \dot{m}_{18} h_{18}$ $\dot{m}_{16} = \dot{m}_{17} + \dot{m}_{18}$	$\dot{E}_{D,FD} = \dot{E}_{16} - (\dot{E}_{17} + \dot{E}_{18})$
4SC	$\dot{W}_{4SC} = \dot{m}_{18}(h_{19} + h_{20} - h_{18})$	$\dot{E}_{D,4SC} = \dot{W}_{4SC} - (\dot{E}_{19} + \dot{E}_{20} - \dot{E}_{18})$
Pump	$\dot{W}_{pum} = \dot{m}_1(h_1 - h_{20})$ $\eta_{is,pum} = (h_{1s} - h_{20})/(h_1 - h_{20})$	$\dot{E}_{D,Pum} = \dot{W}_{pum} - (\dot{E}_1 - \dot{E}_{20})$

Subsequently, the overall system performance is defined in terms of net power output:

$$\dot{W}_{net} = \dot{W}_{EX1} + \dot{W}_{EX2} + \dot{W}_{EX3} - \dot{W}_{Pum} - \dot{W}_{4SC} - \dot{W}_{FC1} - \dot{W}_{FC2} \tag{12.3}$$

Energy efficiency is defined as the ratio of net power output to the chemical energy input of the fuel, representing how effectively the fuel's energy is converted into useful work. Also, exergy efficiency is defined as the ratio of net power output to the exergy content of the fuel, reflecting the maximum theoretical work potential extracted from the fuel. Finally, the energy efficiency and exergy efficiency of the system are expressed as:

$$\eta_{en} = \frac{\dot{W}_{net}}{\dot{m}_{fuel} \cdot LHV_{fuel}} \tag{12.4}$$

$$\eta_{ex} = \frac{\dot{W}_{net}}{\dot{n}_{fuel} \cdot \overline{ex}_{ch}^{CH_4}} \tag{12.5}$$

12.4 Results

12.4.1 Model Validation

To assess the thermodynamic performance of the MATIANT cycle, a comparative analysis was carried out using key operational parameters, including the high-temperature turbine inlet temperature, high-temperature turbine inlet pressure, high-pressure turbine inlet temperature, high-pressure turbine inlet pressure, net output work, and overall energy efficiency. The model was validated against simulation data reported in case studies by Saeidi et al. [17] and Soltanieh et al. [29], which represent previously examined MATIANT cycle configurations. The results of this evaluation show that the relative deviations in output work and energy efficiency, when compared with the benchmark data, range between 0.2–3.45 and 0.21–3.36%, respectively. These minimal deviations confirm the accuracy and consistency of the current simulation with respect to prior validated studies.

Additionally, further thermodynamic and performance parameters obtained from the present simulation are summarized in Table 12.3. The strong agreement with corresponding reference data underscores the reliability of the developed model and its capability to replicate the behavior of the MATIANT cycle under similar operating conditions.

12.4.2 Thermodynamic Evaluation Outcomes

The simulation results of the MATIANT cycle illustrate the detailed thermodynamic and operational performance of its main components, as summarized in Table 12.4. The high-pressure turbine inlet temperature reaches approximately 873 K, reflecting the elevated thermal conditions required for efficient energy conversion within the system.

In the power distribution section, the first and second fuel compressors consume approximately 2,026 and 338 kW, respectively. These values indicate that a moderate portion of the total generated power is utilized by the compression stages to maintain the necessary fuel pressure and ensure stable operation. The expanders, serving as the primary power-generating units, deliver substantial energy output,

Table 12.3 Comparison of the current simulation results with corresponding findings reported in the literature [17, 29].

MATIANT cycle [17–29]	High-temperature turbine inlet temperature (K)	1,573	1,573	1,573	0
	High-temperature turbine inlet pressure (bar)	40	40	40	0
	High-pressure turbine inlet temperature (K)	873.2	873	873	0.02
	High-temperature turbine inlet pressure (bar)	300	300	300	0
	Output work (MWh)	94.29	94.5	97.66	0.2–3.45
	Energy efficiency (%)	47.15	47.25	48.79	0.21–3.36

Table 12.4 Results of the thermodynamic performance analysis of the MATIANT system.

Performance parameters	Value
High-pressure turbine inlet temperature, T_{HPT} (K)	873.15
Fuel compressor 1 power, $\dot{W}_{FC1}$ (kW)	2,026
Fuel compressor 2 power, $\dot{W}_{FC2}$ (kW)	337.8
Expander 1 power, $\dot{W}_{EX1}$ (kW)	45,060
Expander 2 power, $\dot{W}_{EX2}$ (kW)	75,544
Expander 3 power, $\dot{W}_{EX3}$ (kW)	33,154
4SC power, $\dot{W}_{4SC\_com}$ (kW)	57,706
Pump power consumption, $\dot{W}_{Pum}$ (kW)	6,464
Heat exchanger load, $\dot{Q}_{HX}$ (kW)	45.42
MATIANT energy efficiency, $\eta_{en,Matiant}$ (%)	47.15
MATIANT exergy efficiency, $\eta_{ex,Matiant}$ (%)	41.95

with the three expanders producing around 45, 75.5, and 33 MW, respectively. The staged expansion process allows effective energy recovery across multiple pressure levels, thereby improving overall cycle efficiency. The four-stage CO_2 compressor requires nearly 57.7 MW to maintain continuous circulation of the working fluid within the closed-loop system, while the main pump adds approximately 6.5 MW of power consumption, primarily associated with liquid CO_2 pressurization and transfer. The overall performance assessment indicates an energy efficiency of about 47% and an exergy efficiency of approximately 42%. These findings highlight the strong thermodynamic performance of the MATIANT cycle, demonstrating its capability to convert the chemical energy of fuel into useful mechanical power with minimal irreversibilities and energy losses.

12.4.3 Analysis of Exergy in the MATIANT Cycle

Figure 12.2 presents the percentage distribution of exergy destruction across the components of the MATIANT cycle. The second combustion chamber exhibits

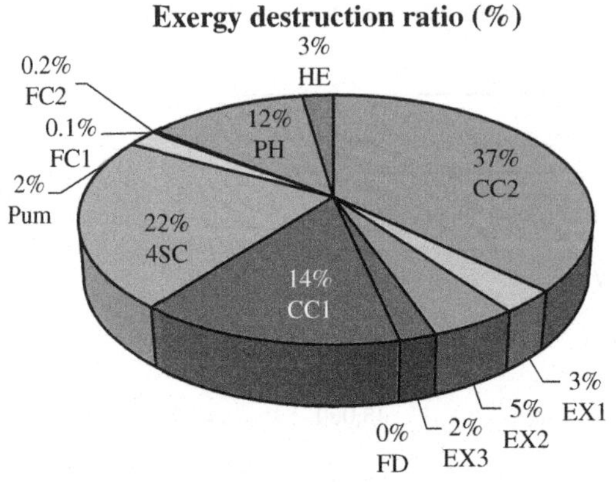

Figure 12.2 Percentage distribution of exergy destruction across the main components of the MATIANT cycle.

the highest exergy loss, accounting for roughly 37% of the total. This significant destruction is mainly due to the high-temperature combustion process and the entropy generated from nonideal mixing between the fuel and oxidizer within a CO_2-rich environment. The large temperature difference between the combustion gases and the working fluid in this region also leads to inefficient heat transfer and increased thermodynamic irreversibility.

The 4SC contributes the second-largest share, approximately 22%, owing to irreversible compression, pressure drops in the intermediate stages, and mechanical inefficiencies. Being one of the most energy-intensive components, this compressor strongly affects overall cycle performance and represents a key area for optimization. The first combustion chamber accounts for around 14% of total exergy destruction, ranking third. Although it operates at slightly lower temperatures than the second chamber, considerable entropy is still generated due to nonideal combustion and large temperature gradients, resulting in notable exergy loss. The preheater contributes about 12% of the total, primarily because of the substantial temperature differences between the hot and cold streams during preheating, which promote nonideal heat transfer and entropy generation. Among the remaining components, the heat exchangers and expanders, expander 1, expander 2, and expander 3, show relatively smaller contributions, approximately 3, 5, and 2%, respectively. These units operate with relatively high thermal effectiveness, and their exergy losses are mainly caused by pressure drops and imperfect temperature matching between the working streams. Minor exergy destruction is also observed in the pump (2%) and main heat exchanger (2%), mainly due to mechanical inefficiencies and hydraulic pressure drops during liquid compression. Finally, the fuel compressors and flash drum exhibit the lowest levels of exergy destruction, around 0.2, 0.1, and 0.05, respectively, indicating high operational efficiency and an optimized design with minimal impact on the overall cycle performance.

12.4.4 Parametric Analysis

This subsection examines the impact of key operational input variables on the overall thermodynamic performance of the MATIANT cycle. The analysis focuses on parameters including the fuel molar flow rate, the inlet temperature of the first expander, the pressure ratio across the four-stage CO_2 compressor, and the outlet pressure of the main pump. Each variable is systematically varied to evaluate its effect on the system's power output and efficiency. Through this approach, the study seeks to determine the relative importance of these factors in influencing both the energetic and exergetic performance of the MATIANT cycle.

12.4.4.1 Effect of Fuel Molar Flow Rate on the Energetic and Exergetic Performance of the MATIANT Cycle

Figure 12.3 illustrates how the energetic and exergetic efficiencies of the MATIANT cycle vary with the fuel molar flow rate at the first fuel compressor. As the fuel flow rate increases from 0.05 to 0.40 kmol/s, both energy and exergy efficiencies show a consistent decline. At lower fuel flow rates, the system operates more efficiently due to more complete energy utilization and reduced irreversibilities during combustion and expansion.

However, as the fuel input gradually rises, the total energy supplied increases faster than the useful output power, resulting in a decrease in both energy and exergy efficiencies. This trend is primarily due to higher thermal losses, elevated exhaust gas temperatures, and a reduced temperature gradient across the expanders, collectively lowering the cycle's overall thermodynamic performance.

Throughout the range of fuel flow rates, the exergy efficiency remains consistently higher than the corresponding energy efficiency. This indicates that, although a portion of the supplied chemical energy is irreversibly lost, the cycle maintains a robust capacity to convert available exergy into mechanical work under optimized operating conditions.

12.4.4.2 Influence of Pump Outlet Pressure on the Energetic and Exergetic Performance of the MATIANT Cycle

Figure 12.4 illustrates how variations in pump outlet pressure influence the energy and exergy efficiencies of the MATIANT cycle. As the outlet pressure rises from approximately 50 to 370 bar, both efficiencies show a progressive upward trend, indicating an overall enhancement in thermodynamic performance.

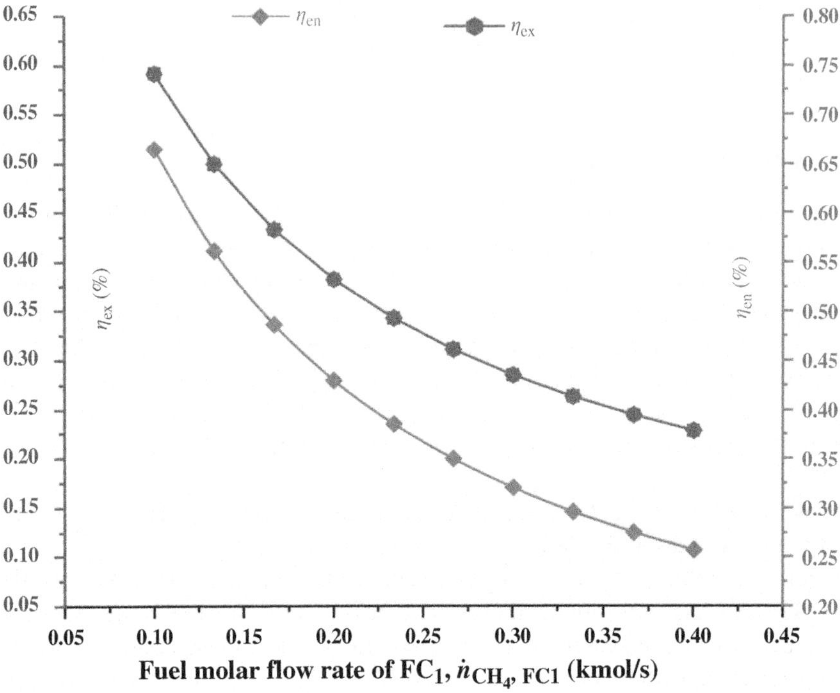

Figure 12.3 Variation of the MATIANT cycle's energy and exergy efficiencies as a function of fuel molar flow rate in FC_1.

Higher pump outlet pressures improve the pressure distribution throughout the system, reducing flow losses and mechanical inefficiencies in downstream components. This leads to more effective compression and expansion processes, thereby increasing the overall cycle efficiency. In addition, elevated liquid pressure enhances heat transfer within the heat exchangers, supporting more favorable combustion conditions and increasing the power output from the expanders. The energy and exergy efficiencies generally rise nearly linearly with increasing pressure up to a certain threshold, after which the rate of improvement slows. This behavior highlights the presence of an optimal pump outlet pressure range, which maximizes thermodynamic performance while avoiding excessive power consumption for pumping. It is also evident that the energy efficiency remains higher than the exergy efficiency across all operating conditions. This difference arises because exergy efficiency accounts for process irreversibilities and thermodynamic degradation, whereas energy efficiency only considers the first law energy balance.

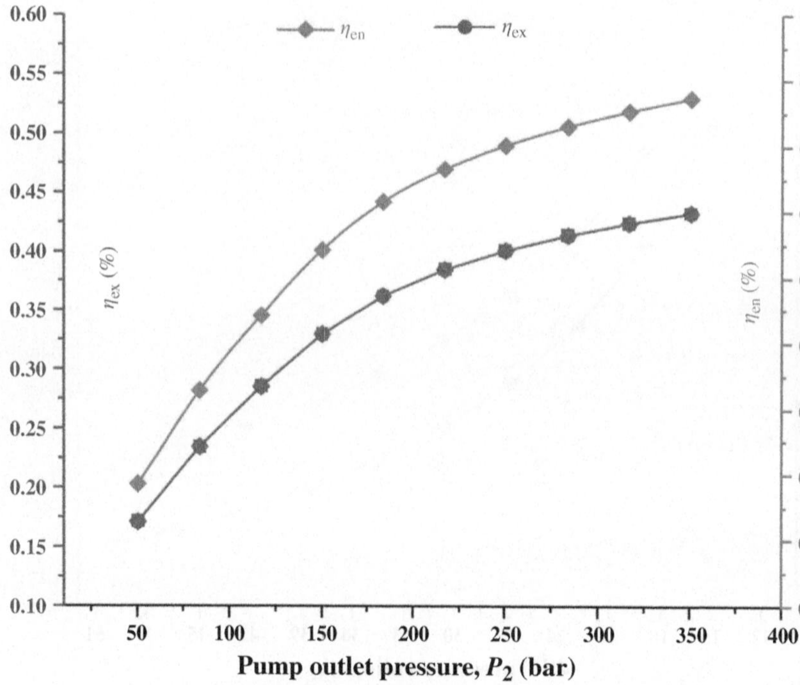

Figure 12.4 The effect of pump outlet pressure on the energy and exergy efficiencies of the MATIANT cycle.

12.4.4.3 Effect of Pressure Ratio of 4SC on the Performance of the MATIANT Cycle

Figure 12.5 presents the impact of the pressure ratio of 4SC on the thermodynamic performance of the MATIANT cycle, expressed through energy efficiency and exergy efficiency. The results indicate that both efficiencies decrease progressively as the 4SC pressure ratio increases. In the MATIANT cycle, an increase in the 4SC pressure ratio intensifies the compressor load and elevates the temperature of the working fluid before entering subsequent heat exchange processes. This leads to higher energy consumption in the compression stage and increased thermal mismatch across the heat exchangers, thereby reducing the net output work of the cycle. Additionally, the elevated pressure causes an increase in exergy destruction within the combustion and heat recovery units due to greater entropy generation and nonideal heat transfer. The decline in exergy efficiency is more pronounced than that of energy efficiency, as exergy efficiency accounts for both the quantity and quality of energy, including irreversibility caused by entropy

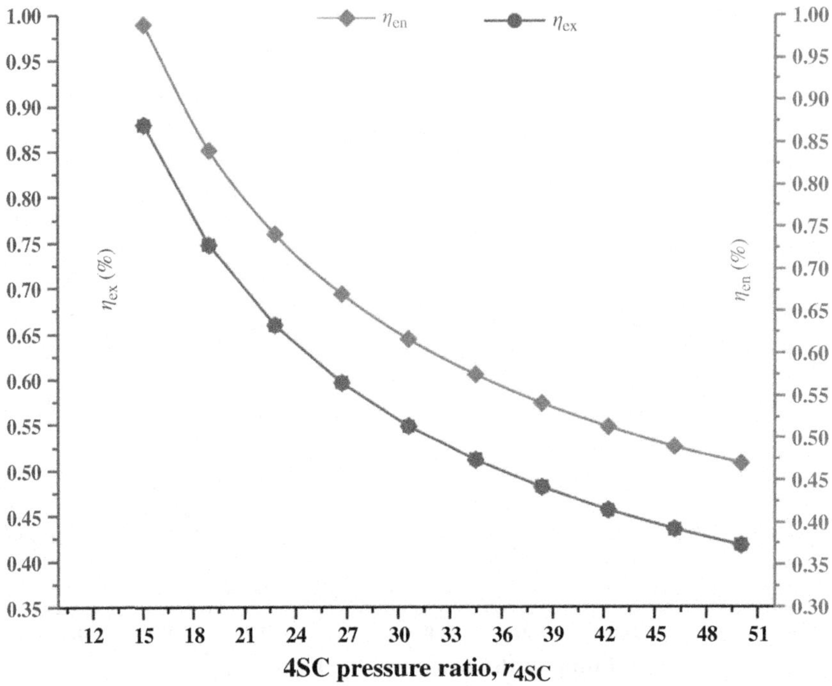

Figure 12.5 The effect of the pressure ratio of 4SC on the energy and exergy efficiencies of the MATIANT cycle.

production. Consequently, at higher pressure ratios, the useful work potential of the stream decreases more rapidly, reflecting a stronger deterioration in exergy performance.

12.4.4.4 Effect of the Expander 1 Inlet Temperature on the Performance of the MATIANT Cycle

Figure 12.6 illustrates the influence of the expander 1 inlet temperature on the energy and exergy efficiencies of the MATIANT cycle. The results demonstrate that both energy and exergy efficiencies increase steadily with rising the expander inlet temperature. In the MATIANT cycle, increasing the expander inlet temperature enhances the specific enthalpy drop during the expansion process, thereby improving the power output delivered by the expander. This directly contributes to a higher energy efficiency, as a greater portion of the supplied thermal energy is converted into mechanical work. From an exergy standpoint, a higher expander inlet temperature increases the specific exergy of the working fluid entering the expander, which elevates the maximum theoretical work potential available. At the same time, the relative magnitude of exergy destruction decreases because

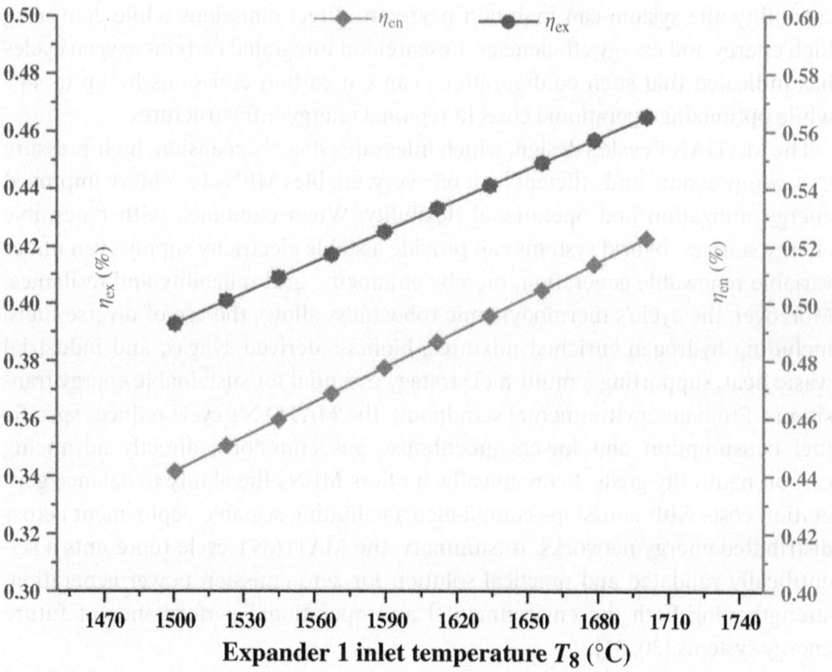

Figure 12.6 Variation of the MATIANT cycle's energy and exergy efficiencies with respect to the inlet temperature of Expander 1.

the ratio of useful work output to total exergy input improves. As the inlet temperature of the expander increases, the exergy efficiency of the MATIANT cycle improves, although it continues to be slightly lower than the corresponding energy efficiency due to inherent irreversibilities arising from combustion, heat transfer, and fluid friction. These findings indicate that elevating the expander temperature enhances the overall thermodynamic performance of the cycle. Nonetheless, practical implementation requires careful consideration of material limitations, turbine blade cooling strategies, and thermal stress constraints to prevent potential efficiency losses or mechanical damage.

12.4.5 Zero-emission Power Effects on Sustainable Deep Decarbonization of MENs

Implementing the oxy-fuel MATIANT cycle offers an effective strategy to facilitate profound and sustainable decarbonization of MENs. Transitioning to oxy-fuel combustion power cycles offers a practical route for substantially reducing carbon dioxide emissions from the energy sector. Thanks to its inherent CO_2 capture capability, the system can maintain near-zero direct emissions while delivering high energy and exergy efficiencies. Research on integrated carbon–oxygen cycles has indicated that such configurations can cut carbon emissions by up to 44% while optimizing operational costs in regional energy infrastructures.

The MATIANT cycle's design, which integrates staged expansion, high-pressure CO_2 compression, and efficient heat recovery, enables MENs to achieve improved energy utilization and operational flexibility. When combined with renewable energy sources, hybrid systems can provide a stable electricity supply even under variable renewable generation, thereby enhancing grid reliability and resilience. Moreover, the cycle's thermodynamic robustness allows the use of diverse fuels, including hydrogen-enriched mixtures, biomass-derived syngas, and industrial waste heat, supporting a multi-fuel strategy essential for sustainable energy transitions. From an environmental standpoint, the MATIANT cycle reduces specific fuel consumption and lowers greenhouse gas emissions, directly advancing carbon neutrality goals. Economically, it offers MENs the ability to balance generation costs with emissions compliance, facilitating scalable deployment across distributed energy networks. In summary, the MATIANT cycle represents a scientifically validated and practical solution for zero-emission power generation, strengthening both the environmental and operational performance of future energy systems [30, 31].

12.5 Conclusion

The thermodynamic analysis of the MATIANT cycle provides a detailed understanding of its performance under various operating conditions. Model validation confirms excellent agreement between the current simulation and benchmark data from previously validated studies, with deviations in power output and overall efficiency remaining below 3.5%. This close correspondence demonstrates that the developed model reliably predicts the actual operational behavior of the MATIANT cycle. The evaluation reveals that the staged expansion design plays a crucial role in achieving high energy and exergy performances. The combined output from the expanders exceeds 150 MW, while the cycle reaches energy and exergy efficiencies of approximately 47 and 42%, respectively. These results indicate that the MATIANT cycle efficiently converts fuel chemical energy into mechanical work, with limited thermodynamic irreversibilities and strong integration across subsystems. Exergy analysis shows that the major losses occur in the combustion and compression stages, where nonideal reactions, high thermal gradients, and pressure drops dominate. Accordingly, optimizing combustion chamber design, improving multistage compression, and enhancing heat recovery are essential for minimizing exergy destruction and further improving cycle performance.

Parametric studies highlight the significant impact of fuel molar flow rate, pump outlet pressure, 4SC pressure ratio, and expander inlet temperature on overall performance. Higher fuel flow rates lead to reduced energy and exergy efficiencies due to increased thermal losses and irreversibilities. Conversely, elevating the pump outlet pressure enhances system efficiency by stabilizing flow and improving pressure balance, up to an optimal point beyond which additional pumping work reduces net gains. Increasing the 4SC pressure ratio can decrease efficiencies because of higher compression work and greater entropy generation within the heat recovery network. In contrast, elevating the expander inlet temperature has a favorable effect, increasing the power output and both efficiency indicators, although practical constraints related to material strength and thermal stress must be carefully considered.

Key findings from this study, emphasizing the role of the oxy-fuel MATIANT cycle in supporting zero-emission and resilient power generation for sustainable decarbonization of MENs, include:

- Near-zero CO_2 emissions are achieved through the cycle's oxy-fuel and carbon capture configuration.
- The high energy efficiency (47.15%) and exergy efficiency (41.95%) indicate strong thermodynamic performance.
- The system exhibits robustness and operational flexibility, maintaining stable performance under varying conditions.
- Compatibility with renewable and low-carbon fuels enables hybrid and sustainable power systems.
- The MATIANT cycle proves to be a dependable and adaptable method for generating low-emission electricity, ensuring both operational stability and sustainable efficiency across different load scenarios.
- This cycle offers a robust and evidence-supported strategy to lower greenhouse gas emissions and enable comprehensive decarbonization of MENs.

References

1 S. F. Ahmadi, A. Minaei, M. Ebadollahi, H. Ghaebi, and M. H. Shahrivar, "Energy management and reducing the environmental impacts of industrial flare gases using a new trigeneration energy system," *Process Safety and Environmental Protection*, vol. 177, pp. 1129–1141, 2023.

2 M. Ebadollahi, M. Amidpour, O. Pourali, and H. Ghaebi, "Development of a novel flexible multigeneration energy system for meeting the energy needs of remote areas," *Renewable Energy*, vol. 198, pp. 1224–1242, 2022.

3 J. Beyrami, M. Jalili, M. Ziyaei, A. Chitsaz, and M. A. Rosen, "A novel system for electricity and synthetic natural gas production from captured CO_2: techno-economic evaluation and multi-objective optimization," *Journal of CO_2 Utilization*, vol. 63, p. 102116, 2022.

4 R. Nazerifard, L. Khani, M. Mohammadpourfard, B. Mohammadi-Ivatloo, and G. G. Akkurt, "Design and thermodynamic analysis of a novel methanol, hydrogen, and power trigeneration system based on renewable energy and flue gas carbon dioxide," *Energy Conversion and Management*, vol. 233, p. 113922, 2021.

5 S. Oleś, J. Mularski, D. Pyka, H. Pawlak-Kruczek, and A. Pozarlik, "Optimization of hydrogen supercritical oxy-combustion in gas turbines," *Fuels*, vol. 6, no. 1, p. 6, 2025.

6 D. Raimi, Y. Zhu, R. G. Newell, and B. C. Prest, *Global Energy Outlook 2024: Peaks or Plateaus*. Resources for the Future, 2024.

7 M. Ebadollahi, M. Amidpour, O. Pourali, and H. Ghaebi, "Flexibility concept in design of advanced multi-energy carrier systems driven by biogas fuel for sustainable development," *Sustainable Cities and Society*, vol. 86, p. 104121, 2022.

8 H. Li, X. Bai, H. Li, and L. Bai, "Optimal scheduling of integrated energy systems considering oxy-fuel power plants and carbon trading," *Energies*, vol. 18, no. 14, p. 3814, 2025.

9 B. Metz, O. Davidson, H. C. De Coninck, M. Loos, and L. Meyer, *IPCC Special Report on Carbon Dioxide Capture and Storage*. Cambridge University Press, 2005.

10 K. Owebor, C. Okereke, O. E. Diemuodeke, A. B. Owolabi, and C. O. Nwachukwu, "A systematic review of literature on the decarbonization of the Nigerian power sector," *Energy, Sustainability and Society*, vol. 15, no. 1, p. 34, 2025.

11 P. K. Pathak, A. K. Yadav, and I. Kamwa, "Green hydrogen: a strategic energy vector for achieving net-zero emissions by 2050," *Sustainable Energy & Fuels*, vol. 9, no. 19, pp. 5218–5226, 2025.

12 H. Jericha, W. Sanz, J. Woisetschläger, and M. Fesharaki, "CO_2-retention capability of CH_4/O_2-fired Graz cycle," in *CIMAC Conference Paper*, Interlaken, Switzerland.

13 H. Jericha and E. Göttlich, "Conceptual design for an industrial prototype Graz cycle power plant," in *Turbo Expo: Power for Land, Sea, and Air*, vol. 3607, ASME, 2002, pp. 413–420.

14 H. Jericha, E. Göttlich, W. Sanz, and F. Heitmeir, "Design optimization of the Graz cycle prototype plant," *Journal of Engineering for Gas Turbines and Power*, vol. 126, no. 4, pp. 733–740, 2004.

15 O. Bolland, H. M. Kvamsdal, and J. C. Boden, "A comparison of the efficiencies of the oxy-fuel power cycles water-cycle, Graz-cycle and MATIANT-cycle," in *Capture and Separation of Carbon Dioxide from Combustion Sources*, vol. 1, Elsevier, 2005, pp. 499–511.

16 P. Mathieu and R. Nihart, "Zero-emission MATIANT cycle," *Journal of Engineering for Gas Turbines and Power*, vol. 121, no. 1, pp. 116–120, 1999.

17 S. Saeidi, S. M. S. Mahmoudi, H. Nami, and M. Yari, "Energy and exergy analyses of a novel near zero emission plant: combination of MATIANT cycle with gasification unit," *Applied Thermal Engineering*, vol. 108, pp. 893–904, 2016.

18 K. Damen, M. van Troost, A. Faaij, and W. Turkenburg, "A comparison of electricity and hydrogen production systems with CO_2 capture and storage. Part A: Review and selection of promising conversion and capture technologies," *Progress in Energy and Combustion Science*, vol. 32, no. 2, pp. 215–246, 2006.

19 Y. Zhao, J. Chi, S. Zhang, and Y. Xiao, "Thermodynamic study of an improved MATIANT cycle with stream split and recompression," *Applied Thermal Engineering*, vol. 125, pp. 452–469, 2017.

20 F. C. Barba, G. M. D. Sanchez, B. S. Segui, H. G. Darabkhani, and E. J. Anthony, "A technical evaluation, performance analysis and risk assessment of multiple novel oxy-turbine power cycles with complete CO_2 capture," *Journal of Cleaner Production*, vol. 133, pp. 971–985, 2016.

21 N. Rogalev, A. Rogalev, V. Kindra, O. Zlyvko, and P. Bryzgunov, "Review of closed sCO_2 and semi-closed oxy-fuel combustion power cycles for multi-scale power generation in terms of energy, ecology and economic efficiency," *Energies*, vol. 15, no. 23, p. 9226, 2022.

22 Y. Ji-chao and B. Sobhani, "Integration of biomass gasification with a supercritical CO_2 and Kalina cycles in a combined heating and power system: a thermodynamic and exergoeconomic analysis," *Energy*, vol. 222, p. 119980, 2021.

23 N. R. Udengaard, "Hydrogen production by steam reforming of hydrocarbons," *American Chemical Society, Division of Fuel Chemistry*, vol. 49, no. 2. p. 906, 2004.

24 J. S. Kang, D. H. Kim, S. D. Lee, S. I. Hong, and D. J. Moon, "Nickel-based tri-reforming catalyst for the production of synthesis gas," *Applied Catalysis A: General*, vol. 332, no. 1, pp. 153–158, 2007.

25 T. Parikhani, J. Jannatkhah, A. Shokri, and H. Ghaebi, "Thermodynamic analysis and optimization of a novel power generation system based on modified Kalina and GT-MHR cycles," *Energy Conversion and Management*, vol. 196, pp. 418–429, 2019.

26 A. Bejan, G. Tsatsaronis, and M. J. Moran, *Thermal Design and Optimization*. John Wiley & Sons, 1995.

27 E. Soleymani, S. G. Gargari, and H. Ghaebi, "Thermodynamic and thermoeconomic analysis of a novel power and hydrogen cogeneration cycle based on solid SOFC," *Renewable Energy*, vol. 177, pp. 495–518, 2021.
28 H. Ghaebi and E. Soleymani, "Investigating the effect of parameters in the thermodynamic analysis of the solid oxide fuel cell cycle using response surface methodology," *Scientific Reports*, vol. 15, no. 1, p. 181, 2025.
29 M. Soltanieh, K. M. Azar, and M. Saber, "Development of a zero emission integrated system for co-production of electricity and methanol through renewable hydrogen and CO_2 capture," *International Journal of Greenhouse Gas Control*, vol. 7, pp. 145–152, 2012.
30 R. Wang, X. Wen, X. Wang, Y. Fu, and Y. Zhao, "Low-carbon economic dispatch of regional integrated energy system based on carbon-oxygen cycle," *Frontiers in Energy Research*, vol. 11, p. 1206242, 2023.
31 F. G. Weiss and C. Bouallou, "Supercritical carbon dioxide power cycles with oxy-combustion capture: towards a new improved cycle layout for energy production," *Chemical Engineering Transactions*, vol. 114, pp. 145–150, 2024.

13

Policy-making Models for Carbon Removal of MENs Under the CESN Framework

*Ali Akbar Rezazadeh and Akram Avami**

Energy Systems Engineering Group, Department of Energy Engineering, Sharif University of Technology, Tehran, Iran

*Corresponding author: avami@sharif.edu

List of Acronyms

a	Output of the hidden layer
a_i and b_i	The respective values of A and B in tuple i
$\overline{A}$ and $\overline{B}$	Mean values of A and B
ABM	Agent-based modeling
AI	Artificial intelligence
ANN	Artificial neural network
b	Bias vector
BNN	Bayesian neural network
CESN	Climate–energy–sustainability nexus
f	Activation function
GDP	Gross domestic product
GHG	Greenhouse gas
LCA	Life cycle assessment
LCCA	Life cycle costing analysis
LP	Linear programing
LSTM	Long short-term memory
m	Momentum term
MENs	Modern energy networks
ML	Machine learning

List of Acronyms	
MLP	Multilayer perceptron
MP	Mathematical programing
n	number of tuples
O_i	The output from the preceding neuron
p	Input vector
R^2	Coefficient of determination
$r_{\mathrm{A,B}}$	Correlation
RF	Random forest
RNN	Recurrent neural network
S-LCA	Social life cycle assessment
SWOT	Strengths, weaknesses, opportunities, and threats
SD	System dynamics
W	The weight matrix
x_{i} and y	Attributes
$\widehat{y_i}$	Prediction of y_i
$\bar{y}$	Average of y
β_i and α	Regression coefficients
η	Learning rate
σ_{A} and σ_{B}	Standard deviations of A and B
δ_j	Local error gradient

13.1 Introduction

In this chapter, the focus is on how the most effective strategies can be identified and selected from among various options, and how these strategies can be modeled in the modern energy networks (MENs) to support informed and optimized decision-making for the climate–energy–sustainability nexus (CESN). This decision is a critical and strategic one, as it significantly affects the survival and long-term trajectory of an organization or society. A strategic decision typically involves high levels of uncertainty, long-term implications, substantial resource commitments, and complex interdependencies. Moreover, such decisions are often irreversible or extremely costly to reverse [1]. Therefore, the selection of an appropriate course of action must be made with great caution and precision.

It should also be recognized that decision-making is rarely influenced by a single factor; rather, it encompasses multiple dimensions and requires a systematic, holistic perspective of the entire issue to ensure that the most suitable solution is identified considering all relevant aspects. Also, it is more important in CESN.

MENs must meet energy demand in a reliable, safe, and economical way. They are strongly linked to complex environmental issues arising from the Earth's ecosystem such as water availability, climate change, and air pollution. Climate change causes an increase in average temperatures, which increases energy demand [2], and thus increases the need for more energy supply. Also, the increase in temperature reduces the efficiency of MEN technologies, which affects both renewable [3] and nonrenewable energy [4], as well as transmission infrastructure [5]. In addition, extreme events such as floods, caused by climate change, pose significant risks to MEN infrastructure [6]. MENs are also the largest contributor to greenhouse gas (GHG) emissions, making them a major focus of climate change mitigation strategies [7]. In the water domain, interdependence is also significant. The energy sector is a major consumer of water (e.g., for cooling in power generation) [4], while water systems require energy for purification, desalination, and distribution [8, 9]. Air pollution is another important issue in MENs. The MENs are the primary source of air pollutants [10], and at the same time, the performance and resilience of energy infrastructure can be adversely affected by air quality [11, 12]. Therefore, holistic decision-making strategies and integrated policies for CESN design must account for all dimensions of the energy system with water, climate, and air quality to ensure sustainable and resilient development of the energy system.

This chapter consists of four other sections. Section 13.2 focuses on the taxonomy of various policy-making and decision-making methods. In this section, models are categorized into quantitative and qualitative approaches, with their respective strengths and weaknesses discussed. Another classification divides the models into predictive models and prescriptive models, highlighting the key characteristics of each. It is also noted that many models do not fit into these categories or are instead hybrids that combine elements from multiple approaches.

Section 13.3 provides a more detailed discussion of several quantitative models used in policy-making in MENs, including system dynamics (SD), mathematical programming (MP), agent-based modeling (ABM), game theory, data-driven models, life cycle assessment (LCA), and an integrated model. This section reviews the advantages and limitations of each model and introduces available tools or software. These approaches cover a wide range; for example, data-driven models range from basic linear regression and econometric techniques to more advanced methods based on machine learning (ML). Section 13.4 introduces two qualitative models, the Delphi method and the strengths, weaknesses, opportunities, and threats (SWOT) analysis. The advantages and limitations of each method are discussed in detail and supported by practical examples to illustrate their applications.

Finally, Section 13.5 provides conclusions. This section also points to potential areas for future research and practical advancements to further enhance decision-making within the CESN framework. The main objectives of this chapter are as follows:

- Classification of different types of policy-making models for CESN of MENs: it includes quantitative and qualitative approaches, predictive and prescriptive, and top-down and bottom-up models.
- Description of the CESN models: they include the nexus of energy, climate, water, food, and land.
- Comparison of various policy-making models according to their pros and cons to facilitate the selection of appropriate models for researchers: includes models such as Delphi, SWOT, ML, MP, and ABM.
- Review important literature and applications on different case studies: for each model, one or more studies are presented to demonstrate the application of the methods. By selecting a diverse range of studies, various types of applications are covered.

13.2 A Taxonomy of Policy-making Models in MENs for CESN

Policy-making models for the MENs can be categorized into two main types: quantitative and qualitative (Figure 13.1). Quantitative models are based on mathematical calculations and algorithms that aim to evaluate policies in measurable terms. For instance, using a set of equations, the cost of each scenario can be determined, and through cost-benefit analysis, a quantitative model can be made. In these models, expert judgment is incorporated not directly, but indirectly, often through mathematical formalization. Sometimes, translating expert knowledge into mathematical terms is straightforward, such as when comparing costs using equations or calculating energy flows within a model. However, in other cases, this process can become more complex, for example, when human behavior is important, such as the tendency of agents to use bicycles in a scenario where people are motivated by environmental concerns. On the other hand, qualitative models stand in contrast to quantitative approaches. Unlike quantitative models, which are based on mathematical relationships, qualitative models seek to incorporate expert judgment directly. These models aim to synthesize the opinions of specialists and, where possible, achieve consensus to propose suitable policy solutions.

A key strength of qualitative models lies in the expert judgment process, which allows for the consideration of indirect or intangible factors that might influence

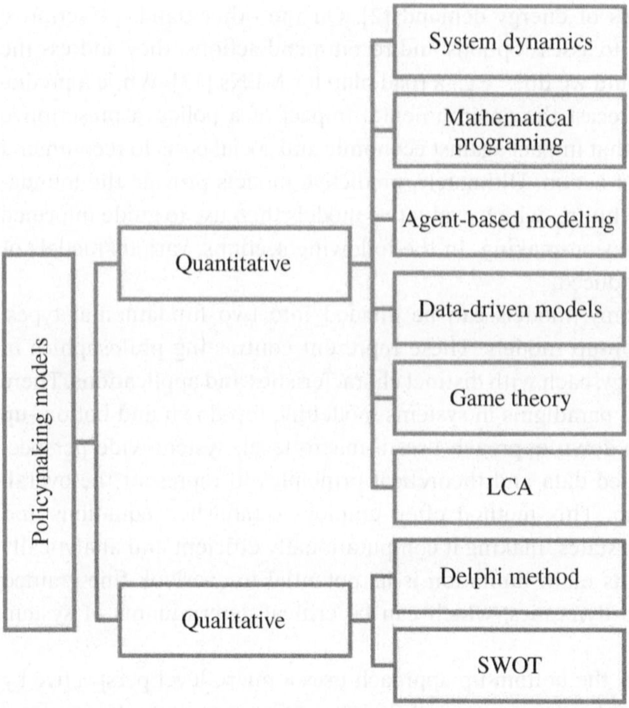

Figure 13.1 The qualitative and quantitative policy-making models.

outcomes. For example, a solution may be technically and economically perfect yet be socially unacceptable or lack public support, which are more important in CESN, leading to its rejection within a qualitative framework. However, qualitative methods also have limitations. They may be influenced by experts' biases, leading to suboptimal or wrong recommendations. Additionally, qualitative models often require some form of weighting or prioritization mechanism to facilitate final results or rank policy alternatives systematically.

There are also hybrid models that combine both approaches. In practice, the results derived from quantitative models often need to be refined through qualitative expert judgment. In other words, the best outcomes are usually achieved when a synergistic relationship between quantitative and qualitative models is established. By iterating between these two approaches, a more comprehensive and robust policy recommendation can be developed.

Beyond the quantitative and qualitative spectrum, models can also be categorized based on their core function into predictive and prescriptive types. Predictive models are primarily concerned with forecasting future outcomes based on historical data and patterns; they answer the question, "What will happen?," e.g., projecting trends of energy demand [2]. On the other hand, prescriptive models are designed to assess options and recommend actions; they address the question, "What should we do?," e.g., a road map for MENs [13]. While a predictive model might forecast the environmental impact of a policy, a prescriptive model would weigh that impact against economic and social costs to recommend the optimal course of action. Ultimately, predictive models provide the foundational data about the future, which evaluative models then use to guide informed and value-driven decision-making. In the following sections, various models of this type will be introduced.

In another taxonomy, models can be divided into two fundamental types: top-down and bottom-up models. These represent contrasting philosophies in modeling methodology, each with distinct characteristics and applications. There are two fundamental paradigms in systems modeling, top-down and bottom-up approaches. The top-down approach uses a macro-level, system-wide perspective. It uses aggregated data and theoretical principles to represent the overall behavior of a system. This method often employs established equations and assumes equilibrium states, making it computationally efficient and analytically tractable. However, its main limitation is its potential to overlook fine-grained details and localized dynamics, which can be critical determinants of system outcomes.

On the other hand, the bottom-up approach uses a micro-level perspective by defining the individual components and their specific parameters. It simulates the system by modeling the interactions and behaviors of these constituent parts. This strength allows it to capture emergent phenomena, nuanced behaviors, and complex interactions that top-down models might miss. Bottom-up models provide greater detail and realism but at the expense of significantly higher computational complexity and data requirements. The selection between these approaches is contingent upon the research objectives, data availability, and the specific characteristics of the system under study. Many contemporary modeling efforts seek to integrate both approaches, leveraging the strengths of each to create more robust and comprehensive representations of complex systems.

13.3 Foundational Quantitative Models for Policy Analysis

13.3.1 Data-driven Models

Today, data can be considered the equivalent of oil, a highly valuable commodity. But in reality, data is even more valuable than oil. Extracting and using it is far more complex, and unlike oil, it never runs out; in fact, it keeps growing every day. Just like crude oil, raw data has limited utility in its initial state. Its true value is realized only after careful extraction and refinement, turning it into actionable and useful knowledge. Data by itself is not useful; what is valuable is the knowledge derived from it [14, 15].

Working with data involves several stages. The first step is preprocessing, which includes tasks such as data cleaning, integrating databases, and selecting relevant features [16]. The resulting database is then fed into various modeling algorithms, which can range from a simple average to more complex structures. Data-driven models can be divided into two main categories: supervised models, where data points have a label, regression and classification models fall into this group, and unsupervised models, where no label is present, and the database is partitioned into clusters based on similarities among data points [2].

In line with common practice in ML literature, the available dataset is partitioned into separate subsets for model development and evaluation. For example, data is allocated as follows: 70, 15, and 15% for training, validation, and test, respectively, after database shuffling. Training data is used to fit the model's parameters. Validation data is employed during the training process to tune hyperparameters, monitor for overfitting, and provide an unbiased assessment of the model's performance on unseen data before final testing. Crucially, the testing set is held out entirely from both the training and validation phases; it is used only for the final, conclusive assessment of the model's generalization ability and predictive performance [2, 16]. In the following subsections, we will explore some of the data-driven models.

13.3.1.1 Correlation

Correlation is a statistical measure that indicates the degree of linear dependence between two features. A large positive value (max +1) represents a direct (positive) relationship, while a large negative value (min −1) indicates an inverse (negative) relationship. In the correlation analysis, input variables are selected based on their linear correlation. Independent input variables (near zero) are removed, and a new model is developed without these variables. The correlation is calculated by Eqs. (13.1) and (13.2), where n is the number of tuples, a_i and b_i are corresponding values of A and B in tuple i, $\overline{A}$ and $\overline{B}$ are the mean values of A and B, and σ_{A} and σ_{B} are the standard deviations of A and B [15],

$$r_{\mathrm{A,B}} = \frac{\sum_{i=1}^{w} (a_i b_i - n\overline{A}\overline{B})}{n\sigma_{\mathrm{A}}\sigma_{\mathrm{B}}} \tag{13.1}$$

$$\begin{cases} r \sim +1 \rightarrow \text{Positively correlated} \\ r \sim 0 \rightarrow \text{Independent} \\ r \sim -1 \rightarrow \text{Negatively correlated} \end{cases} \tag{13.2}$$

Verpeta used correlation to assess the relationships between climate parameters and a solar power plant's efficiency [17]. Rezazadeh et al. used correlation to identify key parameters affecting energy demand to demonstrate the impact of climate change on energy consumption [2]. In addition, another study used correlation to show the co-benefits between climate change mitigation and health due to air pollution in MENs [18]. Canales et al. applied correlation to measure relationships between wind, solar, and hydropower to quantify their relationship in a MEN [19]. These studies successfully used correlation to show the relationship between two parameters in MENs analysis. Therefore, the correlation is useful for detecting relationships between climate, sustainability, and MEN and can also determine the strength of these relationships. Although correlation analysis is necessary and essential for any data analysis, it can be considered the beginning of working with data, not the end. This is because this method is typically used for feature selection, not as the main model itself. After selecting the influential features with its help, a suitable model can then be developed.

13.3.1.2 Association Rules

Association rule mining is an if-then technique for uncovering knowledge and meaningful relationships and patterns hidden in a database [20]. It has been widely applied across diverse fields, including market basket analysis [21], medical disease research [22], industrial quality control [23], alarm analysis [24], and fault diagnosis [20]. A typical association rule might be: "If A occurs, then B is 80% likely to occur." Each rule consists of two components, an antecedent (if clause) and a consequent (then clause). These rules are extracted from the database by analyzing data for frequent if-then patterns and using the criteria support and confidence to identify the most significant relationships. Support measures how frequently the relevant items co-occur in the dataset, while confidence reflects how often the rule holds true when the antecedent is satisfied [15, 25]. For example, an association rule is as follows:

Access Card-Present → Windows-Open

This rule states that when an access card is detected as present in the area of the building, the windows in that same area tend to be open. It is mined with a support of approximately 0.05 (meaning this pattern occurred in 5% of the database) and a confidence of about 0.6 (meaning that when people are present, the windows are open 60% of the time) [26]. The practical significance of this rule is that it reveals a specific occupant behavior leading to energy inefficiency. A decision maker can make the right decision accordingly.

Wang et al. use association rules to extract rules in building operation patterns and the consumption of energy in commercial buildings. The study successfully identifies strong association rules between operational factors like equipment schedules, occupancy levels, and environmental conditions with high energy use. The use of association rules is highly useful because it reveals nonobvious, allowing for the development of targeted energy-saving strategies and data-driven operational improvements [27]. Dolores et al. utilized an association rule for a smart home and implemented its system.

They demonstrated that the discovered rules are highly beneficial, and a control system based on these rules achieved superior performance [26]. Association rules are useful techniques for knowledge discovery in databases, as they quantitatively express the likelihood of co-occurrence between variables.

13.3.1.3 Multiple Linear Regression Model

Regression is used for predicting a continuous value. Multiple linear regression (MLR) is a simple model for predicting a dependent variable by a series of independent variables by fitting a linear equation, Eq. (13.3),

$$Y = \alpha + \beta_1 X_1 + \beta_2 X_2 + \beta_3 X_3 + \dots + \beta_n X_n \tag{13.3}$$

where x_i and y are independent variables and the dependent variable. β_i and α (known as regression coefficients) define the slope of the line and the y-intercept, respectively [15, 25]. Predictive performance of the model is evaluated using the coefficient of determination (R^2) as shown in Eq. (13.4). When R^2 is close to one, the fit of the model is good; when R^2 is close to zero, the fit is poor [15], where $\hat{y}_i$ is the value of the property according to the model, y_i is value according to the original data, and $\bar{y}$ is the average of the attributes,

$$R^2 = \frac{\sum_{i=1}^{m} (\hat{y}_i - \bar{y})^2}{\sum_{i=1}^{m} (y_i - \bar{y})^2} \tag{13.4}$$

Osobajo et al. use MLR to predict GHG emissions by energy demand, gross domestic product (GDP), population, and capital formation across 70 countries. The results demonstrated that all variables had a statistically significant positive relationship with GHG emissions. Specifically, a 1% increase in energy consumption led to a 0.88% rise in GHG emissions, while a 1% increase in economic growth resulted in a 1.02% increase. The high R^2 value (0.91) confirmed that the model effectively explained the data, which is good [28]. Saed et al. used linear regression to quantify the relationships between water, energy, food, GDP, and GHG emissions for Canada and calculated R^2 to measure the strength of each relationship [29].

13.3.1.4 Artificial Neural Network

An artificial neural network (ANN) is a computational model that is a powerful regression for identifying and modeling complex and nonlinear relationships between input and output variables. A common type of ANN is the multilayer perceptron (MLP), which is structured with an input layer, an output layer, and one or more hidden layers. Figure 13.2 illustrates a simple MLP with one hidden layer.

The output of each neuron is calculated using Eq. (13.5) [16]. a is the result of applying an activation function (such as Sigmoid or ReLU) to the weighted sum of the inputs p, plus a bias term b,

$$a = f(Wp + b) \tag{13.5}$$

The network learns by iteratively adjusting its weights (W_i) to minimize prediction error. The weights are initially randomized and then updated in each iteration (often using the backpropagation algorithm) according to Eq. (13.6) [16], where the weights W_i are initially randomized and then iteratively updated using a learning rule that incorporates the learning rate η, the local error gradient δ_j, and the output O_i from the preceding neuron. The weight update uses a momentum

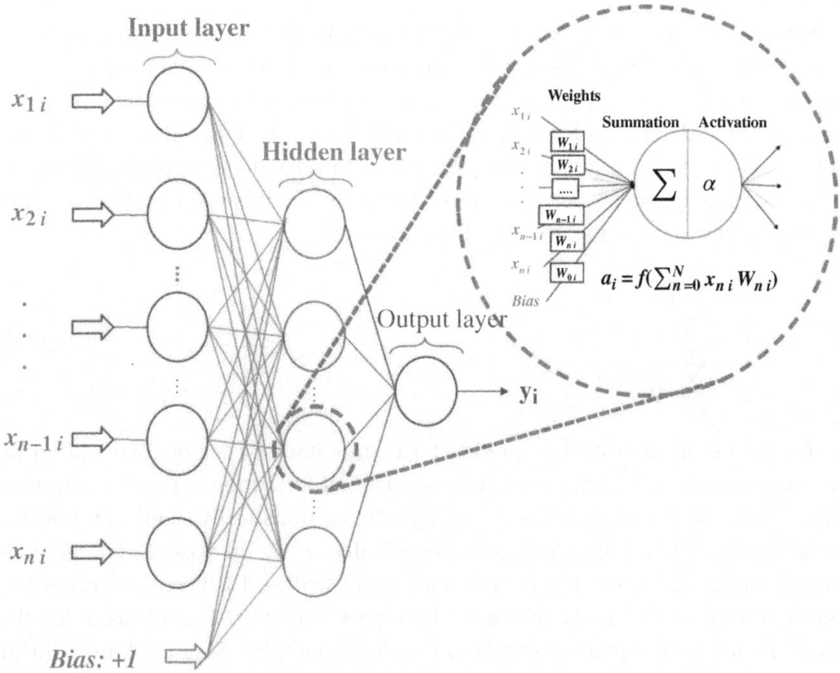

Figure 13.2 Schematic of MLP architecture with a detailed view of a single artificial neuron.

term (m) to speed up training and reduce fluctuations. It does this by adding a fraction of the previous update to the current update,

$$\Delta W_i^{p+1} = \eta \times (\delta_j O_i) + m\Delta W_i^p \tag{13.6}$$

ANN becomes a Bayesian neural network (BNN) when its weights and biases are represented as probability distributions rather than deterministic values. This change allows the model to quantify the uncertainty in its predictions. BNNs are particularly useful in high-risk applications where understanding the certainty of a prediction is as critical as the prediction itself. They help prevent overfitting and improve decision-making under uncertainty.

ANN becomes a recurrent neural network (RNN) when it is designed to process sequential data and time series. Unlike MLP where information flows strictly in one direction, in RNN the information flows from forward and backward. RNNs are essential in areas such as time-series analysis, where future values depend on past observations. Their ability to model dependencies over time makes them an essential tool for any application involving sequential or time-dependent data. When working with recurrent methods such as RNNs, it is essential not to shuffle the dataset.

Long short-term memory (LSTM) is a specialized type of RNN designed to solve the critical problem of vanishing gradients that plague standard RNNs. While a standard RNN has a simple repeating module (like a single tanh layer), an LSTM unit has a more complex, sophisticated structure. Its core innovation is the cell state managed by three regulatory gates, forget gate (to discard information), input gate (to add new information), and output gate (to determine the hidden state) (Figure 13.3). This mechanism enables LSTMs to selectively remember information over long sequences, making them highly effective for tasks with long term like complex time-series forecasting.

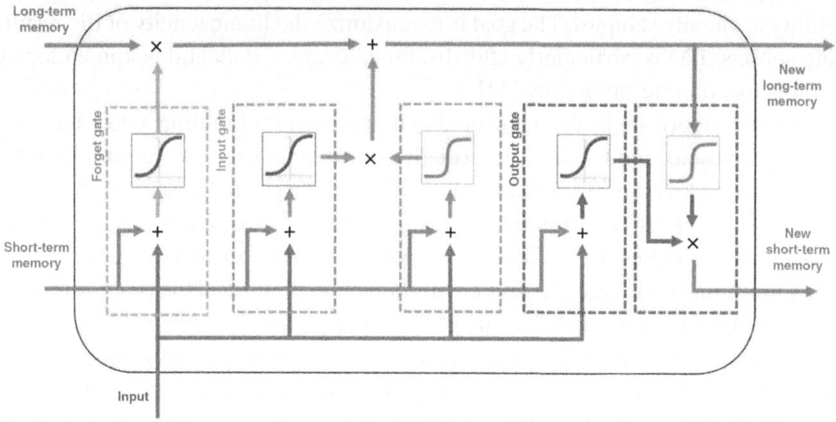

Figure 13.3 Schematic of an LSTM node.

ANNs have been employed for prediction in the MENs. For example, in Europe, inputs such as GDP, population, electricity demand, total primary energy consumption, average temperature, and the share of energy sources are used. The model is trained on historical data and then applied to future data, enabling the estimation of emission levels based on different energy policy scenarios [30].

The BNN algorithm has shown great potential in MEN planning, for example, in vehicle-to-grid planning. It has shown better performance compared to other approaches, making it a powerful tool for CESN [31]. Avami and Boroushaki developed an RNN to predict total energy consumption based on GDP and population data. The model effectively captured the nonlinear temporal dependencies between economic indicators and energy demand, making it particularly suitable for modeling complex and dynamic systems such as MENs [32]. Tang et al. use LSTM as their core predictive model to forecast transportation GHG emissions in China. The choice of an LSTM is crucial due to its inherent ability to capture complex, nonlinear temporal dependencies and long-term patterns within time-series data, which traditional models often miss [33].

13.3.1.5 Classification Models

Classification is a form of data analysis that extracts models to describe important data classes, which are subsequently used to predict categorical (discrete, unordered) class labels [15]. Bayesian classifiers are a family of statistical classifiers rooted in Bayes' theorem. Their key advantage is the ability to predict class membership probabilities, meaning they can quantify the likelihood that a given data sample belongs to a specific class [15]. Iterative Dichotomizer 3 (ID3) is a fundamental algorithm for generating decision trees (DT) from labeled training data. It operates by selecting the most informative attribute at each node to split the dataset, using a metric called information gain to measure an attribute's ability to classify samples. The goal is to maximize the homogeneity of the resulting subsets. ID3 is particularly effective for categorical data and is known for its simplicity and interpretability [34].

The ID3 algorithm is a greedy, top-down approach for building a decision tree. It works by selecting the most informative attribute at each step to partition the data. The process can be summarized as follows: for a given dataset, the algorithm calculates a selection criterion (such as information gain) for all unused attributes. It then chooses the attribute that optimizes this criterion (e.g., provides the maximum information gain) and creates a new node for it. This process repeats for each resulting subset of the data until a stopping condition is met [34].

Venkatesh et al. used Bayesian classifiers for detecting photovoltaic module visual faults [35]. Wu et al. used DT to analyze MENs balance, which integrates energy security, equity, and environmental sustainability. The DT approach helped identify the most influential factors affecting energy system balance and revealed how their importance shifted over time. The method is valuable because it provides clear, interpretable rules for policymakers, enabling data-driven and balanced energy policy design that adapts to dynamic global challenges [36].

13.3.1.6 Clustering Models

Clustering, in contrast to previous methods, is an unsupervised learning algorithm. This means that the dataset contains no pre-assigned labels. The algorithm partitions the dataset into clusters based on the similarity between data points. The core objective is to group data points such that those within the same cluster have the highest degree of similarity to each other, while being as dissimilar as possible to points in other clusters. One of the most common and foundational clustering techniques is the *k*-means algorithm [25].

Naderian et al. applied the *k*-means clustering method to group citizens based on their energy consumption, CO_2 footprint, and sociodemographic attributes. This approach is highly valuable as it successfully identified distinct citizen profiles, providing a critical foundation for policymakers to develop tailored and effective energy plans for each specific cluster of citizens [37]. Wang et al., in order to achieve better results, divided the database into several clusters using clustering techniques prior to executing their main algorithm for association rule mining. They then applied the algorithm separately to each cluster. This approach led to improved outcomes in their study [27].

13.3.1.7 Ensemble Model

Ensemble learning is a powerful approach for improving predictive model accuracy by combining multiple models. The core motivation is to reduce the overall error rate. This method operates on the principle that while individual base classifiers may make errors, an ensemble will only misclassify an instance if a majority of its components are wrong.

Common ensemble methods include bagging, boosting, and random forest (RF). Bagging (bootstrap aggregation) is a foundational technique that creates multiple versions of the training set through bootstrap sampling (sampling with replacement). A separate model is trained on each of these datasets, and their predictions are combined through voting (for classification) or averaging (for regression) to produce a single, robust output [15, 25].

Boosting is a highly powerful and widely used ensemble method. It works by iteratively training a series of classifiers. After each classifier is trained, the weights of the training tuples are updated to force the next classifier to focus more on the instances that are misclassified. The final prediction is a weighted vote of all classifiers, where each classifier's influence is determined by its accuracy. In contrast, RF is particularly well suited for analyzing complex data, especially in datasets with a very high number of features but a relatively small number of records. An RF is an ensemble of multiple DT. Each tree is built on a random subset of the data and features. During classification, every tree casts a vote, and the most popular class is selected as the final output [15, 25].

Rezazadeh et al. employed a recurrent ensemble model to forecast temperature and energy demand. By leveraging the complementary strengths of two powerful RNNs and a recurrent RF for handling complex feature interactions. This integrated approach significantly enhanced prediction accuracy and generalization capability, effectively modeling both the sequential patterns in time-series data and the nonlinear relationships influencing energy demand and temperature fluctuations. The ensemble method proved highly effective in producing reliable forecasts, demonstrating the value of combining deep learning with ensemble techniques for complex spatiotemporal prediction tasks [2].

13.3.2 Mathematical Programming

An MP typically consists of one or more objective functions, subject to constraints that define the feasible solution space. If the objective function and the constraints of such a model are linear, it is referred to as linear programming (LP). LP is highly desirable because its feasible solution space is convex, guaranteeing that the obtained optimal solution is global rather than trapped in local optima. This property is exceptionally valuable, which is why significant effort is often made to linearize nonlinear models when possible.

MP can be formulated as an integer programming problem, meaning that some decision variables are not continuous. For example, variables such as the number of power plant units or other discrete quantities are represented by integer values. Integer programming allows the model to accurately incorporate these discrete decision variables [38].

Rezazadeh and Avami developed an LP to represent MENs from primary sources to useful energy output to study CESN. By conducting extensive sensitivity analyses alongside uncertainty assessments on energy carrier prices and carbon costs, they provided a more precise projection of future scenarios. When the parameters within an LP are deterministic, the approach is referred to as deterministic programming; however, if these parameters are uncertain and modeled probabilistically, the methodology is known as stochastic programming, as exemplified in this study [13]. LP can be a bi-level problem to address hierarchical decision-making processes in complex systems. For instance, at the lower level, total descanted cost can be optimized, while at the upper level, the optimal carbon price to achieve a specific emissions target [39, 40].

13.3.3 Agent-based Modeling

ABM is a computational model that can be used in various disciplines [41]. In this method, the behavior and characteristics of each agent are determined through a set of computations. The interaction of different agents within an environment whose conditions are also defined by constraints and mathematical formulations forms the basis of the problem. The interaction between agents makes it possible to model complex and systemic problems [42]. ABM allows for modeling each agent individually and defining its unique characteristics. This makes ABM powerful for addressing multi-agent problems and effectively capturing the complex and diverse behaviors of agents within MENs. The responsibility and behavior of each agent depend on the case study and its underlying assumptions.

Doraki et al. utilized the power of ABM for transportation modeling. In this study, the transportation network and 100,000 agents are modeled, with each agent optimizing a scoring function based on early arrival, late arrival, and time spent in traffic, which is optimized through iteration. Using this algorithm, they simulated the effects of policies such as flexible working hours and telecommuting on the transportation system. In such models, agents influence each other; for instance, the collective choices of agents cause traffic on a particular route, which in turn affects all agents. This creates a complex, dynamic system where micro-level decisions lead to emergent macro-level phenomena [43]. In another study, ABM is employed to simulate the interactions among agents within the energy supply and demand sectors, with a specific focus on industrial symbiosis. It is designed such that distinct policies, tailored to the behavior of individual agents at the micro-level, generate specific and observable conditions at the macro-level. This emergent system behavior is effectively captured and illustrated by the simulation [44].

13.3.4 Integrated Models

Integrated modeling refers to a computational approach that combines multiple sub-models into a single cohesive framework. These sub-models typically represent different systems or disciplines, such as energy, climate change, economy, water, air pollution, health, and land. The core purpose of integration is to capture the critical feedback loops and dynamic interactions between these systems. For instance, an integrated model might show how a climate policy affects energy and health, which in turn influences GHG and air pollution emissions whose impacts then feed back into MENs [45].

The integration of models is typically categorized as either hard-linked or soft-linked, which differ in their level of coupling and data exchange. In a hard-linked model, all sub-models are formulated and solved within a single, unified computational framework like [45]. On the other hand, soft-linked models maintain separate sub-models that exchange data through input and output. A soft-linked model can be further categorized: if it employs iteration and feedback loops between its components to achieve consistency, it is termed an iterative soft-linked model [46]. Conversely, if data flows in a single pass without any iterative revision, the model is referred to as a one-way soft-linked model [47].

An integrated assessment model can be based on any of the previous models or a combination of them, although LP is also a common approach for this task. The objective function in integrated LP is usually to minimize costs. However, emission penalty, health costs, etc. may also be included in these costs. Additionally, by modifying the objective function and some constraints, different approaches can be incorporated into the model. Studies have shown that greedy and single-objective approaches are not desirable because they can worsen other goals [45]. To diagnose convergence between different objectives in modeling results, specific indices, such as correlation, distance from the origin index (DOI), and co-benefit area index (CBAI), have been introduced [45].

To this end, numerous studies have successfully employed this framework to create integrated hard link land [48], water [49], both land and water [50], and even public health CESN [45]. Several of these integrated models are also open source, often through open-source platforms or user-friendly web interfaces [51–53]. These tightly coupled models facilitate the concurrent modeling and optimization of a single problem with diverse, and often competing, objectives. Studies demonstrate that neglecting these interconnections can lead to significant trade-offs and suboptimal outcomes, where marginal gains in one objective incur substantial, and often unacceptable, costs to others. In some cases, this hard link interconnection might be relaxed. Instead, models can establish a soft link between separately run models to capture complex, cross-sectoral effects. For instance, a soft link could be created to quantify the impact of MEN policies on public health outcomes [54, 55].

A key feature of many integrated models is their open-source availability, a characteristic that fosters transparency, facilitates the reproducibility of results, and encourages collaborative development. Widely used examples span various methodological approaches, including: the global change analysis model (GCAM), which employs a simulation and partial equilibrium framework [56, 57]; the optimization-based open-source energy modeling system (OSeMOSYS) which utilizes LP [58, 59]; the climate, land, energy, and water systems (CLEWs) framework, also based on LP [51, 52]; the integrated MARKAL-EFOM system (TIMES), which utilizes optimization and partial equilibrium principles [60]; and the long-range energy alternatives planning system (LEAP), which uses simulation, accounting, and econometric methods [61].

13.3.5 System Dynamics

SD is a methodology and computer simulation modeling approach used to understand, analyze, and discuss the nonlinear behavior of complex systems over time. It is particularly focused on how the structure of a system (the interconnections between its parts) generates its behavior. Unlike ABM (which focuses on the actions of individual entities), SD takes a top-down, macro-level view. It deals with aggregates and flows, treating the system as a whole. An advantage of SD is its unique ability to quantify qualitative relationships.

This method is particularly effective for capturing causal relationships. For instance, it enables the modeling of cause-and-effect dynamics in the development of MENs. It allows for the integration of various factors such as government support, human capital, and import tariffs within a complex system, facilitating the analysis of each policy's individual impact [62]. Furthermore, this method can be applied to analyze the cause-and-effect relationships within a specific industry, such as cement. It allows for the modeling of how carbon emission mitigation technologies apply in an industrial ecosystem [63].

13.3.6 Game Theory

Game theory is a mathematical framework designed to model and analyze strategic interactions among rational decision makers, often referred to as players. It studies how players choose strategies to maximize their own outcomes in situations where their success depends not only on their own actions but also on the choices of others. Unlike traditional centralized energy systems, modern systems are defined by markets, distributed technologies, and numerous independent players, making game theory more necessary [64]. For example, in the electricity market, every producer, including renewable and nonrenewable ones, would achieve higher profits by colluding and selling at the maximum price. However, each player's incentive to sell more leads them to offer a lower price. The repetition of this behavior causes prices to collapse. This game may have another player, such as the government, which tries to change the macro conditions by imposing tariffs, subsidies, or taxes [65]. Furthermore, game theory can consider critical factors such as resource availability and geopolitical positioning. This is particularly useful in analyzing natural gas transit, where the strategic interdependence between producing, consuming, and transit states creates a complex multi-level game. The geographic distribution of resources and pipelines makes bargaining power and vulnerabilities, thereby influencing strategies and potential equilibria in negotiations over energy security, pricing, and supply [66, 67].

13.3.7 Life Cycle Assessment

LCA, which is primarily used for environmental analyses, offers a comprehensive approach to evaluating processes or policies. In this method, impacts are assessed "from cradle to grave," meaning the entire life cycle of a product is considered. This approach not only allows for comprehensive environmental assessments to include energy and climate change, but also extends to include economic by life cycle costing analysis (LCCA) and social by social life cycle assessment (S-LCA), making it more suitable for policy decision-making. The life cycle approach provides a more comprehensive sustainability analysis by evaluating all stages, rather than focusing solely on the operational phase. LCA typically consists of four main phases: goal and scope definition, inventory analysis, impact assessment, and interpretation of results.

Alizadeh et al. used LCA to investigate the influence of carbon capture in the energy system. Through this study, the advantages and disadvantages of using carbon capture at different stages of MENs are considered [68]. On the other hand, LCCA estimates the costs over the life cycle instead of the environmental impact, which is useful for studying efficient strategies for MENs management over the life cycle [69]. Another important factor for decision-making is the consideration of social issues. This can be investigated through S-LCA, which allows the assessment of the social impacts of a policy or project. For example, the social benefits of implementing a policy, such as the social consequences of developing green hydrogen, are assessed by using this approach [70].

13.4 Foundational Qualitative Models for Policy Analysis

13.4.1 SWOT Analysis

SWOT is a powerful strategic planning tool that helps MEN's policymakers to assess the current situation and make better decisions. It involves four key components: strengths (positive internal characteristics under the policymaker's control), weaknesses (negative internal factors that may hinder progress), opportunities (positive external conditions that the organization can exploit), and threats (negative external factors that may pose risks or challenges). By systematically analyzing these elements, policymakers understand internal capabilities and the external environment. This comprehensive understanding allows them to develop actionable strategies, maximize strengths and opportunities, and minimize weaknesses and threats. For example, this method could be applied to develop a strategic plan for wind energy as a part of MENs in the context of climate change, as shown in Figure 13.4 [71]. In another study, this method is used to assess the transition toward hydrogen in MENs [72].

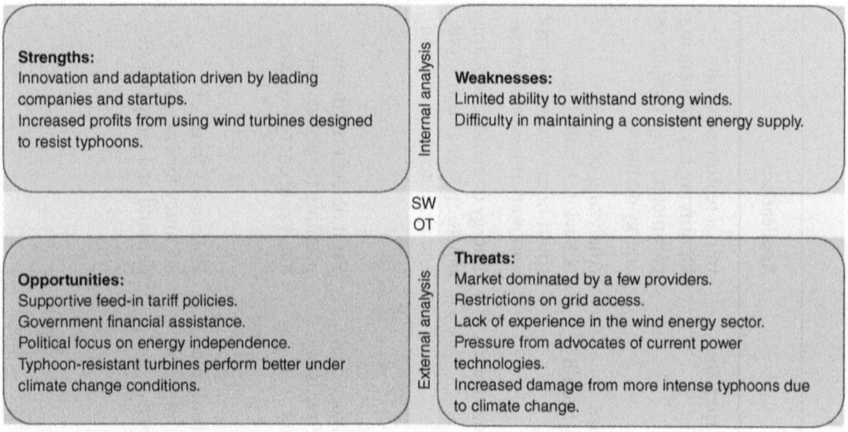

Figure 13.4 SWOT analysis for wind energy depending on climate change.

13.4.2 The Delphi Method

The Delphi method is a policy-making model that relies on a panel of experts. The basic principle of this method is to achieve convergence of opinions on the problem using multiple rounds of questionnaires sent to a panel of experts and providing controlled feedback after each round. The method is designed to overcome the disadvantages of traditional group decision-making.

In this method, a group of experts may come together over a specified period of time, such as a week, to discuss and provide their insights about the MENs transition. The Delphi technique ensures that over time, participants reach consensus on their responses, leading to more refined and reliable results [73]. The Delphi method is also used to identify and select the most suitable areas within a city to achieve a positive energy balance for carbon-neutral cities [74].

13.5 Conclusion and Future Perspective

This chapter enables researchers and decision makers to gain a basic understanding of various policy models and helps them select the most suitable model for their own research.

There is a wide range of approaches for policy modeling, summarized in Table 13.1, and they can be broadly classified into two main groups: quantitative and qualitative. Several quantitative models have been described, including SD, MP, ABM, LCA, game theory, and data-driven methods. SD is a top-down model that focuses on modeling complex systems and their behavior over time using feedback loops, stocks, and flows. It is particularly useful for understanding

Table 13.1 Comparative analysis of policy-making models for MENs under the CESN framework.

Specific model(s)	Advantages	Disadvantages	Features	Applications
Correlation analysis	Simple to compute and interpret. Identifies strength and direction of relationships	Only measures association, not causation. Cannot capture complex, nonlinear relationships	Statistical measure (−1 to +1)	Initial feature selection, identifying key influencing parameters (e.g., climate vs. energy demand)
Association rules	Discover hidden, nonobvious relationships in data. Provide quantitative likelihood	Can generate a large number of trivial rules	If-then rules (e.g., "If A then B")	Analyzing occupant behavior in buildings, smart home system control, and market basket analysis
Regression	Powerful for prediction and forecasting. Captures complex, nonlinear relationships	Requires large, high-quality datasets. Given that it is a black box; it has little interpretability. Risk of overfitting	Predicts continuous values	Energy demand forecasting, emission scenario analysis, and time-series prediction
Classification and clustering	Organizes data into classes or groups. DTs are interpretable. Unsupervised clustering finds hidden patterns	Performance depends on data quality and the chosen algorithm. May struggle with high-dimensional data. Risk of overfitting	Predicts categorical labels or groups of similar data	Fault detection, citizen profiling for policy, and analyzing energy systems
Ensemble methods	Increase predictive accuracy and robustness. Reduce overfitting and variance	Computationally intensive. Less interpretable than single models	Combine multiple models to improve performance	Enhancing forecast accuracy for energy demand and temperature

Mathematical programming	Provides a mathematically optimal solution. LP guarantees a global optimum. Clearly defined constraints and objectives	Can be computationally complex for very large systems. Requires precise formulation; sensitive to input parameters	Objective function minimization/ maximization subject to constraints.	Resource allocation, cost minimization, energy system design, and carbon pricing
ABM	Captures emergent behavior from micro-level interactions. Models heterogeneity and adaptive behavior of agents	Computationally expensive, especially with many agents. Can be difficult to validate and calibrate	Bottom-up simulation of autonomous agents	Modeling transportation networks, market dynamics, technology adoption, and industrial symbiosis
SD	Models feedback loops and long-term, systemic behavior. Effective for policy testing and understanding unintended consequences	Top-down approach may overlook individual heterogeneity. Relies on the accurate definition of causal relationships	Top-down, macro-level view using stocks, flows, and feedback	Analyzing renewable energy development, cause-and-effect in industries, and long-term trend analysis
Game theory	Models strategic interactions between rational decision makers. Analyzes markets, competition, and cooperation	Assumes rational actors, which may not always reflect real-world behavior. Can become very complex with many players	Analyzes strategies and equilibria in multi-agent systems	Electricity market analysis, natural gas transit negotiations, and policy design with subsidies/taxes
LCA	Provides a comprehensive "cradle-to-grave" perspective. Can integrate environmental, economic, and social dimensions	Requires extensive data inventory. System boundary definition can significantly impact results	Evaluates impacts across a product's/ policy's entire life cycle	Evaluating carbon capture technologies, comparing technology sustainability, and assessing the social impacts of policies

(Continued)

Table 13.1 (Continued)

Specific model(s)	Advantages	Disadvantages	Features	Applications
SWOT analysis	Simple, flexible, and intuitive framework. Encourages holistic strategic thinking by considering internal/external factors	Subjective and can be influenced by participant bias. Does not provide prioritization or decision rules on its own	Structured listing of strengths, weaknesses, opportunities, and threats	Strategic planning for technology development (e.g., wind energy, hydrogen transition)
Delphi method	Systematically builds consensus among experts. Groupthink by using anonymous, iterative feedback	Time-consuming and resource-intensive. Success depends on the selection and commitment of experts	Iterative expert opinion elicitation to converge on a consensus	Selecting optimal areas for urban energy projects and refining complex policy questions
Integrated models	Capture critical feedback loops between sectors (energy, water, climate, and health). Avoid suboptimal outcomes from greedy analysis	High complexity in model development and integration. Data requirements can be hard	Combine models from different disciplines into a single framework	Holistic resource planning, analyzing policy co-benefits and trade-offs (e.g., energy–emission–health nexus). Examples: CLEWs, OSeMOSYS, GCAM, and LEAP

long-term, large-scale trends and unintended consequences of policies. MP is designed to identify the best possible solution from a set of options, given specific constraints and objectives. These models are often used in resource allocation, logistics, and policy design where cost minimization is essential. This approach is also useful for integrated modeling. ABM simulates the actions and interactions of agents to assess their effects on the entire system. This approach is valuable for understanding emerging behaviors and testing scenarios where individual decision-making plays a key role. Data-driven methods rely on statistical analysis, ML, and data mining to recognize patterns, make predictions, or extract knowledge. These approaches are increasingly central to evidence-based policy-making in the era of big data. Each of these quantitative techniques offers a structured, analytical framework to support decision-making, though their applicability depends on the problem, data availability, and the level of complexity involved.

Qualitative models represent the other major category of policy modeling. Models such as SWOT provide decision makers with a structured framework for enhanced critical thinking. On the other hand, the Delphi method aims to elicit and converge expert opinions. The outcome of such a model is a synthesis derived from a panel of experts, which can provide significant support to the decision maker. A key advantage is that these models do not suffer from the limitations inherent in quantitative approaches.

The future perspective of energy policy modeling is moving toward higher integration using open-source codes/platforms. Next-generation models will increasingly use artificial intelligence (AI) and ML to enhance predictive accuracy, optimize complex system interactions, and manage large-scale data inputs. These data-driven techniques will be used in combination with other models. There will be a stronger emphasis on open-source and collaborative modeling frameworks, enabling greater transparency, reproducibility, and global cooperation. Additionally, models will incorporate dynamic socioeconomic and behavioral factors to better reflect real-world decision-making and policy impacts. The energy models will evolve into more interactive, accessible, and policy-relevant tools to support decarbonization and sustainable development goals globally. The critical issues of CESN will be highlighted in the future design of the MENs even more.

An energy modeler is like an artist, skillfully using a diverse set of tools and models to paint a complete and accurate picture of the future or recommend a path toward improvement. Such a MEN modeler may use expert judgment to develop future scenarios, use ML to estimate residential energy demand, use ABM to simulate transportation, and employ LP algorithms to optimize MENs using AGB, LCA, or game theory approaches. Therefore, a professional MEN modeler must be able to select and integrate one or more of these methods to suit the specific problem. Through careful analysis, they combine the results to provide a strong and useful basis for effective decision-making.

References

1 R. Grant, *Contemporary Strategy Analysis*, 8th edn. Wiley, 2013.

2 A. A. Rezazadeh, A. Avami, and M. Soleymani Baghshah, "Climatic scenario-based integrated recurrent ensemble model for energy demand forecasting," *Energy and Buildings*, vol. 310, p. 114103, 2024. https://doi.org/10.1016/j.enbuild.2024.114103

3 S. Dubey, J. N. Sarvaiya, and B. Seshadri, "Temperature dependent photovoltaic (PV) efficiency and its effect on PV production in the world—A review," *Energy Procedia*, vol. 33, pp. 311–321, 2013. https://doi.org/10.1016/j.egypro.2013.05.072

4 S. Kahe, A. Sharif, F. Joda, A. Avami, and A. Kianbakhsh, "Water-energy nexus in the combined cycle power plant using energy, exergy, economic and environmental analyses," *Energy Conversion and Management*, vol. 293, p. 117493, 2023. https://doi.org/10.1016/j.enconman.2023.117493

5 M. Rahman, F. Atchison, and V. Cecchi, "Temperature-dependent system level analysis of electric power transmission systems: A review," *Electric Power Systems Research*, vol. 193, p. 107033, 2021. https://doi.org/10.1016/j.epsr.2021.107033

6 A. T. D. Perera, V. M. Nik, D. Chen, J.-L. Scartezzini, and T. Hong, "Quantifying the impacts of climate change and extreme climate events on energy systems," *Nature Energy*, vol. 5, no. 2, pp. 150–159, 2020. https://doi.org/10.1038/s41560-020-0558-0

7 H. Ritchie, "Sector by sector: Where do global greenhouse gas emissions come from?," *Our World in Data*, 2022. Accessed: Aug. 25, 2022. [Online]. Available: https://ourworldindata.org/ghg-emissions-by-sector

8 EPA, "Strategies for saving energy at public water systems," pp. 1–16, 2013. [Online]. Available: https://www.epa.gov/sites/default/files/2015-04/documents/epa816f13004.pdf

9 P. Yousefi and A. Avami, "Optimal configuration of low-grade waste heat driven seawater desalination using data envelopment analysis: A case study of industrial application," *Environmental Progress and Sustainable Energy*, vol. 43, no. 5, p. e14438, 2024. https://doi.org/10.1002/ep.14438

10 European Commission, "Air pollution from key sectors," *European Commission*, 2025. Accessed: Sept. 4, 2025. [Online]. Available: https://environment.ec.europa.eu/topics/air/air-pollution-key-sectors_en

11 M. J. Kim, "Air pollution and solar photovoltaic power generation: Evidence from South Korea," *Energy Economics*, vol. 139, p. 107924, 2024. https://doi.org/10.1016/j.eneco.2024.107924

12 J. Rechlitz, *Air Pollution Impacts and Energy Infrastructure Efficiency: Empirical Evidence with Official Micro Data*. Technischen Universität Berlin, 2023.

13 A. Rezazadeh and A. Avami, "An integrated policy approach for sustainable decarbonization pathways of energy system in a city under climate change scenarios," *Energy Policy*, vol. 195, p. 114394, 2024.

14 C. Stach, "Data is the new oil–sort of: A view on why this comparison is misleading and its implications for modern data administration," *Future Internet*, vol. 15, no. 2, p. 71, 2023. https://doi.org/10.3390/fi15020071

15 J. Han, J. Pei, and M. Kamber, *Data Mining: Concepts and Techniques*, The Morgan Kaufmann Series in Data Management Systems. Elsevier Science, 2011. [Online]. Available: https://books.google.ca/books?id=pQws07tdpjoC

16 P. Movahed, A. A. Rezazadeh, A. Avami, M. Soleymani Baghshah, and M. Mashayekhi, "Modeling and optimization of NO emission for a steam power plant by data-driven methods," *Environmental Progress and Sustainable Energy*, vol. 42, no. 3, p. e14059, 2023. https://doi.org/10.1002/ep.14059

17 V. Verpeta, "Correlation analysis of parameters of climate geoinformation systems for renewable energy," *System Research in Energy*, vol. 2025, no. 2, pp. 91–100, 2025. https://doi.org/10.15407/srenergy2025.02.091

18 A. Rezazadeh, A. Avami, and M. S. Hassanvand, "Assessment of co-benefits exploring the intersections of human health, air pollution, climate change, and sustainable energy systems by rigorous integrated optimization models," 2024.

19 F. A. Canales, J. Jurasz, A. Beluco, and A. Kies, "Assessing temporal complementarity between three variable energy sources through correlation and compromise programming," *Energy*, vol. 192, p. 116637, 2020. https://doi.org/10.1016/j.energy.2019.116637

20 Y. Li et al., "Association rule-based feature mining for automated fault diagnosis of rolling bearing," *Shock and Vibration*, vol. 2019, pp. 1–12, 2019. https://doi.org/10.1155/2019/1518246

21 K.-I. Ahn, "Effective product assignment based on association rule mining in retail," *Expert Systems with Applications*, vol. 39, no. 16, pp. 12551–12556, 2012. https://doi.org/10.1016/j.eswa.2012.04.086

22 J. Nahar, T. Imam, K. S. Tickle, and Y.-P. P. Chen, "Association rule mining to detect factors which contribute to heart disease in males and females," *Expert Systems with Applications*, vol. 40, no. 4, pp. 1086–1093, 2013. https://doi.org/10.1016/j.eswa.2012.08.028

23 B. Kamsu-Foguem, F. Rigal, and F. Mauget, "Mining association rules for the quality improvement of the production process," *Expert Systems with Applications*, vol. 40, no. 4, pp. 1034–1045, 2013. https://doi.org/10.1016/j.eswa.2012.08.039

24 T. Li and X. Li, "Novel alarm correlation analysis system based on association rules mining in telecommunication networks," *Information Sciences*, vol. 180, no. 16, pp. 2960–2978, 2010. https://doi.org/10.1016/j.ins.2010.04.013

25 M. Kantardzic, *Data Mining: Concepts, Models, Methods, and Algorithms*. Wiley, 2011. [Online]. Available: https://books.google.com/books?id=ZZ7l6v0CvRMC

26 M. Dolores, C. Fernandez-Basso, J. Gómez-Romero, and M. J. Martin-Bautista, "A big data association rule mining based approach for energy building behaviour analysis in an IoT environment," *Scientific Reports*, vol. 13, no. 1, pp. 1–18, 2023. https://doi.org/10.1038/s41598-023-47056-1

27 F. Wang et al., "Association rule mining based quantitative analysis approach of household characteristics impacts on residential electricity consumption patterns," *Energy Conversion and Management*, vol. 171, pp. 839–854, 2018. https://doi.org/10.1016/j.enconman.2018.06.017

28 O. A. Osobajo, A. Otitoju, M. A. Otitoju, and A. Oke, "The impact of energy consumption and economic growth on carbon dioxide emissions," *Sustainability*, vol. 12, no. 19, pp. 1–16, 2020. https://doi.org/10.3390/su12197965

29 B. Saed, A. Elshorbagy, and S. Razavi, "Quantifying interactions in the water-energy-food nexus: Data-driven analysis utilizing a causal inference method," *Frontiers in Environmental Science*, vol. 11, pp. 1–19, 2023. https://doi.org/10.3389/fenvs.2023.1328009

30 D. J. Đozić and B. D. Gvozdenac Urošević, "Application of artificial neural networks for testing long-term energy policy targets," *Energy*, vol. 174, pp. 488–496, 2019. https://doi.org/10.1016/j.energy.2019.02.191

31 F. Wang, B. Jiang, L. Hu, H. Jin, Q. Yu, and Y. Shi, "Bayesian deep neural network for low-carbon scheduling strategy planning of V2G," *International Journal of Low-Carbon Technologies*, vol. 19, pp. 642–653, 2024. https://doi.org/10.1093/ijlct/ctad145

32 A. Avami and M. Boroushaki, "Energy consumption forecasting of Iran using recurrent neural networks," *Energy Sources, Part B: Economics, Planning, and Policy*, vol. 6, no. 4, pp. 339–347, 2011. https://doi.org/10.1080/15567240802706734

33 J. Tang, R. Gong, H. Wang, and Y. Liu, "Scenario analysis of transportation carbon emissions in China based on machine learning and deep neural network models," *Environmental Research Letters*, vol. 18, no. 6, 2023. https://doi.org/10.1088/1748-9326/acd468

34 RapidMiner, "Operator manual," RapidMiner Documentation, 2020. Accessed: July 16, 2020. [Online]. Available: https://docs.rapidminer.com/latest/studio/operators/

35 S. N. Venkatesh, V. Sugumaran, B. Subramanian, J. S. F. Josephin, and E. G. Varuvel, "A comparative study on bayes classifier for detecting photovoltaic module visual faults using deep learning features," *Sustainable Energy Technologies and Assessments*, vol. 64, p. 103713, 2024. https://doi.org/10.1016/j.seta.2024.103713

36 S. Wu, Q. Wu, and J. Tan, "Assessing and analysing energy system balance: A decision tree approach," *Energy*, vol. 304, p. 131902, 2024. https://doi.org/10.1016/j.energy.2024.131902

37 S. Naderian, A. Ioannou, and G. Falcone, "A novel approach to deriving energy citizens profiles for effective energy policies: Clustering and statistical analyses applied to UK data," *Journal of Cleaner Production*, vol. 519, p. 146036, 2025. https://doi.org/10.1016/j.jclepro.2025.146036

38 H.-W. Hsu and Z.-W. Fan, "Multi-faceted procurement with mixed integer linear programming for corporate 100% renewable energy goal," *Energy*, vol. 320, p. 135144, 2025. https://doi.org/10.1016/j.energy.2025.135144

39 A. A. Rezazadeh and A. Avami, "Optimal carbon pricing considering human health benefits from reduction in air pollutants emissions in urban energy systems," *Journal of Environmental Studies*, vol. 51, no. 2, 231–253, 2025.

40 A. Shahbazi, A. Avami, M. Moeini-Aghtaie, and H. Tavassoti, "Optimum carbon tax rate for emission targets of electricity generation system by life cycle techno-economic-environmental optimization model," *Renewable Energy*, vol. 237, p. 121877, 2024. https://doi.org/10.1016/j.renene.2024.121877

41 H. J. Scholl, "Agent-based and system dynamics modeling: A call for cross study and joint research," in *Proceedings of the 34th Annual Hawaii International Conference on System Sciences*, IEEE, 2001, pp. 1–8.

42 U. Wilensky and W. Rand, *An Introduction to Agent-Based Modeling: Modeling Natural, Social, and Engineered Complex Systems with NetLogo*. MIT Press, 2015. [Online]. Available: https://books.google.com/books?id=LQrhBwAAQBAJ

43 M. Erfan Doraki, A. Avami, M. Boroushaki, and Z. Amini, "Agent-based modeling for sustainable urban passenger vehicle mobility: A case of Tehran," *Transportation Research Part D: Transport and Environment*, vol. 135, p. 104380, 2024. https://doi.org/10.1016/j.trd.2024.104380

44 Z. Saghafi and R. Roshandel, "Agent-based simulation for technology implementation in an energy-based industrial symbiosis network," *Resources, Conservation and Recycling Advances*, vol. 21, p. 200201, 2024. https://doi.org/10.1016/j.rcradv.2023.200201

45 A. A. Rezazadeh, A. Avami, and M. S. Hassanvand, "Energy planning in urban areas under climate change: An integrated energy-emission-health optimization model," *Energy Reports*, vol. 14, pp. 5922–5940, 2025. https://doi.org/10.1016/j.egyr.2025.08.017

46 M. Rosendal, J. Janin, T. Heggarty, D. Pisinger, R. Bramstoft, and M. Münster, "The benefits and challenges of soft-linking investment and operational energy system models," *Applied Energy*, vol. 385, p. 125512, 2025. https://doi.org/10.1016/j.apenergy.2025.125512

47 M. H. Khoshgoftar Manesh, A. Rezazadeh, and S. Kabiri, "A feasibility study on the potential, economic, and environmental advantages of biogas production from poultry manure in Iran," *Renewable Energy*, vol. 159, pp. 87–106, 2020. https://doi.org/10.1016/j.renene.2020.05.173

48 H. Rahemi, S. A. Torabi, A. Avami, and F. Jolai, "Bioethanol supply chain network design considering land characteristics," *Renewable and Sustainable Energy Reviews*, vol. 119, p. 109517, 2020. https://doi.org/10.1016/j.rser.2019.109517

49 A. HajHasan and A. Avami, "Water and emissions nexus for biodiesel in Iran," *Renewable and Sustainable Energy Reviews*, vol. 93, pp. 354–363, 2018. https://doi.org/10.1016/j.rser.2018.05.052

50 A. Aghadadashi and A. Avami, "Water-land-food-human health nexus in bioenergy supply chain optimization under climatic constraints," *Science of The Total Environment*, vol. 957, p. 177572, 2024. https://doi.org/10.1016/j.scitotenv.2024.177572

51 A. Beltramo, E. P. Ramos, C. Taliotis, M. Howells, and W. Usher, "The global least-cost user-friendly CLEWs open-source exploratory model," *Environmental Modelling and Software*, vol. 143, p. 105091, 2021. https://doi.org/10.1016/j.envsoft.2021.105091

52 E. P. Ramos et al., "The climate, land, energy, and water systems (CLEWs) framework: A retrospective of activities and advances to 2019," *Environmental Research Letters*, vol. 16, no. 3, p. 033003, 2021. https://doi.org/10.1088/1748-9326/abd34f

53 E. P. Ramos et al., "Climate, land, energy and water systems interactions—From key concepts to model implementation with OSeMOSYS," *Environmental Science and Policy*, vol. 136, pp. 696–716, 2022. https://doi.org/10.1016/j.envsci.2022.07.007

54 A. A. Rezazadeh, S. Alizadeh, A. Avami, and A. Kianbakhsh, "Integrated analysis of energy-pollution-health nexus for sustainable energy planning," *Journal of Cleaner Production*, vol. 356, p. 131824, 2022. https://doi.org/10.1016/j.jclepro.2022.131824

55 A. A. Rezazadeh, A. Avami, M. Mashayekhi, and A. Kianbakhsh, "Assessing the contribution of different sources in atmospheric dispersion of $PM_{2.5}$ and related health impact in a region of Qazvin, Iran," *Air Quality, Atmosphere and Health*, vol. 15, no. 8, pp. 1379–1394, 2022. https://doi.org/10.1007/s11869-022-01163-6

56 Joint Global Change Research Institute (JGCRI), "Joint Global Change Research Institute," 2025. Accessed: Sept. 3, 2025. [Online]. Available: https://www.pnnl.gov/joint-global-change-research-institute-research

57 The Joint Global Change Research Institute (JGCRI), "Global change intersectoral modeling system," 2025. Accessed: Sept. 3, 2025. [Online]. Available: https://gcims.pnnl.gov/global-change-intersectoral-modeling-system

58 KTH-dESA, "OSeMOSYS documentation," 2021.

59 D. Dreier and M. Howells, “OSeMOSYS-PuLP: A stochastic modeling framework for long-term energy systems modeling,” *Energies*, vol. 12, no. 7, p. 1382, 2019. https://doi.org/10.3390/en12071382

60 R. Loulou and G. Goldstein, “Documentation for the TIMES PART IV: VEDA2.0,” *IEA Energy Technology Systems Analysis Program*, pp. 1–78, 2016.

61 LEAP. Accessed: Sept. 3, 2025. [Online]. Available: https://leap.sei.org/

62 A. Hashemizadeh, Y. Ju, and F. Z. B. Abadi, “Policy design for renewable energy development based on government support: A system dynamics model,” *Applied Energy*, vol. 376, p. 124331, 2024. https://doi.org/10.1016/j.apenergy.2024.124331

63 O. E. Ige, D. V. Von Kallon, and D. Desai, “Carbon emissions mitigation methods for cement industry using a systems dynamics model,” *Clean Technologies and Environmental Policy*, vol. 26, no. 3, pp. 579–597, 2024. https://doi.org/10.1007/s10098-023-02683-0

64 E. Haghi, “Game theory modeling of energy systems,” in *Handbook of Smart Energy Systems*, M. Fathi, E. Zio, and P. M. Pardalos, Eds., Springer International Publishing, 2023, pp. 941–959. https://doi.org/10.1007/978-3-030-97940-9_117

65 Z. Liu, S. Wang, M. Q. Lim, M. Kraft, and X. Wang, “Game theory-based renewable multi-energy system design and subsidy strategy optimization,” *Advances in Applied Energy*, vol. 2, p. 100024, 2021. https://doi.org/10.1016/j.adapen.2021.100024

66 J. P. Gausmann, *A Game-Theoretic Approach to Cooperation in the European Gas Market.* Institut Barcelona Estudis Internacionals, 2012.

67 S. Baghdadi, “Caspian basin natural gas transfer: A cooperative game theory approach,” *Journal of Energy and Economic Development*, vol. 1, pp. 50–66, 2022.

68 S. Alizadeh, A. A. Rezazadeh, and A. Avami, “A cutting-edge tool for sustainable environmental management through life cycle assessment,” *Renewable and Sustainable Energy Reviews*, vol. 192, p. 114194, 2024. https://doi.org/10.1016/j.rser.2023.114194

69 A. Hajare and E. Elwakil, “Integration of life cycle cost analysis and energy simulation for building energy-efficient strategies assessment,” *Sustainable Cities and Society*, vol. 61, p. 102293, 2020. https://doi.org/10.1016/j.scs.2020.102293

70 M. A. Moktadir and J. Ren, “An innovative social life cycle decision-making model for assessing social sustainability: A case of green hydrogen production,” *Sustainable Development*, vol. 2016, pp. 1–22, 2025. https://doi.org/10.1002/sd.3536

71 L. Bachmann et al., “Climate-resilient strategy planning using the SWOT methodology: A case study of the Japanese wind energy sector,” *Climate Risk Management*, vol. 46, p. 100665, 2024. https://doi.org/10.1016/j.crm.2024.100665

72 D. S. de Oliveira, A. L. Costa, and C. E. Velasquez, "Hydrogen energy for change: SWOT analysis for energy transition," *Sustainable Energy Technologies and Assessments*, vol. 72, p. 104063, 2024. https://doi.org/10.1016/j.seta.2024.104063

73 T. M. Ildefonso, J. Švaikauskaitė, J. A. Tvinnereim, A. Cieśla, E. Tyburn, and V. Laugery, "Urban synergy thinkers utilizing the Delphi technique to optimize energy transition planning," 2024.

74 I. Aparisi-Cerdá, D. Ribó-Pérez, T. Gómez-Navarro, M. García-Melón, and J. Peris-Blanes, "Prioritising positive energy districts to achieve carbon neutral cities: Delphi-DANP approach," *Renewable and Sustainable Energy Reviews*, vol. 203, p. 114764, 2024. https://doi.org/10.1016/j.rser.2024.114764

14

Real-world Case Studies for Climate–Energy–Sustainability Nexus in Deep Decarbonization of Modern Energy Networks

Sina Nabighadim[1], Hamed Tajalli[1], Asma Nasiri[1], Alireza Zare Ansaroudi[1], Kazem Zare[1,2], Behnam Mohammadi-Ivatloo[3], and Sajjad Tohidi[1]*

[1] *Energy Systems Research Institute (ESRI), Faculty of Electrical and Computer Engineering, University of Tabriz, Tabriz, Iran*
[2] *Energy Systems Research Center, Khazar University, Baku, Azerbaijan*
[3] *Department of Electrical Engineering, School of Energy Systems, LUT University of Technology, Lappeenranta, South Karelia, Finland*

*Corresponding author: sinanabi1374@gmail.com

List of Acronyms

CAISO	California independent system operator
CARB	California Air Resources Board
CARE	California Alternative Energy Rates
CCS	Carbon capture and storage
CEC	California Energy Commission
CESN	Climate–energy–sustainability nexus
CfD	Contract for difference
CHP	Combined heat and power
CPUC	California Public Utilities Commission
EEG	Erneuerbare-Energien-Gesetz
EIA	US Energy Information Administration
ETS	Emissions trading system
EV	Electric vehicle

List of Acronyms	
FIT	Feed-in tariff
GHG	Greenhouse gas
IEA	International Energy Agency
MEN	Modern energy network
PtX	Power-to-X
RPS	Renewable portfolio standard
TOU	Time-of-use
UHV	Ultrahigh voltage

14.1 Introduction

Climate change, energy security, and accessibility, as well as long-term environmental and socioeconomic sustainability, are fundamental topics that cannot be addressed separately. These subjects are interconnected and mutually reinforcing. By addressing these challenges in an integrated manner, climate–energy–sustainability nexus (CESN) provides a comprehensive framework that aligns deep decarbonization strategies with broader societal goals. Deep decarbonization, which can be introduced in colloquial language as a deep reduction in greenhouse gas (GHG) emissions in energy systems, has three fundamental pillars that can be presented by Figure 14.1:

- Low-carbon energy sources: the first and perhaps most effective and attainable action is the transition from fossil fuels to renewable energies, such as wind, solar, and hydroelectricity, as well as wind, solar, hydroelectricity, and beyond, such as biomass [1].
- Electrification and efficiency: electrification of end-user sectors (transportation, buildings, industry) that directly use fossil fuels for their operations, and also improving energy efficiency to reduce consumption per unit of service are other main actions in the decarbonization process. This means that the electrification alone cannot be effective; in return for electrification, service efficiency must also be improved so that the amount of electricity consumption does not increase drastically as a result of overall electrification [2].
- System integration and flexibility: renewable energies are the front-runners in decarbonization, with production with very high uncertainty based on highly variable environmental factors. The environmental factors that affect the reliability of producing energy for electricity generation. Adding storage systems [3] and demand-side plans [4] is also the main key to alleviating these issues.

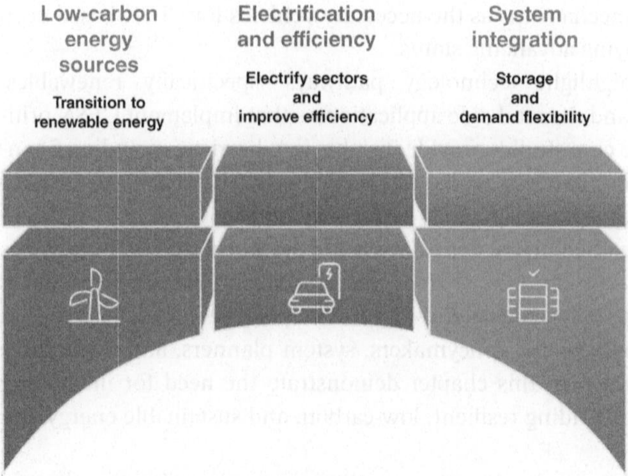

Figure 14.1 Decarbonization pillars.

The aforementioned three pillars can also be viewed as the conceptual foundations of deep decarbonization—however, what the CESN concept provides significance given the challenge of translating them into modern energy networks (MENs)—an integrated network of frameworks recognizing the connections between climate, energy and sustainability objectives, as the conceptual framework of CESN connects and articulates technological innovations related to those aims and stakeholders as well as socioeconomic and environmental and even policy dimensions of energy transformation [5]. In making connections across these domains, the CESN provides the basis for strategies that are more than simply a series of incremental steps forward and provide the potential for greater ability to develop game-changing systems-level transformations within the energy grids, for example. One of the major components of MENs is multi-carrier infrastructures [6] that combine the use of electricity, heating, cooling, hydrogen and other carriers in an integrated manner, it should be noted that decarbonizing is related to the electricity system when defining priorities, and yet can involve more complex inter-sectoral interactions, sector interconnections and coordinated planning across carriers of energy, which again can increase the levels of complexity. This complexity builds a case for an investigation of real-world case studies for the purpose of lowering limitations for identifying practical challenges, trade-offs, and the innovations needed to operationalize the CESN in different contexts.

Actual real-world case studies are important to improve comprehension about the performance of deep decarbonization based on CESN, as they illustrate implementation strategies in local contexts, such as the availability of resources and institutional capability. They also emphasize the importance of supportive policies and market mechanisms as the necessary vehicles for CESN-based deep decarbonization enjoying advancing status.

Moreover, they highlight technology pathways—specifically renewables, storage, smart grids, and demand-side applications—that implement CESN principles [7]. Finally, the case studies provide insights into barriers, including financial barriers, regulatory barriers, and social acceptance. Therefore, this chapter explores real-world case studies of CESN in the deep decarbonization of MENs to identify key lessons learned in practice from around the world, and both successes and limitations. Not only does the chapter intend to document advances, but it also identifies a comparative perspective that can help shape the next generation of decarbonization policies for policymakers, system planners, and researchers alike. The cases revealed in this chapter demonstrate the need for integrating CESN perspectives in building resilient, low-carbon, and sustainable energy systems for the future.

Figure 14.2 shows a word cloud for the core themes represented in this chapter. Words such as "Realworld," "Decarbonization," "Emissions," and "Renewable" feature prominently and support the theme of low-carbon transitions. This word cloud also reflects the case study approach of the chapter, with references to

Figure 14.2 Word cloud illustrating the key themes of this chapter.

state or national jurisdictions such as "Denmark," "California," "Germany," and "China." To develop the most frequently used words throughout this chapter, the authors pulled together about 100 key terms and narrowed down the words to eliminate redundancy, unnecessary text, and theme word. Finally, the remaining words were entered into the online site, "WordClouds.com," and processed into a word cloud. In the word cloud, the size of each word's font represents the frequency with which it has occurred in this chapter. This word cloud has concisely summarized the leading themes featured in this chapter.

The main objectives of this chapter are as follows:

1. To utilize the CESN framework to analyze case studies in Denmark, California, Germany, and China in order to better understand how the coordination of strategies shapes decarbonization outcomes.
2. To analyze the case studies using four indicator categories—technological, socioeconomic, governance and policy-making, and environmental—to identify critical drivers and constraints for transition pathways that support low-carbon development.
3. To contrast similarities and differences between national and regional approaches and emphasize contextual variables that shape decarbonization strategies and outcomes.
4. To support and illustrate that deep decarbonization cannot be achieved straightforwardly or universally but instead is the outcome of approaches that fit policy design, technology deployment, and social engagement to local conditions.

The chapter's structure is organized as follows. Section 14.2 details the methodological approach in analyzing real-world case studies based on CESN, and is also when the indicators pertaining to technology, socioeconomics, governance and policy, and environment are introduced in an organized manner. Section 14.3 presents four case studies that illustrate those based on CESN principles—Denmark, California, Germany, and China. For each case study, sections are devoted to providing the context, strategies, and success of deep decarbonization. Once presented, Section 14.4 analyzes the case studies comparatively by the CESN indicators highlighted in Section 14.2. In light of this, Section 14.5 will synthesize the findings and contributions to strategies, highlighting mutual lessons and recurring struggles across contexts. Future decarbonization pathways that consider CESN will be presented in Section 14.6, where a number of important new features that we expect to see being emphasized include, but are not limited to, digitalization, equity, an integrated multi-sectoral approach, and international engagement. To conclude the chapter, Section 14.7 will summarize the key themes covered in the chapter and reflect on the potential role of CESN for resilient, equitable, and low-carbon energy systems.

14.2 Methodological Approach for Case Study Analysis

It is not sufficient to only outline a number of practical project instances from diverse research in those various nations when examining real cases of deep decarbonization based on CESN. These may appear similar to reports from "corporate-style" presentations, but they serve no function in science and are insufficient for application in the research analysis. To ensure that the analysis is accurate, informative, and valuable for researchers and practitioners, a set of indicators and parameters should be defined that enable case studies to be evaluated in a structured manner and facilitate meaningful comparisons between practical studies. In this regard, the following categories of indicators are proposed as a basis for analyzing and comparing real-world case studies.

14.2.1 Categories of Indicators

To increase analytical depth and precision, this subsection introduces a structured framework of indicators for evaluating and comparing real-world CESN-based decarbonization case studies. In general, four categories of indicators can be defined for examining a real-world case study in this context. The indicators proposed here are adapted from established frameworks for sustainability and energy assessment. These factors are based on the International Energy Agency (IEA) [8, 9], the Intergovernmental Panel on Climate Change [10], and the World Bank's monitoring indicators for sustainable energy [11], and are tailored to CESN's scope and have a specific focus on deep decarbonization of MENs.

14.2.1.1 Technological Indicators

By considering that the energy system modernization and technological advancement are two inseparable factors, technology can be considered the first and most effective indicator in the CESN-based decarbonization process. Technological advances in renewable energy sources, energy storage, digitalization, and sector coupling have expanded the capabilities of MENs and enabled operations that were not previously possible.

- **Renewable energy share:** this index measures the proportion of energy generated from low-carbon sources such as wind, solar, water, biomass, and geothermal. It shows how much of the generation from fossil fuels has been replaced and is a direct measure of decarbonization.
- **Storage and flexibility capacity:** this metric looks at systems' capabilities to accommodate renewable variability and provide reliability. This includes storage and demand-side technologies that contribute to capacity, including

batteries, pumped hydro, and demand-side programs that absorb excess generation and provide capacity when renewable generation output declines.

- **System integration and sector coupling:** this metric captures an improvement in efficiency and emissions in the energy sector—electricity, heating, cooling, transport, and hydrogen. The integration of sectors reduces waste, fossil fuel use, and utilizes excess energy more effectively.

14.2.1.2 Socioeconomic Indicators

Like any major effort, decarbonization approaches need to be economically feasible. Any resources devoted to achieving climate objectives should be recoverable over time and, preferably, even provide additional returns. If not, there is always the chance that even the best projects will stall or simply not occur. Equally important to the economic feasibility is the social dimension, as energy systems ultimately exist to serve people and communities.

- **Affordability and cost-effectiveness:** the index describes whether modern energy systems are financially affordable for households and industry, and whether the savings (in costs or other resources) and benefits over time justify the cost of investment in decarbonization.
- **Job creation and economic development:** this index assesses whether decarbonized activities create jobs and stimulate economic development locally (i.e., at the level of a locality or to a larger economy within a nation) and/or to support the creation of regional economies, develop industries, and secure long-term livelihoods for people.
- **Equity of access:** this index assesses whether clean and reliable energy services are accessible for all groups in society, in particular marginalized societies and rural populations, and are framed in a way that promotes inclusiveness and equitable opportunities for development in society.
- **Public acceptance and social trust:** this index provides an indication of the degree to which communities support and trust decarbonization initiatives and new energy technologies that may be viable from a technical and/or economic perspective but may not progress if the social dimensions of communities are not taken into consideration or public involvement is not harnessed.

14.2.1.3 Governance and Policy Indicators

- **Institutional and regulatory strength:** this index shows the extent to which legislation, policy, and institutions support decarbonization projects. When the rules are clear and stable, and the institutions apply them predictably and impartially, projects can move forward and attract investment. Strong governance reduces the risk of project delays, lack of clarity regarding the rules, and corruption.

14.2.1.4 Environmental Indicators

As the ultimate purpose of decarbonization is to mitigate the environmental consequences, the environmental dimension is incorporated in and is integral to any CESN-based evaluation. Any measure of CESN effectiveness will include the environmental dimension to examine/demonstrate how an energy transition leads to reduced GHG emissions.

- **GHG emissions reduction:** the most straightforward indicator of progress toward decarbonization, it indicates how much a project or system reduces CO_2 and GHG emissions compared to conventional sources of energy.

The indicators mentioned above should not be considered quantitative measures but rather analytical categories for systematic comparison. Each case study is examined within these lenses to discern the characteristics of the technology, select socioeconomic and governance factors, and how these interact with environmental aspects to create the pathway chosen for the study. Figure 14.3 graphically represents the basis for decarbonization.

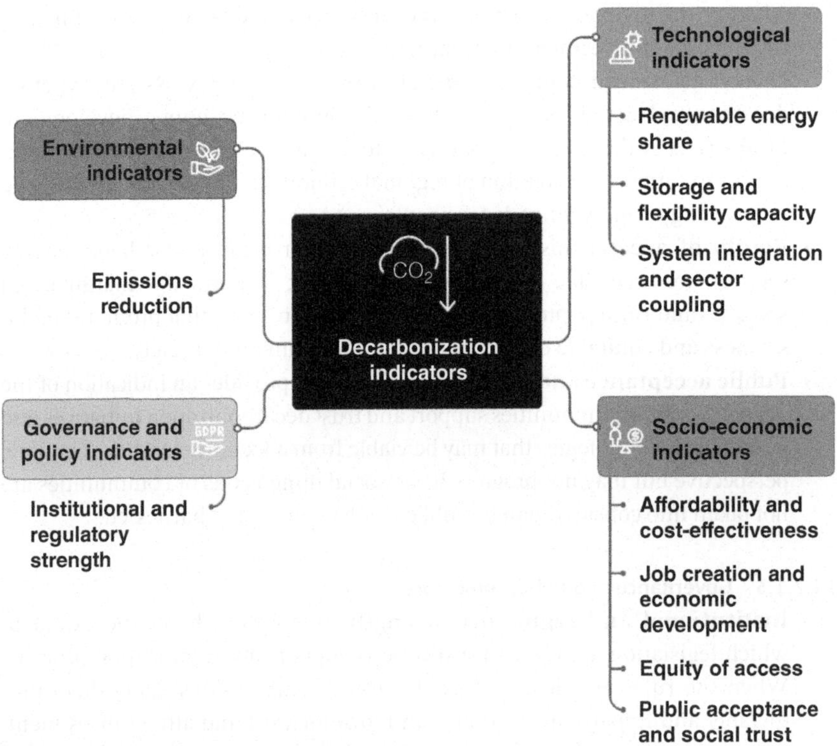

Figure 14.3 Decarbonization indicators.

14.3 Global Trends and Strategic Pathways for Deep Decarbonization

In the last 10 years, there has been an increase in global deep decarbonization as governments, industries, and international institutions recognize the need to curb global temperature rise and shift away from fossil-based systems. Many countries have adopted long-term commitments to target net-zero emissions, generally by 2050 for Europe [12] and North America [9] and 2060 for China [13]. This ambition has resulted in a shift in national energy planning, toward large-scale electrification, renewable energy, sector coupling, and investment in flexible storage systems.

A key characteristic of present-day decarbonization strategies is the increasing focus on sector integration, commonly referred to as sector coupling [14]. This approach attempts to move away from thinking in terms of isolated electricity planning and link more closely power with heating, transportation, and industrial processes into a cohesive energy system. As we electrify end uses, for example, through the installation of heat pumps in buildings and electric vehicles (EVs) in the transportation sector, it enables us to increase demand for electricity that can be supplied through renewable sources [15].

The advancement of this process is being driven more and more by digital energy management tools and worldwide grid coordination platforms. Smart-metering systems, real-time data analysis, and demand response systems enhance system operators' ability to organize supply and demand in a variety of sectors more effectively. Additionally, power-to-X (PtX) modes of operation (including power-to-hydrogen and power-to-heat) are increasingly important for reconciling variable generation of renewable energy with flexible consumption behaviors across industry, buildings, and transport.

Global trends also reveal several benchmark indicators used to evaluate decarbonization progress, including:

- Carbon intensity of electricity generation (gCO_2/kWh) [16].
- Share of renewable energy in final energy consumption (%) [17].
- EV and heat pump adoption rates [15].
- Energy storage deployment capacity (GWh) [18].
- Industrial electrification and hydrogen usage levels [19].

Even with the same overarching commitments, decarbonization pathways differ in their national contexts. For example, states where there is strong policy and social agreement are likely to achieve decarbonization through coordinated integration of sectors, while states with an upward trajectory of energy demand growth or dependence on resources with data will gradually transition. Given this, deep decarbonization is not deterministically singular but is conditional on the context that nations find themselves in. By this, it means that decarbonization pathways are affected by a mix of governance regimes, socioeconomic circumstances, technology suites, and public support.

In numerous situations, the speed and feasibility of decarbonization rely on policy instrument design and implementation. Carbon pricing frameworks, renewable energy subsidies, emissions trading schemes, and long-term planning frameworks for infrastructure in the energy sector are all instrumental in eliciting investment and shaping market incentives. The political sustainability of energy transitions is also influenced by public acceptability and affordability concerns, as energy transitions that result in inequitable economic or social burdens are more likely to face backlash. Thus, successful decarbonization efforts will require sustained technical capacity as well as sustained institutional commitment toward equitable cost distribution across the transition periods and long-term social legitimacy.

14.3.1 Rationale for Deep Decarbonization in Modern Energy Grids

Deep decarbonization has become an essential element of future energy system planning due to several interconnected technical, environmental, and socioeconomic drivers:

- Addressing climate risks: continued dependence on fossil fuels increases global temperature rise, as well as the occurrence and severity of climate-related impacts. In order to meet climate targets agreed upon internationally, GHG emissions must be rapidly reduced.
- Supply security and independence: electrification utilizing renewable and low-carbon energy sources decreases the reliance on imported fuels and provides system resilience to market volatility and geopolitical risks.
- Aging infrastructure and modernization requirements: many national grid systems operate with aging assets. Decarbonization provides an important opportunity to modernize infrastructure and integrate digital control, energy storage, and smart-grid technologies.
- Cost-effectiveness of renewables: the declining cost of wind, solar, and storage technologies has made clean energy both operational and economically viable, while also reducing long-term operations costs of the system.
- Public health and air quality: moving away from combustion-based fuels reduces air pollution and health-related burden from exposure; therefore, it provides a greater public good.

- Policy and regulatory commitments: obligations to national and regional net-zero frameworks require a change in the energy systems, making decarbonization no longer optional but necessary.

14.4 Real-world Case Studies

In this section, four real-life cases of CESN-based decarbonization are introduced. The cases are identified and illustrated in Figure 14.4. The objective is not to explore them in detail here, but to introduce their context, goals, and major outcomes. In each instance, deep decarbonization is featured (using a different trajectory), and this illustrates the variety of contexts in which CESN principles might be applied. This allows this section to segue into the next, where we analyze and contrast the cases using previously outlined indicators.

14.4.1 Denmark—A Blueprint for Integrated and Flexible Energy Systems

Denmark is generally viewed as one of the first movers in the world to fully implement deep decarbonization policies. Based on its geographical advantages and abundance of wind, Denmark has been able to, among other things, meet a remarkable share of its power demand from renewable generation sources, primarily wind. With all that said, Denmark's success is not only based on large volumes of wind turbine installations, but also on organized and deliberate planning of its electrical grid and a socially sophisticated engagement with the energy transition.

From a technical point of view, the nation's electricity grid is connected to the regional Nord Pool market in the neighboring countries, enabling the export (in case of surplus wind energy) of wind power and import (in case of insufficient domestic generation) of hydro power from Norway and Sweden. Through this mechanism, the resource portfolio has provided Denmark with a robust, secure resource under a high variability of renewable energy [20]. Combined heat and

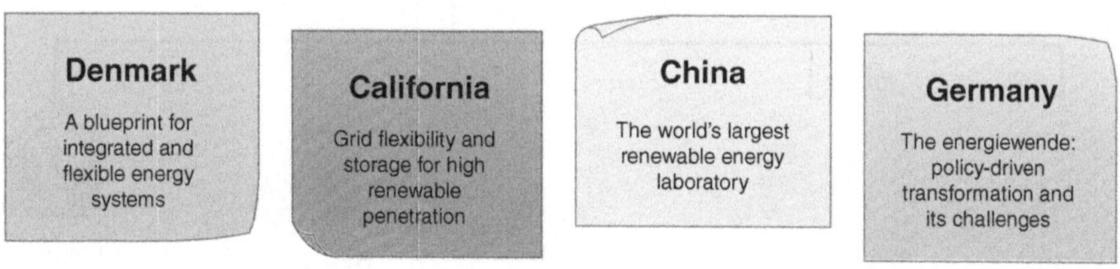

Figure 14.4 Four real-life case studies.

power (CHP) plants are an important component of the Danish energy system, which are widely used in conjunction with district heat networks. It produces heat and electricity instead of limiting its output to electricity, which results in much better energy use in the electricity and heat system than in a conventional power plant. In many cases, the efficiency of CHP plants can exceed 80%, giving them a significant role in reducing fuel consumption and carbon emissions. Over time, Denmark has also connected these heating networks to large thermal storage units and renewable sources, allowing CHP plants to act as a buffer when wind production is too high or too low [21].

This arrangement has not only made the Danish energy system more efficient but also provided the flexibility needed to manage a grid dominated by renewable energy. In addition to the previous ones, the development of PtX and green hydrogen technologies is considered as one of the main pillars for ensuring the long-term sustainability of the Danish energy system [22]. As shown in Figure 14.5, PtX refers to a family of processes in which surplus renewable electrical energy is converted into other energy carriers or valuable chemicals. In the Danish context, the most important applications include the production of green hydrogen, methanol, ammonia, and other synthetic fuels (e-fuels) [23, 24]. These pathways are designed to integrate variable wind energy into broader energy use and enable renewable electricity to decarbonize not only the power and heating sectors, but also industries and transport methods that are difficult to electrify, such as aviation and shipping.

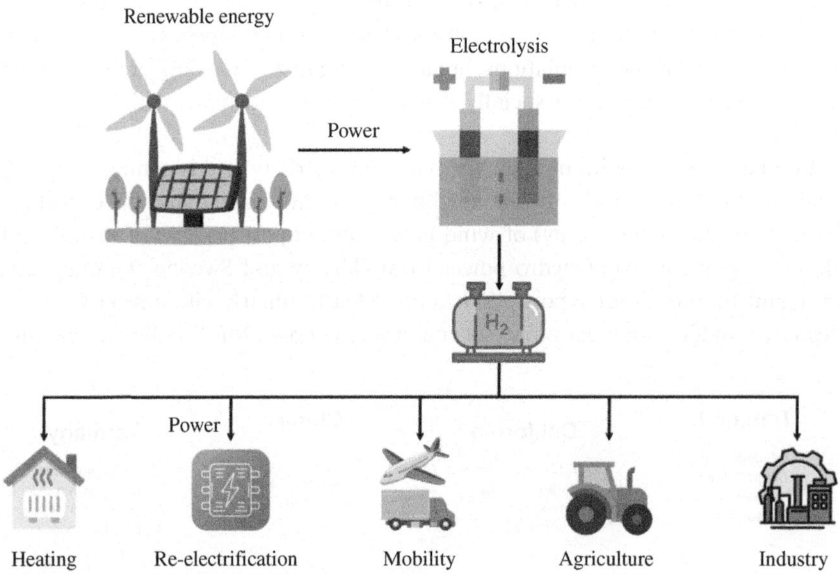

Figure 14.5 Renewable electricity, used for hydrogen production via electrolysis and its distribution across various end-use sectors.

Beyond these structural measures, Denmark is also advancing several flagship projects and policies in deep decarbonization:

- Denmark is implementing a pioneering plan to build the world's first offshore energy islands, islands that will collect and feed electricity generated by offshore wind farms into the grid. One of the islands will be artificially built in the North Sea and will initially have a capacity of around 3–4 GW, with the potential to expand to 10 GW. The second island, located on Bornholm in the Baltic Sea, is designed to produce around 3 GW. Together, the two are expected to provide renewable electricity to millions of households and also serve as a platform for the development of technologies such as PtX [25].
- EcoGrid on Bornholm and EnergyLab Nordhavn in Copenhagen are examples of how Denmark has used its territory as a real laboratory for developing smart grids and increasing the flexibility of the energy system. On Bornholm, EcoGrid projects involved many households in plans to respond quickly to changes in demand and participate in local electricity markets [26]. In Copenhagen, EnergyLab Nordhavn showed how the simultaneous combination and management of technologies such as batteries, EVs, and district heating systems can help to better balance renewable energies in urban environments.
- In Denmark, large thermal storage tanks in the form of open ponds are used; a prominent example is the one in Vojvodina with a capacity of over 200,000 cubic meters. These systems are directly connected to solar thermal power plants and allow heat to be stored from one season to the next. In this way, district heating networks can also benefit from solar energy in the cold months of the year, significantly reducing the need for fossil fuels [27].
- Denmark has started implementing carbon capture and storage (CCS) projects in recent years, including a pilot project to capture carbon dioxide at the Copenhagen waste-to-energy plant. Similar projects are also underway at the Aalborg Portland Cement Plant, one of the country's largest sources of CO_2 emissions. The measures are part of a broader attempt by Denmark to reduce emissions from sectors that are harder to decarbonize.
- In Odense, the heat generated in the Meta (previously Facebook) data center has been captured and fed to a district heating network. The project can offer heating to tens of thousands of homes, and reflects the intersection between digital infrastructure and thermal energy systems [28].

14.4.2 California—Grid Flexibility and Storage for High Renewable Penetration

Traditionally, California has been seen as the leader in the clean energy transition for many reasons in the United States, but mostly a result of the multitude of aggressive policies that the state has developed to support the transition. With the generous solar resource and developed incentives, solar has rapidly expanded in many forms across the state, including large solar farms and increased rooftop solar. Therefore, renewable energy has comprised a significant share of California's electricity load and will continue to do so, but renewable energy has caused significant challenges. One of these challenges is called the "duck curve," which was first presented by California independent system operator (CAISO) [29]. A duck curve is a curve that describes the net demand falling in the early afternoon when solar is at its peak and increasing again later in the day with demand. This process creates a significant imbalance of flexible generation resources that puts additional stress on the system to balance the grid. Figure 14.6 shows this pattern as it appeared in 2016, with a midday "belly" of low demand and a sharp evening "neck" of increased load. California has been investing heavily in battery storage to combat this problem. According to the US Energy Information Administration (EIA), California was projected to install nearly 15,500 MW of battery capacity in May 2025—more than any other state—and the number is expected to reach 55,000 MW by the end of 2045 [30]. These battery systems store excess solar generation occurring in the middle of the day and can discharge into the grid later in the evening during peak times to help smooth the rapid ramps shown in the duck curve. Meanwhile, the CAISO has encouraged the implementation of time-of-use (TOU) tariffs as an additional important demand response mechanism. By inciting consumers to shift electricity consumption to periods of high renewable

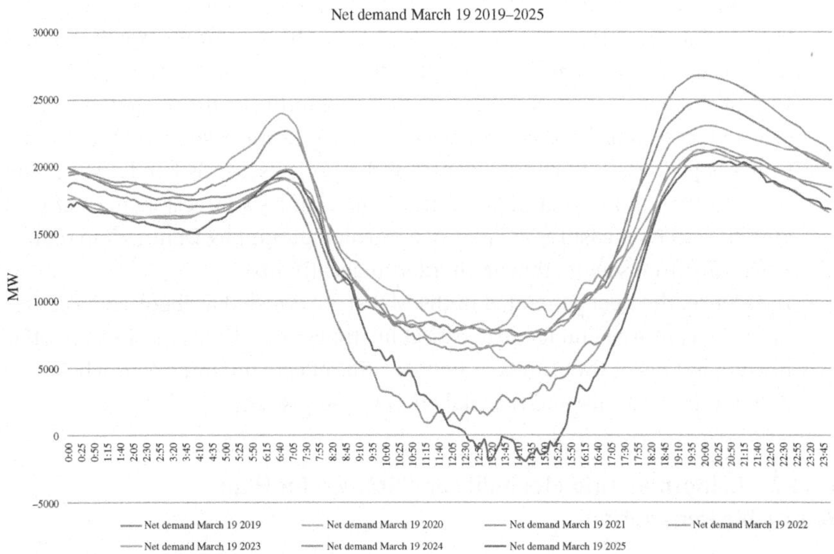

Figure 14.6 Duck curve of California.

generation, TOU pricing helps manage daily surpluses and limits the burden during peak times.

The California decarbonization program is based on a strong set of policies. For example, it is required that the retail electricity sold in California be derived from renewable or carbon-free sources by 2045, and the progress toward this target is assessed with periodic agency reports, which assess the technical feasibility, reliability, and cost [31]. From an electrification approach, the state of California has also sought to promote EVs to create yet another potential flexibility resource. In the second quarter of 2025, approximately 100,671 new EVs were sold in California, which accounts for just over 21.6% of the light-duty vehicle sales for the quarter [32]. These EVs will not only help to reduce transportation emissions, but represent the flexibility resource potential if EVs, including managed charging of EVs and vehicle-to-grid programs, are expanded.

14.4.3 China—The World's Largest Renewable Energy Laboratory

In recent years, China, home to the world's largest electricity system and one of the world's largest CO_2-emitting countries, has emerged as a world leader in renewable energy. China's decarbonization efforts are concentrated on building vast amounts of wind and solar power, rapid development of energy storage, construction of ultrahigh voltage (UHV) transmission lines, deploying EVs, and building a national carbon market. The scale of increased renewable energy in China has been astonishing; from 2021 to 2024, the total capacity for wind and solar was more than double what it was, reaching approximately 1,408 GW of capacity by the end of 2024 [13]. In early 2025, for the first time, the total installed capacity of wind and solar surpassed that of coal, marking a symbolic milestone in China's energy transition. In the first quarter of 2025 alone, China added around 72 GW of solar capacity [33]. To manage such high levels of variable renewable energy generation, China has simultaneously invested in new types of energy storage systems. In 2024 alone, about 42 GW (101 gigawatt-hours [GWh]) of new storage capacity was installed across the country, and by mid-2025, the country had surpassed 100 GW of total storage capacity [34]. Another project of China's energy transition is its UHV transmission grid, which is designed to transport electricity from renewable sources from inland areas where the resources are located to major load centers. For example, the 915-km Qingyang–Tai'an UHV line, completed in 2025, is expected to transfer about 36 TWh of electricity per year, almost half of which will be generated from clean energy sources [35].

China has also expanded its national emissions trading system (ETS). In this process, the government sets a maximum amount of GHGs ("cap") that specific sectors are allowed to emit in a given year. Each of these sectors has its own limit. But if a company (for example, a steel company or a fossil fuel power plant) emits more than the set limit, it must pay a fee to buy new permits (in other words, pay a fine), and if a company emits less than the set limit, it can sell its permits (a financial incentive). This process sets a price for carbon and creates a market. Initially launched in 2021 and limited to the electricity sector, the ETS has become the world's largest carbon market in terms of the amount of GHG emissions covered. In 2025, the scheme was expanded to include the cement, steel, and aluminum industries, covering more than 1,500 companies and almost 3 Gt of CO_2 equivalent. The ETS is designed to gradually reduce GHG emissions in some of China's energy-intensive sectors by providing financial incentives for improved efficiency and cleaner technologies [36].

Transportation electrification represents another key complementary component in China's decarbonization chain. In the first quarter of 2025, global sales of new EVs reached nearly 60% in China, demonstrating that China has a leadership role in the EV market and has further progressed transportation as part of China's decarbonization efforts [15].

These measures show that China is not following a decarbonization pathway solely by heavily relying on the expansion of renewable energy but has also put effort into expanding storage systems, building ultralong transmission lines, building carbon markets, and promoting electric transportation. Although coal remains an important part of the electricity generation mix, the rate of change is extraordinary, and one might reasonably conclude that no other country has the ability to apply such innovation to the globe.

14.4.4 Germany—The Energiewende: Policy-driven Transformation and Its Challenges

Germany is undertaking one of the largest decarbonization efforts in the world, based on its familiar policy of Energiewende (energy transition). The transition began in December 2000 with the establishment of a political framework through the adoption of the Renewable Energy Sources Act (Erneuerbare-Energien-Gesetz [EEG]) [37]. The EEG focused largely on feed-in tariffs (FITs), which were formal agreements by the government to allow its citizens who generated electricity using renewable sources (e.g., solar panel, and wind turbine) to sell this electricity to the grid for a fixed price for several years. Overall, the EEG reduced uncertainty from the market, while providing long-term price certainty for capital markets, resulting in the creation of thousands of energy generation projects of various scales, and Germany quickly became one of the world's leading countries in wind and solar. A major pivot that was a part of Germany's Energiewende was to phase out nuclear energy, referred to as Atomausstieg [38]. After more than two decades of political debate, Germany permanently closed its last three nuclear reactors in April 2023.

This decision was the end point in a long-term strategy to switch from nuclear power generation to renewable energy sources while improving efficiency, storage, and grid infrastructure. The zero nuclear energy is contentious, but it shows that Germany is willing to pursue decarbonization through a system based on renewables instead of low-carbon nuclear power. Germany also committed to phasing out coal, in conjunction with the nuclear power phase-out, with 2038 being the latest possible exit date for removing all generating capacity [39]. Several coal-fired units have been closed as part of this process, and the government has produced a detailed phase-out plan for decommissioning remaining coal capacity. Coal phase-out is critical to the long-term decarbonization strategy in Germany and provides support for an acceleration of the phase-out of nuclear power and the rapid growth of renewable energy generation. The policy has been controversial, and environmentalists are advocating for a much-shortened timeline to achieve the coal exit by 2030, but it does represent a significant step away from fossil fuel use and shows that Germany intends to move its energy system toward a state of climate neutrality target of 2045 [40].

Another important aspect of Germany's Energiewende is cutting GHG emissions outside the electricity sector, especially in the area of heating and in industry. In terms of buildings, the plan is to encourage the diffusion of heat pumps, uphold certain insulation standards, and phase out fossil fuel heating systems from the building stock (see German Energy Policy Review 2025), as these are all important for meeting near-term climate targets. In addition, Germany is making a considerable institutional investment to develop a national hydrogen strategy, the objective of which is to utilize green, low-carbon hydrogen as a clean fuel for heavy industry, long-duration energy storage, and as a flexible energy carrier [37]. Lastly, Table 14.1 highlights the differences between the four cases discussed, with Denmark and Germany dominating the share of RE contributions, California pursuing ambitious clean energy targets, and China attempting to peak emissions before 2030, despite having the highest aggregate emissions overall.

14.5 Comparative Analysis of Case Studies

The aim of this section is to provide a comparative analysis of the case studies described in Section 14.3 using the indicators introduced in Section 14.2. The analysis is organized into four dimensions—technology, socioeconomic, governance and policy, and environment—which together encompass key aspects of CESN-based decarbonization. Rather than presenting each country individually, the discussion is conducted dimension by dimension, where the performance, strengths, and limitations of Denmark, California, China, and Germany are juxtaposed.

Table 14.1 Key indicators of CO_2 emissions, low-carbon electricity share, and 2030 targets in selected CESN cases.

Case	Total CO_2 emissions (Mt)	Low-carbon share in electricity (%)	Key 2030 target(s)	Source
Denmark	42	88.4	70% GHG cut vs. 1990 by 2030 Net-zero 2045	[41]
California	371	57	60% renewable portfolio standard (RPS) by 2030 100% clean electricity by 2045	[42]
China	15,800	38	Peak CO_2 before 2030 Carbon neutrality by 2060	[13]
Germany	649	62.7	80% renewables in electricity by 2030 Net-zero 2045	[37]

14.5.1 Comparative Analysis by Indicator Categories

14.5.1.1 Technological Perspective

Among the four case studies that are presented in the previous section, Denmark stands out globally, with renewables accounting for nearly 88.4% of the country's net electricity generation in 2024 (31,200GWh) [43]. Wind energy, which accounts for approximately 57–60% of the country's total electricity mix, is the dominant source of electricity generation in the country. Bioenergy is in second place with about 15% of total generation, and solar energy is in third place with 13% [44].

California also has a high share of renewable energy in its energy mix, although it does so in a different way, with a large use of solar energy. In 2024, California's electricity generation reached approximately 216,181 GWh. Collectively, renewable energy—including solar, wind, hydro, geothermal, and biomass—would account for just over 57% of California's in-state electricity generation in 2024 [45]. Solar energy, which made up the majority of this generation, accounted for approximately 22.6% of the state's total generation, including photovoltaic systems and solar thermal power plants. In second place was hydroelectric power, with approximately 13.5% of total generation. Wind power also played an important role, accounting for approximately 7.3% of total generation. Geothermal power generation, with 4.8%, and biomass, with 2.2%, complemented these sources. These reflect the state's advanced but still diverse clean energy portfolio [46].

In 2024, China's electricity generation mix saw a considerable increase in renewable sources, with the country having a significant amount of clean electricity generation globally (3,470,000 GWh), but fossil fuels still dominated. Low-carbon sources—including hydropower, wind, solar, and bioenergy—accounted for about 38% of the country's total electricity generation. Hydropower, the largest clean source, accounted for about 13% of China's electricity generation. Wind and solar power generation together accounted for about 18%, with solar alone accounting for about 8% and wind for about 10%. Biomass came in last at about 2%. Despite the increase in renewables, coal and other fossil fuels still provided about 62% of China's electricity in 2024 [47, 48].

Germany is also one of the countries with a high share of renewable energy in its electricity mix, although its production is more evenly distributed between wind, solar, biomass, and hydropower and is not dominated by a single source. In 2024, approximately 59% of Germany's total electricity production (254,900 GWh) came from renewable energy sources. Of this amount, wind energy has the largest share among renewable energy sources, accounting for around 31–33% of total electricity production. In the same year, solar energy accounts for around 14%, biomass for 6–8%, and hydropower for 4–5% of the German energy portfolio [49, 50].

Overall, these four examples present both the diversity of renewable energy portfolios and the varying structural pathways that countries and regions take to pursue deep decarbonization. Denmark demonstrates the highest level of relative penetration, with renewables representing nearly 90% of the electricity it generates, with wind forming the backbone of the system, while California manifests a strong solar pathway, taking advantage of the vast solar resource, with more than half of its renewables being solar, but also employing hydropower and many clean sources as well. Germany presents a balanced energy portfolio model with a strong reliance on wind and solar representing most of their renewables, and biomass and hydropower also acted as critical pillars to support stability needed for phasing out both nuclear power and coal, showing a clear reliance on diversification to provide the stability to phase out nuclear and coal, both of which will depend on stable electrical resources that can provide electrical service. Lastly, China is acknowledged as the global leader in overall renewable electricity generation, generating more clean electricity than any other single source in the world, while also relying to an incredible extent on coal due to the sheer size of the system. Collectively, these examples illustrate that the integration of renewables is as much a reflection of geography, resource and policy choices, as it is about access to technology: some systems seek depth of penetration (Denmark) others seek breadth of a technology mix (Germany), others seek scale of deployment (China), and others seek dominance in a specific sector (California).

In terms of storage capacity, China is the clear leader among the four case studies. Approximately 168 GWh of new energy storage are set to be installed by 2024 [51], making it the world's largest storage system. California is in second place with around 60 GWh of battery storage capacity in 2024 [18], more than half of which will be directly coupled with solar and wind installations to balance variability. Germany is in third place with around 7 GWh of fixed storage in residential, industrial, and grid-scale applications, although this is expected to increase significantly in the coming years with focused development plans [52]. Denmark is currently lagging behind, with its operational and planned projects totaling around 1.5 GWh, reflecting its smaller system size and its firm reliance on sector interconnection rather than large-scale batteries [53].

System integration and sector connectivity are progressing rapidly in three of the four cases, each with distinct priorities. Denmark is advancing the integration of electricity, heat, and hydrogen—notably through PtX plants [54] and a planned hydrogen pipeline to Germany [55]—as part of its path toward a fully renewable electricity system by 2030. China is also investing heavily in connecting electricity with heating, transport, and industrial applications, and is using hydrogen, EVs, and surplus renewable energy for industrial processes to manage variability and demand [13]. Germany follows fully connected system models in its planning scenarios, particularly for electricity and building heating [56], while placing energy conversion to gas and hydrogen as essential for the decarbonization of hard-to-abate sectors [57]. In contrast, California has focused primarily on electricity and storage solutions and has made limited progress in connecting sectors compared to others. Table 14.2 provides a comparative overview of the four case studies based on technology-related indicators, synthesizing the analysis presented in this subsection.

14.5.1.2 Socioeconomic Perspective

In Denmark, citizen ownership has been one of the strongest drivers of public acceptance for wind energy. Participatory models such as the Middelgrunden offshore wind farm (completed in 2000) invited thousands of local citizens to invest directly in turbines, thereby generating economic benefits as well as a sense of shared responsibility among citizens [58, 59]. Since 2009, Danish law has required developers of new onshore projects to offer at least 20% equity to local residents, which has fostered local buy-in [60]. This model has lessened local opposition toward visual and environmental impacts, while also increasing community trust in renewable energy. In parallel, the economic transition in Denmark has been buoyed by the global emergence of businesses like Vestas and Ørsted. Ørsted was originally a fossil fuel (oil and gas) company (DONG Energy), but shifted its entire focus to renewables after 2009, and is now a global leader in offshore wind energy, demonstrating how industrial transition can yield socioeconomic benefits—jobs and exports [61, 62].

In California, policy involving socioeconomic policies is particularly focused on equity (with protections for vulnerable populations). Policymakers have

Table 14.2 Comparative assessment of technological indicators across case studies.

Case	Renewable energy share	Storage and flexibility	System integration and coupling
Denmark	**Leader**—among the highest renewable penetration globally, wind-dominated system	**Weak**—very limited large-scale batteries; relies instead on thermal storage and CHP	**Strong**—advanced sector coupling (electricity, heat, hydrogen), PtX projects
California	**Strong**—high renewable share, solar-dominant portfolio	**Strong**—rapid battery expansion, closely linked with solar and wind	**Weak**—limited sector coupling; progress mostly within electricity sector
China	**Moderate**—lower renewable share, but largest absolute generation worldwide	**Leader**—world's largest storage deployment, rapidly scaling	**Improving**—major push for EVs, hydrogen, and industrial coupling
Germany	**Strong**—diversified renewable mix (wind, solar, biomass, and hydro)	**Moderate**—steady growth in storage, still behind leaders	**Strong**—comprehensive sector-coupling strategies, hydrogen central to long-term plans

underscored that the transition to clean energy should not transfer excessive burden to low-income households or disinvested communities. Programs like California Alternative Energy Rates (CARE) offer discounts of up to one-third on electricity bills and ensure that the cost of renewable energy and infrastructure upgrades does not worsen the lack of energy [63]. Beyond affordability, California has tied clean energy deployment to job creation in the solar, storage, and EV sectors.

In Germany, the Energiewende has had issues with energy affordability. For many years, households have received an extra EEG cost added to their electric bills that financed the renewable energy tariff, which led to conflicts and concerns [64]. As of July 2022, the additional cost of EEG has ended and been moved to federal budgets and GHG emissions trading so that the benefit goes directly to households and small businesses [65]. Concurrently, Germany has created a large renewable job base but argues over regional equity—especially, the siting of wind farms that are not equal to urban demand in rural areas. The German experience demonstrates how socioeconomic success not only depends on the generation of renewable capacity but also on continuing to adjust cost-sharing mechanisms to sustain public participation.

In China, the socioeconomic impacts of the clean energy transition are being defined on a large scale. With over seven million renewable sources jobs in solar, wind, hydroelectric, and EVs, China is a world leader in renewable energy employment. Deliberately, the government has opted to strategically position renewable energy and EVs as key components of its economic policy pillar and is committing substantial effort and resources into supporting industrial capacity and jobs. The government has also made a strategic commitment to advance equity through its national carbon market (ETS), which previously focused on the electricity sector and now incorporates heavy industry such as steel and cement. This illustrates in a gradual way the disparate distribution of and costs of decarbonizing society while also growing clean industries, and while ensuring, in some manner, that the energy transition also aids economic modernization and promotes longer-term social stability [36, 66]. Table 14.3 delivers a comparative overview of the four case studies focused on socioeconomics-related indicators, consolidating the discussion and analysis in this subsection.

14.5.1.3 Governance and Policy Perspective

The Danish governance system relies on legally binding, long-term climate policies with complementary mechanisms. The Climate Act (2020) has committed to reducing GHG emissions by 70% (compared to 1990 emissions) by 2030 and requires annual policy proposals and independent reports, creating a stable environment for investors and businesses [67]. Recent green tax reforms [68], and an economy-wide carbon price—which includes a coordinated path for highly emissive sectors in the transition period (e.g., agriculture)—increase confidence in the longevity of the policy quality and indicate that decarbonization incentives would continue following campaigning. All in all, these actions provide clear targets + sustainable institutions + coherent mechanisms that create lower policy risk and capital lock-in.

California exemplifies some of the most ambitious clean energy targets in the world. Under SB 100, or the 100 Percent Clean Energy Act of 2018, all retail electricity generation must be renewable by 2045. SB 100 also requires the alignment of three principal agencies (the California Energy Commission [CEC], California Public Utilities Commission [CPUC], and California Air Resources Board [CARB]) to issue joint reports regarding progress every four years in addition to creating plans that report under existing law and are accountable to the California State Legislature—an institutional design that establishes locking in goals without sacrificing plan design, market regulation, and oversight [69]. Frequent reporting and proposed course correction signals the effectiveness of regulation, which, alongside the state's expansion of renewable energy, storage, and transmission technologies, increases transparency and confidence with investors.

Germany's Climate Protection Act (Klimaschutzgesetz [KSG]) establishes a legally binding goal of climate neutrality by 2045, reducing GHGs by 65% by 2030 compared to 1990, 88% by 2040, and achieving climate neutrality by 2045. IEA's 2025 Review highlights the need for political stability while reforming the market design. Berlin is doing this by modernizing its support instruments (e.g., by moving

Table 14.3 Comparative overview of socioeconomic indicators.

Case	Affordability and cost-effectiveness	Job creation and economic development	Equity of access	Public acceptance and social trust
Denmark	Generally affordable due to strong policy support and cooperative financing models	Strong job creation and global leadership via companies like Vestas and Ørsted	Equity promoted through mandatory community ownership in wind projects	Very high, due to cooperative models and local investment opportunities
California	Affordability maintained through programs like CARE, protecting low-income households	Significant employment growth in solar, storage, and EV sectors	Strong focus on equity in energy policy, with targeted subsidies for vulnerable groups	Broad public support, though mainly framed as environmental and economic justice
Germany	Initially challenged by high household costs from EEG surcharge, later improved after surcharge removal in 2022	Large renewable employment base, with ongoing debates about regional economic fairness	Access generally equitable, though rural–urban tensions remain in wind siting	Moderate–strong; public support exists, but cost concerns created political debates
China	Large-scale deployment lowers technology costs, but affordability still varies regionally	Global leader in renewable employment (7+ million jobs) and industrial growth	Expanding equity via the national ETS and wider distribution of transition costs	Improving, with strong state-led direction but limited bottom-up participation compared to Europe

away from older fixed-price contracts to approximations of the market with instruments such as contracts for differences [CfDs], and capacity instruments for flexible hydrogen backup, "ready-to-use hydrogen backup") [70]. The outcome develops a governance structure of blending policy certainty and adaptive regulation to mitigate investment risk in an expensive and complicated transition. China's governance framework for decarbonization melds a strong aspect of government planning along with an increasing number of regulatory instruments. The national emissions trading scheme was initially established in the electricity sector, and subsequently was implemented with energy-intensive sectors, including steel and cement, resulting in the largest carbon market in the world, with thousands of companies required to report and trade on an integrated basis. Meanwhile, a renewable energy portfolio quota system mandates provinces and system companies to achieve a minimum share of renewable energy consumption and guarantees that new wind and solar generation facilities are utilized, rather than curtailed, to meet this requirement. The latter is a continuation of longer-term coordination in the Fourteenth Five-Year Plan (2021–2025), which sets binding capacity and limits emissions to minimum targets and directs investment into undersea ultrahigh-voltage transmission lines to receive generation from remote renewable generation bases to coastal demand centers. Together, these instruments—the ETS, renewable energy quotas, and five-year planning—provide a clear regulatory signal, reduce investor uncertainty, and demonstrate strong institutional capacity to drive decarbonization on a national scale [36]. Table 14.4 provides a comparative overview of the four case studies based on governance and policy-related indicators, synthesizing the analysis presented in this subsection.

Table 14.4 Comparative overview of governance and policy (institutional and regulatory strength).

Case	Institutional and regulatory strength	Reference
Denmark	**Strong and stable**—long-term climate law, binding targets, and consistent instruments (carbon taxes, building codes, renewable subsidies) provide high policy certainty and strong institutional credibility	[67]
California	**Strong but adaptive**—ambitious statutory target (SB 100), with clearly defined institutional division of roles (CEC, CPUC, and CARB). Regularly reported goals to ensure accountability; however, political prioritization can cause challenges in implementation	[69]
Germany	**Robust yet evolving**—Climate Protection Act (KSG) is legally binding which anchors target commitments. Institutions are implementing annual monitoring and evaluating market-based instruments. On balance, stability is strong; however, they are politically facing challenges on costs and regional fairness.	[70]
China	**Powerful but centralized**—the expansion of the national ETS, renewable quotas, and Five-Year Plans demonstrate strong institutional capacity and regulatory reach. There is clarity and signals for investment; however, we still remain heavily top-down in decision-making with limited local level participation	[36]

Case	Institutional and regulatory strength	Reference
Australia	**Strengthening through legislative reform**—The Climate Change Act 2022 establishes a legally recognized emissions reduction target. The Climate Change Authority, AEMO, Clean Energy Regulator, and the Safeguard Mechanism support implementation, although historical policy reversals have affected investment certainty.	[71]
Canada	**Federal but multi-level**—The Canadian Net-Zero Emissions Accountability Act provides the legal framework for net-zero by 2050, supported by national carbon pricing. Shared federal-provincial responsibilities create strong oversight but uneven implementation.	[72]
Japan	**Well-established but consensus-driven**—The Act on Promotion of Global Warming Countermeasures and the GX Strategy guide decarbonization. Strong institutions provide continuity, although consensus-based governance can slow reforms.	[73]
Spain	**Strong within the EU framework**—The Climate Change and Energy Transition Law (2021) establish binding climate objectives aligned with EU legislation. National planning and renewable energy policies provide regulatory stability.	[74]
United Kingdom	**Highly institutionalized and accountable**—The Climate Change Act 2008 established legally binding carbon budgets with independent oversight from the Climate Change Committee, providing high policy certainty.	[75]
South Korea	**Ambitious but implementation-focused**—The Framework Act on Carbon Neutrality and Green Growth and the Korean ETS support long-term decarbonization while balancing industrial competitiveness.	[76]

14.5.1.4 Environmental Perspective

Through strong legislation, as well as good policy instruments and third-party participation, Denmark continues to make progress in decreasing its GHG emissions. The IEA estimates that CO_2 emissions from energy have been reduced by over 50% for the last 30years, decreasing from just under 58million tons in 1990 to roughly 27million tons in 2022 [77]. The downward trend in emissions reduction is positively drawn in the downward sloping curve of Figure 14.7. Much of this downward trend is due to the continued phase-out of fossil fuels and the broad adoption of renewable energy sources into the energy system. In 2023, GHG emissions per capita in Denmark totaled approximately 4.5tons per person, again substantially lower than the OECD average of approximately 6.4tons of

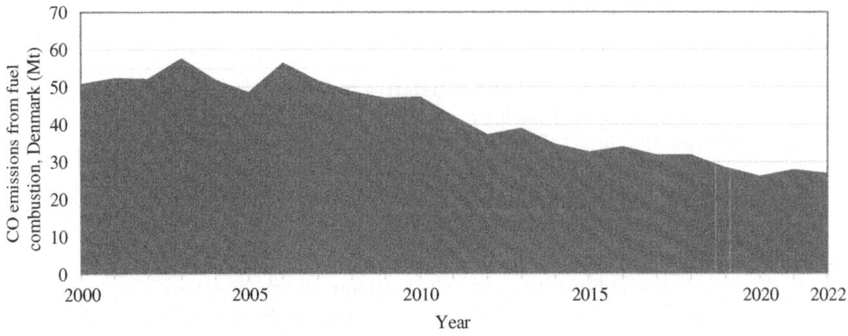

Figure 14.7 CO emissions from fuel combustion in Denmark (2000–2022).
Source: Adapted from [77].

CO_2 emissions per person [78]. Based on current projections, Denmark remains on pace to achieve its 70% reduction target by 2030. Substantial work remains to be done in the areas of heavy transport and agriculture, where emissions reductions are more difficult. Overall, Denmark is an example of a country that is achieving measurable and persistent reductions in GHG emissions supported by well-defined institutional setups, green policies, and acceptance by the populace.

California is a leader in climate policy in the United States. Not only did this state achieve its first legal goal of returning GHG emissions to 1990 levels four years ahead of schedule (in 2016), but it has also expanded its targets to higher levels [79]. Under current law, California must reduce GHG emissions by 40% below 1990 levels by 2030 and achieve carbon neutrality by 2045 [80].

According to the California GHG inventory maintained by CARB, the state's reduction in GHG emissions is primarily due to a cleaner electricity mix—where renewable energy has replaced purchased fossil fuels—along with increased energy efficiency standards and stricter regulations on vehicle GHG emissions [81].

California not only has mandatory long-term carbon reduction goals but also an acceptable short-term path supported by enforceable regulations and sector-specific policies. The challenge ahead involves sectors where reducing GHG emissions is very difficult, such as transportation, heavy industry, and agriculture, but the path so far shows that institutional design and the integration of state policies are already delivering measurable environmental results. As shown in Figure 14.8, California's total GHG emissions have declined steadily since 2007, falling below 1990 levels by 2016.

Germany's achievements in reducing GHG emissions have not been as impressive as those of some of its rivals, but they are not insignificant either. Long-term energy policies and a binding climate protection law (KSG) set clear short-term goals for reducing emissions and dedicate the country to being environmentally friendly by 2045 [82]. The IEA says that CO_2 emissions from energy sources have

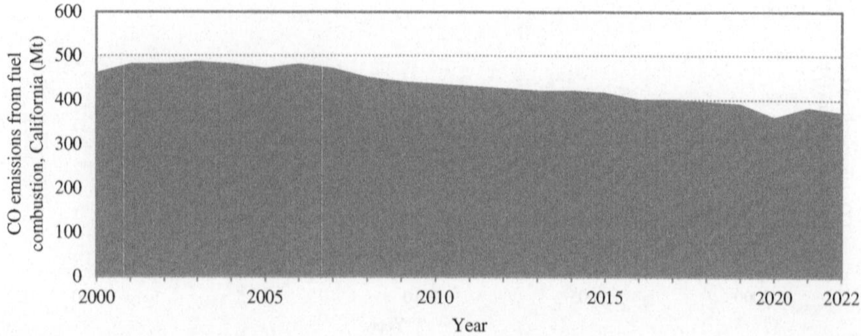

Figure 14.8 California's total GHG gas emissions, 2000–2022. *Source*: Adapted from [81].

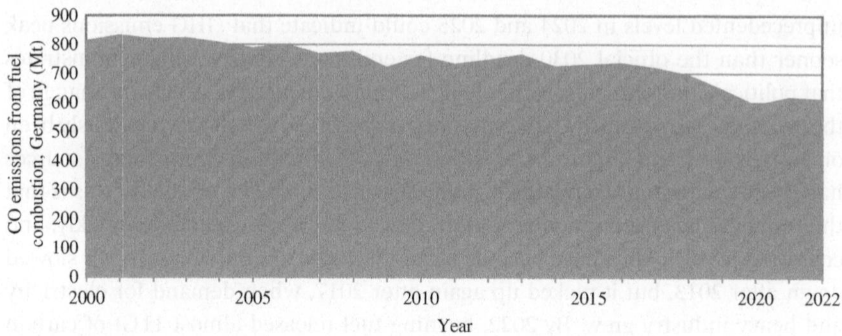

Figure 14.9 CO emissions from fuel combustion in Germany (2000–2022). *Source*: Adapted from [83].

dropped from about 820 Mt in 2000 to about 620 Mt in 2022. Figure 14.9 shows that the overall decline has been slow, and GHG emissions stayed about constant from 2000 to 2015. Reductions have only sped up in the last few years, especially since 2015, when coal production decreased and renewable capacity went up.

Germany's total GHG emissions had dropped to about 649 million tons of carbon dioxide by 2024. This was almost 48% less than in 1990, making it the lowest level in decades.

GHG emissions in China continue to rise, and the country is the world's largest emitter of GHGs, accounting for nearly 31% of global CO_2 [84]. China is a paradox that is important to discuss deep decarbonization: it is the world's largest consumer of coal and the largest investor in clean energy technologies such as solar, wind, EVs, batteries, and hydrogen. That's why the situation in China is a little different. This dual role illustrates the difficulty of being innovative while also responding to major environmental issues. In addition to increases in GHG emissions for decades, more recently, there is evidence that China appears to be reaching an important inflection point. The rapid level of deployment of renewable energy at

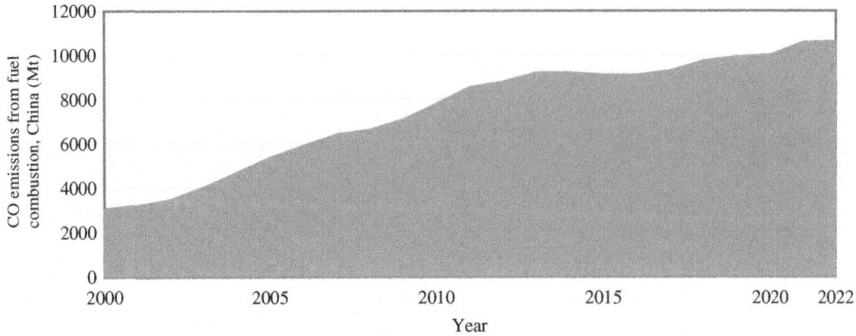

Figure 14.10 CO emissions from fuel combustion in China (2000–2022).
Source: Adapted from [84].

unprecedented levels in 2024 and 2025 could indicate that GHG emissions peak sooner than the official 2030 deadline for emissions control. This demonstrates that political commitments, technological improvement, and structural change of the economy can overturn GHG emissions trajectories, which had previously been on an upward trend. Figure 14.10 shows that CO_2 emissions from energy sources have been increasing at a relatively high rate since 2000. The reason for this is that the economy has been growing rapidly, cities have been expanding quickly, and coal-fired power plants have become larger. The rate of emissions growth slowed down after 2013, but it picked up again after 2017, when demand for electricity and heavy industry grew. By 2022, burning fuel released almost 11 Gt of carbon dioxide, making China the country with the most emissions in the world.

China's emissions are increasing, which has been accompanied by the approval of numerous laws and regulations demonstrating a commitment to advancing its decarbonization agenda. For example, the "1+N" system, established in 2021, organizes national objectives with respect to carbon peaking and neutrality into a coherent set of policies and guidelines [85]. The "1" refers to the main long-term guidance given in 2021, which sets the overall goal of reaching carbon neutrality by 2060 and peaking CO_2 emissions before 2030. "N" indicates a group of supporting programs and actions, such as the National Action Plan for Carbon Peaking, that show how different areas and sectors will help reach this goal. The new energy law, which came into effect in January 2025, brings energy planning, use, storage, and the promotion of renewable energies under stricter government control [86]. The draft Ecological and Environmental Law (2025) also aims to bring together and update current environmental laws. It does this by creating a consistent legal framework that includes climate change, energy conservation, and low-carbon development.

Table 14.5 provides a comparative overview of the four case studies based on environment-related indicators, synthesizing the analysis presented in this subsection.

Table 14.5 Comparative overview of environment (GHG emissions reduction).

Case	GHG emissions reduction—assessment
Denmark	**Leader**—more than 50% reduction since 1990; per-capita emissions (~4.5 t) well below OECD average; on track for 70% cut by 2030. Strong institutional and policy alignment, though agriculture and transport remain difficult sectors
California	**Strong performer**—met 1990-level target early (2016); legally binding 40% reduction by 2030 and carbon neutrality by 2045. Clear downward trajectory since 2007, with major gains from the power sector, efficiency, and vehicle standards. Challenges remain in transport and industry
Germany	**Moderate progress**—~48% reduction vs. 1990 by 2024, but the curve shows slow, plateaued reductions from 2000–2015 with acceleration only after 2015. Climate law (KSG) sets neutrality by 2045, yet hard-to-abate sectors pose risks to future progress
China	**Critical but lagging**—Still rising, ~11 Gt CO_2 in 2022, nearly 30% of global total. Dual paradox: world's largest coal consumer and largest clean energy investor. Legal framework ("1+N" system, 2021; new Energy Law 2025) aims for peaking before 2030 and neutrality by 2060. Early signs suggest emissions may plateau sooner than expected

14.6 Key Findings and Strategic Implications

The cases of Denmark, California, Germany, and China illustrate that deep decarbonization may develop in different ways. The end results of decarbonization emerge from a distinctive configuration of elements—including technologies, institutional capacities, socio-technical context, and political priorities.

To start, there are indications that stable and binding may act as an integral characteristic in decarbonization. In both Denmark and Germany, stable laws and targets have provided local and regional investors with the confidence that policies will not be shifting. This is also the case in California, where periodic reporting requirements (SB 100) increase accountability across sectors. In contrast, where policy frameworks are less organized and/or extremely centralized (e.g., China), while resources may be mobilized quickly and decision-making efficiencies streamlined, rapid activity can be complicated by incomplete implementation and limited degrees of social engagement. Overall, in considering what countries might need to structure sustainable decarbonized outcomes, these countries may need to develop regulations and regulatory processes that are not easily interrupted by political cycles.

The second aspect, social acceptance and justice is factors that increase the speed and acceptance of transitions. For instance, within the Danish model, community involvement by requiring local co-ownership in wind projects could shift any resistance by the public to decarbonization changes, including visual aesthetics or land use, to passive acceptance or even support. Germany's surcharge tax on its energy generation, known as the EEG surcharge, creates a paradoxical risk: if costs are judged excessive or cross-subsidization is poorly applied, even a rich, advanced society like Germany might revolt against renewable energy policies and demand change. California's CARE program shows the merits of equity and is a well-functioning and cost-effective program that can keep rising costs for low-income households in check, thus adding an equity lens to a reliable clean energy transition. China's approach in generating clean energy power through central planning allows for the rapid deployment of clean energy industries, but limited pathways for citizen or community engagement can potentially erode social trust and ultimately long-term legitimacy. In short, it may be suggested that even good technical policies may go unheeded without pathways for cost and equity distribution and some meaningful opportunity for engagement of the citizens.

Third, the success of decarbonization relies upon multiple factors, not simply on the availability of resources, but also on the variety of technology portfolios and the resilience of system design. The Danish model is very effective at a small scale, but persists in connectivity and flexibility, both of which are unlikely to be replicable everywhere. The solar-focused system in California proves the benefit of large storage and sector integration. Germany's complex diversity provides flexibility, yet its adoption is slow, evidence that new balancing will be delayed without sufficient urgency. In China, investing heavily in renewables does not mean decarbonization because coal generation can remain. In the future, systems should seek not to depend on a single source, integrate storage, and proactively interconnect sectors while looking for technology options based on the scale and location of the system.

In conclusion, it should be emphasized that when put together, renewable capacity developed in a country, the level of financial investment in renewables, or the policy language is simply not a reliable indicator of decarbonization. Underpinning this environmental discussion is another significant discussion. Decarbonization policies and programs should be judged not on capacity or financing but instead based on the facilitated and measured sector that will produce reductions in GHG. In the end, these provide a mapping: successful near-term deep decarbonization ultimately hinges on governance structures that bind, equitable and socially just policy frameworks, distributed and resilient technological systems, and, more than anything, measured environmental outcomes. As countries approach these processes in the next decade, they can test out the lessons focus on their own policy realities—learn from Danish social models, consider equity-based frameworks like California, reflect upon Germany's institutional rigor, mobilize, scale the achievability of industrial transformation consistent with their overall political environment (as for example is the case of China), while reflecting upon and weighting against direct costs of other comparative (or prices of decarbonization), lock-in of practices to deliverables in sectors, and coal reliance.

14.7 Future Directions for CESN-based Decarbonization

The experiences of Denmark, California, Germany, and China show that decarbonization requires more than mere technological implementation or significant policy objectives. The management of the CESN—the relationship between climate action, energy transitions, and bigger targets for sustainable development—will determine this. Future progress depends on strategies that simultaneously balance these three dimensions:

- Optimizing the nexus across sectors. To effectively manage the CESN, it is necessary to align climate targets with the design of the energy system and sustainability goals such as health, resilience, and resource efficiency. Provided that electricity, heat, transportation, and industry are all aligned with sustainability strategies, then there is no difficulty in utilizing renewable energies. Also, nexus-based planning enables additional renewable electricity to be employed for heating or other transportation fuels or for green hydrogen, thereby decreasing waste and achieving mutually advantageous climate outcomes.
- Ensuring social equity and sustainability. The sustainability aspect of CESN suggests that just transitions should be socially equitable and carbon-lean. Accordingly, policy responses to climate and energy should not exclude people from decision-making processes. Inclusion may be achieved through community engagement, citizen co-ownership, and targeted programs that do not contribute to increased costs. Denmark and California emphasize that simply adding equity mechanisms helps to achieve green transition by clarifying communities, whereas failing to consider distributive justice can lead to negative outcomes in the transition exemplified by Germany and China. Advancing digitalization for connected governance. Because CESN is so complicated, digital platforms need to be able to combine climate goals, energy flows, and sustainability indicators. AI, big data, and smart grids can help people use resources more effectively while keeping track of their carbon emissions and other sustainability goals. Countries run the risk of only meeting some of their energy goals and missing their greater climate and sustainability goals if they don't use digital monitoring.
- Diversify technology within a resilience framework. The CESN vision calls for a wider range of options, not just in energy technologies but also in strategies for the future, like using circular resources and adapting to climate change. In regions like Denmark, where wind is the only resource, or

California, where solar is the only resource, systems need to be able to store energy, connect to other systems, and manage land and resources in a way that is good for the environment. This reduces your vulnerability to climate, economic, or geopolitical shocks.

- Connect national CESN strategies to global sustainability. No country manages the grid in isolation. Decarbonization needs to stay within a global carbon budget and respect international commitments to sustainability. Smaller economies can facilitate a faster pace of decarbonization while larger producers provide economies of scale, human capital, and innovation. International cooperation will take on an increasingly important role in CESN. This will happen through hydrogen corridors, regional super grids, and shared sustainability models!
- Measure network outcomes instead of just inputs. The effectiveness of CESN should not be measured solely through investment or installed capacity. Effectiveness should measure reductions in GHG emissions, improvements in sustainability indicators, and contributions to global climate objectives. It is important to show that engaging in network management that advances decarbonization is intended for the long-term, and those intentionality's are monitored and reported on clearly.

14.8 Conclusion

This chapter demonstrates that the CESN can be a useful framework for facilitating deep decarbonization in contemporary energy systems. The fact of the matter is, we can summarize four key pillars of deep decarbonization over the long-term by conducting case studies on selected cases:

1. A strong and unified government that ensures decarbonization policies are sustained over time.
2. A diverse, broad, and integrated portfolio of technology that adapts to change and provides the reliability that is necessary.
3. Social and economic measures that keep energy affordable, equitable, and trusted by the public.
4. Verifiable and proven environmental outcomes that contribute to actual and sustained reductions in GHG emissions, not just the addition of new renewable capacity.

The assessment of the analyzed case studies showed both accomplishments and barriers in each reviewed example. Denmark is known as a global leader in wind energy and regional integration, but is reaching its limits in seasonal issues around energy storage; California is also pushing the limits on solar and storage while experiencing a severe duck curve, increasing costs, and limited sector-to-sector connection as it continues to push that limit; Germany's Energiewende program provides the example of expanded renewables, but is progressing with significantly high household energy costs, disparate regional participation of wind energy projects, and slow to decarbonize heating and transportation; and China, while demonstrating incredible scale of renewables, rapid build-out, and collaborative capacity development of infrastructure, faces continued use of coal and a centralized governance model for meaningful social engagement and adoption.

These cases demonstrate that no matter the maturity and scale, each demonstrates some gaps, proving the concept of the need for a CESN-systems thinking-based decarbonization approach. As we consider what is ahead, a CESN-systems thinking approach signifies a recognition of digital governance tools, differentiated approaches to equity, resilient linkages across sectors, and fit with global sustainability agendas. While CESN will further enable low-carbon, inclusive, adaptive, and resilient modern energy systems, it is needed to address the ongoing technological, social, and environmental issues facing our world.

References

1 A. Mahone et al., "Deep decarbonization in a high renewables future," E3 Inc., San Francisco, CA, 2018.

2 K. Holmes, E. Zeitler, M. Kerxhalli-Kleinfield, and R. DeBoer, "Scaling deep decarbonization technologies," *Earth's Future*, vol. 9, no. 11, p. e2021EF002399, 2021.

3 E. Panos et al., "Deep decarbonisation pathways of the energy system in times of unprecedented uncertainty in the energy sector," *Energy Policy*, vol. 180 p. 113642, 2023.

4 E. O'Shaughnessy, M. Shah, D. Parra, and K. Ardani, "The demand-side resource opportunity for deep grid decarbonization," *Joule*, vol. 6, no. 5, pp. 972–983, 2022.

5 X. Pan, T. Shao, X. Zheng, Y. Zhang, X. Ma, and Q. Zhang, "Energy and sustainable development nexus: A review," *Energy Strategy Reviews*, vol. 47 p. 101078, 2023.

6 M. Daneshvar, B. Mohammadi-Ivatloo, and K. Zare, "An innovative transactive energy architecture for community microgrids in modern multi-carrier energy networks: A Chicago case study," *Scientific Reports*, vol. 13, no. 1, p. 1529, 2023.

7 Z. Wang, Y. Sun, H. Kong, and C. Xia-Bauer, "An in-depth review of key technologies and pathways to carbon neutrality: Classification and assessment of decarbonization technologies," *Carbon Neutrality*, vol. 4, no. 1, p. 15, 2025.
8 International Energy Agency, "World energy outlook 2021," 2021. Accessed: Oct. 9, 2025. [Online]. Available: https://iea.blob.core.windows.net/assets/4ed140c1-c3f3-4fd9-acae-789a4e14a23c/WorldEnergyOutlook2021.pdf
9 International Energy Agency, "Net zero by 2050: A roadmap for the global energy sector," 2021. Accessed: Oct. 9, 2025. [Online]. Available: https://iea.blob.core.windows.net/assets/deebef5d-0c34-4539-9d0c-10b13d840027/NetZeroby2050-ARoadmapfortheGlobalEnergySector_CORR.pdf
10 IPCC, "Global warming of 1.5 °C: An IPCC special report on the impacts of global warming of 1.5 °C above pre-industrial levels and related global greenhouse gas emission pathways," 2018. Accessed: Oct. 9, 2025. [Online]. Available: https://www.ipcc.ch/site/assets/uploads/sites/2/2019/02/SR15_Citation.pdf
11 World Bank, Energy Sector Management Assistance Program, "Regulatory indicators for sustainable energy (RISE) 2020—sustaining the momentum," 2020. Accessed: Oct. 9, 2025. [Online]. Available: https://hubs.worldbank.org/docs/imagebank/pages/docprofile.aspx?nodeid=32668096
12 Climate Action Tracker, "EU—Net zero targets/net zero target evaluations (country)," 2025. Accessed: Oct. 9, 2025. [Online]. Available: https://climateactiontracker.org/countries/eu/net-zero-targets
13 M. Yang, B. Yang, S. Butler-Sloss, and E. Graham, "China energy transition review 2025," 2025. Accessed: Oct. 9, 2025. [Online]. Available: https://ember-energy.org/app/uploads/2025/09/China-Energy-Transition-Review-2025.pdf
14 IRENA Coalition for Action, "Sector coupling: A key concept for accelerating the energy transformation," 2022. Accessed: Nov. 8, 2025. [Online]. Available: https://coalition.irena.org/-/media/Files/IRENA/Coalition-for-Action/Publication/IRENA_Coalition_sector_coupling_2022.pdf
15 International Energy Agency, "Global EV outlook 2025: Trends in electric car markets," 2025. Accessed: October 9, 2025. [Online]. Available: https://www.iea.org/reports/global-ev-outlook-2025/trends-in-electric-car-markets-2
16 Ember, "Global electricity review 2024," 2024. Accessed: Nov. 8, 2025. [Online]. Available: https://ember-energy.org/app/uploads/2024/05/Report-Global-Electricity-Review-2024.pdf
17 International Energy Agency, the International Renewable Energy Agency, the United Nations, the World Bank, and the World Health Organization, "Tracking SDG 7: The energy progress report 2024," 2024. Accessed: Nov. 8, 2025. [Online]. Available: https://www.irena.org/-/media/Files/IRENA/Agency/Publication/2024/Jun/IRENA_Tracking_SDG7_energy_progress_2024.pdf

18 California Independent System Operator, Department of Market Monitoring, "2024 special report on battery storage," 2025. Accessed: Oct. 9, 2025. [Online]. Available: https://www.caiso.com/documents/2024-special-report-on-battery-storage-may-29-2025.pdf

19 International Energy Agency, "Global hydrogen review 2024: Executive summary," 2025. Accessed: Oct. 9, 2025. [Online]. Available: https://www.iea.org/reports/global-hydrogen-review-2024/executive-summary

20 Nord Pool AS, "The power market," 2025. Accessed: Oct. 9, 2025. [Online]. Available: https://www.nordpoolgroup.com/en/the-power-market

21 H. Lund, B. Möller, B. V. Mathiesen, and A. Dyrelund, "The role of district heating in future renewable energy systems," *Energy*, vol. 35, no. 3, pp. 1381–1390, 2010.

22 Danish Energy Agency, "Denmark's climate status and outlook 2021," 2021. Accessed: Oct. 9, 2025. [Online]. Available: https://ens.dk/media/4889/download

23 A. Singlitico, "Overview of the power-to-X potential integration within the Danish energy system," 2022. Accessed: Oct. 9, 2025. [Online]. Available: https://orbit.dtu.dk/files/309586496/DTU_Kraka_report_final.pdf

24 I. Skov, N. Schneider, G. Schweiger, J. Schöggl, and A. Posch, "Power-to-X in Denmark: An analysis of strengths, weaknesses, opportunities and threats," *Energies*, vol. 14 p. 913, 2021.

25 Danish Energy Agency, "Denmark's energy islands," 2025. Accessed: Oct. 9, 2025. [Online]. Available: https://ens.dk/en/energy-sources/offshore-wind-power/denmarks-energy-islands

26 C. Heinrich, C. Ziras, A. L. Syrri, and H. W. Bindner, "EcoGrid 2.0: A large-scale field trial of a local flexibility market," *Applied Energy*, vol. 261 p. 114399, 2020.

27 Z. Tian et al., "Large-scale solar district heating plants in Danish smart thermal grid: Developments and recent trends," *Energy Conversion and Management*, vol. 189 pp. 67–80, 2019.

28 R. G. F. Fyn, "Meta: Surplus heat to district heating (Tietgenbyen Energy Centre)," 2019. Accessed: Oct. 9, 2025. [Online]. Available: https://www.ramboll.com/projects/energy/meta-surplus-heat-to-district-heating

29 California Independent System Operator (CAISO), "Flexible resources help renewables: Fast facts," 2016. Accessed: Oct. 9, 2025. [Online]. Available: https://www.caiso.com/Documents/FlexibleResourcesHelpRenewables_FastFacts.pdf

30 A. Bonner, "California's battery storage installations grow to almost 16 GW," Energy-Storage News/Solar Media Ltd., 2025. Accessed: Oct. 9, 2025. [Online]. Available: https://www.energy-storage.news/california-battery-storage-grows-to-15763mw/

31 California Energy Commission, California Public Utilities Commission, and California Air Resources Board, "SB 100 joint agency report," 2025. Accessed: Oct. 9, 2025. [Online]. Available: https://www.energy.ca.gov/sb100
32 Veloz, California Energy Commission, and California Air Resources Board, "California EV market report," 2025. Accessed: Oct. 9, 2025. [Online]. Available: https://www.veloz.org/ev-market-report/#:~:text=Analysis&text=California%20added%20100%2C671%20new%20electric,evidence%20of%20a%20maturing%20market/
33 Ember, "Wind and solar generate over a quarter of China's electricity for the first month on record," Ember, 2025. Accessed: Oct. 9, 2025. [Online]. Available: https://ember-energy.org/latest-updates/wind-and-solar-generate-over-a-quarter-of-chinas-electricity-for-the-first-month-on-record/#:~:text=China%20achieved%20a%20new%20milestone,global%20energy%20think%20tank%20Ember
34 China Energy Storage Alliance and National Energy Administration, "China releases official energy storage report highlighting 130% growth in installed capacity," 2025. Accessed: Oct. 9, 2025. [Online]. Available: https://en.cnesa.org/latest-news/2025/8/3/china-releases-official-energy-storage-report-highlighting-130-growth-in-installed-capacity#:~:text=The%20most%20notable%20finding%3A%20by,in%20the%20international%20NES%20market
35 D. Yiting, H. Linliang, and W. Jintao, "China unveils first integrated wind-solar-thermal UHV power project—Longdong–Shandong ±800 kV DC line," 2025. Accessed: Oct. 9, 2025. [Online]. Available: https://en.people.cn/n3/2025/0523/c98649-20318815.html
36 International Carbon Action Partnership (ICAP), "China officially expands national ETS to cement, steel and aluminum sectors," 2025. Accessed: Oct. 9, 2025. [Online]. Available: https://icapcarbonaction.com/en/news/china-officially-expands-national-ets-cement-steel-and-aluminum-sectors
37 International Energy Agency (IEA), "Germany 2025: Energy Policy Review," 2025. Accessed: Oct. 9, 2025. [Online]. Available: https://iea.blob.core.windows.net/assets/7fea0ad0-1cc1-45e9-810b-2d602e64642f/Germany2025.pdf
38 M. Moraal and L. Shan, "Understanding the German nuclear exit," Heinrich-Böll Stiftung, 2025. Accessed: Oct. 9, 2025. [Online]. Available: https://us.boell.org/en/2023/04/21/understanding-german-nuclear-exit
39 Bloomberg, "Germany—global coal countdown," 2025. Accessed: Oct. 9, 2025. [Online]. Available: https://bloombergcoalcountdown.com/countries/DE
40 Clean Energy Wire, "Germany's finance minister rejects 2030 coal exit without affordable and secure alternatives," 2025. Accessed: Oct. 9, 2025. [Online]. Available: https://www.cleanenergywire.org/news/german-finance-minister-rejects-2030-coal-exit-without-affordable-and-secure-alternatives

41 Danish Energy Agency, "Energy statistics 2023 (Energistatistik—Samlet_2023)," 2023. Accessed: Oct. 9, 2025. [Online]. Available: https://ens.dk/media/6308/download

42 Anaheim.net, "Draft environmental impact report no. 357—section 4.5: Greenhouse gas emissions (OC River Walk Project)," 2025. Accessed: Oct. 9, 2025. [Online]. Available: https://www.anaheim.net/DocumentCenter/View/65776/

43 Invest in Denmark, "Denmark #1 in share of renewables in net electricity generation for 2024 in the EU," 2025. Accessed: Oct. 9, 2025. [Online]. Available: https://investindk.com/insights/denmark-1-in-share-of-renewables-in-net-electricity-generation-for-2024-in-the-eu

44 Low-Carbon Power, "Denmark electricity generation mix 2024/2025," 2025. Accessed: Oct. 9, 2025. [Online]. Available: https://lowcarbonpower.org/region/Denmark

45 U.S. Energy Information Administration, "California: State energy profile & analysis," 2025. Accessed: Oct. 9, 2025. [Online]. Available: https://www.eia.gov/state/analysis.php

46 California Energy Commission, "Electric generation capacity and energy," 2025. Accessed: Oct. 9, 2025. [Online]. Available: https://www.energy.ca.gov/data-reports/energy-almanac/california-electricity-data/electric-generation-capacity-and-energy

47 Ember, "China accounted for more than half of the global increase in wind and solar power in 2024," 2025. Accessed: Oct. 9, 2025. [Online]. Available: https://ember-energy.org/countries-and-regions/china

48 G. Maguire, "China sets new clean electricity milestones during Q1 2025," Reuters, Sustainability/Climate & Energy, 2025. Accessed: Oct. 9, 2025. [Online]. Available: https://www.reuters.com/sustainability/climate-energy/china-sets-new-clean-electricity-milestones-during-q1-2025-maguire-2025-04-22

49 Bundesnetzagentur, "Growth in renewable energy in 2024," 2025. Accessed: Oct. 9, 2025. [Online]. Available: https://www.bundesnetzagentur.de/SharedDocs/Pressemitteilungen/EN/2025/20250108_EE.html

50 SMARD, "Evaluation of last year," Bundesnetzagentur, 2025. Accessed: Oct. 9, 2025. [Online]. Available: https://www.smard.de/page/en/topic-article/5892/215704/evaluation-of-last-year

51 V. Shaw, "China's new energy storage capacity surges to 74 GW/168 GWh in 2024, up 130% YoY," PV-Magazine, 2025. Accessed: Oct. 9, 2025. [Online]. Available: https://www.pv-magazine.com/2025/01/23/chinas-new-energy-storage-capacity-surges-to-74-gw-168-gwh-in-2024-up-130-yoy

52 J. Figgener et al., "The development of battery storage systems in Germany: A market review (status 2023)," arXiv Preprint arXiv:2203.06762, 2022.

53 J. Finney, “How battery storage is powering Denmark’s renewable energy revolution and driving global decarbonization,” EnergyTech, 2025. Accessed: Oct. 9, 2025. [Online]. Available: https://www.energytech.com/renewables/article/55246886/how-battery-storage-is-powering-denmarks-renewable-energy-revolution-and-driving-global-decarbonization

54 Energinet and Energy Systems Integration Group, “P2X and future sector coupling in Denmark (webinar),” 2025. Accessed: Oct. 9, 2025. [Online]. Available: https://www.esig.energy/event/webinar-p2x-future-sector-coupling

55 State of Green, “Green light for cross-border hydrogen pipeline from Denmark to Germany,” 2025. Accessed: Oct. 9, 2025. [Online]. Available: https://stateofgreen.com/en/news/green-light-for-cross-border-hydrogen-pipeline-from-denmark-to-germany

56 R. Dickel, “Decarbonizing Germany’s heating sector,” 2024. Accessed: Oct. 9, 2025. [Online]. Available: https://www.oxfordenergy.org/wpcms/wp-content/uploads/2024/02/ET29-Decarbonising-Germanys-heating-sector.pdf

57 B. van der Zwaan et al., “Electricity- and hydrogen-driven energy system sector-coupling in net-zero CO_2 emission pathways,” *Nature Communications*, vol. 16, no. 1, p. 1368, 2025.

58 State of Green, “Middelgrunden wind turbine co-operative (Middelgrunden Vindmøllelaug),” 2025. Accessed: Oct. 9, 2025. [Online]. Available: https://stateofgreen.com/en/solutions/middelgrunden-wind-turbine-co-operative-middelgrunden-vindmollelaug

59 N. Simcock, R. Willis, and P. Capener, “Cultures of community energy: International case studies,” 2016. Accessed: Oct. 9, 2025. [Online]. Available: https://base.socioeco.org/docs/coce_international_case_studies_online.pdf

60 International Energy Agency, “Denmark: Community ownership of renewables,” 2025. Accessed: Oct. 9, 2025. [Online]. Available: https://www.iea.org/policies/17800-denmark-community-ownership-of-renewables

61 McKinsey & Company, “Ørsted’s renewable-energy transformation,” 2025. Accessed: Oct. 9, 2025. [Online]. Available: https://www.mckinsey.com/capabilities/sustainability/our-insights/orsteds-renewable-energy-transformation

62 M. N. Wachtell and M. B. Tang, “Augmented collaborative advantages: Vestas and the Danish wind energy strategy,” Copenhagen Business School, 2022. [Online]. Available: https://research-api.cbs.dk/ws/portalfiles/portal/76450757/1346038_Augmented_Collaborative_Advantages_Vestas_and_the_Danish_wind_industry_strategy.pdf

63 California Public Utilities Commission, “Low income rates,” 2025. Accessed: Oct. 9, 2025. [Online]. Available: https://www.cpuc.ca.gov/lowincomerates

64 Agora Energiewende, “The German power market 2019: Summary,” 2020. Accessed: Oct. 9, 2025. [Online]. Available: https://www.agora-energiewende.

org/fileadmin/Projekte/2019/Jahresauswertung_2019/A-EW_German-Power-Market-2019_Summary_EN.pdf

65 TransnetBW GmbH, "Charging of levy under German Renewable Energy Act (EEG) from 1 July 2022," 2025. Accessed: Oct. 9, 2025. [Online]. Available: https://www.transnetbw.de/en/energy-market/renewables-and-levies/eeg-support/charging-of-levy-under-german-renewable-energy-act-eeg-from-1-july-2022

66 International Renewable Energy Agency and International Labour Organization, "Renewable energy and jobs: Annual review 2024," 2024. Accessed: Oct. 9, 2025. [Online]. Available: https://www.ilo.org/sites/default/files/2024-10/IRENA-ILO%20Renewable%20energy%20and%20jobs_2024.pdf

67 Danish Ministry of Climate, Energy and Utilities, "Denmark climate programme 2020," 2020. Accessed: Oct. 9, 2025. [Online]. Available: https://unfccc.int/sites/default/files/resource/Denmark%20Climate%20Programme%202020.pdf

68 Climate Change Laws of the World, "Green tax reform agreement," 2020. [Online]. Available: https://climate-laws.org/documents/green-tax-reform-agreement-9372

69 Energy Council of Canada, "2025 SB100 report scoping tribal listening session," 2023. [Online]. Available: https://www.energy.ca.gov/sites/default/files/2023-03/2025_SB100_Report_Scoping_Tribal_Listening_Session_ADA.pdf

70 Federal Ministry for Economic Affairs and Energy, "The 2023 climate action programme of the German Federal Government," 2023. Accessed: Oct. 9, 2025. [Online]. Available: https://www.bundeswirtschaftsministerium.de/Redaktion/EN/Downloads/C/climate-action-programme-2023.pdf

71 Australian Government, "Climate Change Act 2022," 2022. International Energy Agency (IEA), "Australia 2023 Energy Policy Review," 2023. https://www.legislation.gov.au/Details/C2022A00037

72 Government of Canada, "Canadian Net-Zero Emissions Accountability Act (S.C. 2021, c.22)," 2021. OECD, "OECD Environmental Performance Reviews: Canada," 2023. https://laws-lois.justice.gc.ca/eng/acts/C-19.3/

73 Government of Japan, "Act on Promotion of Global Warming Countermeasures," (latest amended version). International Energy Agency (IEA), "Japan 2021 Energy Policy Review," 2021.

74 Government of Spain, "Law 7/2021 on Climate Change and Energy Transition," 2021. European Commission, "Spain's National Energy and Climate Plan (NECP) 2021–2030," 2021.

75 UK Government, "Climate Change Act 2008," (as amended) 2008. Climate Change Committee (CCC), "Progress in Reducing Emissions," 2024. https://www.legislation.gov.uk/ukpga/2008/27/contents

76 Government of the Republic of Korea, "Framework Act on Carbon Neutrality and Green Growth," 2021. International Energy Agency (IEA), "Korea 2020 Energy Policy Review," 2020.

77 International Energy Agency, "Denmark—emissions," 2025. Accessed: Oct. 8, 2025. [Online]. Available: https://www.iea.org/countries/denmark/emissions

78 TheGlobalEconomy.com, "Carbon dioxide (CO_2) emissions per capita—OECD," 2025. Accessed: Oct. 9, 2025. [Online]. Available: https://www.theglobaleconomy.com/rankings/Carbon_dioxide_emissions_per_capita/OECD

79 California Air Resources Board (CARB), "Climate pollutants fall below 1990 levels for first time," 2025. Accessed: Oct. 9, 2025. [Online]. Available: https://ww2.arb.ca.gov/news/climate-pollutants-fall-below-1990-levels-first-time

80 California Environmental Protection Agency, "California climate dashboard," 2025. Accessed: Oct. 9, 2025. [Online]. Available: https://calepa.ca.gov/climate-dashboard

81 California Air Resources Board, "California greenhouse gas emissions from 2000 to 2022: Trends of emissions and other indicators," 2024. [Online]. Available: https://ww2.arb.ca.gov/sites/default/files/2024-09/nc-2000_2022_ghg_inventory_trends.pdf

82 Plan A Earth, "Germany's Climate Protection Act (Klimaschutzgesetz)," 2025. Accessed: Oct. 9, 2025. [Online]. Available: https://plana.earth/policy/reform-germany-climate-protection-act

83 International Energy Agency, "Germany—emissions," 2025. Accessed: Oct. 9, 2025. [Online]. Available: https://www.iea.org/countries/germany/emissions

84 International Energy Agency, "China—emissions," 2025. Accessed: Oct. 9, 2025. [Online]. Available: https://www.iea.org/countries/china/emissions

85 D. De Boer and D. Fan, "Impressive progress in China's 1+N policy framework," CCICED (China Council for International Cooperation on Environment and Development), 2025. Accessed: Oct. 9, 2025. [Online]. Available: https://cciced.eco/climate-governance/how-is-progress-in-chinas-1n-policy-framework/

86 Climate Cooperation China, "China releases its first energy law," 2025. Accessed: Oct. 9, 2025. [Online]. Available: https://climatecooperation.cn/climate/china-releases-its-first-energy-law/

Index

IEEE Press Series on Power and Energy Systems

Series Editor: Ganesh Kumar Venayagamoorthy, Clemson University, Clemson, South Carolina, USA.

The mission of the IEEE Press Series on Power and Energy Systems is to publish leading-edge books that cover a broad spectrum of current and forward-looking technologies in the fast-moving area of power and energy systems including smart grid, renewable energy systems, electric vehicles and related areas. Our target audience includes power and energy systems professionals from academia, industry and government who are interested in enhancing their knowledge and perspectives in their areas of interest.

1. *Analysis of Electric Machinery*
   Paul C. Krause, Oleg Wasynczuk, S. D. Sudhoff
2. *Analysis of Faulted Power Systems*
   Paul M. Anderson
3. *Electrical Power Systems: Design and Analysis, Revised Printing*
   Mohamed E. El-Hawary
4. *Power System Stability*
   Edward W. Kimbark
5. *Inspection of Large Synchronous Machines: Checklists, Failure Identification, and Troubleshooting*
   Isidor Kerszenbaum
6. *Electric Power Applications of Fuzzy Systems*
   Mohamed E. El-Hawary
7. *Power System Protection (1st Edition)*
   Paul M. Anderson
8. *Subsynchronous Resonance in Power Systems*
   Paul M. Anderson, Basant L. Agrawal, J. E. Van Ness
9. *Understanding Power Quality Problems: Voltage Sags and Interruptions*
   Math H. J. Bollen
10. *Analysis of Electric Machinery and Drive Systems, Second Edition*
    Paul C. Krause, Oleg Wasynczuk, S. D. Sudhoff
11. *Principles of Electric Machines with Power Electronic Applications, 2nd Edition*
    Mohamed E. El-Hawary
12. *Electric Power Systems: Analysis and Control*
    Fabio Saccomanno

13. *Electricity Economics: Regulation and Deregulation*
Geoffrey Rothwell, Tomás Gómez, Mohamed E. El-Hawary
14. *Optimization Principles: Practical Applications to the Operation and Markets of the Electric Power Industry*
Narayan S. Rau
15. *Power System Control and Stability, Revised Printing*
Paul M. Anderson, A. A. Fouad
16. *Pulse Width Modulation for Power Converters: Principles and Practice*
D. Grahame Holmes, Thomas A. Lipo
17. *Modeling and High-Performance Control of Electric Machines*
John Chiasson
18. *Operation and Maintenance of Large Turbo-Generators*
Geoff Klempner, Isidor Kerszenbaum
19. *Rating of Electric Power Cables in Unfavorable Thermal Environment*
George J. Anders
20. *Risk Assessment Of Power Systems: Models, Methods, and Applications*
Wenyuan Li
21. *Signal Processing of Power Quality Disturbances*
Math H. J. Bollen, Irene Y. H. Gu
22. *Electric Power System Basics for the Nonelectrical Professional (1st Edition)*
Steven W. Blume
23. *Elements of Tidal-Electric Engineering*
Robert H. Clark
24. *Instantaneous Power Theory and Applications to Power Conditioning (1st Edition)*
Hirofumi Akagi, Edson H. Watanabe, Mauricio Aredes
25. *Maintaining Mission Critical Systems in a 24/7 Environment*
Peter M. Curtis
26. *Modern Heuristic Optimization Techniques: Theory and Applications to Power Systems*
Kwang Y. Lee, Mohamed A. El-Sharkawi
27. *Handbook of Large Turbo-Generator Operation and Maintenance, Second Edition*
Geoff Klempner, Isidor Kerszenbaum
28. *Introduction to Electrical Power Systems*
Mohamed E. El-Hawary
29. *Real-Time Stability Assessment in Modern Power System Control Centers*
S. C. Savulescu
30. *Economic Market Design and Planning for Electric Power Systems*
James A. Momoh, Lamine Mili

31. *Insulators for Icing and Polluted Environments*
Masoud Farzaneh, William A. Chisholm

32. *Introduction to FACTS Controllers: Theory, Modeling, and Applications*
Kalyan K. Sen, Mey Ling Sen

33. *Modeling and Control of Fuel Cells: Distributed Generation Applications*
M. H. Nehrir, C. Wang

34. *Optimization of Power System Operation (1st Edition)*
Jizhong Zhu

35. *Power Distribution System Reliability: Practical Methods and Applications*
Ali Chowdhury, Don Koval

36. *An Introduction to Wavelet Modulated Inverters*
S. A. Saleh, M. Azizur Rahman

37. *Control of Electric Machine Drive Systems*
Seung-Ki Sul

38. *Operation and Control of Electric Energy Processing Systems*
James A. Momoh, Lamine Mili

39. *Restructured Electric Power Systems: Analysis of Electricity Markets with Equilibrium Models*
Xiao-Ping Zhang

40. *Doubly Fed Induction Machine: Modeling and Control for Wind Energy Generation*
Gonzalo Abad, Jesus Lopez, Miguel Rodriguez, Luis Marroyo, Grzegorz Iwanski

41. *Electric Distribution Systems (1st Edition)*
Abdelhay A. Sallam, Om P. Malik

42. *Electricity Power Generation: The Changing Dimensions*
Digambar M. Tagare

43. *High Voltage and Electrical Insulation Engineering (1st Edition)*
Ravindra Arora, Wolfgang Mosch

44. *High Voltage Protection for Telecommunications*
Steven W. Blume

45. *Integration of Distributed Generation in the Power System*
Math H. J. Bollen, Fainan Hassan

46. *Maintaining Mission Critical Systems in a 24/7 Environment, Second Edition*
Peter Curtis

47. *Power Conversion and Control of Wind Energy Systems*
Bin Wu, Yongqiang Lang, Navid Zargari, Samir Kouro

48. *Practical Lighting Design with LEDs (1st Edition)*
Ron Lenk, Carol Lenk

49. *Probabilistic Transmission System Planning*
Wenyuan Li

50. *Electromechanical Motion Devices, Second Edition*
Paul Krause, Oleg Wasynczuk, Steven Pekarek

51. *Smart Grid: Fundamentals of Design and Analysis*
James Momoh

52. *Analysis of Electric Machinery and Drive Systems, Third Edition*
Paul C. Krause, Oleg Wasynczuk, S. D. Sudhoff, Steven D. Pekarek

53. *Electrical Energy Conversion and Transport: An Interactive Computer-Based Approach, 2nd Edition*
George G. Karady, Keith E. Holbert

54. *Electrical Energy Conversion and Transport: An Interactive Computer-Based Approach (1st Edition)*
George G. Karady, Keith E. Holbert

55. *Extruded Cables for High-Voltage Direct-Current Transmission: Advances in Research and Development*
Giovanni Mazzanti, Massimo Marzinotto

56. *Handbook of Electrical Power System Dynamics: Modeling, Stability, and Control*
Mircea Eremia, Mohammad Shahidehpour

57. *Advanced Power Electronics Converters: PWM Converters Processing AC Voltages*
Euzeli dos Santos, Edison R. da Silva, Mohamed E. El-Hawary

58. *Electrical Insulation for Rotating Machines, 2nd Edition*
Greg C. Stone, Ian Culbert, Edward A. Boulter, Hussein Dhirani

59. *Power Magnetic Devices: A Multi-Objective Design Approach (1st Edition)*
S. D. Sudhoff

60. *Practical Power System Operation*
Ebrahim Vaahedi

61. *Principles of Electrical Safety*
Peter E. Sutherland

62. *Risk Assessment of Power Systems: Models, Methods, and Applications, 2nd Edition*
Wenyuan Li

63. *The Selection Process of Biomass Materials for the Production of Bio-Fuels and Co-firing*
N. Altawell

64. *Digital Control of High-Frequency Switched-Mode Power Converters*
Luca Corradini, Dragan Maksimovic, Paolo Mattavelli, Regan Zane

65. *Industrial Power Distribution, 2nd Edition*
Ralph Fehr

66. *Optimization of Power System Operation, 2nd Edition*
Jizhong Zhu

67. *PID and Predictive Control of Electric Devices and Power Converters Using MATLAB®/Simulink®*
Liuping Wang, Shan Chai, Dae Yoo, Lu Gan, Ki Ng
68. *Power System Harmonics and Passive Filter Designs*
J. C. Das
69. *Advanced Solutions in Power Systems: HVDC, FACTS, and Artificial Intelligence*
Mircea Eremia, Chen-Ching Liu, Abdel-Aty Edris, Mohamed E. El-Hawary
70. *Current Signature Analysis for Condition Monitoring of Cage Induction Motors*
William T. Thomson, Ian Culbert
71. *Electric Power System Basics for the Nonelectrical Professional, 2nd Edition*
Steven W. Blume
72. *Energy Production Systems Engineering*
Thomas Howard Blair
73. *High-Power Converters and AC Drives, 2nd Edition*
Bin Wu, Mehdi Narimani
74. *HVDC Grids: For Offshore and Supergrid of the Future*
Dirk Van Hertem, Oriol Gomis-Bellmunt, Jun Liang
75. *Model Predictive Control of Wind Energy Conversion Systems*
Venkata Yaramasu, Bin Wu
76. *Power Grid Operation in a Market Environment*
Hong Chen
77. *Understanding Symmetrical Components for Power System Modeling*
J. C. Das
78. *Advances in Electric Power and Energy Systems: Load and Price Forecasting*
Mohamed E. El-Hawary
79. *Electricity Markets: Theories and Applications*
Jeremy Lin, Fernando H. Magnago
80. *Instantaneous Power Theory and Applications to Power Conditioning, 2nd Edition*
Hirofumi Akagi, Edson Hirokazu Watanabe, Mauricio Aredes
81. *Introduction to AC Machine Design*
Thomas A. Lipo
82. *Introduction to Electric Power and Drive Systems*
Paul C. Krause, Oleg Wasynczuk, Timothy O'Connell, Maher Hasan
83. *Multiphysics Simulation by Design for Electrical Machines, Power Electronics and Drives*
Marius Rosu, Ping Zhou, Dingsheng Lin, Dan M. Ionel, Mircea Popescu, Frede Blaabjerg, Vandana Rallabandi, David Staton

84. *Practical Lighting Design with LEDs, 2nd Edition*
Ron Lenk, Carol Lenk

85. *Advanced Control of Doubly Fed Induction Generator for Wind Power Systems*
Dehong Xu, Frede Blaabjerg, Wenjie Chen, Nan Zhu, Mohamed E. El-Hawary

86. *Electric Distribution Systems, 2nd Edition*
Abdelhay A. Sallam, Om P. Malik

87. *Electrical Railway Transportation Systems*
Morris Brenna, Federica Foiadelli, Dario Zaninelli

88. *Energy Processing and Smart Grid*
James A. Momoh

89. *Handbook of Large Turbo-Generator Operation and Maintenance, 3rd Edition*
Geoff Klempner, Isidor Kerszenbaum

90. *Modular Multilevel Converters: Analysis, Control, and Applications*
Sixing Du, Apparao Dekka, Bin Wu, Navid Zargari

91. *Distributed Fiber Optic Sensing and Dynamic Rating of Power Cables*
Sudhakar Cherukupalli, George J. Anders, Mohamed E. El-Hawary

92. *Power Electronics in Renewable Energy Systems and Smart Grid*
Bimal K. Bose

93. *Power System Control and Stability, 3rd Edition*
Vijay Vittal, James D. McCalley, Paul M. Anderson, A. A. Fouad

94. *A New Swing-Contract Design for Wholesale Power Markets*
Leigh Tesfatsion

95. *Advances in Electric Power and Energy: Static State Estimation*
Mohamed E. El-Hawary

96. *Applications of Modern Heuristic Optimization Methods in Power and Energy Systems*
Kwang Y. Lee, Zita A. Vale

97. *Arc Flash Hazard Analysis and Mitigation, 2nd Edition*
J. C. Das

98. *Distributed Energy Management of Electrical Power Systems*
Yinliang Xu, Wei Zhang, Wenxin Liu, Wen Yu

99. *Electromechanical Motion Devices, 3rd Edition*
Paul C. Krause, Oleg Wasynczuk, Steven D. Pekarek, Timothy O'Connell

100. *Handbook of Large Hydro Generators: Operation and Maintenance*
Glenn Mottershead, Stefano Bomben, Isidor Kerszenbaum, Geoff Klempner

101. *Maintaining Mission Critical Systems in a 24/7 Environment, 3rd Edition*
Peter M. Curtis

102. *Reference Frame Theory: Development and Applications*
Paul C. Krause

103. *Alternative Liquid Dielectrics for High Voltage Transformer Insulation Systems*
U. Mohan Rao, Issouf Fofana, Ramanujam Sarathi

104. *Interval Methods for Uncertain Power System Analysis*
Alfredo Vaccaro

105. *Introduction to the Analysis of Electromechanical Systems*
Paul C. Krause, Oleg Wasynczuk, Timothy O'Connell

106. *Optimal Coordination of Power Protective Devices with Illustrative Examples*
Ali R. Al-Roomi

107. *Power Flow Control Solutions for a Modern Grid Using SMART Power Flow Controllers*
Kalyan K. Sen, Mey Ling Sen

108. *Power Magnetic Devices, 2nd Edition*
Scott D. Sudhoff

109. *Power System Protection, 2nd Edition*
Paul M. Anderson, Charles F. Henville, Rasheek Rifaat, Brian Johnson, Sakis Meliopoulos

110. *Power System Protection: Fundamentals and Applications*
John Ciufo, Aaron Cooperberg

111. *Probabilistic Power System Expansion Planning with Renewable Energy Resources and Energy Storage Systems*
Jaeseok Choi, Kwang Y. Lee

112. *Real-Time Electromagnetic Transient Simulation of AC-DC Networks*
Venkata Dinavahi, Ning Lin

113. *Resilient Control Architectures and Power Systems*
Craig Rieger, Ronald Boring, Brian Johnson, Timothy McJunkin

114. *Soft-Switching Technology for Three-phase Power Electronics Converters*
Dehong Xu, Rui Li, Ning He, Jinyi Deng, Yuying Wu

115. *Control of Power Electronic Converters with Microgrid Applications*
Arindam Ghosh, Firuz Zare

116. *Coordinated Operation and Planning of Modern Heat and Electricity Incorporated Networks*
Mohammadreza Daneshvar, Behnam Mohammadi-Ivatloo, Kazem Zare

117. *Electromagnetic Analysis and Condition Monitoring of Synchronous Generators*
Hossein Ehya, Jawad Faiz

118. *High Voltage and Electrical Insulation Engineering, 2nd Edition*
Ravindra Arora, Wolfgang Mosch

119. *Intelligent Data Mining and Analysis in Power and Energy Systems*
Zita A. Vale, Tiago Pinto, Michael Negnevitsky, Ganesh Kumar Venayagamoorthy
120. *Introduction to Modern Analysis of Electric Machines and Drives*
Paul C. Krause, Thomas C. Krause
121. *Modeling and Control of Modern Electrical Energy Systems*
Masoud Karimi-Ghartemani
122. *Smart Energy for Transportation and Health in a Smart City*
Chun Sing Lai, Loi Lei Lai, Qi Hong Lai
123. *Transportation Electrification: Breakthroughs in Electrified Vehicles, Aircraft, Rolling Stock, and Watercraft*
Ahmed A. Mohamed, Ahmad Arshan Khan, Ahmed T. Elsayed, Mohamed A. Elshaer
124. *Wireless Power Transfer: Principles and Applications*
Zhen Zhang, Hongliang Pang
125. *Graph Database and Graph Computing for Power System Analysis*
Renchang Dai, Guangyi Liu
126. *Modular Multilevel Converters: Control, Fault Detection, and Protection*
Fujin Deng, Chengkai Liu, Zhe Chen
127. *Practical Partial Discharge Measurement on Electrical Equipment*
Greg C. Stone, Andrea Cavallini, Glenn Behrmann, Claudio Angelo Serafino
128. *Stability-Constrained Optimization for Modern Power System Operation and Planning*
Yan Xu, Yuan Chi, Heling Yuan
129. *The Power of Artificial Intelligence for the Next-Generation Oil and Gas Industry*
Pethuru Raj Chelliah, Venkatraman Jayasankar, Mats Agerstam, B. Sundaravadivazhagan, Robin Cyriac
130. *Interconnected Modern Multi-Energy Networks and Intelligent Transportation Systems*
Mohammadreza Daneshvar, Behnam Mohammadi-Ivatloo, Amjad Anvari-Moghaddam, Reza Razzaghi
131. *Microgrids: Theory and Practice*
Peng Zhang
132. *Analysis of Electric Machinery and Drive Systems, 4th Edition*
Paul C. Krause, Oleg Wasynczuk, Steven D. Pekarek, Timothy O'Connell
133. *Communities for Clean Energy Justice and Equity in Grid Modernization*
Mohammadreza Daneshvar, Behnam Mohammadi-Ivatloo, Amjad Anvari-Moghaddam

134. *Electric Machinery and Drives: An Electromagnetics Perspective*
Nabeel A. O. Demerdash, JiangBiao He, Hao Chen

135. *Electric Power System Basics for the Nonelectrical Professional, 3rd Edition*
Steven W. Blume

136. *Hidden Risk in Occupational Electrical Safety*
H. Landis Floyd, Ganesh Kumar Vanayagamoorthy

137. *Modern Power System*
Arindam Ghosh

138. *Smart Cyber-Physical Power Systems: Challenges and Solutions, 2 Volume Set*
Ali Parizad, Hamid Reza Baghaee, Saifur Rahman

139. *Understanding Electromagnetic Transients in Power Systems*
Luiz Cera Zanetta Jr.

140. *Artificial Intelligence Empowered Smart Energy Systems*
Qiang Yang, Gang Huang

141. *Measuring the Electric Grid: Mysteries Explained*
Harold Kirkham, Jan-Philipp Kitzig, David Laverty, Artis Riepnieks, Dani Strickland, Rod White

142. *Deep Decarbonization of Modern Energy Networks for Climate–Energy–Sustainability Nexus*
Mohammadreza Daneshvar, Amjad Anvari-Moghaddam, Mahdieh S. Sadabadi

Wiley The manufacturer's authorized representative according to the EU General Product Safety Regulation is Wiley-VCH GmbH, Boschstr. 12, 69469 Weinheim, Germany, e-mail: Product_Safety@wiley.com.

Printed and bound by CPI Group (UK) Ltd, Croydon, CR0 4YY
19/09/2026
02219301-0002